2197504 C 54.95 67 I

δ_q quenching distance (Ch. 7)
γ_T, δ_h temperature (thermal) BL thickness
δ^* displacement thickness (Ch. 4)
$\delta\{\ \}$ Dirac impulse function (Ch. 6)
δ_m mass-transfer BL thickness (Ch. 6)
$\Delta\{\ \}$ change in ( ), (Chs. 2, 3)
ξ dimensionless axial variable (duct flow) (Ch. 5)
ε_{ij} depth of energy "well" for interaction of molecules of species i and j (Chs. 3, 7)
ε turbulent dissipation per unit mass (Ch. 5)
ε emittance or emissivity (Ch. 5)
κ bulk viscosity (Ch. 3)
κ equilibrium partition coefficient (Ch. 6)
κ_t turbulent kinetic energy per unit mass (Ch. 5)
Λ fin parameter governing heat transfer effectiveness (Ch. 5)
η_{cap} capture fraction (Ch. 6)
η efficiency or effectiveness factor (Chs. 5, 6)
μ viscosity (Newtonian fluid) (Ch. 2)
ν kinematic viscosity (momentum diffusivity), μ/ρ
ν_O, ν_F reaction orders (with respect to Oxidizer, Fuel), Chs. 6, 7
ρ fluid density (Ch. 2)
σ_B Stefan-Boltzmann constant (Ch. 5)
σ molecular-collision diameter (Chs. 1, 3)
$\mathbf{T}$ viscous (extra-) stress tensor (Ch. 2)
$\mathbf{\Pi}$ total stress tensor (Ch. 2)
Φ mixture ratio parameter (Chs. 6, 7)
ϕ specific gravitational potential energy (Ch. 2), intermolecular potential (Ch. 7), or volume fraction (Ch. 4)
ϕ Thiele modulus governing η_{cat} (Ch. 6)
ϕ meridional angle (Ch. 2)
ϕ_T heat flux potential (Chs. 3, 5)
Φ fuel/air ratio (Ch. 7)
θ polar angle (Ch. 2)
ψ stream function (Ch. 4)
$\boldsymbol{\omega}$ fluid rotational velocity (Ch. 3)
ω_i mass fraction of chemical species i (ρ_i/ρ) in mixture
Ω dimensionless normalized dependent variable (Ch. 6)
Ω_μ, Ω_D molecular potential collision integral functions (Chs. 3, 7)
Ω solid angle (Ch. 5)
$\boldsymbol{\zeta}$ vorticity vector (Ch. 4)

Dimensionless Groups

Arr Arrhenius parameter: $(E/(RT)_{ref}$ (Chs. 6, 7))
Bi Biot number (Chs. 5, 6)
C catalytic parameter (Ch. 6)
C_D drag coefficient (Chs. 4, 7)
C_f skin-friction coefficients (Chs. 4, 5, 6, 7)
C_p pressure coefficient (Ch. 4)
Ec Eckert number (Chs. 5, 7)
Dam Damköhler number (t_{flow}/t_{chem}) (Ch. 7)
$\mathcal{Q}$ chemical heat release parameter (Chs. 6, 7)
Fr Froude number (Ch. 7)
Gr Grashof number (Chs. 5, 6, 7)
Kn Knudsen number (Ch. 2)
Le Lewis number (Chs. 5, 6)
Ma Mach number (Chs. 4, 7)
Nu Nusselt number (Chs. 5, 6)
Pe Peclet number (Chs. 5, 6)
Pr Prandtl number (C... 7)
Ra Rayleigh numb... 7)
Re Reynolds' nu...
Sc Schmidt n...
St Stanto...
Stk Stok...

Sub...

b
B
bo ...off" (Ch. 7)
c ...odynamic critical state (Ch. 7)
cf "cold" flow (nonreactive) (Ch. 6)
CF chemically frozen
chem chemical contribution (Ch. 3)
comb combustion
conv convection
elem pertaining to chemical elements
e BL edge
f flame (Chs. 2, 6), formation (Ch. 2), friction (Ch. 4)
f film (Ch. 8)
F fuel
g gas
g gas
h heat (energy) transfer (Ch. 5)
i interstitial (Ch. 4) i
i species i i ($i = 1, 2, \ldots, N$) (Chs. 2, 6)
j jet (primary)
LTCE local thermochemical equilibrium (Ch. 3)
m model (Ch. 7)
m model
m mass transfer (Ch. 6)
n normal
nc natural convection
p prototype (Ch. 7)
p particles (Chs. 6, 8)
$proj$ projected (Ch. 6)
rad radiation (Ch. 5)
r recovery (Chs. 5, 6)
s pertaining to the secondary jet flow (Ch. 5)
sat saturation (LTE) value
sep separation point (Ch. 4)
t turbulent; tangential
t time
u unburned (Chs. 1, 7)
vap vaporization
w wall
x at position x
$\alpha, \beta, \ldots$ pertaining to phases $\alpha, \beta, \ldots$
∞ far-field (upstream infinity)
0 "stagnation" value (Ch. 4)
0 initial value

D1255629
6583
WITHDRAWN FSEC SOLAR ENERGY CENTER LIBRARY

Transport Processes in Chemically Reacting Flow Systems

BUTTERWORTHS SERIES IN CHEMICAL ENGINEERING

SERIES EDITOR

HOWARD BRENNER
Massachusetts Institute of Technology

ADVISORY EDITORS

ANDREAS ACRIVOS
Stanford University

JAMES E. BAILEY
California Institute of Technology

MANFRED MORARI
California Institute of Technology

E. BRUCE NAUMAN
Rensselaer Polytechnic Institute

ROBERT K. PRUD'HOMME
Princeton University

SERIES TITLES

Heterogeneous Reactor Design *Hong H. Lee*

Phase Equilibria in Chemical Engineering *Stanley M. Walas*

Transport Processes in Chemically Reacting Flow Systems *Daniel E. Rosner*

RELATED TITLES

Solid Liquid Separation *Ladislav Svarovsky*

Liquid and Liquid Mixtures *J. S. Rowlinson and F. L. Swinton*

Fundamentals of Fluidized Beds *John G. Yates*

Enlargement and Compaction of Particulate Solids *Nayland Stanley-Wood*

Mixing in the Process Industries *N. Harnby, M. Edwards, and A. W. Nienow*

Supercritical Fluid Extraction *Mark A. McHugh and Val J. Krukonis*

Transport Processes in Chemically Reacting Flow Systems

Daniel E. Rosner
Professor of Chemical Engineering and
Applied Science
Department of Chemical Engineering
Yale University

LIBRARY
Florida Solar Energy Center
1679 Clearlake Road
Cocoa FL 32922-5703

Butterworths
Boston London Durban Singapore Sydney Toronto Wellington

LIBRARY
Florida Solar Energy Center
300 State Road 401
Cape Canaveral, Florida
32920-4099

6583

Copyright © 1986 by Butterworth Publishers.
All rights reserved.
No part of this publication may be reproduced, stored in a retrieval system, or transmitted, in any form or by any means, electronic, mechanical, photocopying, recording, or otherwise, without the prior written permission of the publisher.

Library of Congress Cataloging in Publication Data
Rosner, Daniel E.
Transport processes in chemically reacting flow systems.
(Butterworths series in chemical engineering)
Includes bibliographies and index.
1. Mass transfer. 2. Transport theory. 3. Fluid dynamics.
I. Title. II. Series.

TP156.M3R67 1986 660.2′842 85-16634
ISBN 0-409-95178-1

Butterworth Publishers
80 Montvale Avenue
Stoneham, MA 02180

10 9 8 7 6 5 4 3 2 1

Printed in the United States of America

To my wife, Susan

Contents

2 Governing Conservation Principles 27

List of Primary Figures

List of Primary Tables

Preface

I have succumbed to the temptation to write the book for which I searched in vain while I was a student and young lecturer! This text has evolved from my teaching of Yale University undergraduate and graduate courses dealing with the transport of energy, mass, and momentum in chemically reacting fluids, to students of engineering (chemical, mechanical, aeronautical, etc.) and applied science (e.g., materials, geophysics/geochemistry, medicine). The manuscript was put into its present form with the partial support of EXXON Research and Engineering Co., in connection with my teaching of the short course: "Introduction to the Fluid Mechanics of Combustion" (part of the EXXON R & E—Technical Education Program) and Olin Corporation (in connection with a short course presenting chemical engineering concepts to chemists). Accordingly, it is written in such a way as to be accessible to students and practicing scientists whose background has until now been confined to physical chemistry, classical physics, and/or applied mathematics. Indeed, the basic principles of these underlying fields are here generalized and reformulated so as to be able to deal with chemically reacting flow systems of current and future engineering interest. It is not necessary that the student have a previous course in fluid mechanics (i.e., momentum transfer by convection and diffusion); however, in that case the material presented here should certainly be covered over a period of more than one semester. Reflecting my own interdisciplinary background and involvement in ME, AeroE, and ChE, a special effort has been made to write the book in such a way as to make accessible to engineers educated in one area (say, ME or AeroE) the fruitful approaches and results of engineers in adjacent disciplines (especially chemical engineering, as in my treatment of the topics of momentum/energy/mass transport in packed bed exchangers, and also residence-time distribution analysis). While this is undoubtedly not the first such attempt at unifying these engineering fields under one cover, it may be the first having some of the advantages associated with being written by a single author. Inevitably, portions of my notation will, at first, appear unfamiliar and, perhaps, downright cumbersome, but, in most cases, it possesses a certain logic and suggestiveness which does not tax one's memory. Thus, it will not take the reader long to identify a quantity like $\dot{m}''_{3\textcircled{2}}$ as the convective *mass* flow rate per unit area (say, kg/m^2-s) of chemical species 3 evaluated at station (location) ②, etc. (see Nomenclature).

J. W. Gibbs remarked that the role of theory in any science is to find the perspective from which the subject appears in its simplest form. My purpose is to present in a simple language but rather general form, principles and approaches that have proven to be very fruitful, and that will doubtless remain so in solving the challenging problems still ahead of us. Thus, while our perspective and scope is broader than that found in many previous transport textbooks (especially those intended for undergraduates), the presentation here is deliberately concise and very selective, leaving many "details" for student exercises. I hope the result provides the dedicated reader with the fundamentally oriented yet up-to-date background needed to tackle more advanced, specialized topics. In any event, I am confident it will put the reader in a position to properly formulate and solve many important problems involving *rates* of energy, mass, or momentum transport in fluids that may be reacting chemically.

The pedagogical choice of *combustion* for many of the examples is not merely the result of my own research background. For the reasons outlined below I am convinced that combustion is an excellent "prototype" for presenting the important concepts of transport in chemically reacting fluid flows. First, it is perhaps the only area of chemically reacting flows not only common to chemical engineering, mechanical engineering, and aeronautical engineering, but also familiar in the daily experience of all applied scientists. Second, while avoiding the dazzling variety of phases, states, and chemical species encountered in present-day ChE reactor applications, combustors exhibit all of the important qualitative features of nonideal, transport-limited, nonisothermal reactors used to synthesize valuable chemicals—indeed, many chemicals (C_2H_2, HCl, P_2O_5, TiO_2, etc.) *are* routinely produced in "flame" reactors. Finally, it should not be necessary to remind the reader of the economic importance of the efficient use of our remaining fossil fuels, and the prevention of combustion-related accidents. Since one of my primary objectives is to lay a proper foundation for subsequent study and R & D, in this introductory treatment I have deliberately avoided many topics, more heavily dependent on empiricism, associated with interacting multiphase transport (e.g., boiling, bubbling fluidized bed dynamics, etc.). However, as indicated in Section 2.6.4, the *macroscopic* conservation conditions (see the Introduction to Chapter 2) on which we systematically build our understanding of *single-phase* flow systems also provide the starting point for rational pseudo-continuum theories of dispersed *multiphase* situations. Therefore, it is appropriate that these underlying principles first be mastered in the context of either single-phase flows, or simple limiting cases of two-phase flows (e.g., steady flow through isothermal porous media) or packed beds (Sections 4.7, 5.5.5, and 6.5.1) and diffusion with chemical reaction in porous solid media (Section 6.4.4). [Study of Section 2.5 can be postponed without a loss in continuity; however, several of the derived forms of the conservation equations given here will prove useful in Chapters 4, 5, and 6.]

Also deliberately excluded is explicit material on what might be called the "systems" aspects of heat/mass exchangers, chemical reactors, and networks thereof. Thus, while we formulate and exploit the principles on which individual exchangers and chemical reactors are selected and designed (e.g., sized), explicit consideration of the economic optimization of specific devices, or the integration of many separate devices (as in multistage arrangements, or chemical "plants") would take us too far from our central themes.

While Chapters 4, 5, and 6 deal successively with momentum, energy, and mass transport, we clearly develop, state, and exploit useful quantitative "analogies" between these transport phenomena, including interrelationships that remain valid even in the presence of homogeneous or heterogeneous chemical reactions (Sections 6.5.3 and 6.5.5). Moreover, we include a separate chapter (7) on the use of transport theory in the systematization and generalization of experimental data on chemically reacting systems, emphasizing "similitude" methods that go far beyond ordinary "dimensional analysis." Because of our present emphasis on the transport mechanisms of *convection* and *diffusion*, which operate for momentum, energy, and (species) mass, the somewhat "singular" subject of *radiative* energy transport (Section 5.9) is only briefly included. While some chemical reactors are *intended* to produce photons (e.g., combustion-driven furnaces or chemical lasers), radiation is often an incidental "by-product." These factors, together with the "one-way" nature of the fluid dynamics–radiative energy coupling in most engineering devices (i.e., the fluid-momentum, energy, and species "density" fields are needed to predict the radiative behavior, but not vice versa), account for the brevity of this section. Nevertheless, what little is included is intended to indicate the nature of the radiative transport problem, and to suggest fruitful alternative approaches to deal with it.

Following a concise "overview" (Chapter 7, Summary) of the main points of each chapter, many of these principles and methods are then brought together in a comprehensive numerical example (Chapter 8) intended to also serve as a prototype (see Appendix 8.1, Recommendations on Problem-Solving) for student solutions to the novel problems posed at the end of each chapter. These "exercises," which are an extremely important part of this textbook from the viewpoint of a student's education, have been designed to bring out important qualitative and quantitative engineering implications of the topics treated in each chapter. Unless otherwise specified they were developed by the author in connection with his previous teaching, research, and consulting; however, in some cases (clearly cited), they are elaborations or revisions of similar problems included in earlier textbooks or treatises. Several complete solutions are provided to demonstrate the specific use of seemingly "abstract" concepts, mathematical formulae, and/or graphical or tabular data provided in each chapter. While our preference is for metric units (m-kg-s, or cm-g-s), some examples are deliberately included in other commonly used engineering unit systems (for conversion factors, see Appendix 8.6). Most equations derived or quoted herein are either dimensionless or, if dimensional, stated in a form in which they are valid in any self-consistent unit set.

In summary, the principles developed and often illustrated here for combustion systems are important not only for the rational design and development of engineering equipment (e.g., chemical reactors, heat exchangers, mass exchangers) but also for scientific research involving coupled transport processes and chemical reaction in flow systems. Moreover, the groundwork is laid for the systematic further study of more specialized topics (chemical reactor analysis/design, separation processes, multiphase transport, radiative energy transport, computational fluid mechanics, combustion science and technology, etc.). Indeed, while developed primarily for use as a graduate (and undergraduate) textbook in transport processes (energy, mass, and momentum), our emphasis on fluids containing molecules capable of undergoing chemical reaction (e.g., combustion) should make this book useful in more specialized engineering courses, especially chemical reaction engineering and combustion fundamentals. Specific sequences of topics in each of these possible courses are identified in Tables P1 and P2. In each case it is assumed that the relevant background in the underlying sciences of chemical thermodynamics and chemical kinetics can be provided *via* readily available texts in these classical areas.

By this time the reader will have noted that this text is concerned with the *principles* underlying the development of comprehensive rational computer models of chemically reacting flow systems, rather than the description of recently developed computer aids to engineering design. Thus, our emphasis is on the use of fundamental laws in the clever exploitation of a judicious blend of experiment, analysis, and numerical methods to first develop the requisite understanding, and, ultimately, to develop mathematical models for the essential portions of engineering problems involving energy, mass, and/or momentum exchange. In this respect, the particular problems and solutions I have chosen to explicitly include here should be regarded merely as instructive "prototypes" for dealing with the challenging new engineering problems that face us.

Much of my own learning occurs in the process of doing research in the general area of transport processes in chemically reacting systems. For this reason I wish to acknowledge the Office of Scientific Research of the U.S. Air Force and NASA–Lewis Research Laboratories for their financial support of research that has strongly influenced the orientation and content of this book. I am also indebted to

Table P1 Chemical Reaction Engineering

Topic(s)	*Textbook Section(s)*
Introduction	Ch. 1, Section 6.1
Conservation (Balance) Laws	Ch. 2
Transport (Diffusion) Laws	Ch. 3, Section 6.2
Ideal Plug-Flow Reactors: Empty and Packed (Fixed) Beds	Ex. 2.14, 5.11; Sections 6.1.3.1, 6.4.4, 6.7
Ideal (Well-) Stirred Tank Reactors	Sections 6.1.3.2, 6.7
Nonideal Reactors: PDE Models/Solution Methods	Section 7.4.3; Appendix 8.2
Modular Models of Real Flow Reactors; Stability and Parametric Sensitivity	Sections 6.7.4, 6.7.6, 6.7.7; Ex. 6.12
Diffusion and Chemical Reaction in Porous Media	Sections 3.4.4, 6.4.4; Ex. 6.6
Fixed-Bed Mass Transfer	Section 6.5.1; Ex. 6.10
Unpacked Duct Wall Reactors	Sections 6.1.3.1, 6.5.3; Ex. 6.9, 6.10
Similitude Methods in Systems with Chemical Reaction	Sections 6.4.4, 7.2.3.2, Ex. 6.8, 6.9
Chemical Reactor Scale Model Theory	Section 7.2.3
Summary	Section 7.4.2

Table P2 Combustion Fundamentals

Topic(s)	*Textbook Section(s)*
Introduction to Combustion: Scope, Importance	Ch. 1
Conservation (Balance) Laws	Ch. 2
Transport (Diffusion) Laws	Ch. 3
Thermodynamics of Combustion	Section 2.5.4; Ex. 2.9 (Solution)
Chemical Kinetics of Combustion	Sections 3.1.2, 6.7.6, 7.2.2.2
Premixed Flames (Deflagration Waves)	Sections 4.3.2, 6.5.5.8, 7.2.2.2
Detonation Waves	Section 4.3.2; Ex. 4.5
Flame Stabilization (Flashback, Blow-off)	Section 7.2.3.2; Ex. 7.4; Fig. 1.2-4
Ignition Energy	Section 1.1.1; Ex. 7.5
Diffusion Flames (Laminar and Turbulent)	Sections 1.1.2, 1.1.3, 6.5.5, 7.2.3.2c
Surface-Catalyzed Combustion/Incineration	Ex. 5.11; Section 6.9.1; Ex. 6.9; Section 6.5.3
Fuel Droplet Vaporization and Combustion Theory	Sections 1.1.3, 6.4.3.3, 6.5.5.7, 7.2.3.2d; Ex. 6.11
Stability and Parametric Sensitivity of Combustors	Sections 6.7.6, 7.2.3.1
Modular Mathematical Models of Combustor Performance	Sections 6.7.4, 6.7.6; Ex. 6.12
PDE Models of Combustion/Numerical Methods	Section 7.4.2; Appendix 8.2
Scale Model and Similitude Theory in Combustion Engineering	Section 7.2.3; Appendix 7.1
Heat and Mass Transfer from Combustion Gases	Sections 5.8, 5.9.3, 6.5.4; Ch. 8, Ex. 7.1, 7.6
Summary	Section 7.4.2

many colleagues at Yale University and EXXON Corporation for their helpful comments, and to the members of Technion–Israel Institute of Technology for their hospitality during the Fall of 1982, when this manuscript was essentially put into its present form. However, the author takes full responsibility for any errors of commission or omission associated with this first edition, and will welcome the written feedback of students, faculty, and practicing engineers and applied scientists who use this book.

Daniel E. Rosner
New Haven

1

Introduction to Transport Processes in Chemically Reactive Systems

INTRODUCTION

The information needed to design and control engineering devices for carrying out chemical reactions will be seen below to extend well beyond the obviously relevant underlying sciences of:

i. thermochemistry/stoichiometry, and
ii. chemical kinetics,

already encountered in each student's preparatory courses. Even the basic data of these underlying sciences are generated by using idealized laboratory configurations (e.g., closed calorimeters ("bombs"), well-stirred reactors, etc.) bearing little outward resemblance to practical chemical reactors.

This book deals with the role, in chemically reacting flow systems, of *transport processes*—particularly the transport of *momentum*, *energy*, and (chemical species) *mass* in fluids (gases and liquids). The laws governing such transport will be seen to influence:

i. the local rates at which reactants encounter one another;
ii. the ability of the reactants to be raised to a temperature at which the rates of the essential chemical reactions are appreciable;
iii. the volume (or area) required to carry out the ensuing chemical reaction(s) at the desired rate; and
iv. the amount and the fate of unwanted (by-) products (e.g., pollutants) produced.

For systems in which only *physical* changes occur (e.g., energy and/or mass exchange, perhaps accompanied by *phase* change), the same general principles can be used to design (e.g., size) or analyze equipment, usually with considerable simplifications. Indeed, we will show that the transport laws governing nonreactive systems can often be used to make rational predictions of the behavior of "analogous" chemically reacting systems. For this reason, and for obvious pedagogical reasons, the simplest illustrations of momentum, energy, and mass transport (in Chapters 4, 5, and 6, respectively) will first deal with nonreacting systems, but using an approach and a viewpoint amenable to our later applications or extensions to chemically reacting systems. This strategy is virtually an educational necessity, since chemical reactions are now routinely encountered not only by chemical engineers, but also by many mechanical engineers, aeronautical engineers, civil engineers, and researchers in the applied sciences (materials, geology, etc.).

Engineers frequently study momentum, energy, and mass transport in three separate, sequential, one-semester courses, as listed in the accompanying table.

Course	*Topics Usually Covered*
1. Fluid mechanics	Transport of linear momentum in incompressible fluid flow
2. Heat transfer	Energy transport in quiescent and flowing nonreactive media
3. Mass transfer[1]	Species transport in quiescent and flowing media (solute exchange between two solvents, evaporation, condensation, etc.)

Here the essential viewpoints and features of these *three* subjects will be concisely presented from a unified perspective (Chapters 2 through 7), with emphasis on their relevance to the quantitative understanding of *chemically reacting flow systems* (Chapters 6 and 7). Our goal is to complete the *foundation* necessary for dealing with modern engineering problems and more specialized topics useful to:

a. chemists, physicists, and applied mathematicians with little or no previous experience in the area of transport processes, and/or
b. engineers who *have* studied certain aspects of transport processes (e.g., including fluid mechanics, and heat transfer) but in isolation and/or divorced from their immediate application to chemically reacting (e.g., combustion) flow systems.

For several reasons (see the Preface) we will illustrate the principles of chemically reacting flows and reactor design and analysis using examples drawn primarily from the field of *combustion*—i.e., that branch of the engineering of chemical reactions in which the net exoergic chemical change accompanying the mixing and reaction of "fuel" and "oxidizer"[2] is exploited for such specific purposes

---

[1] Regrettably, many recent mechanical engineering students have completed curricula without a course in mass transfer.

[2] It is interesting to note that in 1816 the young Michael Faraday cited, as evidence of French chemists' prejudice, the fact that they refused to call the rapid exo-ergic reaction of iron with sulfur vapor by the name of *combustion* because the process involved no oxygen! Indeed, today, "combustion" embraces all fuels and oxidizers (chemical type and physical state), as well as exo-ergic decompositions of oxygen-free compounds (e.g., monopropellants such as hydrazine).

as power generation, propulsion, heat exchange, photon production, chemical synthesis, etc. Combustion examples have the merits that:

i. The field is encountered by virtually all engineers and applied scientists;
ii. These examples exhibit most of the important features of nonideal, transport-limited, nonisothermal chemical reactors; and
iii. They deal with an interdisciplinary subject of enormous industrial and strategic importance.

Convenient chemical fuels (easy to store, clean burning, energetic per unit volume, etc.) are today an especially vital commodity, and the need to efficiently synthesize and utilize such fuels continues to provide a powerful incentive to the study and advancement of combustion science and technology.

In the remainder of this introductory chapter we will first briefly illustrate the role of physical transport processes in several reasonably familiar combustion applications to which we will return in Chapters 6 and 7. We then introduce the basic strategy we will adopt to formulate and solve technologically important transport problems with (or without) chemical reactions.

1.1 PHYSICAL FACTORS GOVERNING REACTION RATES AND POLLUTANT EMISSION: EXAMPLES OF PARTIAL OR TOTAL "MIXING" RATE LIMITATIONS

The following specific but representative examples illustrate the important role of *physical* rate processes, and will serve to motivate our subsequent treatment of the quantitative laws of momentum, energy, and mass transport in flow systems with chemical reaction. For definiteness (see Preface) they deal, respectively, with a *premixed* (fuel + oxidizer) system, an initially unmixed *gaseous* fuel + oxidizer system, and an initially unmixed *condensed* fuel ("heterogeneous" combustion) system.

1.1.1 Flame Spread across IC Engine Cylinder

Consider events in a well-carbureted internal combustion (IC or "piston") engine cylinder following the firing of the spark plug (Figure 1.1-1). Of interest is the adequacy of the spark for ignition, and the time required for the combustion reaction to consume the fuel vapor + air mixture in the cylinder space (defined, in part, by the instantaneous piston location). One finds that if the spark energy is adequate the localized combustion reaction is able to spread outward, consuming the fresh reactants in a propagating *combustion "wave,"* which usually appears to be a "wrinkled" discontinuity in a high-speed photograph. What factors govern the

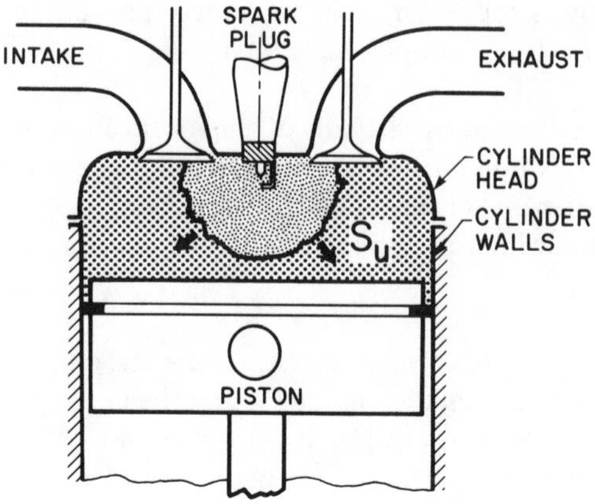

Figure 1.1-1 Flame spread across a carbureted internal combustion (piston) engine.

necessary spark energy deposition rate per unit/volume, and "flame" propagation rate across the chamber? Certainly the rates and exo-ergicities of the participating chemical reactions play a role, however; so do the *transport processes* which participate in determining the local compositions and temperatures. Thus:

a. If the spark energy deposition rate is too small, or the gas velocities past the gap too large, no local region of the gas mixture will be heated to a temperature high enough to allow the combustion region to continuously spread into the unburned gases—i.e., the fledgling combustion zone will "extinguish" in response to an unfavorable ratio of the rate of heat loss to that of heat generation.
b. Even where the unburned premixed gas is locally motionless, the rate at which it is heated to temperatures at which combustion reactions become appreciable is determined in part by the rate of energy diffusion (*conduction*) from the already burned gas (on the "hot" side of the wave).
c. Chaotic (turbulent) gas motion may augment the local rates of energy transfer to the unburned gas by suddenly projecting pockets of burned gas into the unburned region. This effect, combined with an augmentation in the (now wrinkled) flame *area*, act to increase the rate of propagation of the combustion wave across the unburned gas space.
d. In the immediate vicinity of the water-cooled cylinder walls the premixed gas loses energy to the wall, becoming considerably more difficult to ignite. This can cause local extinction without the combustion wave being able to consume all of the unburned fuel vapors originally in the chamber.

Clearly, each of these important facets of internal-combustion engine performance involves transport phenomena in a central way.

1.1.2 Gaseous Fuel Jet

Figure 1.1-2 shows an ignited, horizontal, turbulent gaseous fuel jet introduced into a surrounding air stream at a pressure level near 1 atm. Here, in contrast to Figure 1.1-1, the fuel and oxidizer vapors do not coexist initially, but must first "find each other" in a narrow reaction zone that separates the unreacted fuel region from the surrounding air. Hot combustion products, generated in this narrow reaction zone, mix in both directions, locally "diluting" but heating both fuel and oxidizer streams.

Initially unmixed fuel/oxidizer systems of this type have two principal advantages over their "pre-mixed" counterparts:

1. There is little explosion hazard involved in "recycling" energy (that might have been wasted in the effluent stream) into one or both of the reactant feed streams.
2. In "furnaces" (where the goal is to transfer as large a fraction as possible of the reaction energy to "heat sinks" placed within the reactor), such ("diffusion") flames are found to be better heat *radiators* (owing to the transient presence of hot *soot* particles) when carbonaceous fuels are burned (see Section 5.9.2).

The shape and the length of flame required to completely burn the fuel vapor are here dominated by turbulent transport, rather than chemical kinetic factors. This is suggested by the following interesting behavior:

a. An increase in the fuel jet velocity does *not* appreciably lengthen the flame!
b. Large changes in the chemical nature of the fuel have only a small influence on flame length and shape characteristics.

Thus, while (as is often the case) chemical kinetic factors may play an important role in pollutant emission (e.g., NO(g), soot), physical *transport processes* control the overall volumetric energy release rate (flame length, etc.).

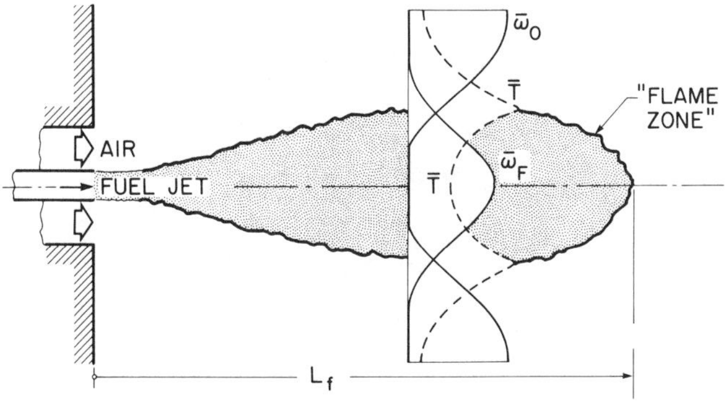

Figure 1.1-2 Horizontal gaseous fuel jet issuing into a turbulent, co-flowing, oxidizer-containing gas stream.

1.1.3 Single Fuel Droplet and Fuel Droplet Spray Combustion

Many useful fuels are not only conveniently *stored* as liquids, but they can be effectively *burned* (without complete "pre-vaporization") by spraying them directly into the combustion space (e.g., oil-fired furnaces, diesel-engine cylinders). If the ambient oxidizer concentration is adequate and the *relative* velocity between the droplet and gas is sufficiently small, such a droplet may be surrounded by an "envelope" diffusion flame (Figure 1.1-3) in which fuel *vapors* generated at the droplet surface meet inflowing oxygen. The situation is reminiscent of Section 1.1.2 in that the fuel and oxidizer vapor are separated from one another, meeting only at a thin reaction "front." However, here the energy generated at the flame zone must also be fed back to sustain the endothermic fuel *vaporization* process itself. Again, these *physical* processes usually control the overall combustion rate, as evidenced by the following behavior (cf. Rosner (1972) and Section 6.5.5.7):

a. The time to completely consume a fuel droplet depends *quadratically* on the initial droplet diameter;
b. The droplet lifetime is only weakly dependent on ambient gas temperature and pressure level, and even chemical characteristics of the fuel.

Actually, a fuel droplet usually finds itself in a local environment that cannot support an individual envelope diffusion flame. Rather, the conditions of oxidizer transport into the droplet cloud are such that most droplets collectively vaporize in fuel-rich environments which then supply a single vapor-phase *jet* diffusion flame, much like that shown in Figure 1.1-2. Again, this overall behavior is quite insensitive to the intrinsic *chemical kinetic* properties of these fuel/air systems. These considerations not only pertain to conventional (hydrocarbon) liquid fuels but also

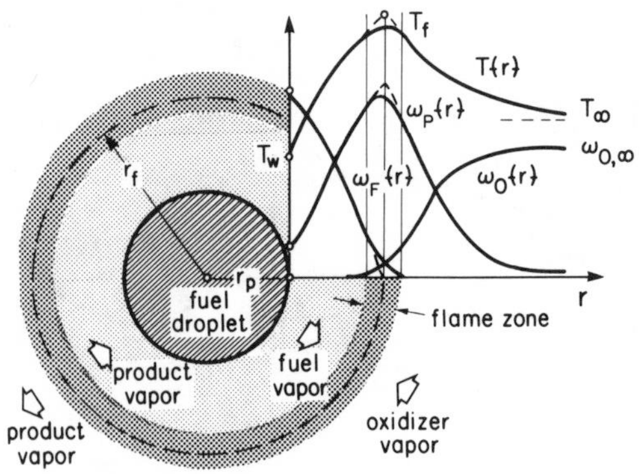

Figure 1.3-1 Envelope flame model of isolated fuel droplet combustion in an ambient gas containing oxidizer (adapted from Rosner (1972)).

to "fuels" such as liquid sulfur or liquid phosphorus burned in spray devices like that described here to produce, respectively, SO_2(g) (one step in the production of sulfuric acid), or P_2O_5 for fertilizer production.

1.2 CONTINUUM (*VS.* MOLECULAR) VIEWPOINT: LENGTH AND TIME SCALES OF FLUID-DYNAMIC INTEREST

Recall that from an experimental (phenomenological) point of view the laws of thermochemistry and chemical kinetics can be developed without postulating a molecular model of matter. This "macroscopic" point of view can be extended to deal with continuously deformable media (fluids) encountered in engineering applications. The resulting subject is then called "continuum" mechanics, or fluid mechanics, except that we must deal with fluids whose composition (and other state properties) change from point to point and with time.

Historically, the laws of mechanics and thermodynamics were first developed for discrete amounts of matter—e.g., a mass "point," an artillery projectile, the moon, etc. These same laws can, however, be extended to apply to fluids *that appear continuous on a macroscopic level* (e.g., the entire gas phase in Figures 1.1-1, 1.1-2), as shown in Chapter 2. This program, initiated by Cauchy, Euler, Lagrange, and Fourier, among others, provides the basis for quantitatively understanding and even predicting complex fluid motions, without or with simultaneous chemical reaction.

More generally, the combustion space is filled with more than one phase (e.g., Figure 1.1-3) and hence is discontinuous on a scale that is coarser than microscopic. Even such flows can be treated as continuous on the length scales of combustor interest (e.g., many centimeters or meters). However, to complete such a *pseudo-continuum* formulation, information is required on the local interactions between the "co-existing phases." Often this is provided from an intermediate scale analysis in which the region is considered piecewise continuous, and conservation laws are imposed within each continuum.

For definiteness, consider the gaseous-fuel jet situation sketched in Figure 1.2-1. A number of *macroscopic* dimensions are indicated, e.g.:

$r_j \equiv$ fuel jet radius,
$r_f \equiv$ radial location of the flame zone,
$L_f \equiv$ total flame length,
$L_c \equiv$ characteristic dimension of the combustion space.

Typically $r_j < r_f < L_f < L_c$; however, what is more important here is that each such *macroscopic* dimension is very large compared to the following "*microscopic*"

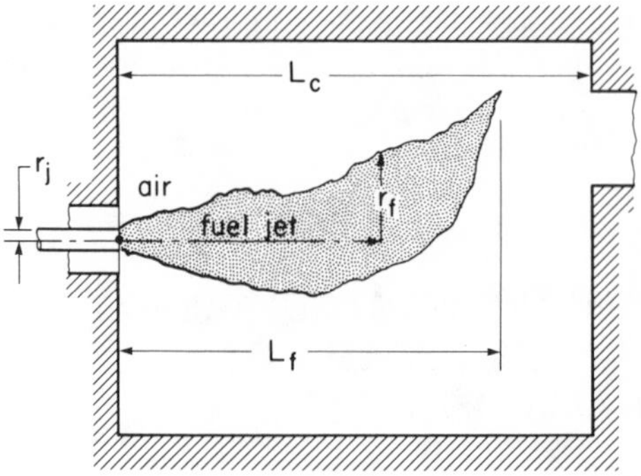

Figure 1.2-1 Gaseous fuel jet in confined furnace space: characteristic macroscopic lengths.

lengths:

$\sigma \equiv$ the characteristic diameter of a single molecule,
$n^{-1/3} \equiv$ the average distance between molecules (where $n \equiv$ number density),
$\ell \equiv$ the average distance traveled by a single molecule before it encounters another molecule (the "mean-free-path").

This disparity makes it possible to use a continuum formulation to treat *macroscopic* problems, explicitly ignoring the "molecularity" or "granularity" of the (gaseous) medium (e.g., at a point where $p = 1$ atm and $T \cong 1000$ K we have, approximately, $\sigma \approx 4 \times 10^{-1}$ nm, $n^{-1/3} \approx 5$ nm, and $\ell \approx 200$ nm (where 1 nm $\equiv 10^{-9}$ m)).

An equivalent statement can be made in terms of characteristic *times*, about which more will be said in Chapter 7. Thus, we have the macroscopic characteristic *times*:

$t_{\text{flow}} \equiv$ time for a representative fluid parcel to traverse the combustor length ($\approx L_c/U$, where U is the characteristic axial air velocity);
$t_{\text{diff}} \equiv$ time for a representative fluid parcel (or tracer constituent) to diffuse from the jet centerline to the flame zone,

etc. These *macroscopic* times are usually much longer than 1 ms ($\equiv 10^{-3}$ seconds). In contrast, consider

$t_{\text{interaction}} \equiv$ characteristic time during which a molecule interacts with a collision partner during a single binary encounter,
$t_{\text{collision}} \equiv$ average time elapsed before a molecule encounters another molecule,
$t_{\text{chem}} \equiv$ average time between encounters that are successful in bringing about reaction.

Ordinarily $t_{\text{interaction}} \ll t_{\text{collision}} \ll t_{\text{chem}}$; however, we will be concerned here with flows for which the first two of these times are very small compared to t_{chem} *and* the macroscopic t_{diff} and t_{flow}. (For a gaseous system with $p \approx 1$ atm, $T \approx 1000$ K, ordinarily $t_{interaction} \approx 0.5$ ps and $t_{collision} \approx 0.2$ ns (1 ps $\equiv 10^{-12}$ s).)

Once we suppress the "granularity" of real matter by treating it as continuous on the scale of our interest, we simplify enormously the local description of the fluid. For example, rather than attempting to describe the instantaneous translational, rotational, and vibrational motion of *each* of the 7×10^{18} molecules in a typical cubic centimeter (!) of gas mixture (at 1 atm, 1000 K, say), we merely have to consider certain *average* properties per unit volume of the prevailing fluid mixture, such as:

$\rho \equiv$ prevailing gas-mixture mass "density" (mass per unit volume)
$\rho\mathbf{v} \equiv$ prevailing gas-mixture linear momentum "density" (linear momentum/volume),

etc. Here ρ can be regarded as the limiting value of mass/volume when the volume about point $\mathbf{x}$ is shrunk to a value small compared to the cube of any *macroscopic* length, but kept much larger than the cube of the microscopic dimensions σ, $n^{-1/3}$, or ℓ, etc. (cf. Figure 1.2-2b). Because such quantities can be defined at each point, $\mathbf{x}$, in space at each time, t, they define "fields." While a complete description of an exchanger or reactor flow might require the simultaneous specification of many such fields (pressure, mass density, momentum density or velocity, temperature, individual species densities, etc.) in a certain spatial and temporal domain, this task is not only tractable (compared to that of keeping track of individual molecules), but also more than sufficient for most engineering purposes.

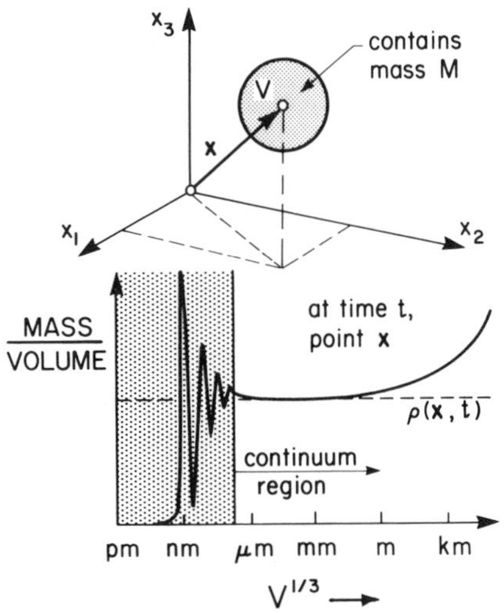

Figure 1.2-2 Operational definition of continuum density "at a point"; illustrative case of total mass density.

It is instructive to note, in passing, that it is sometimes necessary to define and compute "field densities" in a space that transcends physical space. In such cases the concept of a continuum "fluid" filling the resulting "hyperspace" is retained, and the procedures for deriving conservation equations governing such densities are identical to those to be detailed in Chapter 2 for field densities (mass, momentum, energy, entropy) in physical space. Perhaps the best example of the need for an "augmented" space arises in deriving the equation governing *radiative energy transport* (Section 5.9.4). In that case the minimum level of detail required usually includes not only position and time, but also frequency and direction (in space).

Unsteady scalar fields are often displayed at each instant by giving "snapshot" contours of constant field values; for example,

isobars ≡ contours of constant pressure (cf. meteorological map);
isotherms ≡ contours of constant temperature;
isopleths ≡ contours of constant species density, etc.

Often flows are locally chaotic ("turbulent") but the *time-average field variables* may be steady at each point. Steady *velocity* fields, or time-averaged velocity fields are usually displayed by giving their "stream-lines"—i.e., fluid parcel trajectories relative to fixed coordinates (experimentally realized in particle track photographs). Figures 1.2-3 through 1.2-8 collect some representative "fields" of combustion interest. They show, respectively:

1.2-3 Temperature distributions and streamlines in a natural gas/air laminar premixed flame.

1.2-4 Time-averaged temperature distributions and streamlines in the turbulent wake of a flame stabilizer (see also Section 7.2.3.2, Part b).

1.2-5 Species and temperature distributions in a one-dimensional premixed methane/air laminar flame (see Section 7.2.2).

1.2-6 Time-averaged NO profiles (ppm by volume) in a turbulent diffusion flame furnace.

1.2-7 Time-averaged streamlines in a swirling turbulent jet.

1.2-8 Time-averaged streamlines in a gas-turbine combustion chamber (schematic).

Note that these flow fields can be quite complicated,[3] and, hence, difficult to measure[4] or predict. Yet the field viewpoint and description is conceptually valuable in dealing quantitatively with reacting fluid-flow problems, including combustion.

---

[3] Figures 1.2-4, 1.2-7, and 1.2-8 reveal the existence of *closed* streamlines—i.e., important regions of fluid *recirculation.* Were it not for molecular and/or turbulent mixing, fluid parcels within these regions would never escape, but would merely cycle around forever!

[4] It is interesting to note that, in effect, Figure 1.2-5 displays the "internal" *structure* of a reaction zone so thin that it appears to be a "discontinuity" on the scale of Figure 1.2-3. Necessary relations between field variables across such "fluid dynamic discontinuities" will be discussed in Chapters 2 and 4.

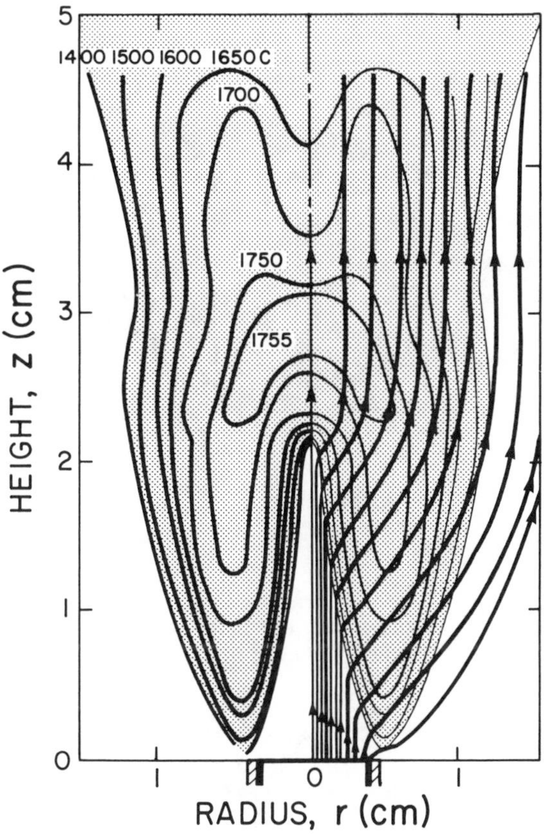

Figure 1.2-3 Temperature distributions and streamlines in a natural gas/air premixed flame (adapted from Lewis and von Elbe (1956)).

In the continuum description of a multiphase reacting flow, e.g., a spray fuel combustor flame, one can also define contours of constant time-average *liquid* fuel "density," i.e., mass of liquid per unit volume of space. A more complete description would include additional local information, such as:

a. the statistics of turbulent fluctuations about the local mean,
b. the local distribution of liquid fuel with respect to droplet size at each position **x** and time *t*, etc.

The habit of thinking in terms of coupled, coexisting "fields" does not imply that such a description is necessary, or even achievable, in every important engineering problem. As in all branches of applied science, it is the task of the engineer to select that level of approximation that is *sufficient* to deal with the problem at hand. Some problems can be solved by considering the entire device

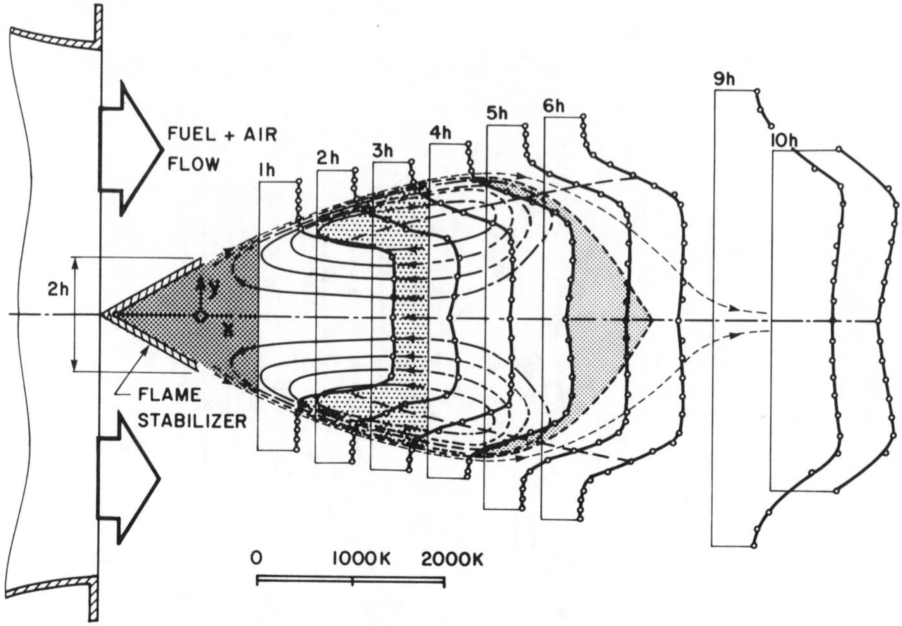

Figure 1.2-4 Time-averaged temperature distributions and streamlines in the wake of a premixed flame "gutter"-type stabilizer (adapted from Bespalov, *et al.* (1967)).

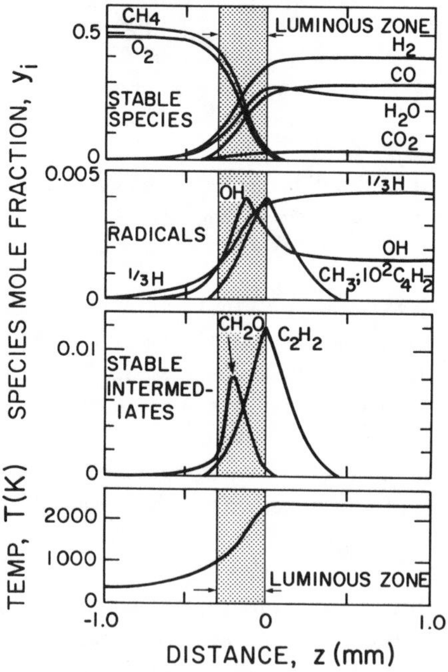

Figure 1.2-5 Species and temperature distributions in a one-dimensional methane/air flame (adapted from Goodings, *et al.* (1976)).

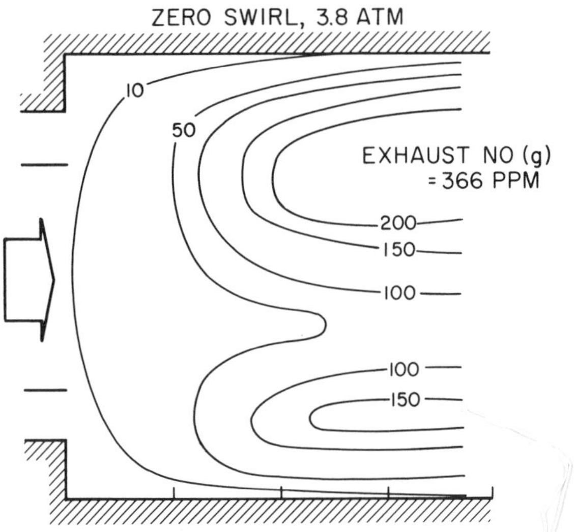

Figure 1.2-6 Time-averaged NO profiles (ppm by volume) in a turbulent diffusion flame furnace (adapted from Owen, *et al.* (1976)).

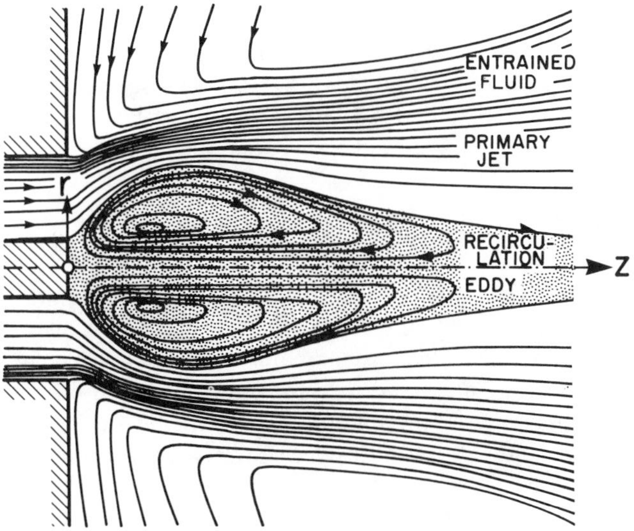

Figure 1.2-7 Time-averaged streamlines in a swirling turbulent annular jet containing a recirculation zone (adapted from Beer and Chigier (1974)).

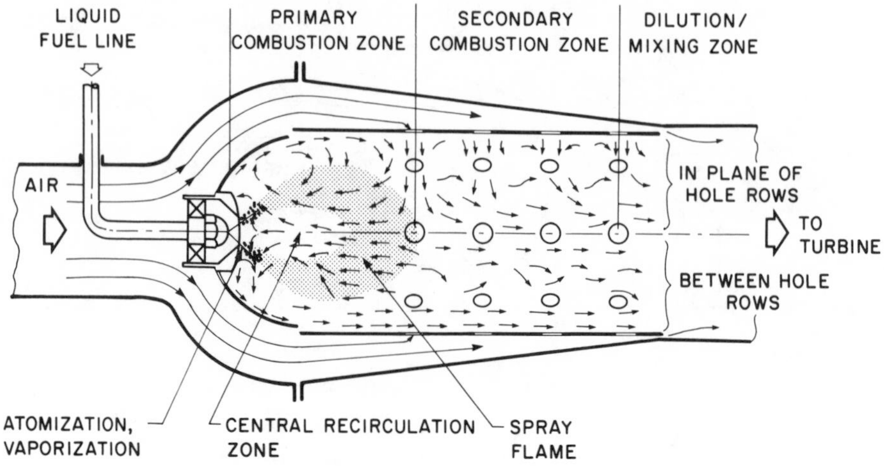

Figure 1.2-8 Sketch of time-averaged streamlines in a gas turbine combustion chamber.

as a single "black box," or as a network of smaller interconnected "black boxes." However, many current and future reactor problems, especially those involving new environmental constraints, will require a level of understanding and control attainable only through a *field description.*

We conclude this section with the following remark: The fact that we exploit a *macroscopic, continuum* mechanics viewpoint that is *phenomenological* in nature does not mean that we reject the insights available from *microscopic* theories of matter in various states of aggregation (e.g., vapor, liquid, solid)! Indeed, this information on the nature of the laws governing individual substances, the so-called *constitutive laws* (equations of state, flux laws, and coefficients, etc.), is especially valuable for interpolating or extrapolating the inevitably incomplete base of thermochemical, chemical kinetic, and transport property data.

1.3 TYPES/USES OF "CONTROL" VOLUMES

If one were to apply the conservation laws of classical mechanics to the single encounter between *two* discrete molecules, there would be no question as to the appropriate amount of matter to consider in our analysis. However, in applying the conservation laws of mechanics and thermodynamics to fluid matter *treated as a continuum,* questions arise as to the amount of matter to be considered. We shall see that in any particular flow problem this decision is often based on convenience or level of detail sought; moreover, there is no single choice—rather, many possibilities exist that can lead to the same useful predictions.

Indeed, the conservation laws of continuum dynamics can be applied to the fluid contained in a volume of arbitrary size, shape, and state of motion (cf. Figure 1.3-1). The volume considered is called a "control" volume (CV), and, conceptually, the simplest control volume is perhaps one that moves at every point on its surface with the local fluid velocity. Such a control volume is called a "*material*" *control volume* since, in the absence of molecular- (and/or eddy-) diffusion, it always retains the material originally present within its "control" surface (CS).

While the conservation laws of fluid dynamics are readily *stated* for material control volumes they are not so readily *used* in this form. This is because material control volumes:

i. move through space,
ii. change their volume,
iii. deform in shape.

As one simple example of material control volumes, while the valves of an IC piston engine are closed, the volume defined by the cylinder head, walls, and moving piston face is a material control volume of macroscopic dimensions (Figure 1.3-2).

An analysis of the motion of *material* control volumes is usually termed "*Lagrangian.*" Time derivatives calculated for material control volumes are sometimes called "*material derivatives,*" or "substantial" derivatives.

Another simple class of control volumes is the class of "*fixed*" *control volumes*—i.e., those defined by surfaces *fixed* in physical space, *through which the*

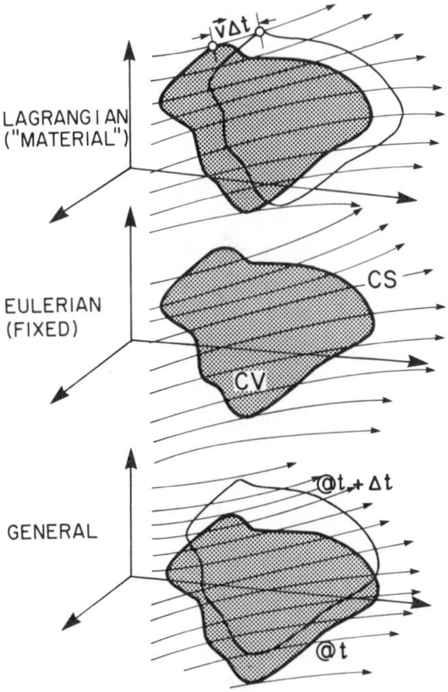

Figure 1.3-1 Types of fully macroscopic control volumes.

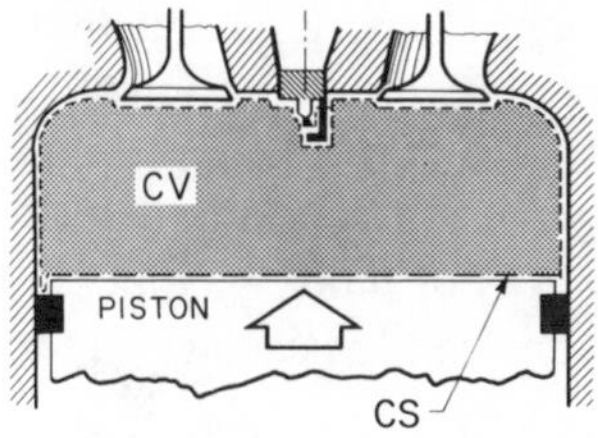

Figure 1.3-2 Total IC-engine combustion space as a "material" control volume (fully macroscopic).

fluid flows. Such control volumes are called "Eulerian," and may also be:

i. macroscopic in all directions,
ii. semi-"differential" (incremental),
iii. "differential" (incremental) in all directions.

The fluid contained within such fixed (Eulerian) CVs is said to be an "open" (flow) system in the thermodynamic sense. In the limit of ever-decreasing incremental dimension (subject again to the continuum restriction), conservation equations for cases (ii) and (iii) will be seen to take the form of *differential equations*, as discussed and displayed in Chapter 2. Most reactive fluid-dynamics problems are in fact solved using fixed (Eulerian) control volumes, and most diagnostic measurements are performed across fixed control surfaces. An example would be the volume defined by the internal walls of an entire reactor space (Figure 1.3-3, including surfaces that span (cut across) the inlet/exit ducts. A fully incremental (differential) representative fixed control volume would be the volume $\Delta x\,\Delta y\,\Delta z$ near the fixed point defined by the cartesian coordinates x, y, z (Figure 1.3-3). While the material CV shown in Figure 1.3-1 moves and deforms in a time increment Δt, by definition *fixed* (*Eulerian*) CVs initially coinciding with those above will remain in these positions, experiencing instantaneous fluid inflows and outflows.

The most general type of control volume is defined by surfaces that are neither fixed nor moving with the local fluid velocity—i.e., surfaces that move "arbitrarily." Such control volumes are used to analyze the behavior of non-material "waves" in fluids—such as flame sheets, detonations, etc., as well as moving phase boundaries in the presence of mass transfer across the interface (see, e.g., Section 2.6.1).

Occasionally the most convenient control volume to consider is a "hybrid" in the sense that some of its surfaces are fixed, some move with the local fluid velocity ("material"), and some may even move at a velocity that is neither of the above. For example, the macroscopic control volume defined by the liquid contained in an open-ended funnel is a "hybrid" CV (why?).

The conservation laws discussed in the next section, and more explicitly in Chapter 2, can be applied to the fluid contained in any of the above-mentioned types and "sizes" of control volume(s). In general, each resulting conservation ("balance") equation will contain time derivatives and, depending upon the number of incre-

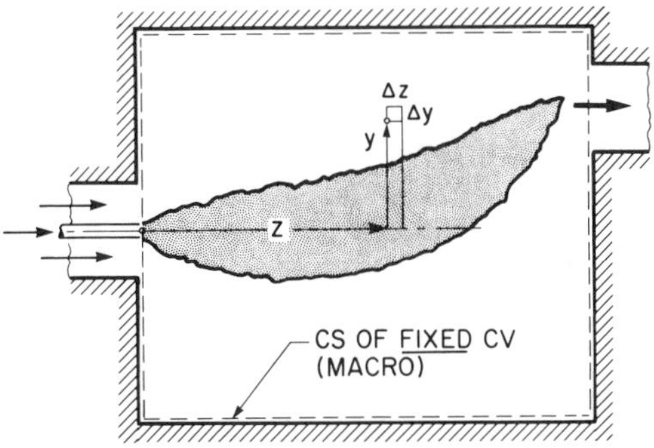

Figure 1.3-3 Fixed (Eulerian) control volumes for the quantitative description of furnace combustion. (a) Macroscopic control volume enveloping entire space within furnace; (b) representative fully incremental control volume at point x, y, z within the furnace space.

mental dimensions of the CV, will generate either:

a. integral equations governing single-phase or multiphase regions;
b. differential (or difference) equations in 1, 2, or 3 *spatial* dimensions within each single-phase subregion (see also Section 2.6.3); or
c. "jump" conditions across the interfaces (which separate adjacent continua).

1.4 NOTION OF CONSERVATION PRINCIPLES AND THEIR APPLICATION TO MOVING CONTINUA

Experience teaches us that there are certain laws which must be valid irrespective of the type of fluid mixture being considered. These laws, loosely called "conservation laws" or "balance" principles, will be seen to impose constraints upon the coexisting fields (pressure, temperature, species composition, etc.) describing instantaneous motions of a chemically reacting fluid mixture.

One such law is that, for ordinary chemical reactions: (a) total *mass* is "conserved." This is actually a corollary of another important conservation law governing all chemical reactions, *viz.*, each chemical element is "conserved." This means that ordinary chemical reactions involve the elements merely "changing partners," rather than their production or destruction. Individual chemical *species* are not really "conserved" (since they can be generated or consumed in chemical reactions), but one can nevertheless write an appropriate "balance" equation for each chemical species (Chapter 2).

Additional balance equations that must apply to *all* fluids deal with the *mixture*:

b. momentum (classical mechanics of Newton),

c. energy (First Law of Thermodynamics),
d. entropy (Second Law of Thermodynamics),

and will be discussed in detail in Chapter 2.

In elementary courses of physics, chemistry, and thermodynamics these principles are usually stated and illustrated for:

i. *discrete* systems (e.g., colliding spheres),
ii. *single* stoichiometric chemical reactions—e.g., the gas reaction $H_2 + \frac{1}{2}O_2 \rightarrow H_2O$, or
iii. *closed* systems (communicating with their "surroundings" *via* heat exchange and work terms).

To understand and predict exchanger (energy, mass) or reactor behavior, these same physicochemical laws must be cast in a form applicable to *multiphase, multicomponent, chemically reacting continua* in "open" flow systems defined by any of the control volume types discussed in Section 1.3. For mass and momentum conservation (e.g., classical fluid mechanics), this program was initiated some two centuries ago, with the primary contributions being associated with Newton, Euler, Cauchy, and Lagrange. Energy and entropy conservation principles for flowing continua were formulated somewhat later, with important contributions made by Fourier, Helmholtz, Clausius, Gibbs, Meixner, Prigogine, deGroot, and Kirkwood.

Depending on the type of control volume considered, these laws take rather different-looking mathematical forms (Chapter 2), but their simple ("primitive") physical content remains unaltered in each case. The many ways in which these laws can be (are) *used* will be discussed in Section 1.6 and further in the Introduction to Chapter 2.

1.5 NOTION OF "CONSTITUTIVE" LAWS (AND COEFFICIENTS) FOR PARTICULAR SUBSTANCES

All supplementary "laws," not necessarily valid for *all* substances but required to complete the quantitative description of any particular reactive flow, will be termed "*constitutive.*" While this term implies a dependence on the nature of the fluid (its "constitution"), such laws can in principle be completely "phenomenological"—i.e., directly measurable, and hence independent of any particular molecular-level model of the fluid.

Even completely phenomenological laws, however, are not free of *general* constraints imposed by the principles of continuum mechanics. One such constraint is set by the property that the resulting balance laws must retain the same form despite certain differences in "vantage point" (e.g., relative to observers who are translating with respect to one another). Such "invariance" requirements, to be considered briefly in Chapter 3, provide important guidance in the selection of relevant constitutive relations, including those governing turbulent flows.

Examples of "constitutive" relations are equations relating:

a. local "state variables" describing the mixture ("equations of state" (EOS));
b. the local *diffusion fluxes* of momentum, energy, and species mass to appropriate properties of the field variables (Section 2.5);
c. the local net rates of species production (due to chemical reactions) to the local field variables (concentration, temperature).

As will be seen in Chapter 3, these constitutive equations contain parameters (or "coefficients") that take on particular ranges of values for particular substances. In practice, these coefficients (chemical rate constants, thermal conductivities, etc.) are derived either from:

a. direct experiments on the same mixture but under extremely simple conditions (e.g., no flow, isothermal system, etc.—see also Section 1.6.1); or
b. a molecular model (e.g., kinetic theory of a "low-density" gas) which relates these coefficients with other, more readily measurable and more fundamental properties of the mixture or all of its constituents.

Approach (b), above, which draws heavily on the science of physical chemistry, is actually indispensable since the alternative of *measuring* all of the parameters appearing in the above-mentioned phenomenological constitutive reactions, under all possible local environmental conditions, would be quite out of the question.

1.6 USES OF CONSERVATION/CONSTITUTIVE PRINCIPLES IN SCIENCE AND TECHNOLOGY

To motivate a more detailed study of conservation (balance) principles and constitutive laws/coefficients for chemically reacting continua (Chapter 2), it is helpful to briefly consider some of the diverse *uses* to which these equations can be and will be put (see Figure 1.6-1).[5]

1.6.1 Inference of Constitutive Laws/Coefficients Based on Analysis and Measurement of Simple ("Canonical") Flow/Transport Situations

Constitutive laws and coefficients are, in fact, *determined* by comparing experimental measurements with predictions based on these equations for particularly

[5] We will return to this figure in Section 7.4, giving specific examples that are explicitly covered in this book.

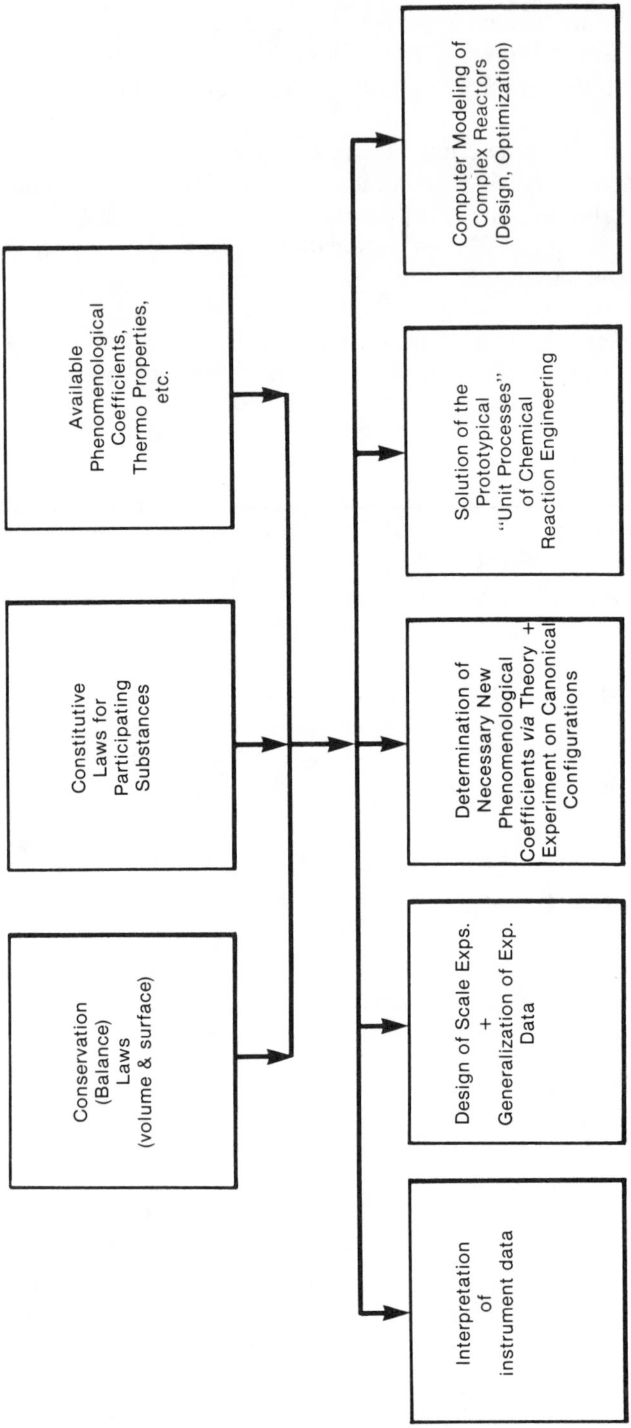

Figure 1.6-1 Uses of the conservation and constitutive relations.

simple geometric and flow situations with a minimum number of participating physicochemical phenomena. An example is the use of steady flow through straight channels (Chapter 4) or circular ducts, to determine the viscosity coefficient μ of so-called Newtonian fluids (see Chapters 2 and 4).

1.6.2 Solution of Simpler "Prototype" Problems Illustrating Effects of the Basic Interacting Phenomena

To gain an *overall* understanding of transport processes in practical devices (chemical reactors, furnaces, engines, etc.), it is necessary to gain an understanding of the relevant "unit" processes—i.e., how chemical and physical processes *interact* in representative simpler, usually smaller-scale, circumstances. Examples would be simple quantitative *theories* of:

a. the ignition of a planar or spherical pocket of premixed gas following thermal-energy deposition;
b. the steady-state propagation of a planar flame through a uniform, premixed, nonturbulent gas;
c. the oxidation rate of an isolated spherical fuel droplet (Figure 1.1-3) or particle.

Often the emphasis of these analyses is not on accuracy but rather on an understanding of important relevant groupings of property, geometric, and environmental parameters, and functional dependencies. For this reason apparently drastic simplifying assumptions are introduced; however, the essential interacting physicochemical phenomena must, of course, be retained. Apart from the understanding gained, these results are usually also needed as input information in the solution of more complex, larger-scale reaction-engineering problems (cf. Section 1.6.4).

1.6.3 Guide Design of Small-Scale or Full-Scale Experiments, and the Interpretation and Generalization of Experimental Results

In designing a set of experiments on, say, a small-scale ("model") reactor, one must realize that:

a. Measurements of *all* properties at *all* points in the reactor, separator, etc., are never feasible, or even necessary, for the purpose at hand;
b. Only if operating conditions and geometry are selected in a particular way (see Chapter 7) will it be simple to quantitatively predict full-scale behavior on the basis of small-scale model test results.

Quantitative analyses based on the conservation and constitutive laws of transport provide the information necessary to minimize the required number of measurements and make maximum use of whatever measurements *are* made.

1.6.4 Comprehensive Exchanger/Reactor Design Predictions via Computer Modeling

Now that high-speed, large-memory, electronic digital computers are available, it is becoming feasible to assemble basic data, conservation and constitutive laws, and our understanding of the elementary phenomena and "unit processes" into comprehensive computer codes to *predict* and optimize new exchanger or reactor performance. This approach, seriously initiated in the 1970s, should ultimately reduce the costs of new reactor system design and development. It has already had an effect on reaction-engineering *research* strategy, helping to direct attention to areas of uncertainty having the greatest influence on device performance.

1.6.5 Interpretation of Instrument Measurements Made in the Laboratory or Field

In validating proposed mathematical models or making tests to determine full-scale or model system performance, many *measurements* (temperatures, velocities, compositions, flow rates, etc.) are required, using new or "classical" immersion ("probe") techniques, optical (nonintrusive) techniques, etc. To properly interpret these measurements, as well as to make rational corrections for the effects of unwanted but inevitable systematic errors, it is always necessary to invoke the above-mentioned conservation and constitutive equations, or their immediate consequences. A simple but representative and important example is the use and proper interpretation of thermocouple temperature probe measurements (i.e., correcting for the effects of "stagnation" temperature rise in high-velocity streams (Section 4.3.1.2), thermal-radiation transfer (Section 5.9), conduction heat loss along the thermocouple wire leads and/or "well" walls (Section 5.4.5), and/or surface-catalyzed chemical reactions (of unburned fuel vapor and/or nonequilibrium molecular fragments (Section 6.5.4). (Can you supply another example from your own experience?)

With this background we now turn to the requisite underlying conservation laws (Chapter 2) and constitutive laws (Chapter 3), and their interesting consequences for the transport of *momentum* (Chapter 4), *energy* (Chapter 5), and (species) *mass* (Chapter 6).

SUMMARY

- Most chemical reactor applications involve momentum, energy, and mass-transport processes in a central way; useful predictions of time and/or space requirements *cannot* be made solely on the basis of thermodynamics and chemical kinetics.

- Most flows of engineering interest are fortunately well into the continuum range; this means that we can (a) circumvent the need to treat the motion of individual molecules, (b) define field densities that are continuous functions of $\mathbf{x}, t$ (some fields are scalars, some vectors, and some tensors, defined by six independent scalars at each $\mathbf{x}, t$), and (c) develop a *phenomenological* theory, independent of any postulates about the molecular nature of the fluid.
- The field densities are not independent of one another but are linked by *conservation constraints*. These can be expressed as:

  integral constraints (any macroscopic control volume (CV));

  differential (local) constraints (differential CV);

  "jump" constraints (across apparent discontinuities).

  The most convenient CVs are usually fixed (Eulerian), except in the case of CVs that "straddle" moving discontinuities (shocks, phase boundaries, etc.), or those defined relative to coordinate systems attached to moving objects (e.g., a turbine rotor blade).
- The individuality of substances shows up in their *constitutive* laws; the conservation (balance) laws apply to *all* substances.
- These quantitative laws summarize the cumulative experience of centuries, have many uses, and are the logical starting point in grappling with novel transport problems with chemical reactions (see Figure 1.6-1 for uses) *via* both experimental and theoretical methods.

TRUE/FALSE QUESTIONS

1.1 T F The individuality of substances shows up in their *constitutive* laws; in contrast, the *conservation* (balance) laws apply to *all* substances, irrespective of their state, chemical composition, etc.

1.2 T F The success of the continuum approximation is a consequence of the fact that characteristic flow times and diffusion times are always much larger than the characteristic time between "successful" (chemically reactive) molecular encounters.

1.3 T F The fundamental conservation laws applied to the fluid within an Eulerian control volume necessarily lead to partial differential equations (PDEs).

1.4 T F A continuum formulation cannot be used to predict the behavior of flow fields that contain local "discontinuities," such as shock waves, detonation waves, or deflagration waves.

1.5 T F A detailed local "field description" is now needed to solve all transport problems of engineering interest.

1.6 T F While the partial differential equations and boundary/initial conditions governing all coupled fields can always be written, they cannot be solved, even numerically, in any case of practical interest.

1.7 T F "Jump" conditions can be derived only across discontinuities that are so thin that they preclude a continuum analysis of their structure (e.g., strong detonations, shock waves, phase boundaries, etc.).

1.8 T F An empirical approach, based on the use of extensive probe measurements, circumvents the need to understand and apply the principles of transport processes.

1.9 T F The final answer one gets in solving an engineering transport problem should be independent of the particular control volume(s) used in the analysis, but some choices of control volume(s) will inevitably be much more convenient than others.

1.10 T F Control volumes that are *neither* fixed in space (Eulerian) nor moving with the fluid (material, or Lagrangian) are not useful in the analysis of chemically reacting fluid mixtures.

REFERENCES

1. Bespalov, I.V., Raushenbakh et al., translated from Russian, Clearinghouse for Federal Scientific and Technical Information, USA, AD 658372, p. 366 (May, 1967).
2. Goodings, J.M., D.K. Bohme, and T.M. Sugden, *Sixteenth Symp. (Int.) on Combustion*, The Combustion Inst., Pittsburgh, PA. (1976), pp. 891–902.
3. Lewis, B., and G. von Elbe, in Vol. 2, Section G, *High-Speed Aerodynamics and Jet Propulsion*, pp. 216–311, Princeton Univ. Press, Princeton, NJ (1956).
4. Owen, F.K., L.J. Spadaccini, C.T. Bowman, in *Sixteenth Symp. (Int.) on Combustion*, The Combustion Inst., Pittsburgh, PA (1976), pp. 105–117.
5. Rosner, D.E., in *Liquid Propellant Rocket Combustion Instability* (D.T. Harrje, ed.) NASA SP-194, Section 2.4, pp. 74–100 (1972).

BIBLIOGRAPHY

Barnard, J.A., and J.N. Bradley, *Flame and Combustion.* London: Chapman and Hall Science Paperback (1985).

Beer, J.M., and N.A. Chigier, *Combustion Aerodynamics.* New York: Halstead Press, Div., J. Wiley (1972).

Bird, R.B., W.E. Stewart, and E.N. Lightfoot, *Transport Phenomena.* New York: J. Wiley (1960).

Chigier, N.A. *Energy, Combustion and Environment.* New York: McGraw-Hill (1981).

Faraday, M. *The Chemical History of a Candle* (6 Lectures 1860–1861). New York: Collier Books (Paperback AS348) (1962).

Fristrom, R., and A.A. Westenberg, *Flame Structure.* New York: McGraw-Hill (1965).

Gaydon, A.G., and H.G. Wolfhard, *Flames, Their Structure, Radiation, and Temperature* (4th Ed.). London: Chapman and Hall (1979).

Glassman, I., *Combustion.* New York: Academic Press (1977).
Hirschfelder, J.O., C.F. Curtiss, and R.B. Bird, *The Molecular Theory of Gases and Liquids.* New York: J. Wiley (1954).
Jost, W. *Explosion and Combustion Processes in Gases.* New York: McGraw-Hill (1946).
Lewis, B., R.N. Pease, and H.S. Taylor, Eds., *Combustion Processes* (Vol. II), High-Speed Aerodynamics and Jet Propulsion. Princeton, NJ: Princeton Univ. Press (1956).
Lewis, B., and G. von Elbe, *Combustion, Flames, and Explosions of Gases.* New York: Academic Press (1951).
Rosner, D.E., *Chem. Engrg. Ed.* (ASEE) **14**, 193–196, 209–212 (1980).
Spalding, D.B., *Some Fundamentals of Combustion.* London: Butterworths (1955).
Strehlow, R.A., *Combustion Fundamentals.* New York: McGraw-Hill (1984).
Williams, F.A., *Combustion Theory.* Reading, MA: Addison-Wesley (1965); Second Ed. Benjamin-Cummings (1985).

Additional Sources of Information

Symposia (Int.) on Combustion (20 Proceedings through 1985).
Progress in Energy and Combustion Sci. (Journal, Pergamon Press).
Combustion Sci. and Technology (Journal, Gordon and Breach, Inc.; printed in Great Britain).
Combustion and Flame (Journal, the Combustion Institute).
J. Propulsion and Power (Amer. Inst. Aeron. Astronautics).
J. Inst. Energy (formerly, *Fuel*; Journal, London).

SOLAR ENERGY CENTER LIBRARY

2

Governing Conservation Principles

INTRODUCTION

Approach

Chemically reacting fluids, including gas mixtures, can be quantitatively understood in terms of two types of "laws" applied to matter treated as a continuum:

a. Conservation laws, which summarize the experience of the last three centuries on the behavior of *all* forms of matter;
b. "Constitutive" laws, which quantitatively describe the behavior of certain subclasses of "fluids" (e.g., perfect gas mixtures, elastic solids, etc.).

The conservation principles, to which we first direct our attention, ensure that, *for any fluid, in any state of motion*, the following quantities are either conserved or, more generally, "balanced":

Mass	Total mixture mass Individual chemical species (e.g., CH_4, O_2, CO_2, etc.) mass Individual chemical element (e.g., C, O, S, N, etc.) mass
Momentum	Total linear momentum of the mixture (a vector) Total angular momentum of the mixture (a vector)
Energy	Total energy (thermodynamic + kinetic) (First Law of Thermodynamics) Kinetic (mechanical) energy
Entropy[1]	Total entropy of mixture (describing the consequences of transformations between forms of energy that are not thermodynamically equivalent (Second Law) and identifying all sources of irreversibility and the nature of entropy transport by diffusion).

[1] From the Greek: εντρωπη, meaning "evolution."

Each of these quantities is associated with a "field density," i.e., a spatial concentration of that quantity, related, in turn, to the local material mass density, as shown in Table 2-1. Note that while most of these field densities are scalars (defined by one number at each point/instant), $\rho\mathbf{v}$ is a *vector* field (defined by three numbers at each point/instant).

These conservation principles apply not only to any fluid, but to any region of space (subject to the continuum restriction). Thus, we state them below in three important forms, applicable to a control volume (CV) that is:

i. of arbitrary size/shape (leading to integral equations);
ii. fully differential (leading to partial differential equations (PDEs) relating the above-mentioned *local* field densities at each point/instant);
iii. straddling a region of rapid spatial change (discontinuous on a macroscopic level, e.g., a phase boundary, thin flame, etc.).

For cases (i) and (ii) we consider the CV to be fixed in space (Eulerian). In case (iii) the most convenient CV is one that moves so that it always "straddles" the relevant "interface."

We make the following basic assumptions; some of which can be relaxed, as outlined below (see Section 2.6):

1. Certain material properties are functions of local state alone and are interrelated by the laws of equilibrium thermodynamics, even though the fluid may be undergoing rapid spatial and temporal change.

Table 2-1 Local Field Densities of Interest in a Continuum

Quantity	*Local Field Density*	*Meaning*
Total mass	ρ	Total mass per unit volume
Species i mass ($i = 1, 2, \ldots, N$)	$\rho_i = \rho\omega_i$	Total mass of species i per unit volume ($\omega_i \equiv$ *mass* fraction)
Element k mass ($k = 1, 2, \ldots, N_{\text{elem}}$)	$\rho_{(k)} = \rho\omega_{(k)}$	Total mass of element k ($k = 1, 2, 3, 4, 5, \ldots$ representing C, O, H, N, S, etc.) per unit volume
Linear momemtum	$\rho\mathbf{v}$	Linear momentum of mixture per unit volume (a vector)
Total energy	$\rho\left(e + \frac{v^2}{2}\right)$	Total (internal + kinetic) energy of material mixture per unit volume
Mechanical (kinetic) energy	$\rho(v^2/2)$	Local kinetic energy per unit volume
Entropy	ρs	Entropy of local mixture per unit volume
etc.		

2. Each subregion to which we apply the above-mentioned conservation principles contains only a single phase, and may be treated as a *continuum*.
3. Only extranuclear (ordinary chemical) reactions occur, so that chemical elements are conserved. Thus, we preclude fission and fusion reactions, which are associated with (a) the production or destruction of certain chemical elements (U, Ba, He, D, etc.), and (b) mass "defects" (converted into energy *via* the Einstein $(\Delta m)c^2$ law).
4. All speeds and thermodynamic state conditions are such that the laws of classical mechanics apply to the motion of the continuum. This precludes relativistic effects (near the speed of light) and quantum effects (superfluidity, etc.) at very low absolute temperatures.

The individuality of substances shows up through their "constitutive" laws, i.e., explicit laws governing:

a. interrelations between local state variables (p, T, e, s, ρ, etc.);
b. interrelations between local diffusional fluxes (mass, momentum, energy, entropy) appearing in the conservation equations and local field densities; and
c. local rates of production of the individual chemical species (chemical kinetic "net-source" terms).

Without these "constitutive" laws, the conservation equations themselves are insufficient to completely solve most practical problems; constitutive laws are needed to "close" the above-mentioned set of quantitative relations.

While most easily *stated* for a control volume (CV) locally moving with the fluid (material or "Lagrangian"), the conservation equations below will be written in their most *useful* form, *viz.*, for a fixed (Eulerian) CV through which the fluid flows (these two alternate forms of the conservation principles are kinematically interrelated). When the CV is differential, we will first state the governing PDE in a form having the same physical meaning irrespective of the particular coordinate system used to locate points in space (cartesian, cylindrical, spherical, etc.). The implications of local conservation are then made explicit for one particular choice of coordinate system; *viz.*, cylindrical polar coordinates (the choice of coordinate system to describe any particular flow is, of course, quite arbitrary, being governed largely by convenience).

All of the fixed CV conservation equations can be cast in the same form; i.e., at each instant:

$$\left\{\begin{array}{l}\text{Rate of}\\ (\quad)\text{ accumu-}\\ \text{lation}\\ \text{in CV}\end{array}\right\} + \left\{\begin{array}{l}\text{Net outflow rate}\\ \text{of }(\quad)\text{ by}\\ \text{convection}\\ \text{across CS}\end{array}\right\} = \left\{\begin{array}{l}\text{Net inflow rate}\\ \text{of }(\quad)\text{ by}\\ \text{diffusion}\\ \text{across CS}\end{array}\right\} + \left\{\begin{array}{l}\text{Net source}\\ \text{of }(\quad)\\ \text{within}\\ \text{CV}\end{array}\right\} \tag{2-1}$$

where ( ) applies to mass, momentum, energy, or entropy (see Section 2.5.2).

When the control volume is differential, each PDE stated below is obtained by simply *dividing* this standard equation by the magnitude of the volume V (e.g., $\Delta x\,\Delta y\,\Delta z$ in a cartesian coordinate system) and passing to the limit $V \to 0$. In stating each PDE we will adopt the following "language":

$$\lim_{V\to 0}\left[\frac{1}{V}\begin{Bmatrix}\text{Net outflow}\\ \text{associated}\\ \text{with (\quad)}\\ \text{across CS}\end{Bmatrix}\right] \equiv \begin{matrix}\text{Local}\\ \text{"divergence" of (\quad)} \equiv \operatorname{div}\{\quad\}\end{matrix} \tag{2-2}$$

Thus, whenever div{ } appears in a subsequent PDE, it simply means the local net outflow associated with the flux ( ), calculated on a per-unit-volume basis. Conversely, −div{ } is the net *inflow* per unit volume.

The conservation equations in their various forms (given below) have many uses indeed. Briefly, the *macroscopic* CV (integral) equations are used:

a. to test predictions or measurements for overall conservation,
b. to solve "black box" problems (without inquiring into internal details of the flow),
c. to derive finite difference (element) equations using arbitrary, coarse meshes, Appendices 8.2 and 8.3, and
d. as a starting point for deriving *multiphase* flow conservation equations (Section 2.6.4).

The PDEs (local field equations) are often used to do the following:

a. Predict the detailed distribution of flow properties within the region of interest.
b. Extract flux laws/coefficients from measurements in simple flow systems.
c. Provide a basis for estimating the important dimensionless parameters governing a chemically reacting flow (Section 7.2.2, Appendix 7.1).
d. Derive finite difference (algebraic) equations for numerically approximating the field densities (Appendix 8.2).
e. Derive (in the case of the entropy conservation equation) the *entropy production* expression (Sections 2.4, 2.5.6) and provide guidance as to the proper choice of flux-driving force expressions (constitutive laws).

Conservation statements at a surface of discontinuity are used to provide a rational basis for the selection of appropriate boundary conditions or "jump" conditions across real or apparent surfaces of "discontinuity" (e.g., shock waves, flames, etc.) separating adjacent continuum regions within a flow field.

The principles treated in Sections 2.1 through 2.5 will:

a. summarize all of the relevant laws of nature in a form useful for the quantitative treatment of engineering problems, and
b. not only provide the foundation for all subsequent sections of this course, but also subsequent courses in all branches of engineering and applied science.

Accordingly, for the present we concentrate on a concise yet general statement of all of the important laws. In Chapters 3 through 8, applications (special cases) of most of these laws will be encountered, and will warrant reference to this chapter to fully understand the nature of the approximations subsequently introduced.

2.1 CONSERVATION OF MASS

2.1.1 Total Mass Conservation

This is the simplest of the conservation principles, since total mass cannot be created by chemical reactions, and it can only be transported by convection, not diffusion. [The latter is a consequence of our definition of $\mathbf{v}$ as the momentum per unit mass of mixture. This means that $\rho\mathbf{v}$ is not only the linear momentum density, but also the total *mass flux vector* $\dot{\mathbf{m}}''$ (that is, the total amount of mass that passes through a fixed unit area perpendicular to $\mathbf{v}$ in a unit of time).] Accordingly, we can write that, for any fixed CV and at any instant, we must have:

$$\begin{Bmatrix}\text{Rate of}\\ \text{mass}\\ \text{accumu-}\\ \text{lation in CV}\end{Bmatrix} + \begin{Bmatrix}\text{Net outflow rate}\\ \text{of mass by}\\ \text{convection}\\ \text{across CS}\end{Bmatrix} = 0. \tag{2.1-1}$$

Mathematically, Eq. (2.1-1) can be expressed as an *integral constraint* on the ρ and $\rho\mathbf{v}$ fields, i.e.:

$$\frac{\partial}{\partial t}\int_V \rho\, dV + \int_s \rho\mathbf{v}\cdot\mathbf{n}\, dA = 0, \tag{2.1-2}$$

where the "dot product" $\rho\mathbf{v}\cdot\mathbf{n}\,dA$ symbolizes the amount of mass flow through the area $\mathbf{n}\,dA$ per unit time (even when the vectors $\rho\mathbf{v}$ and $\mathbf{n}\,dA$ are not colinear) and $\int_s$ symbolizes summation over *all* such control surface elements on the overall CS (cf. Figure 2.1-1).[2]

Around any typical (nonsingular) point within the fluid we can divide by $V \equiv \int_V dV$ and pass to the limit $V \to 0$. This provides the following local (PDE) constraint on the instantaneous ρ and $\rho\mathbf{v}$ "density" fields:

$$\frac{\partial\rho}{\partial t} + \operatorname{div}\{\rho\mathbf{v}\} = 0, \tag{2.1-3}$$

sometimes (loosely) called the equation of "continuity."

[2] Everywhere the angle between $\rho\mathbf{v}$ and $\mathbf{n}\,dA$ exceeds $\pi/2$ radians (90°) $\rho\mathbf{v}\cdot\mathbf{n}\,dA$ changes sign, automatically accounting for the *inflow* contributions.

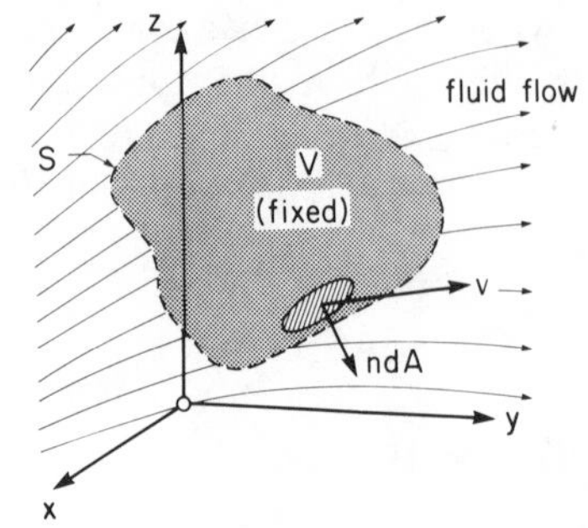

Figure 2.1-1 Arbitrary shape/size fixed (Eulerian) control volume V, bounded by a fixed control surface through which the fluid flows. Note representative area element $d\mathbf{A}$ whose direction is denoted by the unit vector $\mathbf{n}$ (outward normal).

The implications of the PDE (2.1-3) are, perhaps, clearest when examined in any particular coordinate system. For example, if we consider the cylindrical polar coordinate system in which:

i. a point in space is defined by its r, θ, z coordinates (Figure 2.1-2),
ii. the relevant element of volume ΔV is defined by the region of space "trapped" inside of the coordinate surfaces:

$$\left.\begin{array}{lll} r & \text{and} & r + \Delta r = \text{const}, \\ \theta & \text{and} & \theta + \Delta\theta = \text{const}, \\ z & \text{and} & z + \Delta z = \text{const}, \end{array}\right\} \qquad \text{i.e., } \Delta V = r\,\Delta r\,\Delta\theta\,\Delta z,$$

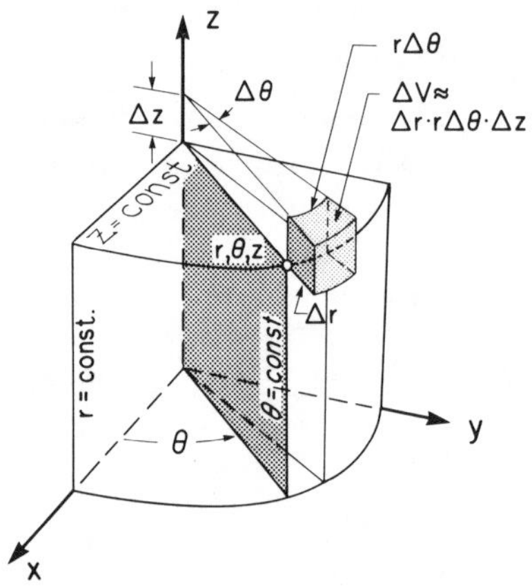

Figure 2.1-2 Coordinate surfaces in the cylindrical polar coordinate system. Here the position of each point in space is defined by the three numbers: r, θ, z. The volume element about this point is $(\Delta r) \cdot (r\,\Delta\theta) \cdot (\Delta z) \equiv \Delta V$. (See, also, Figure 2.5-4(a).) Each local vector (say, the velocity $\mathbf{v}$) can, appropriately, be resolved into its r, θ, z components (v_r, v_θ, v_z, respectively).

and

iii. the net outflow of mass is due to the algebraic sum of the outflows across each of these surfaces,

then it is easy to prove that Eq. (2.1-3) becomes the single explicit PDE:

$$\frac{\partial \rho}{\partial t} + \left\{ \frac{1}{r} \cdot \frac{\partial}{\partial r}(r \rho v_r) + \frac{1}{r} \cdot \frac{\partial}{\partial \theta}(\rho v_\theta) + \frac{\partial}{\partial z}(\rho v_z) \right\} = 0, \tag{2.1-4}$$

involving spatial gradients (derivatives) of the three relevant components of the mass flux vector $\rho\mathbf{v}$ in this particular coordinate system. Similarly, Eq. (2.1-3) can be expressed in any other coordinate system (see Section 2.5.7) but its physical content remains that of Eq. (2.1-1) divided by V in the limit $V \to 0$.

The principle of total mass conservation, Eq. (2.1-1), also applies across a "surface of discontinuity," which may be itself moving (e.g., a premixed flame "sheet"). In such cases the accumulation term is usually negligible and Eq. (2.1-1) takes the degenerate form:

$$\left\{ \begin{matrix} \text{Net outflow rate} \\ \text{of mass by convec-} \\ \text{tion relative} \\ \text{to the CS} \end{matrix} \right\} = 0. \tag{2.1-5}$$

It is usually expressed *per unit area* of the surface (see Figure 2.1-3 and Section 2.6.1).

2.1.2 Individual Species Mass Balance

Any individual species, designated by the subscript $i (i = 1, 2, \ldots, N)$, also is governed by a simple "conservation" (balance) principle, except we must now allow for

a. mass transport by *diffusion* (as well as convection), and
b. net production (production − consumption) as a result of all *homogeneous* reactions.

The standard form of species i mass conservation for a fixed CV can therefore be expressed in words as follows: For any fixed CV, at any instant:

$$\left\{ \begin{matrix} \text{Rate of accu-} \\ \text{mulation of} \\ \text{species } i \text{ mass} \\ \text{within CV} \end{matrix} \right\} + \left\{ \begin{matrix} \text{Net outflow rate} \\ \text{of species } i \\ \text{mass by convec-} \\ \text{tion across CS} \end{matrix} \right\} = \left\{ \begin{matrix} \text{Net inflow rate} \\ \text{of species } i \\ \text{mass by diffu-} \\ \text{sion across CS} \end{matrix} \right\} + \left\{ \begin{matrix} \text{Net chemical} \\ \text{source strength} \\ \text{of species } i \\ \text{mass with CV} \end{matrix} \right\}. \tag{2.1-6}$$

2.1.2.1 Definitions

Since the *convective* flux of mass, $\rho_i \mathbf{v} = \omega_i \rho \mathbf{v}$ does not, in general, account for the *total* local species i flux $\dot{\mathbf{m}}_i''$, we define:[3]

$$\mathbf{j}_i'' \equiv \dot{\mathbf{m}}_i'' - \rho_i \mathbf{v} \equiv \text{“diffusion” flux of species } i.$$

Moreover, let:

$\dot{r}_i'''$ {local state of mixture} ≡ net rate of production of species i mass per unit volume (*via* homogeneous chemical reactions).

In terms of these quantities Eq. (2.1-6) takes the following explicit form as an integral constraint on the local $\rho_i, \rho\mathbf{v}$ and $\mathbf{j}_i''$ fields:

$$\frac{\partial}{\partial t}\int_V \rho_i \, dV + \int_S \rho_i \mathbf{v} \cdot \mathbf{n} \, dA = -\int_S \mathbf{j}_i'' \cdot \mathbf{n} \, dA + \int_V \dot{r}_i''' \, dV. \qquad (2.1\text{-}7a)$$

Division by V and passing to the limit $V \to 0$ about some typical (nonsingular) point in the flow field immediately provides the following *local* PDE constraint on the ρ_i, $\rho\mathbf{v}$ and $\mathbf{j}_i''$ fields:

$$\frac{\partial \rho_i}{\partial t} + \text{div}\{\rho_i \mathbf{v}\} = -\text{div}\{\mathbf{j}_i''\} + \dot{r}_i''' \qquad (i = 1, 2, \ldots, N). \qquad (2.1\text{-}7b)$$

The explicit form of Eq. (2.1-7b) in cylindrical polar coordinates is immediately evident from a comparison of Eqs. (2.1-3) and (2.1-4), which reveals that in this coordinate system the “divergence” (net outflow/unit volume) of *any* vector $\mathbf{B}$ (with components B_r, B_θ, B_z) is given by:

$$\text{div}\,\mathbf{B} = \frac{1}{r}\cdot\frac{\partial}{\partial r}(rB_r) + \frac{1}{r}\cdot\frac{\partial}{\partial \theta}(B_\theta) + \frac{\partial}{\partial z}(B_z). \qquad (2.1\text{-}8)$$

This result, of course, applies to the vectors $\rho_i \mathbf{v}$ and $\mathbf{j}_i''$ appearing in Eq. (2.1-7).

The species mass balance applied to each unit area of a “surface of discontinuity” (which may itself be moving) provides a “jump” condition of the form:

$$\left\{\begin{matrix}\text{Net outflow rate}\\ \text{of species } i\\ \text{mass by con-}\\ \text{vection rela-}\\ \text{tive to the CS}\end{matrix}\right\} \approx \left\{\begin{matrix}\text{Net inflow rate}\\ \text{of species } i\\ \text{mass by}\\ \text{diffusion}\\ \text{across CS}\end{matrix}\right\} + \left\{\begin{matrix}\text{Net chemical}\\ \text{source of}\\ \text{species } i \text{ per}\\ \text{unit area of}\\ \text{discontinuity}\end{matrix}\right\}, \qquad (2.1\text{-}9)$$

[3] Throughout this text we adopt the convenient notation that a triple-primed, double-primed or single-primed quantity refers to that quantity reckoned per unit volume, per unit area, or per unit length, respectively.

where we have (again) neglected the accumulation and surface flow terms (cf. Figure 2.1-3).

It should be remarked that all but one of the N species mass balance equations are independent of the total mass balance. This is because the latter equation must be recoverable from the *sum* of the species mass balances. This is indeed the case, since:

$$\sum_{i=1}^{N} \rho_i = \rho, \qquad \sum_{i=1}^{N} \mathbf{j}_i'' = 0, \qquad \sum_{i=1}^{N} \dot{r}_i''' = 0.$$

Perhaps the second of these equations deserves special comment. While it is certainly possible to define "diffusion" relative to other (e.g., molar-averaged) reference velocities, we have chosen to define the *diffusion* mass flux $\mathbf{j}_i''$ as the vector difference between the total species i mass flux vector $\dot{\mathbf{m}}_i''$ $(= \rho_i \mathbf{v}_i)$ and the species i mass flux due to mixture *convection*, $\omega_i \rho \mathbf{v}$, where $\mathbf{v}$ is the so-called mass-averaged mixture velocity (since $\sum_{i=1}^{N} \rho_i \mathbf{v}_i = \rho \mathbf{v}$). The species i diffusion flux vector $\mathbf{j}_i''$ can therefore also be written as $\rho_i(\mathbf{v}_i - \mathbf{v})$. Summing either of these relations over all species present in the local mixture immediately reveals that, since:

$$\sum_{i=1}^{N} \dot{\mathbf{m}}_i'' = \dot{\mathbf{m}}'' = \rho \mathbf{v}, \qquad \text{then} \qquad \sum_{i=1}^{N} \mathbf{j}_i'' = 0.$$

The $\sum_i \dot{r}_i'''$ equation above, which interrelates all of the chemical source-strength expressions $\dot{r}_i'''$ (however complex the reaction mechanism), expresses the simple fact that the total *mass* (not total particle number) is conserved in ordinary chemical reactions.

In many applications (solution electrochemistry, electrical discharges in gases, etc.), at least some of the chemical species are *ionic* in nature; that is, they possess an

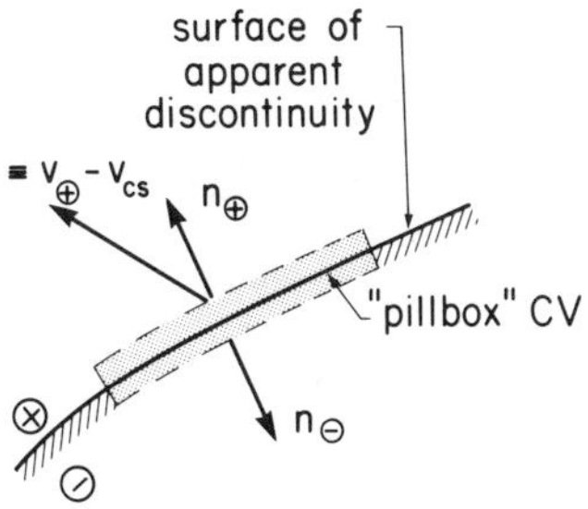

Figure 2.1-3 "Pillbox" control volume and corresponding control surface chosen so as to always "straddle" a unit area of the surface of discontinuity separating adjacent (⊕ and ⊖) continuum regions. Also shown are the unit normal vectors $\mathbf{n}_\oplus$, $\mathbf{n}_\ominus$, and the relative velocity $\mathbf{v}_{\text{rel}}$ of the fluid on the + side of the pillbox, as measured by an observer moving with the pillbox control volume.

integral number of elementary charges of positive sign (protons or the absence of electrons) or negative sign (electrons). In such situations it is appropriate to consider the relationship between the local species balance constraint (Eq. (2.1-7b)) and the principle of *electrical charge conservation.* Suppose each molecule of species i contains an integral number, z_i, of elementary charges (e.g., for the positive ion species ("cation") Na^+, $z_i = 1$; for the negative ion ("anion") species $SO_4^=$, $z_i = -2$, etc.). In such a mixture, species movement is inevitably associated with the flow of electrical charge—i.e., electrical currents. Reconsider the species i local mass balance equation:

$$\frac{\partial \rho_i}{\partial t} + \operatorname{div}(\dot{\mathbf{m}}_i'') = \dot{r}_i''' \qquad (i = 1, 2, \ldots, N). \tag{2.1-7c}$$

Suppose we now multiply both sides of this equation by $z_i \mathrm{F}/m_i$ (where F is *Faraday's constant*, equal to the charge (coulombs) associated with a mole of univalent cations) and then *sum* the result *over all chemical species* $i = 1, 2, \ldots, N$. Since electrical charge is conserved in each elementary chemical reaction, and each $\dot{r}_i'''/m_i$ is the algebraic resultant of many such elementary chemical reactions, one can conclude:

$$\sum_{i=1}^{N} z_i \mathrm{F}\left(\frac{\dot{r}_i'''}{m_i}\right) = 0 \qquad \text{(charge conservation)},$$

(which should be compared to the simpler *mass* conservation relation:

$$\left.\sum_{i=1}^{N} \dot{r}_i''' = 0\right).$$

As to the remainder of the equation, note that:

$$\sum_{i=1}^{N} \left(\frac{\dot{\mathbf{m}}_i''}{m_i}\right) z_i \mathrm{F} \equiv \mathbf{i}'',$$

where $\mathbf{i}''$ is the *total current density* (flux vector for charge transport), and

$$\sum_{i=1}^{N} \left(\frac{\rho_i}{m_i}\right) z_i \mathrm{F} \equiv n_{\text{charge}}$$

is the *net charge density* of the local fluid mixture. Thus, the species balance equation, combined with charge conservation, leads us to the important local interrelation (Newman (1973)):

$$\frac{\partial}{\partial t}(n_{\text{charge}}) + \text{div}\{\mathbf{i}''\} = 0. \quad \text{(2.1-7d)}$$

Note that in regions where the mixture is *locally electrically neutral*, i.e., where $n_{\text{charge}} = 0$, then charge conservation leads to the frequently quoted conclusion: div $\mathbf{i}'' = 0$. In such special cases $\mathbf{i}''$ is associated exclusively with the *diffusion* of charged species, that is,

$$\mathbf{i}'' = \sum_{i=1}^{N} \left(\frac{\mathbf{j}_i''}{m_i}\right) z_i \text{F},$$

since $n_{\text{charge}}\mathbf{v} = 0$.

2.1.3 Individual Chemical Element Conservation

Any individual chemical element, denoted by the subscript *k in parentheses* (where $k = 1, 2, \ldots, N_{\text{elem}} (\leqslant N)$) is also governed by a conservation principle. It will be similar in structure to Eq. (2.1-7a), but we note that, for conventional (extranuclear) chemical reactions, elements change partners but no element can be locally produced, however complex the reaction; that is, for *each* k,

$$\dot{r}'''_{(k)} = 0 \qquad (k = 1, 2, \ldots, N_{\text{elem}}). \quad \text{(2.1-10)}$$

The kth element conservation equation for a fixed macroscopic CV, therefore, takes the "source-free" form:

$$\frac{\partial}{\partial t}\int_V \rho_{(k)}\, dV + \int_S \rho_{(k)}\mathbf{v} \cdot \mathbf{n}\, dA = -\int_S \mathbf{j}''_{(k)} \cdot \mathbf{n}\, dA, \quad \text{(2.1-11)}$$

where $\mathbf{j}''_{(k)}$ is the diffusion flux vector for the element denoted by k (which receives contributions from the diffusion flux $\mathbf{j}_i''$ of each chemical species containing the kth chemical element) and $\rho_{(k)}$ is the partial density of the element denoted by k (which is comprised of contributions from the density of each chemical species i that contains the element denoted by k). (See Section 3.4.2).

The kth element local PDE likewise takes the "source-free" form (cf. Eq. (2.1-7)):

$$\frac{\partial \rho_{(k)}}{\partial t} + \text{div}\{\rho_{(k)}\mathbf{v}\} = -\text{div}\{\mathbf{j}''_{(k)}\}, \quad \text{(2.1-12)}$$

which can be expressed, say, in cylindrical polar coordinate form using Eq. (2.1-8) for the local divergence of the flux vectors $\rho_{(k)}\mathbf{v}$ and $\mathbf{j}''_{(k)}$.

The kth element mass balance applied to each unit area of a surface of discontinuity (which may itself be moving) provides a "jump" condition of the form:

$$\begin{Bmatrix} \text{Net outflow rate} \\ \text{of the } k\text{th} \\ \text{element by con-} \\ \text{vection rela-} \\ \text{tive to CS} \end{Bmatrix} \approx \begin{Bmatrix} \text{Net inflow rate} \\ \text{of the } k\text{th} \\ \text{element by dif-} \\ \text{fusion} \\ \text{across CS} \end{Bmatrix} \tag{2.1-13}$$

Element balances are widely used in the analysis of chemically reacting flows since:

a. there are fewer chemical elements than chemical species (consider all of the molecules that can be constructed from just C and H!);
b. even in the presence of ordinary chemical reactions, the element conservation equations are identical in form to those governing inert ("tracer") species.

If there are N_{elem} elements present in the system, only $N_{\text{elem}} - 1$ element conservation equations are independent of the *total* mass conservation equation (Section 2.1.1) since the latter equation must be obtainable by summing the element equations over all k-values. This is, indeed, the case, since:

$$\sum_{k=1}^{N_{\text{elem}}} \rho_{(k)} = \rho \qquad \text{and} \qquad \sum_{k=1}^{N_{\text{elem}}} \mathbf{j}''_{(k)} = 0.$$

Note: in view of the definition of $\mathbf{j}''_{(k)}$, the latter condition ensures that $\rho\mathbf{v}$ accounts for the *total* mass flux vector.

2.2 CONSERVATION OF MOMENTUM (MIXTURE)

The laws governing conservation of momentum for a single-phase fluid mixture were first derived by Euler and Cauchy, and represent the laws of Newtonian mechanics applied to a "continuum" of mass points. They are expressed here for a fixed or translating (nonaccelerating or "inertial") coordinate system (but they may be transformed, using kinematic relations, to apply to accelerating coordinate systems, as discussed in Section 2.6.2).

2.2.1 Linear Momentum Conservation

Our "standard form" (Eq. (2-1)) applies also to linear momentum conservation. In this case we make the identifications:

$$\left\{\begin{array}{l}\text{Net inflow rate}\\ \text{of linear momentum}\\ \text{by } \textit{diffusion} \text{ across CS}\end{array}\right\} \equiv \begin{array}{l}\textit{Net surface}\\ \text{(contact) } \textit{force}\\ \text{acting on CS}\end{array} \equiv \int_S \mathbf{\Pi} \cdot \mathbf{n}\, dA \tag{2.2-1}$$

and

$$\left\{\begin{array}{l}\text{Net source of}\\ \text{linear momentum}\\ \text{within CV}\end{array}\right\} \equiv \begin{array}{l}\text{Net } \textit{body force}\\ \text{acting on all}\\ \text{species within CV}\end{array} \equiv \sum_{i=1}^{N} \int_V \rho_i \mathbf{g}_i\, dV. \tag{2.2-2}$$

Thus, we define the local stress operator $\mathbf{\Pi}$ such that $\mathbf{\Pi} \cdot \mathbf{n}\, dA$ is the (vector) element of *surface force*, and $\mathbf{g}_i$ is the local "body" force acting on each unit mass of species *i*. More will be said about the nature of $\mathbf{\Pi}$, which is also a "field," in Section 3.2.1.

With these identifications, linear momentum conservation applied to the fluid mixture in an arbitrary fixed CV implies the following integral constraint on the $\rho\mathbf{v}$, $\mathbf{g}_i$, and $\mathbf{\Pi}$ fields:

$$\frac{\partial}{\partial t}\int_V \rho\mathbf{v}\, dV + \int_S \rho\mathbf{v}\mathbf{v} \cdot \mathbf{n}\, dA = \int_S \mathbf{\Pi} \cdot \mathbf{n}\, dA + \sum_{i=1}^{N} \int_V \rho_i \mathbf{g}_i\, dV. \tag{2.2-3}$$

This vector equation is equivalent to three (scalar) equations—one for each component (direction), regardless of the type of coordinate system chosen.

If (2.2-3) is applied to a typical differential region of space, we obtain the following *local* (PDE) constraint on the $\rho\mathbf{v}$, $\mathbf{g}_i$, and $\mathbf{\Pi}$ fields:

$$\frac{\partial}{\partial t}(\rho\mathbf{v}) + \operatorname{div}\{\rho\mathbf{v}\mathbf{v}\} = \operatorname{div}\{\mathbf{\Pi}\} + \sum_{i=1}^{N} \rho_i \mathbf{g}_i. \tag{2.2-4}$$

This (vector) PDE is equivalent to three (scalar) PDEs, as explicitly outlined below for the particular case of cylindrical polar coordinates.

The local, instantaneous, convective momentum flux $\rho\mathbf{v}\mathbf{v}$ and the surface (contact) stress $\mathbf{\Pi}$ appearing in Eq. (2.2-4) each involve no less than nine local scalar quantities for their complete specification (cf. vectors, which are specified by "only" three scalars). For $\rho\mathbf{v}\mathbf{v}$ in cylindrical polar coordinates, the nine quantities (components of the so-called momentum flux "tensor") are evidently:

$$\begin{array}{lll}\rho v_r v_r & \rho v_r v_\theta & \rho v_r v_z,\\ \rho v_\theta v_r & \rho v_\theta v_\theta & \rho v_\theta v_z,\\ \rho v_z v_r & \rho v_z v_\theta & \rho v_z v_z.\end{array} \tag{2.2-5}$$

However, it is clear that, because of symmetry (that is, $\rho v_\theta v_r = \rho v_r v_\theta$, etc.), only six are independent.

Calculation of each of the three components of the net outflow/volume (div) of such quantities requires considerable care, especially in curvilinear (nonplanar coordinate surface) coordinate systems. This is clear from the z-component of Eq. (2.2-4), which is found to take the form:

$$\frac{\partial}{\partial t}(\rho v_z) + [\operatorname{div}\{\rho \mathbf{vv}\}]_z = [\operatorname{div}\{\mathbf{\Pi}\}]_z + \sum_{i=1}^{N} \rho_i \mathbf{g}_{i,z}, \tag{2.2-6}$$

where

$$[\operatorname{div}\{\rho \mathbf{vv}\}]_z = \frac{1}{r}\frac{\partial}{\partial r}(r\rho v_r v_z) + \frac{1}{r}\frac{\partial}{\partial \theta}(\rho v_\theta v_z) + \frac{\partial}{\partial z}(\rho v_z v_z), \tag{2.2-7}$$

and an analogous expression can be written for $[\operatorname{div}\mathbf{\Pi}]_z$, the net *surface* z-force acting on each unit volume, involving the corresponding independent components of the local stress operator $\mathbf{\Pi}$.

The principle of linear momentum conservation, Eq. (2.2-3), also applies across a "surface of discontinuity," which may itself be moving (e.g., a premixed flame sheet, detonation, shock wave, phase boundary, etc.). In such cases the accumulation and "body" force terms are usually negligible, and Eq. (2.2-3) takes the simplified form (cf., e.g., Section 2.6):

$$\begin{Bmatrix}\text{Net outflow rate}\\ \text{of linear mo-}\\ \text{mentum by con-}\\ \text{vection relative to}\\ \text{the moving CS}\end{Bmatrix} \approx \begin{Bmatrix}\text{Net surface}\\ \text{force due to}\\ \text{stresses acting}\\ \text{on both sides}\\ \text{of CS}\end{Bmatrix}. \tag{2.2-8}$$

The nature of the contact stress operator $\mathbf{\Pi}$ may differ from fluid to fluid (a "constitutive" relation). However, we will see that it is convenient to decompose $\mathbf{\Pi}$ into two parts: One is the local thermodynamic *pressure* (the normal stress that would exist in an *equilibrium* fluid with the same local *state* variables); and the remainder is the so-called "extra" (viscous) stress, written $\mathbf{T}$.

2.2.2 Angular Momentum Conservation

Just as each force acting through a moment arm gives rise to a "torque," similarly, each parcel of linear momentum $\rho\mathbf{v}\,dV$, acting through some moment arm (position vector $\mathbf{x}$ relative to a chosen origin) gives rise to a "moment of momentum," called its "angular" momentum, and formally written[4] $(\mathbf{x})\times(\rho\mathbf{v}\,dV)$. For a fluid with

[4] This type of product, called the *vector* cross product, maximizes when the two individual vectors involved are mutually perpendicular, and *vanishes* when they are *colinear* (cf. the scalar or inner product). In general, for any two vectors **A** and **B** subtending an angle θ_{AB} between them, $\mathbf{A}\times\mathbf{B}$ is a vector with magnitude $AB\sin\theta_{AB}$ and a *direction* defined by the direction a right-hand screw would turn if A were rotated into B.

negligible net *internal* angular momentum, experiencing ordinary (contact and body) forces,[5] one can then write a vector angular-momentum conservation (balance) equation (Euler), valid for arbitrary motion through an inertial coordinate frame, by simply making the following identifications:

$$\left\{\begin{matrix}\text{Net inflow rate}\\ \text{of angular mo-}\\ \text{mentum by dif-}\\ \text{fusion across CS}\end{matrix}\right\} \equiv \left\{\begin{matrix}\text{Net surface}\\ \text{torque acting}\\ \text{on CS}\end{matrix}\right\} = \int_S (\mathbf{x}) \times (\mathbf{\Pi} \cdot \mathbf{n}\, dA) \qquad (2.2\text{-}9a)$$

and

$$\left\{\begin{matrix}\text{Net ``source'' of}\\ \text{angular momentum}\\ \text{within CV}\end{matrix}\right\} \equiv \left\{\begin{matrix}\text{Torque due to}\\ \text{net body forces}\\ \text{acting within CV}\end{matrix}\right\} = \int_V (\mathbf{x}) \times \sum_i \rho_i \mathbf{g}_i\, dV. \qquad (2.2\text{-}9b)$$

Thus, in any region of space the motion of such a fluid necessarily satisfies the integral vector constraint:

$$\frac{\partial}{\partial t}\int_V (\mathbf{x}) \times \rho\mathbf{v}\, dV + \int_S (\mathbf{x}) \times \rho\mathbf{v}(\mathbf{v}\cdot\mathbf{n}\, dA) = \int_S (\mathbf{x}) \times (\mathbf{\Pi}\cdot\mathbf{n}\, dA) + \sum_i \int_V (\mathbf{x}) \times (\rho_i \mathbf{g}_i\, dV), \qquad (2.2\text{-}10)$$

which, while not independent of the macroscopic CV linear-momentum conservation constraint (Eq. (2.2-3)), is particularly convenient for analyzing fluid flow in "rotating" (turbo) machinery (centrifugal or axial flow compressors, turbines; see, e.g., Vavra (1960), cyclones, etc.

In view of our previous discussion of the consequences of macroscopic conservation principles when applied "at the local level" (to a typical differential CV), it is natural to ask if Eq. (2.2-10) leads to an independent angular momentum *PDE* analogous to Eq. (2.2-4). The answer is *no*. Instead, we are led to a theorem[6] about the local state of stress, $\mathbf{\Pi}$, *viz*., that the matrix formed from its nine components (in *any* particular coordinate system) must be everywhere *symmetrical* about the diagonal. Hence $\mathbf{\Pi}$ is specifiable at each point in space by "only" six independent scalar quantities, like the *convective* momentum flux $\rho\mathbf{vv}$ already discussed. Indeed, in accord with Eqs. (2.2-1) and (2.2-9a), $-\mathbf{\Pi}$ may be fruitfully viewed as the local *diffusive* linear momentum flux. The "symmetry" of $\mathbf{\Pi}$ for all ordinary fluids will be exploited in Chapter 3, where we relate $\mathbf{\Pi}$ to the other relevant local fields (in particular, the scalar thermodynamic pressure $p(\mathbf{x}, t)$, and spatial nonuniformities in the fluid velocity $\mathbf{v}(\mathbf{x}, t)$).

---

[5] Not direct contact and body *torques* (couples) irrespective of moment arm $\mathbf{x}$.

[6] One can show that if the local contact stress operator $\mathbf{\Pi}$ were *not* symmetrical, then the rate of change of angular momentum of a moving fluid parcel would be infinite (see Exercise 2.21).

2.3 CONSERVATION OF ENERGY (FIRST LAW OF THERMODYNAMICS)

While in classical fluid mechanics it is often sufficient to consider only "mechanical" energy and work, in chemically reacting systems one must consider transformations involving thermal and chemical as well as mechanical energy, and both heat addition and work. There are many forms of the law of energy conservation even when applied to the continuum within a fixed CV. Here we discuss the most basic or "primitive" form, from which all other forms can be derived by straightforward manipulations (the fact that certain forms of energy (e.g., mechanical) are "higher" than others ("heat") in terms of the possibility for conversion to do work will not be contained in the *first* law of thermodynamics, but rather in the *second* law, which will be discussed in Section 2.4).

Our standard form of conservation equation (2-1) for a fixed macroscopic CV applies also to total (thermal + chemical + kinetic) energy of the local material mixture provided we identify:

$$\left\{\begin{array}{l}\text{Net inflow rate}\\ \text{of total energy}\\ \text{by diffusion}\\ \text{across CS}\end{array}\right\} = -\int_S \dot{\mathbf{q}}'' \cdot \mathbf{n}\, dA + \left\{\begin{array}{l}\text{Rate at which work}\\ \text{is done by (surface)}\\ \text{stresses along CS}\\ \text{on fluid within CV}\end{array}\right\}, \qquad (2.3\text{-}1)$$

and

$$\left\{\begin{array}{l}\text{Net "source" of}\\ \text{total energy}\\ \text{within CV}\end{array}\right\} = \int_V \dot{q}'''\, dV + \left\{\begin{array}{l}\text{Rate at which work}\\ \text{is done by each of}\\ \text{the local body forces } \mathbf{g}_i\\ \text{acting on the corre-}\\ \text{sponding moving species } i\end{array}\right\}. \qquad (2.3\text{-}2)$$

Here we have introduced the definitions:

$\dot{\mathbf{q}}'' \equiv$ total energy diffusion flux vector in the prevailing material mixture,
$\dot{q}''' \equiv$ volumetric energy source for the material mixture, typically derived from interaction with a local electromagnetic field ("photon-phase"),

and the work terms appearing, respectively, in Eqs. (2.3-1) and (2.3-2) may be formally written:

$$\int_S (\mathbf{\Pi} \cdot \mathbf{n}\, dA) \cdot \mathbf{v} \qquad \text{and} \qquad \sum_{i=1}^{N} \int_V \dot{\mathbf{m}}_i'' \cdot \mathbf{g}_i\, dV.$$

The macroscopic form of total energy conservation, therefore, implies the following integral constraint on the local field densities:

$$\frac{\partial}{\partial t}\int_V \rho\left(e+\frac{v^2}{2}\right)dV + \int_S \rho\left(e+\frac{v^2}{2}\right)\mathbf{v}\cdot\mathbf{n}\,dA = -\int_S \dot{\mathbf{q}}''\cdot\mathbf{n}\,dA + \int_V \dot{q}'''\,dV$$
$$+\int_S (\mathbf{\Pi}\cdot\mathbf{n}\,dA)\cdot\mathbf{v} + \sum_{i=1}^{N}\int_V \dot{\mathbf{m}}_i''\cdot\mathbf{g}_i\,dV, \tag{2.3-3}$$

where $e \equiv$ specific "internal" energy of the *mixture* (function of local thermodynamic state), including chemical contributions,

$v^2/2 \equiv$ specific kinetic energy possessed by each unit mass of the *mixture* as a consequence of its ordered motion.

Applying Eq. (2.3-3) to a typical differential CV leads to the following *local* PDE constraint on the $\rho(e + (v^2/2))$, $\rho\mathbf{v}$, $\dot{\mathbf{q}}''$, etc., fields:

$$\frac{\partial}{\partial t}\left[\rho\left(e+\frac{v^2}{2}\right)\right] + \operatorname{div}\left\{\left[\rho\left(e+\frac{v^2}{2}\right)\mathbf{v}\right]\right\} = -\operatorname{div}\{\dot{\mathbf{q}}''\} + \dot{q}''' + \operatorname{div}\{\mathbf{\Pi}\cdot\mathbf{v}\}$$
$$+\sum_{i=1}^{N}\dot{\mathbf{m}}_i''\cdot\mathbf{g}_i, \tag{2.3-4}$$

from which all other forms (e.g., those involving mechanical energy, enthalpy, or temperature, etc.) of the local energy balance may be derived.

Application of Eq. (2.1-8) for the divergence of a vector (net outflow rate/volume) in cylindrical coordinates then yields the explicit form of the energy conservation PDE in *cylindrical polar coordinates.* In this connection it is important to note that $\mathbf{\Pi}\cdot\mathbf{v}$ is a vector with the r, θ, z components:

$$\begin{aligned}
[\mathbf{\Pi}\cdot\mathbf{v}]_r &= \Pi_{rr}v_r + \Pi_{r\theta}v_\theta + \Pi_{rz}v_z,\\
[\mathbf{\Pi}\cdot\mathbf{v}]_\theta &= \Pi_{\theta r}v_r + \Pi_{\theta\theta}v_\theta + \Pi_{\theta z}v_z,\\
[\mathbf{\Pi}\cdot\mathbf{v}]_z &= \Pi_{zr}v_r + \Pi_{z\theta}v_\theta + \Pi_{zz}v_z,
\end{aligned} \tag{2.3-5}$$

and the local "divergence" of this vector (calculated *via* Eq. (2.1-8)) is, physically, the *net rate at which work is done* by all of the *surface stresses* acting on the moving *fluid for each unit volume of space.*

The total energy balance can be applied to each unit area of a "surface of discontinuity" (which may itself be moving). In such cases the accumulation and volumetric (energy addition + body force work) terms are usually negligible and one writes:

$$\begin{Bmatrix}\text{Net outflow rate of}\\ \text{total energy by con-}\\ \text{vection relative to}\\ \text{the moving CS}\end{Bmatrix} \cong \begin{Bmatrix}\text{Net inflow rate of}\\ \text{total energy due}\\ \text{to } \dot{\mathbf{q}}'' \text{ at both}\\ \text{surfaces of CS}\end{Bmatrix} + \begin{Bmatrix}\text{Rate at which work}\\ \text{is done by stress}\\ \text{system at both}\\ \text{surfaces of CS}\end{Bmatrix}. \qquad (2.3\text{-}6)$$

(cf. Section 2.6.)

Equations involving only the thermodynamic internal energy or the *mechanical energy* explicitly can be derived by combining the energy equations written above with the corresponding linear-momentum conservation equation (after the latter has been suitably multiplied by the local velocity vector) and subtracting those two equations. This not only provides the explicit "sources" for the $\rho(v^2/2)$-field, but also allows the above-mentioned equations for the $\rho(e + v^2/2)$-field to be converted into those governing ρe (by subtracting the kinetic-energy contribution). Equations of this type will be discussed in Section 2.5.4.

It is interesting to note that, when the "body" forces $\mathbf{g}_i$ (per unit mass) are the same for all chemical species (as in the case of gravitation),[7] then:

$$\sum_{i=1}^{N} \dot{\mathbf{m}}_i'' \cdot \mathbf{g}_i = \left(\sum_{i=1}^{N} \dot{\mathbf{m}}_i''\right) \cdot \mathbf{g} = \dot{\mathbf{m}}'' \cdot \mathbf{g}, \qquad (2.3\text{-}7)$$

a result consistent with the constraints:

$$\sum_{i=1}^{N} \omega_i = 1, \qquad \sum_{i=1}^{N} \mathbf{j}_i'' = 0. \qquad (2.3\text{-}8)$$

Moreover, if $\mathbf{g}$ is associated with a time-independent potential energy field, ϕ, then this body-force work term can be taken over to the left-hand side of the total energy equation, which then applies to the *total energy density field*:

$$\rho\left(e + \frac{1}{2}v^2 + \phi\right).$$

Thus, if the energy equation is written for this kind of an energy density, this type of body-force work has already been included and should not also appear on the right-hand side.

[7] Electrostatic body forces (that depend *on ionic charge*) are *not* the same for each species i. They are important in the fields of electrochemistry and ionized gas flows. (See, also, Section 6.2.1.) In such cases the *net* force $\mathbf{g}_i - \sum_j \omega_j \mathbf{g}_j$ "drives" species i "drift" in the prevailing fluid mixture (see Sections 3.4.3, 6.2.1.2, 6.2.5).

2.4 "CONSERVATION" OF ENTROPY (SECOND LAW OF THERMODYNAMICS)

One usually thinks of the Second Law of Thermodynamics[8] as providing an *inequality*, expressing the familiar observation that *irreversible* phenomena (diffusive momentum transfer, energy transfer, mass transfer, and chemical reactions) lead to entropy "production." However, by combining the above-mentioned laws with simple thermodynamic interrelations between state variables (including the local mixture entropy), one can write a conservation or "balance" equation that must be satisfied by the entropy density field ρs. This equation can be formally expressed in our "standard," fixed CV form; i.e., at each instant:

$$\begin{Bmatrix}\text{Accumulation rate}\\ \text{of entropy}\\ \text{within CV}\end{Bmatrix} + \begin{Bmatrix}\text{Net outflow rate}\\ \text{of entropy by}\\ \text{convection across CS}\end{Bmatrix} = \begin{Bmatrix}\text{Net inflow rate}\\ \text{of entropy by}\\ \text{diffusion across CS}\end{Bmatrix} + \begin{Bmatrix}\text{Net volumetric}\\ \text{source of entropy}\\ \text{within CV}\end{Bmatrix}. \qquad (2.4\text{-}1)$$

Defining:

$\mathbf{j}_s'' \equiv$ diffusion flux vector for entropy,
$\dot{s}''' \equiv$ local volumetric rate of entropy production due to all irreversible processes occurring within the fluid mixture,

Equation (2.4-1) can be written:

$$\frac{\partial}{\partial t}\int_V \rho s\, dV + \int_S \rho s \mathbf{v} \cdot \mathbf{n}\, dA = -\int_S \mathbf{j}_s'' \cdot \mathbf{n}\, dA + \int_V \dot{s}'''\, dV. \qquad (2.4\text{-}2)$$

For any typical (nonsingular) subregion of space, the formal entropy balance (Eq. (2.4-2)) implies the existence of the corresponding local PDE:

$$\frac{\partial}{\partial t}(\rho s) + \operatorname{div}\{\rho s \mathbf{v}\} = -\operatorname{div}\{\mathbf{j}_s''\} + \dot{s}''', \qquad (2.4\text{-}3)$$

interrelating the ρs, $\rho \mathbf{v}$, and $\mathbf{j}_s''$ fields. This local entropy balance can be expressed in any convenient coordinate system. Thus, by introducting Eq. (2.1-8) for the "divergence" terms, one obtains the local entropy balance for a cylindrical polar system (r, θ, z).

[8] There are two remaining laws of thermodynamics which will not *explicitly* concern us here. The so-called *Zeroth Law* (according to which two distinct systems in thermochemical equilibrium (TCE) with a third system must be in TCE with each other) and the *Third Law* (the entropy function vanishes as the temperature absolute zero is approached, thereby fixing *absolute* entropies).

Applied to a moving "surface of discontinuity," the formal entropy balance yields:

$$\left\{\begin{matrix}\text{Net outflow rate of}\\ \text{entropy by convec-}\\ \text{tion relative to CS}\end{matrix}\right\} \cong \left\{\begin{matrix}\text{Net inflow rate of}\\ \text{entropy by diffu-}\\ \text{sion across CS}\end{matrix}\right\} + \left\{\begin{matrix}\text{Rate of entropy}\\ \text{production per}\\ \text{unit area of surface}\end{matrix}\right\}, \quad (2.4\text{-}4)$$

where we have again neglected the accumulation term (see Eqs. (2.6-1 through 11)).

These equations can be used to express important necessary *inequalities* (i.e., Clausius' form of the Second Law of Thermodynamics for "open" flow systems). These inequalities are simply restatements of the fact that the net effect of all transfer processes must be to yield a *positive entropy production.* Thus, we must have:

$$\int_V \dot{s}''' \, dV \geqslant 0 \qquad \text{(for any macroscopic CV)}, \quad (2.4\text{-}5)$$

or, locally:

$$\dot{s}''' \geqslant 0, \quad (2.4\text{-}6)$$

$$\dot{s}'' \geqslant 0 \qquad \text{(for a discontinuity)}. \quad (2.4\text{-}7)$$

For example, the first of these (combined with (Eq. 2.4-2)) implies that for any fixed macroscopic region of space and at any instant:

$$\frac{\partial}{\partial t}\int_V \rho s \, dV + \int_S (\rho s \mathbf{v} + \mathbf{j}_s'') \cdot \mathbf{n} \, dA \geqslant 0. \quad (2.4\text{-}8)$$

By combining the above-mentioned entropy balance equations with thermodynamic interrelations between state variables and the remaining conservation equations (say, for $\dot{q}''' = 0$), we can derive:

a. an explicit expression for the local volumetric entropy production, which is of the instructive *additive* form:[9]

$$\dot{s}''' = \underset{\substack{\text{transport by}\\ \text{diffusion}}}{(\dot{s}''')_{\text{Momentum}}} + \underset{\substack{\text{mass transport}\\ \text{by diffusion}}}{(\dot{s}''')_{\text{Species}}} + \underset{\substack{\text{transport by}\\ \text{diffusion}}}{(\dot{s}''')_{\text{Energy}}} + \underset{\substack{\text{chemical}\\ \text{reactions}}}{(\dot{s}''')_{\text{Nonequilibrium}}}; \quad (2.4\text{-}9)$$

b. an explicit expression for the local entropy flux vector $\mathbf{j}_s''$ in terms of $\dot{\mathbf{q}}''/T$ and the individual $\mathbf{j}_i'' s_i$-values, where s_i is the specific entropy of species i in the mixture.[10]

---

[9] See, also, Eq. (2.5-28), exploited in Chapter 3 (Selection of General Constitutive Laws).
[10] See, also, Eq. (2.5-27), which includes the important "mixing-entropy" contribution.

Applications of these entropy balance equations and examples of both isentropic and nonisentropic fluid flows will be given in Sections 2.5.6, 2.6.1, and Chapters 4 through 7 below.

Interestingly enough, in many cases nonequilibrium systems subjected to time-independent boundary conditions (BCs) evolve toward a nonequilibrium steady state of *minimum entropy production* (Prigogine (1961))—i.e., toward that steady-state for which $\int_V \dot{s}''' \, dV$ (or, equivalently,

$$\left. \int_S (\rho s \mathbf{v} + \mathbf{j}_s'') \cdot \mathbf{n} \, dA \right)$$

is a *minimum* compared to all otherwise "eligible" steady states subject to the imposed BCs (see Section 2.5.6).

The Second Law of Thermodynamics, as reformulated here, will be seen to be useful to:

a. set important constraints on "otherwise possible" physicochemical processes (e.g., minimum work to separate a mixture, impossibility of an expansion shock wave, maximum possible efficiency of "heat" engines, minimum electrical energy to obtain aluminum via electrolysis of molten salt containing Al_2O_3, etc.);
b. provide the basis for *variational methods* to numerically solve nonequilibrium problems within the domain of "linear irreversible thermodynamics" (principle of minimum entropy production; see above and Section 2.5.6).
c. guide the selection of general constitutive laws governing the *diffusion* of momentum, energy, and/or species mass (see Chapter 3) in nonequilibrium chemically reacting mixtures;
d. pinpoint the sources of entropy production and, hence, inefficiency in proposed or actual engineering devices, and provide insights useful in the optimization of engineering devices involving simultaneous transport phenomena and chemical reaction.

2.5 ALTERNATIVE ("DERIVED") FORMS OF THE CONSERVATION (BALANCE) EQUATIONS[11]

2.5.1 Introduction

The PDEs we have written, governing each of the field densities ρ, ρ_i, $\rho_{(k)}$, $\rho(e + (v^2/2))$, ρs, are central to the analysis of all problems involving chemically reacting continua, past, present, and future. Each term has an obvious physical

---

[11] Study of this section can be postponed without a loss of continuity; however, several of these derived forms of the conservation equations will prove useful in Chapters 4, 5, and 6.

meaning, irrespective of the user's particular choice of coordinate system, and for this reason, we consider these PDEs, and their macroscopic counterparts, our "battle-tested" starting point.

Nevertheless, for the solution of particular engineering problems, it is often more convenient to work directly with dependent variables other than the particular densities written above; e.g., variables such as[12] $\omega_i(\equiv \rho_i/\rho)$, $\omega_{(k)} \equiv (\rho_{(k)}/\rho)$, $\rho(v^2/2)$, $\rho(e + (p/\rho))$, etc. Therefore, in this section we develop several important "derived" forms of these conservation (balance) equations, which we will have occasion to use in Chapters 4, 5, and 6. Some of these "derived" equations will contain terms whose physical origin is less clear than the corresponding "primitive" equations, making them more difficult to remember; however, we show that these terms are straightforward consequences of the changes in dependent variables effected.

Of course, many factors dictate the most convenient choices of dependent variables in any given problem. For example,

a. Which field variables are of greatest practical interest?
b. Which are known (specified) at the boundaries of the domain or initially ($t = 0$)?
c. Which could be directly measured in verification experiments?
d. Which satisfy the simplest partial differential equations?

Thus, while the same solution can ultimately be found using any of a number of alternate choices, some choices might be more expeditious than others.

Similar remarks apply to the particular *coordinate frame* chosen to express conservation—i.e., to the choice of *independent* (Eulerian) variables for the problem at hand. We have deliberately stated all conservation equations in a vector form having the same physical meaning for *any* particular *inertial* coordinate frame. However, in Section 2.6.2, below, we display the important consequences of selecting any convenient *accelerating* (noninertial) *coordinate frame* to express the same conservation constraints at the local level.

---

[12] Chemical engineers frequently work with the "hybrid" composition variable ω_i/m_i, readily seen to be the same as n_i/ρ, that is, the local number of molecules (or moles) of chemical species i per unit *mass* of mixture. Another common composition variable is the *mole fraction* (x_i (condensed phase) or y_i (vapor phase)). These are therefore related to the corresponding ω_i-values by:

$$y_i(\text{or } x_i) = \frac{(\omega_i/m_i)}{\sum_{j=1}^{N} (\omega_j/m_j)};$$

therefore

$$m_{\text{mix}} = \sum_{i=1}^{N} y_i m_i = \left(\sum_{i=1}^{N} (\omega_i/m_i) \right)^{-1}.$$

2.5.2 Origin of the "Accumulation Rate + Net Convective Outflow Rate" Structure of all Conservation Equations for a Fixed Control Volume

The left-hand side of each macroscopic *fixed CV* conservation equation (i.e., (accumulation rate) + (net convective outflow rate)) is shown here to be nothing more than the instantaneous rate of change of the total mass, momentum, energy, or entropy of a volume initially the same size and shape as the *fixed* CV but whose CS instead moves with the local fluid velocity: $\mathbf{v}(\mathbf{x}_{CS}, t)$, that is, a "material" control volume. This is seen by applying Leibnitz's theorem for differentiating an integral:[13]

$$\frac{d}{dt}\left(\int_{\substack{\text{CS moving with}\\ \text{velocity}\\ \mathbf{v}_{CS}(\mathbf{x}_{CS},t)}} f(\mathbf{x},t)\,dV\right) = \frac{\partial}{\partial t}\int_{\text{fixed CV}} f(\mathbf{x},t)\,dV + \int_{\text{fixed CS}} f(\mathbf{x}_{CS},t)\mathbf{v}_{CS}\cdot\mathbf{n}\,dA \tag{2.5-1}$$

to the special case[14] $\mathbf{v}_{CS}(\mathbf{x}_{CS},t) = \mathbf{v}(\mathbf{x},t)$. Hereafter, denoting "material" derivatives by D/Dt, we see that:

$$\frac{\partial}{\partial t}\int_V \rho\,dV + \int_S \rho\mathbf{v}\cdot\mathbf{n}\,dA = \frac{D}{Dt}\left(\int_V \rho\,dV\right), \tag{2.5-2}$$

$$\frac{\partial}{\partial t}\int_V \rho_i\,dV + \int_S \rho_i\mathbf{v}\cdot\mathbf{n}\,dA = \frac{D}{Dt}\left(\int_V \rho_i\,dV\right), \tag{2.5-3}$$

$$\frac{\partial}{\partial t}\int_V \rho\mathbf{v}\,dV + \int_S \rho\mathbf{v}(\mathbf{v}\cdot\mathbf{n}\,dA) = \frac{D}{Dt}\left(\int_V \rho\mathbf{v}\,dV\right), \quad \text{etc.} \tag{2.5-4}$$

Thus, each of the above-mentioned *macroscopic* balance equations for a fixed (Eulerian) CV follows immediately from a balance between the accumulation rate for a *material* CV and its only possible "causes," *viz*., an inflow by diffusion (across the CS) and "sources" (within the CV).

An important (material CV $\leftrightarrow$ fixed CV) interrelation of nearly identical structure to that of Eqs. (2.5-2 and 3.4) can be derived for our *differential* equations,

---

[13] This is simply the multidimensional analog of the following well-known result from one-variable calculus:

$$\left(\frac{d}{dt}\int_{a(t)}^{b(t)} f(x,t)\,dx\right)_{t=t_0} = \left(\frac{\partial}{\partial t}\int_{a(t_0)}^{b(t_0)} f(x,t)\,dx\right)_{t=t_0} + \left[f(b,t)\frac{db}{dt}\right]_{t=t_0} - \left[f(a,t)\frac{da}{dt}\right]_{t=t_0}.$$

[14] When $\mathbf{v}_{CS} = \mathbf{v}$, this result is sometimes called "Reynolds' transport theorem," but it is a kinematic theorem, *not* a law of fluid mechanics.

viz., for any[15] $f(\mathbf{x}, t)$ (including vector fields),

$$\frac{\partial}{\partial t}(\rho f) + \operatorname{div}(\rho \mathbf{v} f) = \rho \frac{Df}{Dt}, \tag{2.5-5}$$

where properties of the *local material derivative* Df/Dt required for the derivation of Eq. (2.5-5) will be displayed below (see Eq. (2.5-13c)).

2.5.3 Material Derivative Form of the Conservation PDEs

With $f = 1$, the theorem in Eq. (2.5-5) is in fact immediately identical to our earlier statement of conservation of total mass at the local level (Eq. (2.1-3)). Setting $f = 1/\rho = v$ (specific volume), we obtain an alternative (material derivative) form of conservation of total mass, *viz.*:

$$\operatorname{div} \mathbf{v} = \frac{1}{v} \cdot \frac{Dv}{Dt} = -\frac{1}{\rho}\frac{D\rho}{Dt} \tag{2.5-6a}$$

or

$$\frac{D\rho}{Dt} = -\rho \operatorname{div}(\mathbf{v}). \tag{2.5-6b}$$

A fluid is said to be locally *incompressible* if $D\rho/Dt = 0$,[16] that is, if the rate of change of density of each moving fluid parcel vanishes. In such a case, Eq. (2.5-6) immediately reveals that the instantaneous velocity field $\mathbf{v}(\mathbf{x}, t)$ must then satisfy the local condition: $\operatorname{div}(\mathbf{v}) = 0$. Equation (2.5-6a) also shows that in general the local, instantaneous value of $\operatorname{div}(\mathbf{v})$ is a measure of the *volumetric rate of fluid deformation*, a result which will be used in Section 3.3.2, devoted to the constitutive law governing momentum diffusion.

Considering conservation of chemical species and chemical elements (with $f = \omega_i$ and $\omega_{(k)}$, respectively) we immediately see that:

$$\rho \frac{D\omega_i}{Dt} = -\operatorname{div} \mathbf{j}_i + \dot{r}_i''' \qquad (i = 1, 2, \ldots, N) \tag{2.5-7}$$

and

$$\rho \frac{D\omega_{(k)}}{Dt} = -\operatorname{div} \mathbf{j}_{(k)}'' \qquad (k = 1, 2, \ldots, N_{\text{elem}}) \tag{2.5-8}$$

---

[15] In what follows the arbitrary function f should not be confused with h-Ts, the specific Gibbs free energy of the fluid mixture.

[16] This does *not* necessarily imply *spatially* uniform density. Thus, the steady flow of a stratified $Hg(\ell)$, $H_2O(\ell)$, oil(ℓ) mixture would be "incompressible" despite the spatial nonuniformity of density.

Thus, only an inflow by diffusion and/or local chemical reaction can cause the local species mass fraction ω_i (and hence ω_i/m_i) to change for each moving fluid parcel. Moreover, chemical reactions are incapable of causing the *element* mass fractions $\omega_{(k)}$ $(k = 1, 2, \ldots, N_{\text{elem}})$ to change within each moving fluid parcel, and hence *along* any streamline in a steady flow.

The material control volume form of *linear momentum conservation* for the mixture follows by setting $f = \mathbf{v}$, whereupon Eqs. 2.2-4 and 2.5-5 lead immediately to:

$$\rho \frac{D\mathbf{v}}{Dt} = \operatorname{div}\{\mathbf{\Pi}\} + \sum_{i=1}^{N} \rho_i \mathbf{g}_i, \tag{2.5-9}$$

showing that differences in contact stresses and/or net body forces cause velocity changes (magnitude and/or direction) of a moving fluid parcel.

The material CV form of *energy conservation* follows from Eqs. (2.3-4) and (2.5-5) with the identification $f = e + (v^2/2)$. Then:

$$\rho \frac{D}{Dt}\left(e + \frac{v^2}{2}\right) = -\operatorname{div}\dot{\mathbf{q}}'' + \dot{q}''' + \operatorname{div}\{\mathbf{\Pi}\cdot\mathbf{v}\} + \sum_{i=1}^{N} \dot{\mathbf{m}}_i'' \cdot \mathbf{g}_i. \tag{2.5-10}$$

Finally, the formal entropy balance (Eq. 2.4-3), and Eq. (2.5-5) with $f = s$, lead to:

$$\rho \frac{Ds}{Dt} = -\operatorname{div}\mathbf{j}_s'' + \dot{s}''', \tag{2.5-11}$$

which, with the help of supplementary thermodynamic interrelations, will allow us to derive later (in Section 2.5.6) explicit expressions for $\mathbf{j}_s''$ and $\dot{s}'''$ in terms of the participating diffusion fluxes and irreversible phenomena (e.g., nonequilibrium chemical reaction).

Before using the material CV forms of the conservation PDEs to derive equations explicit in ρe, $\rho v^2/2$ themselves, and the important combinations $\rho(e + (p/\rho))$, $\rho(e + (p/\rho) + (v^2/2))$, we show that the mass balance equations (2.5-7 and 8) are equivalent to Eulerian PDEs governing the ω_i and $\omega_{(k)}$ fields.

To express this relation, we introduce here the notion that each field quantity $f\{\mathbf{x}, t\}$ (including vectors) possesses a local spatial *gradient* defined such that the projection of the *vector* **grad** f in any direction gives the *spatial derivative* of that scalar f in that direction; thus,

$$(\mathbf{grad}\, f)\cdot\mathbf{e}_\xi \equiv \frac{1}{h\{\xi,\ldots\}}\cdot\frac{\partial f}{\partial \xi}, \tag{2.5-12}$$

where $\mathbf{e}_\xi$ is the *unit* vector in the direction of increasing coordinate ξ and $h\{\xi,\ldots\}\,\Delta\xi$

is the *length* increment associated with an increment $\Delta\xi$ in the coordinate ξ (see Section 2.5.7).

If we define the *local material derivative* of f in the following reasonable way:

$$\frac{Df}{Dt} \equiv \lim_{\Delta t \to 0} \left\{ \frac{f\{\mathbf{x} + \mathbf{v}\,\Delta t, t + \Delta t\} - f\{\mathbf{x}, t\}}{\Delta t} \right\}, \tag{2.5-13a}$$

and expand $f\{\mathbf{x} + \mathbf{v}\,\Delta t, t + \Delta t\}$ in terms of $f\{\mathbf{x}, t\}$ using a Taylor series about $\mathbf{x}, t$, that is,

$$f\{\mathbf{x} + \mathbf{v}\,\Delta t, t + \Delta t\} \approx f\{\mathbf{x}, t\} + (\mathbf{grad}\, f)\Big|_{\mathbf{x},t} \cdot \mathbf{v}\,\Delta t + \left(\frac{\partial f}{\partial t}\right)\Big|_{\mathbf{x},t} \Delta t + \cdots, \tag{2.5-13b}$$

then it follows from Eq. (2.5-13a) that an observer moving in the local fluid velocity $\mathbf{v}\{\mathbf{x}, t\}$ will record:

$$\frac{Df}{Dt} = \frac{\partial f}{\partial t} + \mathbf{v} \cdot \mathbf{grad}\, f \tag{2.5-13c}$$

This valuable kinematic interrelation[17] now allows each of the above-mentioned primitive conservation equations to be re-expressed in an equivalent Eulerian form. In particular, each of the local species mass and element mass balance equations (2.5-7 and 8) can now be expressed:

$$\rho\left(\frac{\partial \omega_i}{\partial t} + \mathbf{v} \cdot \mathbf{grad}\, \omega_i\right) = -\operatorname{div} \mathbf{j}_i'' + \dot{r}_i''' \qquad (i = 1, 2, \ldots, N), \tag{2.5-14}$$

$$\rho\left(\frac{\partial \omega_{(k)}}{\partial t} + \mathbf{v} \cdot \mathbf{grad}\, \omega_{(k)}\right) = -\operatorname{div} \mathbf{j}_{(k)}'' \qquad (k = 1, 2, \ldots, N_{\text{elem}}). \tag{2.5-15}$$

Note that there is a nonzero species i mass *convective* term only when the vectors $\rho\mathbf{v}$ and $\mathbf{grad}\,\omega_i$ are not locally perpendicular to one another; that is, when $\rho\mathbf{v}$ and $\mathbf{grad}\,\omega_i$ *are* both perpendicular, there is no convective contribution to the species i mass balance despite the presence of total mass convection.

2.5.4 Alternate Forms of the Energy Conservation PDE

The local material derivative also provides a convenient "shorthand" for making changes in dependent variables, as shown below.

[17] When combined with the Eulerian form of total mass conservation (Eq. (2.1-3)) and the rule for differentiating products, Eq. (2.5-5) is readily derived.

The distributive property of differentiation makes it clear that if we derive an equation for $\rho\, D(v^2/2)/Dt$ and subtract it from the equation for $\rho\, D(e + v^2/2)/Dt$ we can construct an equation for $\rho\, De/Dt$. The latter could then be used to generate an equation for $\rho\, Dh/Dt$ by the addition of $\rho\, D(p/\rho)/Dt$, etc.

If the linear-momentum conservation (balance) Eq. (2.5-9) is multiplied, term by term, by $\mathbf{v}$ (scalar product), we obtain the following equation for $\rho\mathbf{v} \cdot D\mathbf{v}/Dt = \rho\, D(v^2/2)/Dt$:

$$\rho \frac{D}{Dt}\left(\frac{v^2}{2}\right) = \mathbf{v} \cdot \operatorname{div} \mathbf{\Pi} + \sum_{i=1}^{N} \rho_i \mathbf{v} \cdot \mathbf{g}_i. \tag{2.5-16}$$

Subtracting this from the equation of energy conservation, Eq. (2.5-10), permits us to write[18]

$$\rho \frac{De}{Dt} = (-\operatorname{div} \dot{\mathbf{q}}'' + \dot{q}''') + (\mathbf{\Pi}\!:\! \mathbf{grad}\, \mathbf{v}) + \sum_{i=1}^{N} \mathbf{j}_i'' \cdot \mathbf{g}_i, \tag{2.5-17}$$

where we have introduced the short-hand notation:

$$\mathbf{\Pi}\!:\! \mathbf{grad}\, \mathbf{v} \equiv \operatorname{div}(\mathbf{\Pi} \cdot \mathbf{v}) - \mathbf{v} \cdot \operatorname{div} \mathbf{\Pi}. \tag{2.5-18}$$

Equation (2.5-17) therefore governs the rate of change of the specific internal energy of a fluid parcel (see, also, Section 3.2.3).

By adding $\rho D(p/\rho)/Dt (= Dp/Dt - (p/\rho)(D\rho/Dt) = Dp/Dt + p \operatorname{div} \mathbf{v})$ to both the RHS and LHS of this equation, we obtain an equation governing the rate of change of specific enthalpy $h(\equiv e + (p/\rho))$ of a fluid parcel:

$$\rho \frac{Dh}{Dt} = (-\operatorname{div} \dot{\mathbf{q}}'' + \dot{q}''') + \left[\frac{Dp}{Dt} + p \operatorname{div} \mathbf{v}\right] + (\mathbf{\Pi}\!:\! \mathbf{grad}\, \mathbf{v}) + \sum_{i=1}^{N} \mathbf{j}_i'' \cdot \mathbf{g}_i. \tag{2.5-19}$$

This equation can be simplified by rewriting it in terms of that part of the contact stress ($\mathbf{T}$) left after subtracting the local thermodynamic pressure (see Section 3.2.1). In terms of this so-called "extra stress," Eq. (2.5-19) becomes:[19]

$$\rho \frac{Dh}{Dt} = (-\operatorname{div} \dot{\mathbf{q}}'' + \dot{q}''') + \frac{Dp}{Dt} + (\mathbf{T}\!:\! \mathbf{grad}\, \mathbf{v}) + \sum_{i=1}^{N} \mathbf{j}_i'' \cdot \mathbf{g}_i. \tag{2.5-20}$$

---

[18] Note that if $\mathbf{g}_i$ is the same for each species (e.g., gravity), then $\sum_{i=1}^{N} \mathbf{j}_i'' \cdot \mathbf{g}_i$ vanishes by virtue of $\sum_{i=1}^{N} \mathbf{j}_i'' = 0$.

[19] It is readily shown that $\mathbf{\Pi}\!:\! \mathbf{grad}\, \mathbf{v} = -p \operatorname{div} \mathbf{v} + \mathbf{T}\!:\! \mathbf{grad}\, \mathbf{v}$. Thus, for incompressible flow the scalars $\mathbf{\Pi}\!:\! \mathbf{grad}\, \mathbf{v}$ and $\mathbf{T}\!:\! \mathbf{grad}\, \mathbf{v}$ are everywhere equal.

At this point we reiterate that the *specific enthalpy*, $h \equiv e + (p/\rho)$, of the mixture *includes* chemical (bond-energy) contributions, and must be calculated from a *constitutive* relation of the general form:

$$h = \sum_{i=1}^{N} \omega_i \cdot \bar{h}_i(T, p, \omega_1, \omega_2, \ldots), \tag{2.5-20a}$$

where the $\bar{h}_i$-values are the *partial specific enthalpies* in the prevailing mixture. For a mixture of *ideal gases* this relation simplifies considerably to:

$$h = \sum_{i=1}^{N} \omega_i \cdot h_i(T) = \sum_{i=1}^{N} \omega_i \cdot \frac{H_i(T)}{M_i}. \tag{2.5-20b}$$

Alternatively, in terms of mole fractions,

$$h = \frac{H}{M} = \frac{\sum_{i=1}^{N} y_i H_i(T)}{\sum_{i=1}^{N} y_i M_i}. \tag{2.5-20c}$$

Here the $H_i (i = 1, 2, \ldots, N)$ are the "absolute" molar enthalpies of the pure constituents (see Figure 2.5-1), that is,

$$H_i(T) = \Delta H_{f,i}(T_{\text{ref}}) + \underbrace{\int_{T_{\text{ref}}}^{T} C_{p,i}\, dT}_{H_i(T) - H_i(T_{\text{ref}})}, \tag{2.5-20d}$$

where $\Delta H_{f,i}(T_{\text{ref}})$ is the molar "heat of formation" of species i; that is, the enthalpy change across the stoichiometric reaction in which one mole of species i is formed from its constituent chemical elements in some (arbitrarily chosen) reference states (e.g., H_2(g), O_2(g), and C(graphite) @ $T_{\text{ref}} = 298$ K, as in Figure 2.5-1). Because of the (implicit) inclusion of the "heat of formation" in h, the local energy addition term $\dot{q}'''$ appearing on the RHS of the PDE (Eq. 2.5-20) is *not* associated with chemical reactions (this would give rise to a "double-counting" error). An explicit chemical-energy generation term enters energy equations only when expressed in terms of a "sensible-" (or thermal-) energy density dependent variable, such as

$$\int_{T_{\text{ref}}}^{T} c_{p,\,\text{mix}}\, dT$$

(or T itself). Using Eqs. (2.5-19b) and (2.5-20b), a PDE for $\rho c_{p,\,\text{mix}}(DT/Dt)$ is readily derived (Eq. (5.3-10)), and its RHS indeed contains (in addition to $\dot{q}'''$) the explicit

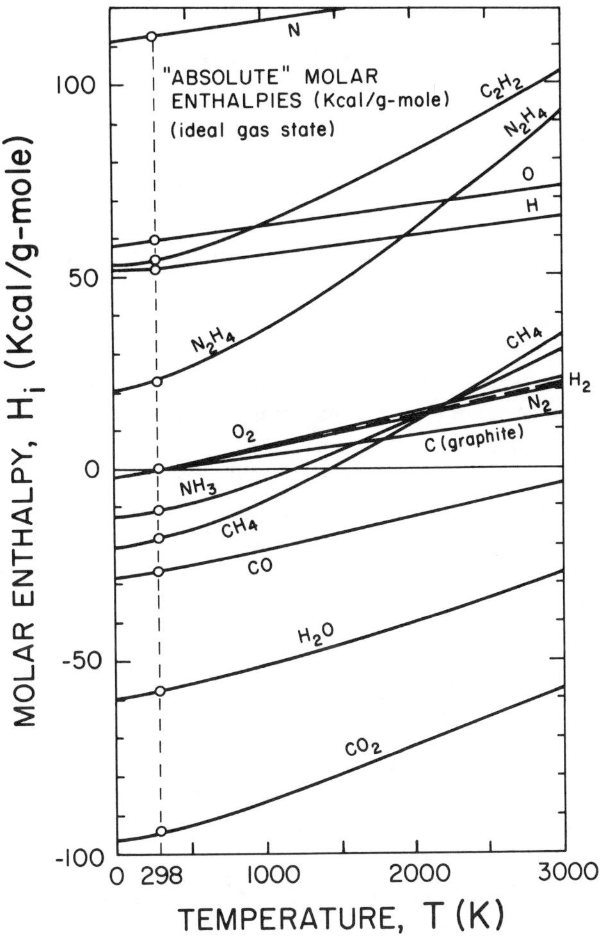

Figure 2.5-1 "Absolute" molar enthalpies for several important ideal gases containing the chemical elements C, H, O, and N. Circular points (at 298 K) are the values of the corresponding heats of formation for one mole of the indicated species from appropriate amounts of H_2, O_2, N_2, and/or C(graphite) at 298 K. Note: To this scale the molar enthalpies of O_2, H_2, and N_2 are almost coincident (based on tabular data from *JANAF Thermochemical Tables*, 2nd Ed. (1971)).

"chemical"-energy source term:

$$-\sum_{i=1}^{N} \dot{r}_i''' h_i,$$

where $h_i(T) = H_i(T)/M_i$ (see Figure 2.5-1).

Finally, addition of the equation for $\rho D(v^2/2)/Dt$ allows us to construct the following PDE for the "total" enthalpy $h_0 \equiv e + (p/\rho) + (v^2/2)$

$$\rho \frac{Dh_0}{Dt} = (-\operatorname{div} \dot{\mathbf{q}}'' + \dot{q}''') + \frac{\partial p}{\partial t} + \operatorname{div}(\mathbf{T} \cdot \mathbf{v}) + \sum_{i=1}^{N} \dot{\mathbf{m}}_i'' \cdot \mathbf{g}_i. \tag{2.5-21}$$

Of course, the same result could also have been obtained from Eq. (2.5-10) by simply adding $\rho D(p/\rho)/Dt$ to both sides.

It is interesting to note that:

a. in contrast to our $\rho D(e + v^2/2)/Dt$ starting point, Eqs. 2.5-17, 20, and 21 contain terms whose physical interpretation is not at all obvious;
b. since the *chemical* energy of the mixture is already included in the specific internal energy e, Eqs. (2.5-17, 20, and 21) do not contain an explicit term involving the energy transformations associated with chemical reactions[20] (such terms appear only when these equations are rewritten in terms of only the "sensible" (thermal) energy or enthalpy, or the mixture *temperature* itself (see Eq. (5.3-10)).

Again, each of these equations can be written in an explicitly Eulerian form by replacing the material derivative on the LHS with its equivalent, using Eq. (2.5-13).[21]

2.5.5 Macroscopic Mechanical-Energy Equation (Generalized Bernoulli Equation)

A widely used macroscopic "mechanical" energy balance can be derived from our equation for $\rho D(v^2/2)/Dt$ (Eq. (2.5-16)) by combining term-by-term volume integration, Gauss' theorem,[22] and the rule for differentiating products. We state the result for the important special case of incompressible flow subject to "gravity" as the only body force, being expressible in terms of the spatial gradient of a time-independent potential function $\phi(\mathbf{x})$, that is,

$$\mathbf{g} = -\mathbf{grad}\, \phi. \tag{2.5-22}$$

[20] In these equations $\dot{q}'''$ allows for *other* kinds of volumetric energy deposition—e.g., net radiation absorption.

[21] The steady-flow form of Eq. (2.5-21) is seen to degenerate to $\rho \mathbf{v} \cdot \mathbf{grad}\, h_0 = 0$ in the absence of local energy addition, viscous stress work, and body force work. For such situations, h_0 will be constant along each streamline even if the fluid is locally brought to rest. Therefore, h_0 is sometimes called the "stagnation" enthalpy.

[22] As can be appreciated from the definition of div **B** (see the chapter Introduction), over any macroscopic volume, $\int_V \operatorname{div} \mathbf{B}\, dV = \int_S \mathbf{B} \cdot \mathbf{n}\, dA$.

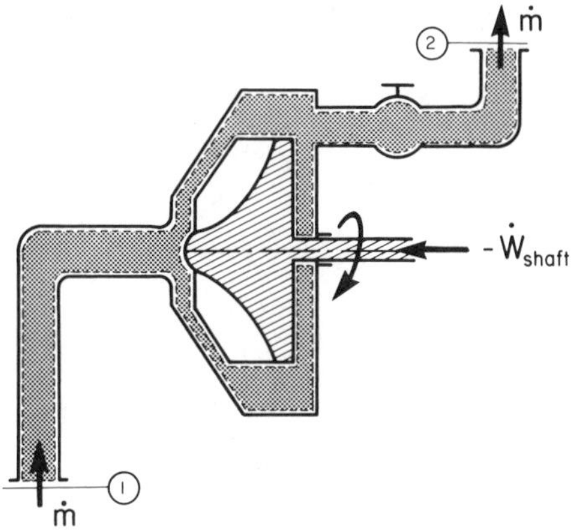

Figure 2.5-2 Control volume for application of macroscopic mechanical energy balance (generalized Bernoulli equation) to a single-inlet, single-outlet, steady-flow system containing a pump (or turbine).

Then for any fixed macroscopic CV:

$$\frac{\partial}{\partial t}\int_V \rho\left(\frac{v^2}{2} + \phi\right)dV + \int_S \rho\left(\frac{p}{\rho} + \frac{v^2}{2} + \phi\right)\mathbf{v}\cdot\mathbf{n}\,dA$$

$$= -\int_V (\mathbf{T}\!:\!\mathbf{grad}\,\mathbf{v})\,dV + \int_S \mathbf{v}\cdot(\mathbf{T}\cdot\mathbf{n}\,dA). \qquad \text{(2.5-23a)}$$

The corresponding results for a variable-density fluid (flow) (Bird, *et al.* (1960)) are rather more complicated than Eq. (2.5-23a), and not, exclusively, "mechanical" in nature.

In contrast to Eq. (2.3-3) note that Eq. (2.5-23) makes no reference to changes in thermodynamic internal energy, nor surface or volume heat addition ($-\dot{\mathbf{q}}''\cdot\mathbf{n}\,dA$, $\dot{q}'''\,dV$)—hence the name "mechanical" energy equation.

Common applications of Eq. (2.5-23) are to the cases of:

a. Passive steady-flow component (pipe length, elbow, valve, etc.) on the control surfaces of which the work done by the extra stress can be neglected. Then there must be a net inflow of $((p/\rho) + (v^2/2) + \phi)$ to compensate for the volume integral of $\mathbf{T}\!:\!\mathbf{grad}\,\mathbf{v}$, a positive quantity[23] shown in Section 2.5.7 to be the local irreversible dissipation rate of mechanical energy (into heat).
b. Steady-flow liquid pumps, fans, and turbines, relating the work required for unit mass flow to the net outflow of $(p/\rho) + (v^2/2) + \phi$.

For (b), in cases with a single inlet and single outlet (see Figure 2.5-2), Eq. (2.5-23a)

[23] For high Re-flow ($Re > 10^5$) of Newtonian fluids, values (mostly experimental) of $(\int \mathbf{T}\!:\!\mathbf{grad}\,\mathbf{v}\,dV)/(\dot{m}v_{②}^2/2)$ are given in Figure 2.5-3.

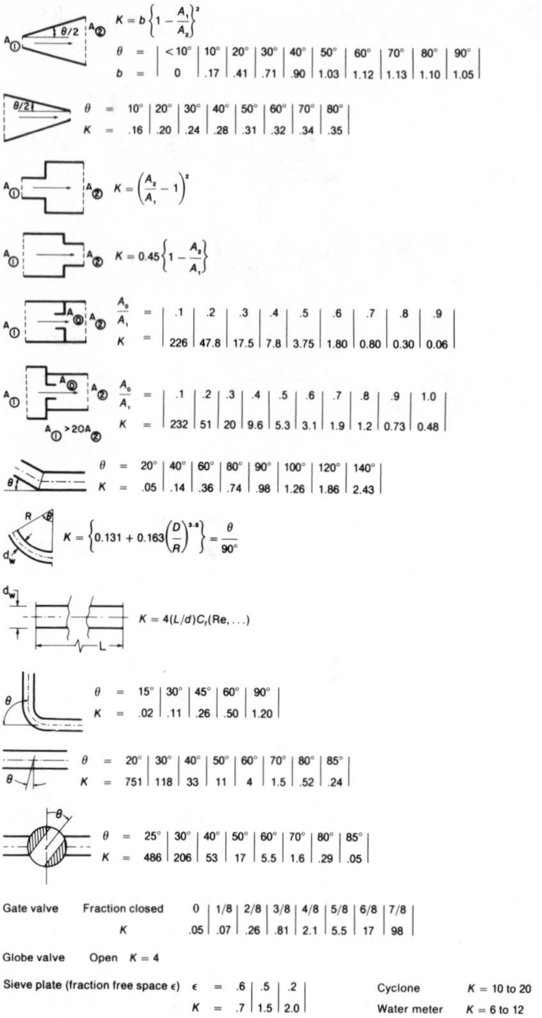

Figure 2.5-3 Summary of friction-loss factors ($\text{Re} > 10^5$) for common fluid-filled system components (after Beek and Muttzall (1975)).

may be rewritten in the "engineering form":[24]

$$\frac{-\dot{W}_{\text{shaft}}}{\dot{m}} = \left\{\left(\frac{p}{\rho} + \frac{v^2}{2} + \phi\right)_{①} - \left(\frac{p}{\rho} + \frac{v^2}{2} + \phi\right)_{②}\right\} + \sum_{\substack{\text{all fluid-}\\\text{filled portions}}} \left\{\frac{v_{\text{exit}}^2}{2} \cdot K(\text{Re}, \text{Shape})\right\}_{\substack{\text{each}\\\text{section}}}. \qquad (2.5\text{-}23b)$$

[24] Inspection of Eq. (2.5-23a) reveals that, strictly speaking, the average fluid velocities $v_①$ and $v_②$ explicitly appearing in Eq. (2.5-23b) are *not* those defined by $\dot{m}/(\rho A_①)$ and $\dot{m}/(\rho A_②)$ respectively, but rather those defined in such a way that $\dot{m}\, v_{\text{avg}}^2/2$ is the actual flow rate of mechanical kinetic energy through each cross-section. While this distinction is not important for highly *turbulent* duct flow conditions (consistent with the friction-loss-factor-values assembled in Figure 2.5-3), it should be made for *laminar* flow situations (see, also, Exercise 4.6).

Here the indicated sum (RHS) accounts for all viscous dissipation losses in fluid-containing portions of the system (other than those contained in the "excluded" "pumping-device" shown in Figure 2.5-2), and $-\dot{W}_{\text{shaft}}$, given by:

$$-\dot{W}_{\text{shaft}} \cong \int_{\substack{\text{portion of fixed}\\ \text{CS enveloping the}\\ \text{pumping device}}} \mathbf{v}\cdot[(-p\mathbf{I}+\mathbf{T})\cdot\mathbf{n}\,dA], \qquad (2.5\text{-}23c)$$

is the rate at which mechanical work is done on the fluid by the indicated pumping device. Note that the work requirement per-unit-mass-flow is the sum of that required to change $[(p/\rho) + (v^2/2) + \phi]$ and that required to overcome the prevailing viscous dissipation losses throughout the system. With a suitable change in signs, this equation can clearly also be used to predict the output of a turbine system for power *extraction* from the fluid.

2.5.6 Explicit Form of the Differential Entropy Balance when $\dot{\mathbf{q}}''' = 0$

The material derivative provides the basis for invoking useful thermodynamic interrelations between small changes in local state variables. For example, the well-known thermodynamic interrelations (J. W. Gibbs):

$$T\,ds = de + p\,d\left(\frac{1}{\rho}\right) - \sum_{i=1}^{N} \hat{\mu}_i\,d\omega_i \qquad (2.5\text{-}24a)$$

or

$$T\,ds = dh - \left(\frac{1}{\rho}\right)dp - \sum_{i=1}^{N} \hat{\mu}_i\,d\omega_i \qquad (2.5\text{-}24b)$$

relate changes in specific entropy, specific volume or pressure, energy, or enthalpy with changes in chemical composition[25] for a given fluid parcel. It follows that in the present continuum flow context:

$$T\frac{Ds}{Dt} = \frac{De}{Dt} + p\frac{D(1/\rho)}{Dt} - \sum_{i=1}^{N} \hat{\mu}_i \frac{D\omega_i}{Dt} \qquad (2.5\text{-}25a)$$

[25] Here the functions $\hat{\mu}_i$ are the partial specific Gibbs free energies, defined such that $h - Ts = \sum \omega_i \hat{\mu}_i$ for the mixture. Moreover, in general, $h = \sum \omega_i \bar{h}_i$, where the $\bar{h}_i$ are called partial specific enthalpies (equal to $h_i(T,p)$ for thermodynamically ideal systems.

and

$$T\frac{Ds}{Dt} = \frac{Dh}{Dt} - \frac{1}{\rho}\frac{Dp}{Dt} - \sum_{i=1}^{N} \hat{\mu}_i \frac{D\omega_i}{Dt}. \tag{2.5-25b}$$

Invoking the previously derived equations for De/Dt or Dh/Dt and the $D\omega_i/Dt$, clearly allows an explicit *balance* equation for $\rho\, Ds/Dt$ to be derived. Comparison with the formal entropy balance (Eq. (2.5-11)) then allows the identification of the entropy flux vector $\mathbf{j}_s''$, as well as the local entropy production rate $\dot{s}'''$. This lengthy but straightforward procedure leads to the interesting identification:

$$\mathbf{j}_s'' = \frac{\dot{\mathbf{q}}''}{T} - \frac{\sum_{i=1}^{N} \mathbf{j}_i'' \hat{\mu}_i}{T} \tag{2.5-26}$$

or, for a thermodynamically ideal mixture:

$$\mathbf{j}_s'' = \frac{1}{T}\left(\dot{\mathbf{q}}'' - \sum_{i=1}^{N} \mathbf{j}_i'' h_i\right) + \sum_{i=1}^{N} \mathbf{j}_i'' \underbrace{\left(s_i + \frac{R}{M_i}\cdot \ln\frac{1}{y_i}\right)}_{\bar{s}_i}. \tag{2.5-27}$$

Moreover, one finds for the product of local absolute temperature and entropy production rate:

$$T\dot{s}''' = (\mathbf{T}\!:\!\mathbf{grad}\,\mathbf{v}) + \left(\dot{\mathbf{q}}'' - \sum_{i=1}^{N} \mathbf{j}_i'' \bar{h}_i\right)\cdot\left(\frac{-\mathbf{grad}\, T}{T}\right)$$
$$+ \sum_{i=1}^{N} \mathbf{j}_i'' \cdot (-\mathbf{grad}_T\, \hat{\mu}_i + \mathbf{g}_i) - \sum_{i=1}^{N} \dot{r}_i''' \hat{\mu}_i. \tag{2.5-28}$$

These equations are the basis for our earlier claim (Section 2.4) that:

a. both energy diffusion and species diffusion contribute to the diffusion flux of entropy; and
b. the local entropy production rate, which must be positive or zero, receives additive contributions from the diffusion of momentum,[26] energy, and mass, as well as nonequilibrium chemical reactions.

Further implications of these results will be discussed in Chapters 3, 4, and 5.

[26] Recall that the "viscous dissipation" rate term $(\mathbf{T}\!:\!\mathbf{grad}\,\mathbf{v})$ appears as a sink in the mechanical-energy equation and as a source in the equation governing changes in e or h.

When steady boundary conditions are imposed that prevent the achievement of local equilibrium everywhere, it has often been observed that fluids tend toward a particular *nonequilibrium steady state* which comes as close as possible to equilibrium while satisfying the imposed constraints. Indeed, Prigogine and co-workers (1961) have shown that, under the restrictive conditions listed as (a)–(c) below, fluids will seek a steady state for which the volume integral of $\dot{s}'''$ (that is,

$$\left.\int_V \dot{s}''' \, dV\right)$$

is a *minimum* compared to all otherwise eligible "neighboring" steady states compatible with the same boundary conditions. This principle of *minimum entropy production* can be used as the basis of a class of variational methods (see, e.g., Schechter (1967) and Section 4.4.4) to approximately (e.g., numerically) solve nonequilibrium steady-state problems involving diffusion (momentum, energy, and/or species mass) as well as nonequilibrium chemical reaction.

However, for a general fluid mixture, satisfaction of this principle can be shown to be equivalent to satisfaction of the previously stated conservation PDEs only if $\dot{q}''' = 0$ and:

a. the momentum, energy, and mass diffusion flux-driving force and chemical kinetic (constitutive) laws are *linear* (see Chapter 3);
b. the transport coefficients appearing in these constitutive laws are constants and satisfy the Onsager reciprocity relationships (see Chapter 3);
c. nonlinear convective terms (momentum) are absent or negligible (e.g., as in "creeping" [$\mathrm{Re} \ll 1$] flow; Section 4.4.1).

While the principle of minimum entropy production is of considerable fundamental interest, unfortunately many important engineering problems do not satisfy these restrictive conditions, especially (b) and (c).

2.5.7 Explicit Form of the Conservation PDEs in Alternate Orthogonal Coordinate Systems

The nature of the above-mentioned symbolic PDEs in any particular inertial coordinate system is briefly explored in this section. We begin by first considering a general curvilinear coordinate system in which the position of each point is defined by the intersection of three nonplanar *surfaces*, say: $\xi_1 = \mathrm{const}_1$, $\xi_2 = \mathrm{const}_2$, and $\xi_3 = \mathrm{const}_3$. However, if the coordinate system chosen is *orthogonal*, the *unit vectors* $\mathbf{e}_1$, $\mathbf{e}_2$, and $\mathbf{e}_3$ (locally in the directions of increasing ξ_1, ξ_2, and ξ_3, respectively) are mutually perpendicular at each such point, and any vector, $\mathbf{B}$, can be expressed in terms of its three local scalar components B_1, B_2, B_3:

$$\mathbf{B} = \mathbf{e}_1 B_1 + \mathbf{e}_2 B_2 + \mathbf{e}_3 B_3. \tag{2.5-29}$$

Consider now the adjacent coordinate surfaces $\xi_1 + \Delta\xi_1$, $\xi_2 + \Delta\xi_2$, and $\xi_3 + \Delta\xi_3$: Then (combined with the original surfaces) we enclose a small control volume ΔV (one "corner" of which is at the point ξ_1, ξ_2, ξ_3). In general, the coordinates themselves do not have the units of length, and the *displacements* associated with the coordinate increments $\Delta\xi_1, \Delta\xi_2, \Delta\xi_3$ are therefore written, respectively, $h_1(\xi_1, \xi_2, \xi_3)\,\Delta\xi_1$, $h_2(\xi_1, \xi_2, \xi_3)\,\Delta\xi_2$, and $h_3(\xi_1, \xi_2, \xi_3)\,\Delta\xi_3$, where the h-functions are the so-called *scale-factor functions*. If $\Delta\xi_1$, $\Delta\xi_2$, and $\Delta\xi_3$ are sufficiently small, then, to a first approximation, the incremental *volume* ΔV is simply:

$$\Delta V \approx (h_1\,\Delta\xi_1)\cdot(h_2\,\Delta\xi_2)\cdot(h_3\,\Delta\xi_3). \tag{2.5-30}$$

We are now in a position to calculate the local "divergence" of any vector $\mathbf{B}$ in this curvilinear system, that is (see Introduction):

$$\operatorname{div}\mathbf{B} \equiv \operatorname*{Lim}_{\Delta V\to 0}\frac{1}{\Delta V}\cdot\int_S \mathbf{B}\cdot\mathbf{n}\,dA. \tag{2.5-31}$$

Including the contributions from each of the six surfaces of ΔV and carrying out the indicated operations, we readily find:

$$\operatorname{div}\mathbf{B} = \frac{1}{h_1 h_2 h_3}\cdot\left\{\frac{\partial}{\partial\xi_1}(h_2 h_3 B_1) + \frac{\partial}{\partial\xi_2}(h_1 h_3 B_2) + \frac{\partial}{\partial\xi_3}(h_1 h_2 B_3)\right\}. \tag{2.5-32}$$

Similarly, for any (nine-component) tensor $\mathbf{T}$ (such as the viscous stress), we find that:

$$(\operatorname{div}\mathbf{T})_1 = \frac{1}{h_1 h_2 h_3}\left\{\frac{\partial}{\partial\xi_1}(h_2 h_3 \tau_{11}) + \frac{\partial}{\partial\xi_2}(h_1 h_3 \tau_{21}) + \frac{\partial}{\partial\xi_3}(h_1 h_2 \tau_{31})\right\}; \tag{2.5-33a}$$

$$(\operatorname{div}\mathbf{T})_2 = \frac{1}{h_1 h_2 h_3}\left\{\frac{\partial}{\partial\xi_1}(h_2 h_3 \tau_{12}) + \frac{\partial}{\partial\xi_2}(h_1 h_3 \tau_{22}) + \frac{\partial}{\partial\xi_3}(h_1 h_2 \tau_{32})\right\}; \tag{2.5-33b}$$

$$(\operatorname{div}\mathbf{T})_3 = \frac{1}{h_1 h_2 h_3}\left\{\frac{\partial}{\partial\xi_1}(h_2 h_3 \tau_{13}) + \frac{\partial}{\partial\xi_2}(h_1 h_3 \tau_{23}) + \frac{\partial}{\partial\xi_3}(h_1 h_2 \tau_{33})\right\}. \tag{2.5-33c}$$

This result follows from the convention that the first subscript deals with the relevant *coordinate surface*, whereas the second subscript refers to *direction*. For example, the *stress* component τ_{12} would be the 2-direction force (per unit area) on the surface $\xi_1 = \text{const}$ in the vicinity of the point ξ_1, ξ_2, ξ_3. (See Section 3.2.1 for further details.)

Table 2.5-1 Scale Factors for the Three Most Commonly Used Orthogonal Coordinate Systems

Coordinate System	ξ_1	h_1	ξ_2	h_2	ξ_3	h_3
Cartesian	x	1	y	1	z	1
Cylindrical Polar	r	1	θ	r	z	1
Spherical Polar[a]	r	1	θ	r	ϕ	$r\sin\theta$

[a] Here r is the radial distance of the point measured from the origin, θ is the complement of the latitude angle, and ϕ is the meridional angle.

It remains[27] for us to explicitly express the *spatial gradient* operation in this general system. In view of the above-mentioned definition of the *scale factors* and *unit vectors*, we see that, if $f(\xi_1, \xi_2, \xi_3, t)$ is any *scalar* function, then:

$$\mathbf{grad}\, f = \frac{1}{h_1}\frac{\partial f}{\partial \xi_1}\mathbf{e}_1 + \frac{1}{h_2}\frac{\partial f}{\partial \xi_2}\mathbf{e}_2 + \frac{1}{h_3}\frac{\partial f}{\partial \xi_3}\mathbf{e}_3. \tag{2.5-34}$$

To appreciate the content of these equations, consider the following simple special cases. Table 2.5-1 gives the ξ and scale-factor assignments for the three most commonly used orthogonal coordinate systems, that is, *cartesian* (x, y, z), *cylindrical polar* (r, θ, z) (Figure 2.5-4)[28] and *spherical polar* (r, θ, ϕ). Note that the cartesian system is singular in that *none* of its coordinate surfaces is "curvilinear" (each of its scale factors is a constant (unity)). If any other orthogonal coordinate system is of interest for a particular class of transport problems, it is only necessary to first determine the appropriate scale factors h_1, h_2, h_3 for that system. Then Eqs. 2.5-32, -33, -34 yield the terms required to express the consequences of local conservation in that system. The properties of additional commonly used curvilinear orthogonal coordinate systems (including spheroidal, parabolic, toroidal, and bi-spherical coordinates) may be found in Morse and Feshbach (1953), or Moon and Spencer (1961). Numerical computations on complicated domains are often facilitated by the

---

[27] The *vector* $\mathbf{T}\cdot\mathbf{v}$ appearing in the energy equation has the following three components:

$$(\mathbf{T}\cdot\mathbf{v})_1 = \tau_{11}v_1 + \tau_{12}v_2 + \tau_{13}v_3,$$
$$(\mathbf{T}\cdot\mathbf{v})_2 = \tau_{21}v_1 + \tau_{22}v_2 + \tau_{23}v_3,$$
$$(\mathbf{T}\cdot\mathbf{v})_3 = \tau_{31}v_1 + \tau_{32}v_2 + \tau_{33}v_3.$$

Note also that, in view of orthogonality ($\mathbf{e}_1\cdot\mathbf{e}_1 = 1$, $\mathbf{e}_1\cdot\mathbf{e}_2 = 0$, $\mathbf{e}_1\cdot\mathbf{e}_3 = 0$, etc.), the scalar $\mathbf{v}\cdot\mathbf{grad}\, f$ appearing in the material derivative will be given by

$$\mathbf{v}\cdot\mathbf{grad}\, f = \frac{v_1}{h_1}\frac{\partial f}{\partial \xi_1} + \frac{v_2}{h_2}\frac{\partial f}{\partial \xi_2} + \frac{v_3}{h_3}\frac{\partial f}{\partial \xi_3}.$$

[28] The student should derive Eq. (2.5-32) and verify that Eqs. (2.1-4, 2.1-8, and 2.2-7 are consistent with Eqs. (2.5-32, -33, -34) and the entries in row 2 of Table 2.5-1.

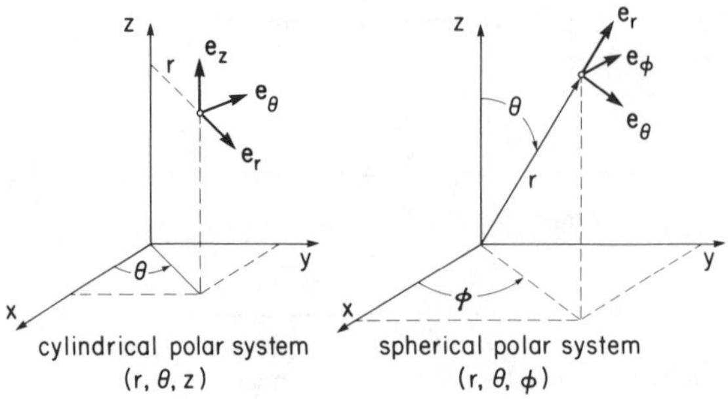

Figure 2.5-4 Orthogonal polar coordinate systems: (a) cylindrical (r, θ, z); (b) spherical (r, θ, ϕ).

numerical computation of orthogonal coordinate surfaces (one of which is the domain boundary), scale factors and mesh point locations (see Chapter 8, Appendices 8.2, 8.3).

It is also interesting to note here that many convective-diffusion problems which involve negligible momentum diffusion (i.e., "inviscid" flows; see Section 2.7.3) but appreciable *energy*- (and/or *species*-) diffusion can be conveniently solved using "streamline coordinates." In such cases the stream surfaces (contours of constant "stream function") constitute one important set of coordinate surfaces for expressing the consequences of local energy (and/or species mass) conservation.

2.6 REMARKS ON IMPORTANT GENERALIZATIONS

2.6.1 Moving Control Volumes and "Jump" Conditions across Moving (or Fixed) Discontinuities (Shock Waves, Phase Boundaries, etc.)

It is often useful to consider the conservation equations for a control volume which is itself moving with respect to both the fluid *and* fixed space. An example might be a CV enveloping the space between two turbine (rotor) blades, or a CV straddling a thin premixed flame propagating across an engine cylinder (Figure 1.1-1). If velocities are still computed in an inertial coordinate system, then the *moving* CS/CV equation introduces one important effect; i.e., the net convective outflow is dictated by the product of the *relative* velocity $\mathbf{v}_{rel} \equiv \mathbf{v} - \mathbf{v}_{CS}$ with each area element $\mathbf{n}\, dA$. Thus, net convective outflows are calculated relative *to an observer moving with the local CS-element*, as implied in Eqs. (2.1-5), (2.1-9), (2.1-13), (2.2-8), (2.3-6), and (2.4-4).

The reason conservation equations applied to control volumes straddling interfaces (phase boundaries or fluid dynamic "discontinuities") yield relationships known as "jump" conditions can be illustrated using total mass conservation (Eq. (2.1-5)) as a simple, but typical, example. Consider a "pillbox" control volume straddling a thin fluid dynamic discontinuity (Figure 2.1-3). Then the total mass balance, Eq. (2.1-5), gives, for a unit area of interface:

$$\rho_{\oplus}\mathbf{v}_{\mathrm{rel}_{\oplus}} \cdot \mathbf{n}_{\oplus} + \rho_{\ominus}\mathbf{v}_{\mathrm{rel}_{\ominus}} \cdot \mathbf{n}_{\ominus} = 0, \tag{2.6-1}$$

where $\mathbf{n}_{\oplus}$ and $\mathbf{n}_{\ominus}$ are the unit outward normal vectors in the + and − sides of the interface, respectively. However, in the discontinuous (thin transition) limit, $\mathbf{n}_{\ominus} = -\mathbf{n}_{\oplus}$, so that Eq. (2.6-1) becomes:

$$[(\rho\mathbf{v}_{\mathrm{rel}_{\oplus}}) - (\rho\mathbf{v}_{\mathrm{rel}_{\ominus}})] \cdot \mathbf{n}_{\oplus} = 0. \tag{2.6-2}$$

If we define $[\rho\mathbf{v}_{\mathrm{rel}}]$ as the "jump" in $\rho\mathbf{v}_{\mathrm{rel}}$ across the interface—that is, if we introduce the "jump" (operator) notation:

$$[\quad] \equiv (\quad)_{\oplus} - (\quad)_{\ominus}, \tag{2.6-3}$$

then conservation of mass at an interface can be restated in the compact form:

$$[\rho\mathbf{v}_{\mathrm{rel}}] \cdot \mathbf{n}_{\oplus} = 0. \tag{2.6-4}$$

This means that neither ρ nor $\mathbf{v}_{\mathrm{rel}} \cdot \mathbf{n}_{\oplus}$ (the normal component of the fluid velocity relative to the surface) need be continuous across such a transition; i.e., *only their product* (physically, the mass flux *through* the interface, written, $\dot{m}''_{\pm}$) *must be continuous.*

Similar explicit "jump" forms of the conservation equations can now be written for chemical species, chemical elements, linear momentum, total energy, and entropy, as outlined in Section 2.1.

The explicit "jump" forms of conservation of each chemical species i $(i = 1, 2, \ldots, N)$ and chemical element $k (k = 1, 2, \ldots, N_{\mathrm{elem}})$ are, respectively (cf. Eqs. (2.1-9) and (2.1-13)):

$$\dot{m}''_{\pm} \cdot [\omega_i] = -[\mathbf{j}''_i] \cdot \mathbf{n}_{\oplus} + \dot{r}''_i \tag{2.6-5}$$

and

$$\dot{m}''_{\pm} \cdot [\omega_{(k)}] = -[\mathbf{j}''_{(k)}] \cdot \mathbf{n}_{\oplus}. \tag{2.6-6}$$

These equations are seen to impose necessary conditions on the "jump" in the *normal component* of the chemical species and element *diffusion fluxes.* A common application of Eqs. (2.6-5 and 2.6-6) will be seen to be at a fluid/catalyst interface. There, in the absence of *net* interfacial mass transfer ($\dot{m}''_{\pm} = 0$), Eq. (2.6-5) specializes to:

$$[\mathbf{j}''_i] \cdot \mathbf{n}_{\oplus} = \dot{r}''_i \qquad (i = 1, 2, \ldots, N); \tag{2.6-7}$$

that is, if there is an (instantaneous) production rate of chemical species i per unit area of interface, this must be balanced by a net diffusional *outflow* of that species from the "pillbox" CV straddling the interface (see, e.g., Section 6.5.3). In this same situation Eq. (2.6-6) implies that, since each element production rate $\dot{r}''_{(k)} = 0$, then the normal component of the *diffusion flux* of each *chemical element* "k" ($k = 1, 2, \ldots, N_{\text{elem}}$) must be continuous across the interface. It should also be remarked that phase boundaries which are the site of nonzero net reaction rates ($\dot{r}''_i \neq 0$) will "violate" the Gibbs *thermochemical equilibrium* constraint: $[\hat{\mu}_i] = 0$, where $\hat{\mu}_i$ is the *chemical potential* (partial specific Gibbs free energy) of chemical species i.

In the case of *linear momentum* it is convenient to consider both the normal (n) and tangential (t) components, and split off the thermodynamic pressure part of the total stress $\mathbf{\Pi}$ (see Eq. (2.5-17)). Then the *normal* component momentum balance at an interface (cf. Eq. (2.2-8)) becomes explicitly:

$$\dot{m}''_{\pm} \cdot [v_{\text{rel,n}}] = -[p] + ([\mathbf{T}] \cdot \mathbf{n}_{\oplus})_{\text{n}} \tag{2.6-8}$$

a result which will be useful when applied to normal shock waves and detonations in the absence of appreciable viscous stress effects (i.e., when the normal linear-momentum outflow is balanced by the $-[p]$ force alone (cf. Section 4.3.2)). The *tangential* component of Eq. (2.2-8) explicitly gives (in the absence of spatial gradients in surface tension):

$$\dot{m}''_{\pm} \cdot [v_{\text{rel,t}}] = ([\mathbf{T}] \cdot \mathbf{n}_{\oplus})_{\text{t}}. \tag{2.6-9}$$

Note that, if there is no velocity "slip" (i.e., if $[v_{\text{rel,t}}] = 0$), then, even if there *is* interfacial mass transfer ($\dot{m}''_{\pm} \neq 0$), the *tangential* component of the extra (viscous) stress can be nonzero but must be *continuous* across the interface.

The energy conservation condition at an interface is, perhaps, most usefully expressed in terms of the jump in the "total" enthalpy $h_0 \equiv e + (p/\rho) + (v^2/2)$. Then Eq. (2.3-6) takes the explicit form:

$$\dot{m}''_{\pm} \cdot [h_0] = \dot{q}'' - [\dot{\mathbf{q}}''] \cdot \mathbf{n}_{\oplus} + [\mathbf{T} \cdot \mathbf{v}] \cdot \mathbf{n}_{\oplus} - [p] v_{\text{CS,n}} \tag{2.6-10}$$

where $\dot{q}''$ (first term on the RHS) accounts for any radiative contributions to/from the interface (Section 5.9), and the second term is the jump in the normal component of the material phase energy flux vector. A common "degenerate" form of Eq. (2.6-10) for a phase boundary without mass transfer and radiative loss is simply that $\dot{q}''_{\text{n}}$ must be continuous. However, when there is a net rate of phase change, and $[h_0] \neq 0$, then $\dot{q}''_{\text{n}}$ must exhibit a corresponding "jump" (see Exercise 5.9).

The *entropy balance* equation applied to a "pillbox" CV straddling an interface (Eq. (2.4-4)) explicitly yields:

$$\dot{m}''_{\pm} \cdot [s] = -[\mathbf{j}''_s] \cdot \mathbf{n}_{\oplus} + \dot{s}'', \tag{2.6-11}$$

where the entropy diffusion flux vector, $\mathbf{j}''_s$, on either side of the interface is given by

Eq. (2.5-26) or (2.5-27) in the absence of radiative energy transport. *Clausius' inequality* for an interface then takes the form $\dot{s}'' \geqslant 0$ or, explicitly:

$$\dot{m}''_{\pm} \cdot [s] + [j''_{s,n}] \geqslant 0. \tag{2.6-12}$$

This is a result we will have occasion to exploit in Section 4.3, dealing with gas-dynamic discontinuities.

We can summarize the results of this section by stating that discontinuities are *not* prohibited within the *continuum* framework; i.e., a flow can be self-consistently treated as *piecewise continuous, provided the field variables across the surfaces of discontinuity "jump" in such a way as to satisfy the fundamental conservation principles.* This is the basis of Eqs. (2.6-4, -5, -6, -8, -9, -10, and -11), special cases of which will be illustrated in Chapters 4, 5, and 6. A still more general viewpoint is that the "jump-conditions" stated in this section are themselves degenerate forms of the corresponding conservation equations governing the behavior of an approximately two-dimensional ("interfacial") phase which, in effect, separates the ordinary three-dimensional phases on either side of it. In their complete forms, such *interfacial phase balance equations* would also include terms that allow for:

accumulation rates per unit area of interface,
net outflow rates by convection in the surface (tangent) plane,
net inflow rates by diffusion in the surface (tangent) plane.

(See, e.g., Rosner (1976), and Brenner (1984).)

2.6.2 Conservation Equations Using an Accelerating (Noninertial) Coordinate Frame

To predict fluid flow near or produced by an *accelerating* object (such as a rotating turbine or compressor blade, a contoured piston face, a propeller, helicopter blade, missile, or the surface of the earth) it is usually most convenient to adopt a coordinate frame fixed to the accelerating object itself. Such a coordinate frame can otherwise be chosen to suit the individual situation (curvilinear, nonorthogonal, etc.), but it will necessarily be noninertial—hence the momentum conservation equations must be expressed with special care. This is because any "apparent" fluid parcel acceleration recorded by an observer in the accelerating coordinate frame (ACF) will generally be rather different from the "true" (or "total") acceleration recorded for the same fluid parcel by an observer using an inertial coordinate frame (ICF).

In formally making any coordinate (independent variable) transformation we require, of course, quantitative interrelationships between the two coordinate systems of interest, as well as corresponding interrelationships between "apparent" and "true" velocities and accelerations. Let us consider that the previously written conservation equations are valid statements for any particular ICF. To express the corresponding statement for a convenient ACF (which may be rotating and/or accelerating away from the origin of the ICF; see Figure 2.6-1), we recall that true and apparent (subscript a) positions, velocities, and accelerations are, respectively, related by:

$$\mathbf{x} = \mathbf{x}_a + \mathbf{x}_0, \tag{2.6-13a}$$

$$\mathbf{v} = \mathbf{v}_a + \dot{\mathbf{x}}_0 + (\boldsymbol{\Omega} \times \mathbf{x}_a), \tag{2.6-13b}$$

$$\mathbf{a} = \mathbf{a}_a + \ddot{\mathbf{x}}_0 + (2\boldsymbol{\Omega} \times \mathbf{v}_a) + (\boldsymbol{\Omega} \times \boldsymbol{\Omega} \times \mathbf{x}_a) + (\dot{\boldsymbol{\Omega}} \times \mathbf{x}_a), \tag{2.6-13c}$$

where the dots pertain to time differentiation.

The "kinematic" interrelations (Eqs. (2.6-13b and c)) are familiar from classical treatments of rigid body motion, except that here $\boldsymbol{\Omega}$ is the instantaneous rotational (angular) velocity *of the accelerating coordinate frame* (ACF). The third and fourth terms in the expression for the total acceleration (**a**) are known as the *Coriolis* (after G. G. Coriolis [1743]) and *centripetal* accelerations, respectively. The second and fifth terms clearly vanish when the ACF is "noninertial" only due to steady rotation (in which case $\dot{\boldsymbol{\Omega}} = 0$, and $\ddot{\mathbf{x}}_0 = 0$).

Since the acceleration vector **a** can be identified with the material derivative $D\mathbf{v}/Dt$ appearing in Eq. (2.5-9), we see from Eq. (2.6-13c) that the linear-momentum conservation equation for the chemically reacting fluid relative to the ACF becomes:

$$\rho \frac{D\mathbf{v}_a}{Dt} = -\operatorname{div} \boldsymbol{\Pi} + \sum_i \rho_i \mathbf{g}_i - \rho\{\ddot{\mathbf{x}}_0 + (2\boldsymbol{\Omega} \times \mathbf{v}_a) + (\boldsymbol{\Omega} \times \boldsymbol{\Omega} \times \mathbf{x}_a) + (\dot{\boldsymbol{\Omega}} \times \mathbf{x}_a)\}. \tag{2.6-14}$$

Thus, to an observer on the accelerating coordinate frame (ACF), it will appear that

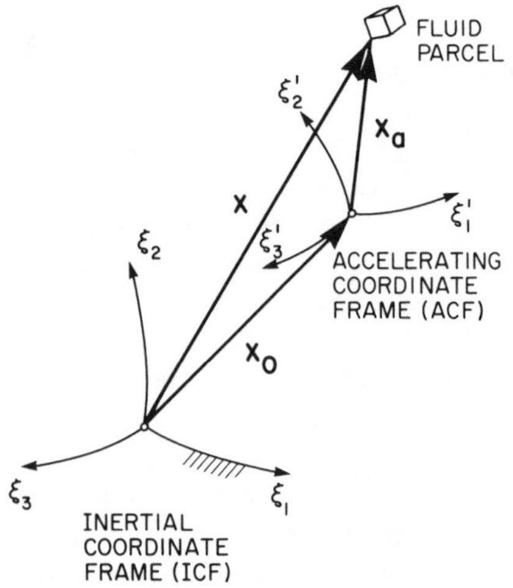

Figure 2.6-1 Relation between accelerating (noninertial) and inertial coordinate frames.

each unit volume of fluid is being subjected to forces other than the "real" contact force ($-\text{div}\,\boldsymbol{\Pi}$) and "real" net body force $\sum \rho_i \mathbf{g}_i$. These additional "apparent forces" (sometimes called "forces of transport") are merely consequences of the fact that $D\mathbf{v}_a/Dt$ is not the complete (true) acceleration vector for each fluid parcel. It is particularly interesting that one of these apparent forces (the Coriolis "force") is itself dependent on the magnitude and direction of the apparent fluid velocity, $\mathbf{v}_a$. The others can be lumped into time- and/or position-dependent *effective body force per unit mass of mixture.*

Note, also, that, as an immediate consequence of Eq. (2.6-14), the PDE governing the apparent local mechanical kinetic energy per unit mass, $v_a^2/2$ (that is, the equation for $\rho D(v_a^2/2)/Dt$) will also contain many "new" terms associated with the ACF coordinate transformation. This observation is important in developing mathematical models for complex *turbulent* flows, in which the *local mechanical kinetic energy associated with the turbulence* (see Sections 2.6.3 and 5.7.3) will be seen to play an important role.

2.6.3 Approach (Reynolds') to Treatment of Turbulence via Time-Averaging the Conservation Equations

A technologically important type of continuum flow not explicitly discussed thus far is "turbulent" fluid flow, in which (a) all local field densities undergo temporal fluctuations about some (time-averaged) values, and (b) spatial correlation lengths (eddy "scales") are more than adequate to ensure the validity of our previous transient *continuum* equations. However, these fluctuations are usually on such a fine temporal and spatial scale as to preclude a completely deterministic description of them *via* these equations. Since we are usually interested in the time-averaged field variables anyway (indeed, these are measured with normal "sluggish" instrumentation), O. Reynolds reasoned that this information is contained in and can be extracted from the conservation equations if we *time-average* them. This procedure indeed reveals the cause of enhanced transport (of mass, momentum, energy, and entropy) in turbulent flows, as well as other "novel" features attributed to the turbulence (augmented dissipation of mechanical kinetic energy into heat, an influence on the apparent time-averaged kinetic source strengths $\bar{\dot{r}}_i'''$, etc.). We illustrate this briefly below, for the case of, say, axial momentum transfer due to radial convection. Since v_z is the axial momentum *per unit mass*, and ρv_r is the mass flux across r = constant surfaces, the corresponding *momentum flux* is $\rho v_r v_z$.

Consider the $\rho v_r v_z$ component of the total convective momentum flux, whose spatial gradient appears in the PDE expressing momentum conservation. Suppose v_r and v_z are each fluctuating, but, for simplicity, consider ρ as constant. Then, at each instant and each point,

$$v_r \equiv \bar{v}_r + v_r', \qquad \text{(2.6-15a)}$$

$$v_z \equiv \bar{v}_z + v_z', \qquad \text{(2.6-15b)}$$

where the overbar defines the time-averaged (mean) value and a prime denotes the fluctuation (Figure 2.6-2). Then the fluctuations will have the property that their time-averages individually vanish; that is,

$$\overline{v_r'} = 0, \qquad v_z' = 0. \tag{2.6-16}$$

However, if we consider the time-average of the axial momentum transfer due to radial convection we find:

$$\overline{\rho v_r v_z} = \rho\overline{(v_r + v_r')(v_z + v_z')} = \rho \bar{v}_r \bar{v}_z + \rho\overline{v_r' v_z'}; \tag{2.6-17}$$

that is, it is different from $\rho \bar{v}_r \bar{v}_z$ (corresponding to the time-average motion) by the amount $\rho\overline{v_r' v_z'}$, which is nonzero as a result of the *time-correlation between* v_r*- and* v_z*-fluctuations at the point in question.* This "additional" momentum transfer (over that associated with the mean motion) is "responsible" for the apparently non-Newtonian behavior of Newtonian fluids in "steady" turbulent flow! Indeed, in cylindrical polar coordinates, for example, a turbulent time-average flow will appear to be governed by an "additional" local *stress*, with components:

$$\begin{pmatrix} -\rho\overline{v_r' v_r'} & -\rho\overline{v_r' v_\theta'} & -\rho\overline{v_r' v_z'} \\ -\rho\overline{v_\theta v_r} & -\rho\overline{v_\theta' v_\theta'} & -\rho\overline{v_\theta' v_z'} \\ -\rho\overline{v_z' v_r'} & -\rho\overline{v_z' v_\theta'} & -\rho\overline{v_z' v_z'} \end{pmatrix}, \tag{2.6-18}$$

called the *Reynolds' stress.* While one can discuss this stress in terms of some "effective" (turbulent) viscosity μ_t, unfortunately, μ_t is not a simple *fluid* property and a general "constitutive" law (analogous to Eq. (3.2-4)) to relate this stress to local time-average field variables (so that turbulent flows can be predicted, at least in a time-average sense) has eluded researchers because it does not exist in principle. This is especially true since many "turbulent" flows are now known to contain underlying large-scale coherent features (e.g., periodicities associated with vortex "shedding") which cannot be adequately described using this simple Reynolds' decomposition ($\bar{\mathbf{v}} + \mathbf{v}', \ldots$) scheme. In any case, if the Reynolds' decomposition scheme is formally adopted, it is clear that such "double-(time-) correlations" (at each point) will arise in calculating the time-average of *any* flux that is *nonlinear* in *fluctuating field* variables. This occurs for the *convective* flux of species *mass* and *energy*, as well as *momentum.*

In a fluid that experiences appreciable *density* fluctuations the time-averaged linear momentum flux would be $\overline{\rho\mathbf{v}\mathbf{v}}$, which now involves the *triple* correlations $\overline{\rho'\mathbf{v}'\mathbf{v}'}$, as well as new double-correlations of the form $\overline{\rho'\mathbf{v}'}$. For this reason an alternative time-averaging procedure (due to Favre (1969)) is actually more convenient for treating the turbulent flow of such fluids. In "Favre-averaging" all fluid mechanical field quantities (except the pressure) are "mass"-averaged; that is,

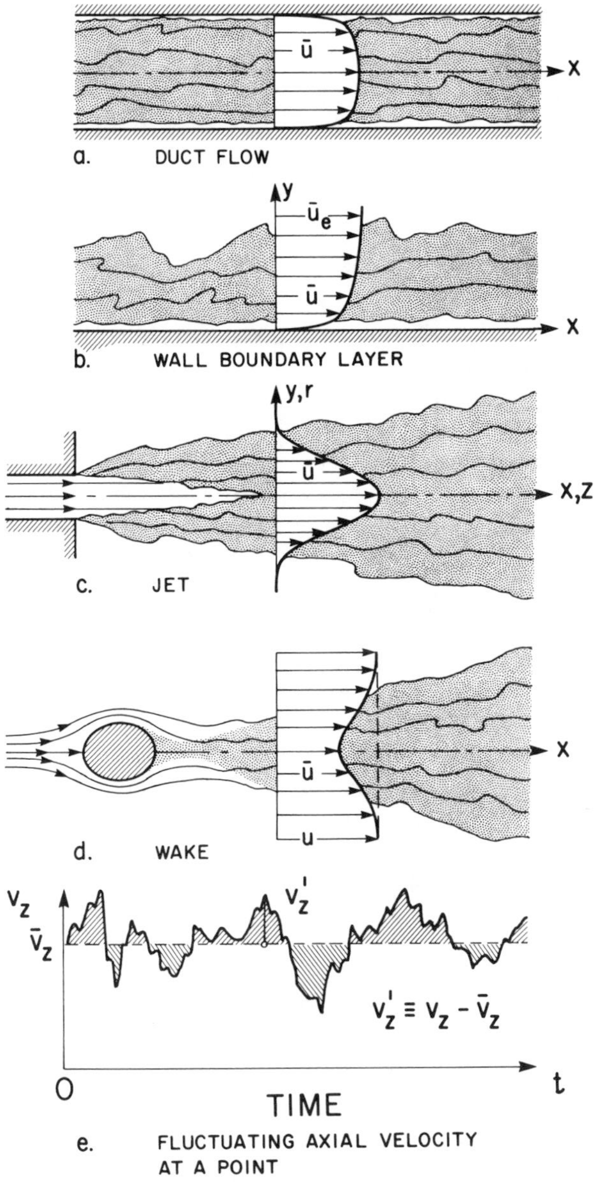

Figure 2.6-2 Typical "steady" turbulent flows (adapted from Kuethe and Chow (1976)), showing instantaneous streamlines, time-averaged (mean) velocity distributions and field boundaries of turbulent shear flows: (a) fully developed flow through tube or channel; (b) boundary layer; (c) jet; (d) wake (N.B.: Turbulent field boundaries are not streamlines); and (e) time-averaged and fluctuating axial (z-component) velocities at a typical point in flow-field (c).

we define a new (Favre-averaged-) velocity by:

$$\tilde{\mathbf{v}} \equiv \frac{\overline{\rho \mathbf{v}}}{\bar{\rho}} \qquad \text{(etc.)}. \tag{2.6-19}$$

Velocity fluctuations about *this* mean are then considered; i.e., the instantaneous local velocity is decomposed into:

$$\mathbf{v} = \tilde{\mathbf{v}} + \mathbf{v}'', \qquad \text{(etc.)}. \tag{2.6-20}$$

(Note the use of $\sim$ and $''$ to distinguish these means and fluctuations from the conventional ones.) In this scheme $\overline{\mathbf{v}''}$ does *not* automatically vanish, but $\overline{\rho \mathbf{v}''}$ does. But, most important, the time-averaged linear-momentum flux can now be expressed in the concise form $\bar{\rho}\tilde{\mathbf{v}}\tilde{\mathbf{v}} + \overline{(\rho \mathbf{v}'')\mathbf{v}''}$, somewhat simplifying the formidable problem of predicting its value in novel situations.

Turbulent fluctuations in the fluid's macroscopic velocity are, of course, also associated with mechanical kinetic energy. Clearly,

$$\frac{\overline{\mathbf{v} \cdot \mathbf{v}}}{2} = \frac{\overline{(\bar{\mathbf{v}} + \mathbf{v}') \cdot (\bar{\mathbf{v}} + \mathbf{v}')}}{2} = \frac{\bar{\mathbf{v}} \cdot \bar{\mathbf{v}}}{2} + \frac{\overline{\mathbf{v}' \cdot \mathbf{v}'}}{2}. \tag{2.6-21}$$

The first term on the RHS is the kinetic energy per-unit-mass associated with the time-averaged motion; the second term (abbreviated κ_t in what follows) is the *mechanical kinetic energy per-unit-mass associated with the local turbulence.* It will be seen to play an important role in characterizing the local turbulence, and estimating its effective "diffusivities" for momentum, energy, and species mass transport (Section 5.7.3).

We will not explicitly write out the time-averaged or Favre-averaged form of each of the conservation equations discussed here; however, it is clear from the above discussion that they contain many new terms that must be "modeled" in order to provide the basis for a semi-empirical "predictive" scheme for new turbulent flows. Particularly for compressible, reactive gas flows, the treatment of these "correlations" is at the forefront of current research (see Section 5.7.3).

2.6.4 Approach to the Treatment of Multiphase Continua via Volume-Averaging the Conservation Equations (See, e.g., Slattery (1972), and Boure and Delhaye (1982))

Despite the disclaimer (see the Introduction to this chapter, assumption 2) some multiphase flows can be treated by the above-mentioned methods merely by considering the second ("dispersed") phase a component (or components) of the mixture. An example might be the flow of a soot-laden combustion gas in a large furnace, in which case the soot (small aggregates of primary carbonaceous particles of some 0.025 μm diameter) could be considered as a high-molecular-weight "gaseous" constituent. In general, however, this approach *cannot* be used, because the "inertia" of the dispersed or continuous phase produces such large velocity differences between the local dispersed and continuous phases as to fall outside the domain of validity of the locally linear "diffusion" flux laws to be discussed in Chapter 3.[29] This would be the case for a fuel-oil spray (containing droplets between 10–100 μm diam.) vaporizing and burning in swirling air, or for gas flow through a coarse-grained porous medium. While it may be true that the above-mentioned continuum equations apply *within* each phase, it would be quite impractical to try to keep track of, say, *each* fuel droplet and its immediate surroundings (this would require a computational resolution of fractions of a micrometer when the overall furnace flow might take place in a space of some 10 meters on a side). The problem is somewhat analogous to keeping track of individual turbulent eddies, an impractical task circumvented by the above-mentioned device of time-averaging, or "time-smoothing." In the present case a similar device can be invoked to yield continuum-like equations for each of the two (or more) phases as though they coexisted in the same space. Rather than time-averaging, the equations governing each phase are *volume-averaged*, taking advantage of the fact that, in such flows, one typically has:

$$\ell \ll d_p < N_p^{-1/3} \ll L; \tag{2.6-22}$$

that is, the individual particle sizes d_p, and average interparticle distances, $N_p^{-1/3}$ (where $N_p \equiv$ particle number density) are very small on the scale of the overall flow field. In effect, instead of dealing each *actual* property P_α of phase α at a point, we deal with the *volume*-averaged property $\langle P_\alpha \rangle^{(\alpha)}$ near each point, defined by (Figure 2.6-3) as:

$$\langle P_\alpha \rangle^{(\alpha)} \equiv \frac{1}{V_\alpha} \cdot \int_{V_\alpha} P_\alpha \, dV, \tag{2.6-23}$$

where the averaging volume is large compared to $N_p^{-1/3}$ but small compared to L. The value of $\langle P_\alpha \rangle^{(\alpha)}$ will be insensitive to the actual value of V_α chosen, for volumes in this size range. The PDEs expressing conservation and governing such *volume-averaged* fields contain two essential novel features:

a. Source terms arising from the interchange of mass, momentum, energy, and entropy *between* the phases across their common interfacial area within the averaging volume.
b. *Spatial* correlation terms (quite analogous to the time-correlations described in Section 2.6.2) involving volume averages of products like $(P_\alpha - \langle P_\alpha \rangle^{(\alpha)}) \cdot (Q_\alpha - \langle Q_\alpha \rangle^{(\alpha)})$, etc. (These spatial correlations represent the "price" paid for the microscale information "lost" when volume averages were taken.)

---

[29] See also the numerical example discussed in Section 8.3.

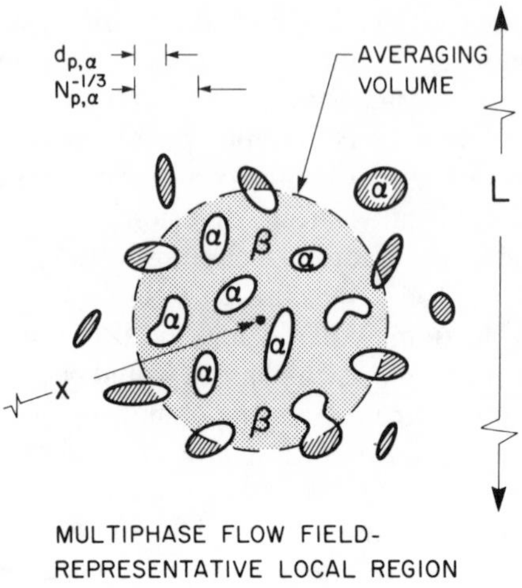

Figure 2.6-3 Volume over which averages are computed in a multiphase region ($\alpha + \beta$) to arrive at continuum-like equations for conservation (mass, momentum, energy) for each "co-existing" phase.

When these terms are empirically or heuristically "modeled," multiphase problems can be solved just like multicomponent single-phase problems, *via* strongly coupled continuum-like PDEs. By essentially dealing with each size range (in a distribution of droplet sizes) as if it were a separate phase, it is also possible in principle to predict the spatial and temporal evolution of a droplet spray in a complex flow, etc.

Clearly, the most challenging problems, unfortunately commonplace in applications, are those involving *turbulent multiphase flows*, which require *both* time- and space-"smoothing." Again, this brings us to the forefront of current chemically reacting flow mathematical modeling.

2.7 COMMENTS ON THE MATRIX OF FLUID MECHANICS

2.7.1 Continuum/Molecular

Even a gas flow can be considered using the continuum viewpoint if the gas kinetic mean-free-path, ℓ (the average distance a molecule travels before its next encounter), is much smaller than all macroscopic distances of interest. For the flow past an object of characteristic dimension L, it is *necessary* that the Knudsen number, Kn, defined by:

$$\mathrm{Kn} \equiv \ell / L, \qquad (2.7\text{-}1)$$

be small compared to unity. If $\mathrm{Kn} \gg 1$, the flow about the object is said to be "free-molecular," and cannot be understood using continuum concepts. Flows with Kn near 1, said to be "transitional," actually present the greatest computational difficulty.

2.7.2 Compressible/Incompressible

Most gas flows involve appreciable fractional changes in the mixture density ρ along streamlines. This is common in cases involving:

a. chemical heat release, or chemically induced changes in mean molecular weight,
b. steady flows at speeds comparable to, or exceeding, the local speed of sound,
c. unsteady flows involving wave propagation (e.g., a "blast" wave),
d. natural ("free") convection in a body force field, such as gravity.

Such *flows* are called "compressible." In other cases, the assumption that the density remains constant (e.g., along a streamline in a steady flow) may provide an adequate quantitative description, especially for *liquids*. Such *flows* are called "incompressible." It is interesting that low speed (Mach number) flows of *gas* (e.g., air) in the absence of appreciable) heat release/transfer are often well approximated as "incompressible" even though the equation of state for such a gas clearly reveals a pressure-dependent density.

2.7.3 Viscous/Inviscid

A region of fluid flow is said to be *inviscid* if linear-momentum *diffusion* (local shear stresses) plays a role subordinate to linear momentum *convection*, normal pressure, and body forces. The importance of viscosity is usually described *via* the *Reynolds' number*:

$$\mathrm{Re} \equiv \frac{(\rho U U)}{(\mu U/L)} = \frac{UL}{\nu}, \tag{2.7-2}$$

which is seen to be the ratio of the characteristic linear-momentum flux by *convection*, ρUU, to the characteristic linear momentum flux by *diffusion*, $\mu U/L$, where μ is the viscosity coefficient discussed in detail in Chapter 3. Alternatively, Re can also be regarded as the ratio of two characteristic *times*, that is, $(L^2/\nu)/(L/U)$, where $\nu \equiv \mu/\rho$ is the fluid's momentum diffusivity, L^2/ν is the characteristic *momentum diffusion time*, and L/U is the *characteristic flow* (or "residence") *time*. When $\mathrm{Re} \gg 1$, the flow-field is predominantly "inviscid" (free of the influence of viscosity), although local regions of the flow (near walls, inside shock waves, etc.) will be viscous (cf. Figure 2.7-1). When $\mathrm{Re} \ll 1$, the entire flow field is "viscous."

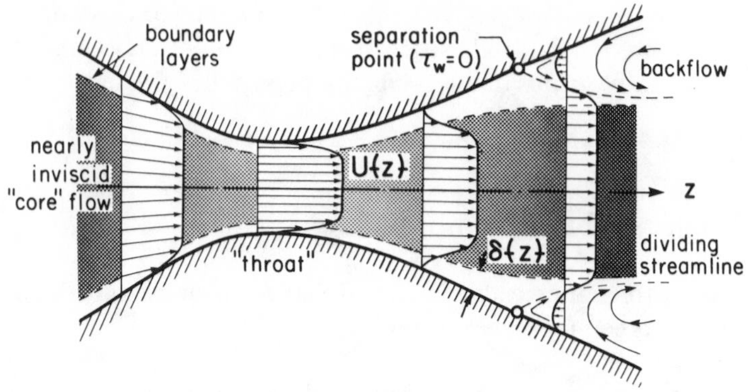

Figure 2.7-1 High-Reynolds'-number steady flow of a nearly incompressible fluid in a nozzle, showing both the "inviscid" core and the effects of fluid viscosity in the immediate vicinity of the duct walls.

2.7.4 Newtonian/Non-Newtonian

Fluids governed by a *linear* stress *vs.* rate of deformation law (Eq. 3.2-4) are said to be "Newtonian," with all *other* types of fluid being "non-Newtonian." Examples of the latter are polymer melts, and slurries (e.g., paint, mud, applesauce, etc.). Fortunately, reacting gas mixtures exhibit Newtonian behavior when in streamline (laminar) flow, but they appear to be non-Newtonian when such flows become unstable (see Sections 2.7.6 below and 2.6.2 above).

2.7.5 Steady/Unsteady

Flows in which the "accumulation rate" terms in each of the above conservation equations are important are called "unsteady," with the remainder being either strictly steady (time-independent) or "quasi-steady" (slowly varying). In the latter case the accumulation term is not identically zero, but rather the small difference between large terms. In some cases, unsteady flows are (a) *periodic* (but not necessarily harmonic), such as the flow in an internal combustion engine cylinder, or (b) nearly *chaotic* (such as turbulent flows at the local level—see Section 2.6.2 above and 2.7.6 below).

2.7.6 Laminar/Turbulent

Rapid response instrumentation or dye tracer experiments reveal that viscous streamline flows (called "laminar") become dynamically unstable above a threshold (transition) Reynolds' number (for forced convection), and exhibit the apparently chaotic motion of parcels of fluid ("eddies") superimposed on a mean (time-averaged) motion which may itself be steady. Such fluid flows, which exhibit enhanced time-averaged rates of mass, momentum, and energy transport, are said to be "turbulent" and require additional empirical information for their adequate description and economical prediction.

2.7.7 Multidimensional/One-Dimensional

In most flows all quantities (fields) vary in each of the three spatial directions (e.g., r, θ, z). Such flows are said to be "three-dimensional," requiring for their description *partial differential equations* containing three independent *space* variables. Often, quantities vary primarily in two directions, being nearly constant in the third direction (e.g., along the span of a wing or gas turbine blade). Such flows, said to be two-dimensional, are described by PDEs in two independent spatial variables. When quantities depend primarily on only *one* spatial variable (e.g., cylindrical radius, or distance along an axially symmetric nozzle, etc.), such flows are termed "one-dimensional" and, if steady, can be adequately described by *ordinary* differential (ODEs) (see Chapter 4). Many flows (called "quasi"-one-dimensional) can be approximated as one-dimensional even though flow properties, strictly speaking, do vary in more than one direction. This is accomplished by averaging dependent variables in all directions but one (e.g., area-averaging across each cross section of a variable-area nozzle (see Figure 2.7-1 and Section 4.3.1).

While there are many other categorizations of our subject (see, e.g., Section 4.1.3), within the realm of *continuum* flows (Section 2.7.1) the integral, differential, and jump conditions developed and expressed in this chapter will be seen to provide the basis for *all* engineering design and analysis techniques, no matter what limiting cases may be useful approximations. However, these general conservation (balance) principles must now be supplemented by the *constitutive* laws (Chapter 3) appropriate to the substances of interest.

SUMMARY

- All of the conservation (balance) laws for Eulerian CVs can be cast in the same standard form, involving:

$$\begin{pmatrix}\text{Accumu-}\\ \text{lation rate}\end{pmatrix} + \begin{pmatrix}\text{Net outflow rate}\\ \text{by convection}\end{pmatrix} = \begin{pmatrix}\text{Net inflow rate}\\ \text{by diffusion}\end{pmatrix} + \begin{pmatrix}\text{Net source}\\ \text{strength}\end{pmatrix}.$$

PDEs are obtained by dividing by the volume V and then passing to the limit $V \to 0$. They can be stated in a symbolic form applicable in *any* coordinate system (the coordinate system chosen to proceed with the solution is a matter of convenience).

- Not all of the conservation equations are independent of one another; the set actually used to solve particular problems is also a matter of convenience: e.g.,

  All chemical species equations and all chemical-element conservation equations are not independent of the total mass-conservation equation;

  Angular momentum and linear momentum constraints are not independent for the flow of a simple fluid (e.g., gas mixture);

  The entropy *balance* equation is not independent of the remaining conservation equations.

- The energy equation can be written in many forms, the simplest being in terms of the specific "total" energy ($e + (v^2/2)$), from which all other forms can be derived. The *enthalpy* function ($e + (p/\rho)$) enters (ultimately in the net-outflow-by-convection term) as a result of the rate at which work is done in overcoming the thermodynamic pressure at those portions of the control surface through which the fluid flows.
- Even the Second Law of Thermodynamics can be written as a *balance* principle, revealing the nature of:

  the local rate of entropy production associated with species mass-, momentum-, and energy-*diffusion*, as well as nonequilibrium chemical reaction, and
  the local diffusion flux of entropy.

- While the continuum equations are, in principle, applicable, the "temporal granularity" of the fluid associated with local *turbulence* forces us to *time-average* these equations, and focus our attention on the time-averaged fields and certain statistical correlations of fluctuations of the various field densities about these means. *Turbulence* (time-averaged effects of nonsteady convection) introduces enhanced ("eddy"-) transport of momentum, energy, and mass, as well as other effects: augmented rate of viscous dissipation, corrections to time-averaged chemical reaction rates, etc.
- Similarly, the spatial granularity of *multiphase* systems can be "suppressed" by the device of local *volume-averaging*, to arrive at a continuum formulation for coexisting, interacting phases such as occur in a fuel droplet spray, or a pulverized coal-dust cloud.

TRUE/FALSE QUESTIONS

2.1 T F The conservation equations expressed in vector form have a physical meaning that is the same no matter what the choice of coordinate system used to define the positions of points in space.

2.2 T F Complete specification of the local *convective* flux of linear momentum requires six independent scalar quantities.

2.3 T F Linear momentum is not conserved in systems described by using accelerating coordinate frames (systems).

2.4 T F The specific enthalpy, $h = e + (p/\rho)$, is a particularly convenient dependent variable for treating energy transfer in time-dependent, nonflow ("batch") systems.

2.5 T F While it is possible to derive useful equations governing the total (thermal, chemical, and mechanical) energy, it is not possible to derive useful equations governing only the "mechanical" portion of the mixture energy.

2.6 T F In a steady flow, nothing short of a nuclear transformation (e.g., fission) can change the mass fraction of each chemical element along a streamline.

2.7 T F In an N-component reacting mixture, all N-species conservation equations are independent of the equation governing total mixture mass conservation.

2.8 T F Whereas each mass-balance equation is a *scalar* equation, the mixture *linear-momentum* equation is a *vector* equation, equivalent to *three* scalar equations (in any coordinate system).

2.9 T F "Viscous dissipation" constitutes a local "sink" for mechanical energy, but a "source" for thermal energy and enthalpy.

2.10 T F Even though the transient continuum equations are still directly applicable, time-averaging and volume-averaging procedures are often used in the treatment of turbulent and multiphase flows to circumvent the need for impractically fine temporal and spatial resolution.

2.11 T F If the body forces $\mathbf{g}_i$ are the same for each species i present in an N-component mixture, then the net body-force term necessarily drops out of the mixture momentum equation.

2.12 T F When the only operative body force is that due to gravity, then it is useful to introduce the notion of gravitational potential energy in the energy equation, to take into account the work done by the gravitational body force on the flowing mixture.

2.13 T F For a chemically reacting mixture in the absence of momentum diffusion, energy diffusion, and species mass diffusion, the local rate of entropy production, $\dot{s}'''$, would necessarily be zero.

2.14 T F Even for the same temperature and pressure (e.g., 298.15 K, 1 atm) all gases do *not* have the same *molar* enthalpy.

2.15 T F The Second Law of Thermodynamics provides a qualitatively useful *inequality* (Clausius') but does not yield a quantitatively useful *balance* (conservation) equation.

EXERCISES

2.1 Why are conservation principles expressed in their *fixed* control volume form ordinarily more useful in problem-solving than those expressed in "material" control volume form? Why is a "material" control volume of interest?

2.2 If a fluid mixture is comprised of many individual species $i\,(i = 1, 2, \ldots, N)$, each of which is drifting through space with absolute velocities $\mathbf{v}_i$, then what is the meaning of the average fluid *mixture* velocity $\mathbf{v}$ appearing in the total mass, momentum, and energy equations?

2.3 If chemical *element* k $(k = 1, 2, \ldots, N_{\text{elem}})$ is found in many chemical species i $(i = 1, 2, \ldots, n)$, then relate the diffusion flux vector for *element* k, $\mathbf{j}''_{(k)}$, to the individual *species* diffusion flux vectors $\mathbf{j}''_i$. Suppose the element C appeared only in the gaseous molecules CO, CO_2, C_2H_6, and CH_4 (for a particular problem). Express the element mass flux $\mathbf{j}''_{(C)}$ in terms of $\mathbf{j}''_{CO}$, $\mathbf{j}''_{CO_2}$, $\mathbf{j}''_{C_2H_6}$, and $\mathbf{j}''_{CH_4}$ and the appropriate molecular masses. Similarly, express the carbon *element* mass fraction, $\omega_{(C)}$ in terms of the *species* mass fractions ω_{CO}, ω_{CO_2}, $\omega_{C_2H_6}$, ω_{CH_4}. Note that $\omega_{(C)}\dot{\mathbf{m}}''$ will be the local *convective mass flux vector* for this chemical element.

2.4 Given the conditions of Problem 2.3, why can we say that no matter what (or how many) chemical reactions are proceeding in the gas phase, $\dot{r}'''_{(O)} = 0$? $\dot{r}'''_{(C)} = 0$?, $\dot{r}'''_{(N)} = 0$?, etc. When would $\dot{r}'''_{CH_4} = 0, \dot{r}'''_{CO_2} = 0$, etc.?

2.5 For systems that need not be in mechanical, thermal, or chemical equilibrium:

a. Would a discontinuity in tangential mass-averaged velocity, v_t, across an interface (e.g., phase boundary) violate any basic *conservation* principle?

b. Would a discontinuity in temperature, T, across an interface violate any *conservation* principle?

c. Would a discontinuity in chemical potential across an interface violate any *conservation* principle?

d. What kind of restrictions *do* the *conservation* equations impose in such cases?

2.6 Under what conditions is the notion of specific "potential energy" (ϕ) useful? Is the work done on a fluid by the body force due to the earth's gravitation already included in ϕ, or must it be included separately in the basic energy equation (Eq. (2.5-23))?

2.7 We have stated that a moving fluid element has a "kinetic" energy equal to $v^2/2$ per unit mass. Verify that this is the amount of work that must be done to accelerate a unit mass from rest to velocity v. (Use Newton's second law and the concept that work equals force times displacement (in the direction of the force)).

2.8 Numerically, compute and compare the following energies (after converting all to the same units, say, calories):

a. The kinetic energy of a gram of water moving at 1 m/s.

b. The potential energy change associated with raising one gram of water through a vertical distance of one meter against gravity (where $g = 0.9807 \times 10^3\ cm/s^2$).

c. The energy required to raise the temperature of one gram of liquid water from 273.2 K to 373.2 K.

d. The energy required to melt one gram of ice at 273.2 K.

e. The energy released when one gram of $H_2O(g)$ condenses at 373 K.

f. The energy released when one gram of liquid water is formed from a stoichiometric mixture of hydrogen ($H_2(g)$) and oxygen ($O_2(g)$) at 273.2 K.

What do these comparisons lead you to expect regarding the relative importance of changes of each of the above-mentioned types of energy in applications of the law of conservation of energy? Is H_2O "singular," or are your conclusions likely to be generally applicable?

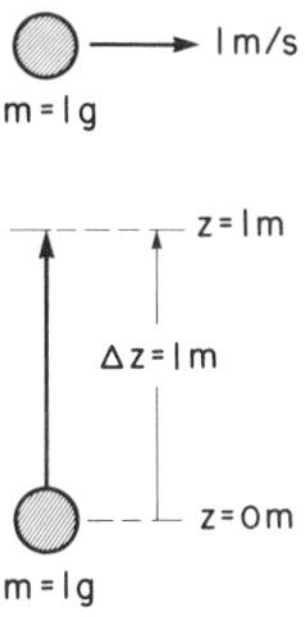

Figure 2.8E

2.9 Apply the macroscopic CV energy equation (Eqs. (2.3-3 and 3.2-6a)) to the following simple special case (sketched in Figure 2.9E):

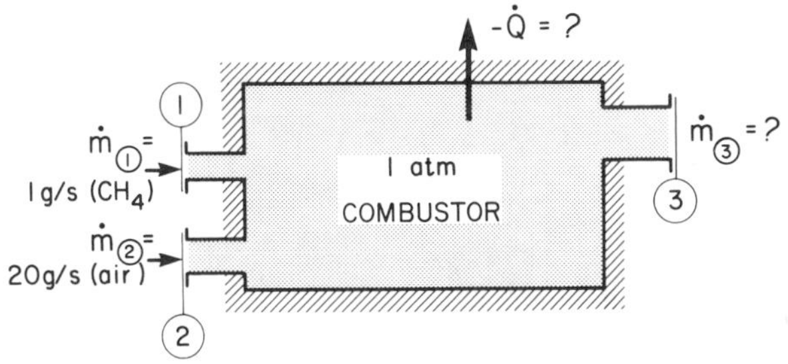

Figure 2.9E

Table 2.9E Gaseous Species data[a] (molar basis; $p = 1$ atm)

Species	*i*	M_i	$\Delta H_{f,i}(298\ K)$	$H_i(1000\ K)-H_i(298\ K)$	$S_i(1000\ K)$
O_2	1	32.00	0.000	5.427	58.192
N_2	2	28.01	0.000	5.129	54.507
H_2O	3	18.016	−57.798	6.209	55.592
CO_2	4	44.01	−94.054	7.984	64.344
CH_4	5	16.04	−17.895	9.125	59.141
Units:		g/(g-mole)	kcal/(g-mole)	kcal/(g-mole)	cal/(g-mole · K)

[a] JANAF Thermochemical Tables, Second Ed. (NSRDS-NBS 37) (1970).

Pure methane gas at 300 K, 1 atm, and pure air at 300 K, 1 atm, steadily flow into a combustor from which a single stream of product gas (CO_2, H_2O, O_2, N_2) emerges at 1000 K, 1 atm. Using the appropriate balance equations and the data assembled here, answer the following questions:

a. Using *total mass conservation* (Eq. (2.1-2)), what is the mass flow rate of the product stream out of the combustor?
b. Using chemical *element* conservation (Eq. (2.1-11)) calculate the chemical composition of the product gas mixture (expressed in *mass* fractions).
c. Using the *energy* balance (Eq. (3.2-6)), calculate the rate of energy extraction, $-\dot{Q}$, necessary to bring the combustion product gas to 1000 K.
d. Calculate the net convective outflow rate of entropy from the combustor—i.e., the surface integral of $\rho s \mathbf{v} \cdot \mathbf{n}\, dA$. What role does this term play in Clausius' (inequality) form of the *Second* Law of Thermodynamics applied to this macroscopic, nonisolated, "open" flow system?

Defend your important assumptions and, for simplicity, treat "air" as having the nominal composition: $\omega_{O_2} = 0.23$, $\omega_{N_2} = 0.73$.

2.10 *Design of a "Pebble Bed" Regenerative Gas Heater.* One way of obtaining hot process gas is to intermittently pass the desired gas through a preheated packed bed of sufficient thermal capacity to obtain an adequate interval of hot gas supply before the bed has to be reheated (say, using combustion products). Consider the design of such a "pebble bed" heater system to supply 1000 lb_m/hr of compressed air at temperatures between 500° and 1000° F for an interval of 0.5 hr. A possible system is sketched in Figure 2.10-1E and -2E below:

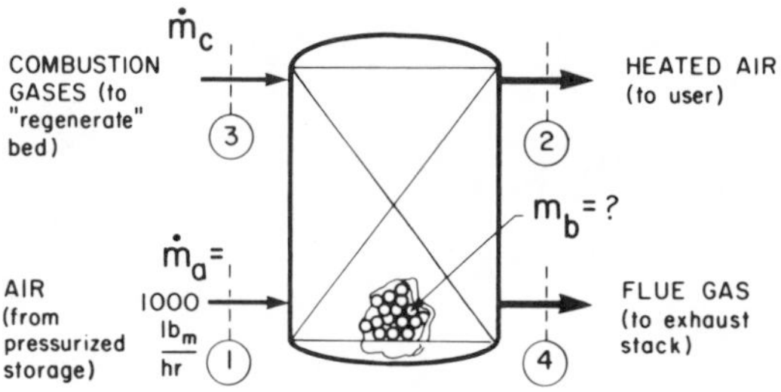

Figure 2.10-1E

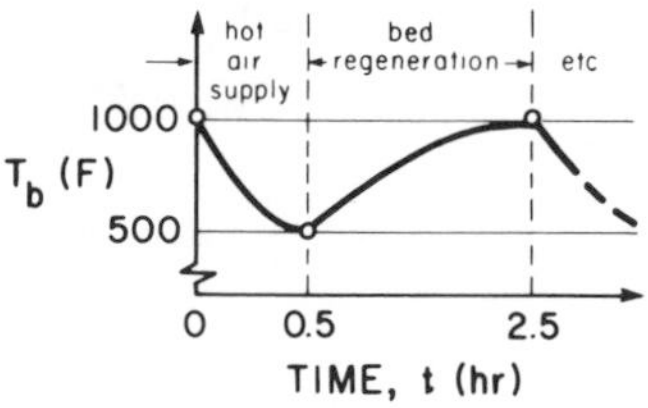

Figure 2.10-2E

Basic Data

Substance	*Average Heat Capacity*	$\left(\frac{BTU}{lb_m\,°F}\right)$	*Remarks*
Compressed air (a)	0.245	$(=c_{p,a})$	Require 0.5 hr. at $500 \leqslant T_2 \leqslant 1000°$ F
Combustion products (c)	0.26	$(=c_{p,c})$	available at 2300° F for two hr.
Bed material (b)	0.13	$(=c_{p,b})$	iron-silicon alloy spheres (1.25 in. (diam); alloy density = 425 lb_m/ft^3; able to withstand 1000° F in oxidizing environment)

Basic Design Questions

1. What total mass of "pebbles" will be needed to allow for a 0.5 hr. supply of air at temperatures between 500° F and 1000° F?
2. What mass flow rate of 2300° F combustion products will be needed to restore the bed temperature to 1000° F in 2 hrs.?

Tentative Approximations

A1. The bed temperature can be considered spatially uniform at each instant—i.e., T_b is a function of time alone.

A2. Thermal contact between the gases (air or combustion products) and the deep pebble bed is sufficiently good that the *outlet* gas temperature at each instant can be taken to be the instantaneous bed temperature.

A3. At any instant the energy content of the vessel is dominated by the energy content of the bed itself (cf. gas).

A4. The bed vessel is well insulated, so that parasitic heat losses are negligible.

A5. Gas temperatures and flow rates at stations 1 and 3 are constant during their respective "on" intervals.

Supplementary Questions

3. Approximately how large a heater vessel will be required? (The void fraction of randomly packed spheres of uniform size is about 0.4.) Can you now check A3 under "Tentative approximations" *a posteriori* (consider the "worst case")?
4. Would gas pressure drop "across" the bed[30] cause your earlier estimates to be seriously in error (explain reasoning)?
5. Briefly consider the possible need for "safety factors" to satisfy the design requirements, i.e., can Assumptions 1 and 2 be relaxed within the confines of a "black box" (macroscopic control volume) approach? What factors do you suspect will govern bed temperature nonuniformities, and the efficiency of gas/bed thermal contact?[31] Could a more accurate design be accomplished based on the *differential* equations of energy conservation (for the gases and bed) within the vessel (perhaps combined with some experimental data)?
6. State one obvious drawback of the above pebble-bed heater concept for supplying hot process gas, and briefly indicate a possible way to overcome it. State one possible *advantage* of the "pebble-bed" concept (cf. combustion-driven continuous heat-exchanger).

2.11 A well-stirred vessel contains 10,000 kg of solution of a dilute methanol–water solution ($\omega_A = 0.05$ mass fraction alcohol). A constant flow rate of 500 kg/min of pure water is suddenly introduced into the tank, and a constant rate of withdrawal of 500 kg/min of solution is simultaneously initiated. These two flows are continued and remain constant during the

[30] These matters will be dealt with quantitatively in Exercise 5.13.

[31] E.g., expressed in $(ft)^3$.

dilution process. Assuming that the solution density does not change appreciably, and the total content (inventory) of the tank remains constant (at 10,000 kg of solution), predict the time that will be required for the alcohol content to drop to $\omega_A = 0.01$ (that is, 1.0 wt%). *Ans.* 32.2 min (after C. Geankoplis (1978), Ex. 2.6-5).

2.12 a. Using a macroscopic CV species mass-balance (Eq. (2.1-7a)), calculate the output response of a perfectly ("well") stirred vessel of total volume V to a step-function of inert, dilute tracer (composition, ω_I) in a steady feed $\dot{m}$ of constant density ρ (N.B.: a perfectly well-stirred [WS] vessel has the property that, at each instant, the outlet (station ②) stream composition is identical to that prevailing *everywhere within the vessel*; it represents an ideal to which real vessels are often compared). The normalized response function:

$$F(t) \equiv \frac{\omega_{I,②}(t) - \omega_{I,①}(0^-)}{\omega_{I,①}(0^+) - \omega_{I,①}(0^-)}$$

and its first (time) derivative will be seen to play an important role in diagnosing and modeling the performance of such vessels when used as continuous flow mixers or "well-stirred" chemical reactors (WSRs, Section 6.7.2);

b. For $\dot{m} = 10^3$ kg/h, $\rho = 10^3$ kg/m^3, $V = 10$ m^3, how long will it take F to rise to the value 0.5?

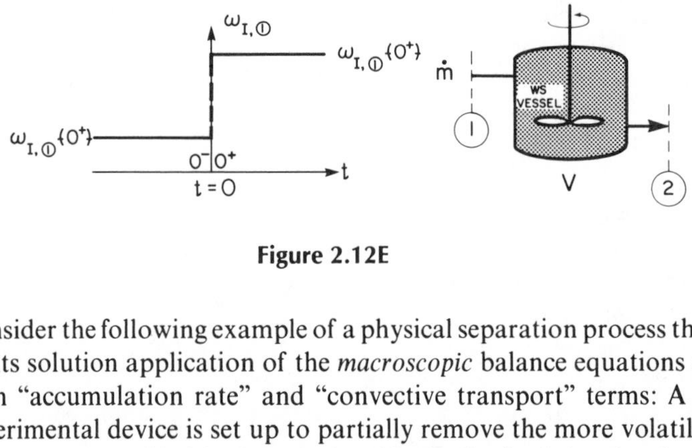

Figure 2.12E

2.13 Consider the following example of a physical separation process that requires for its solution application of the *macroscopic* balance equations containing both "accumulation rate" and "convective transport" terms: A pilot-scale experimental device is set up to partially remove the more volatile contaminant water from liquid ethylene glycol by "stripping" the glycol with dry air. For this purpose a fixed batch of impure glycol is first charged to the vessel and then desiccated air is continuously supplied through a "sparger" (bubbler) beneath the surface of the glycol. For simplicity, assume that the glycol solution is "well stirred" by the bubbling action. Assume, further, that the air leaves the vessel in equilibrium with the prevailing well-stirred solution. Surprisingly, the water-glycol system obeys Raoult's law closely.

Given the following conditions and data, find the *time* for which the air must be passed through the solution to dry the glycol to a water content of

0.1 mole percent (i.e., 1×10^{-3} mole fraction) if its initial water content is 2 mole percent (i.e., 2×10^{-2} mole fraction):

System isothermal at 60° C
H_2O Vapor pressure = 149 mm Hg at 60°C
Relative volatility of water to glycol = 98
Initial liquid inventory = 10 kg-moles glycol
Air flow rate = 5 kg-moles/h
Pressure level ≈ 1 atm.

(after C. J. King (1971), Exercise 3.L_2)

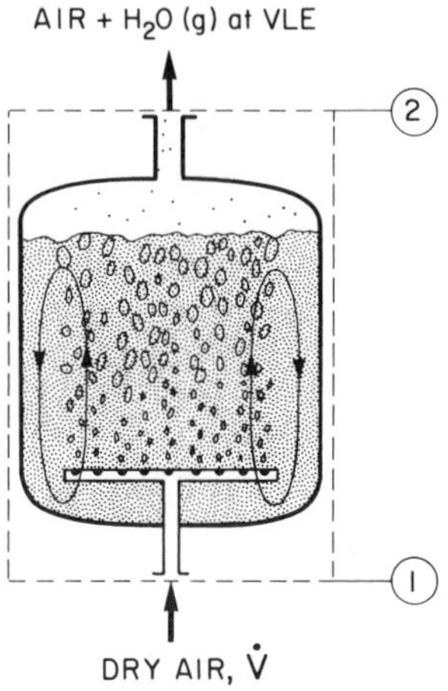

Figure 2.13E

2.14 Consider the use of the species and element mass-balance equations (2.1-7a) to design a *steady-flow* chemical reactor for carrying out the homogeneous thermal decomposition of a feed vapor under conditions such that the kinetics are simple and well established from previous measurements.

"Phosphine," $PH_3(g)$ (the phosphorous analog of ammonia, with a molecular weight, m_{PH_3}, of 34 kg/kg-mole) under the conditions of interest here, is known to thermally decompose according to the stoichiometry:

$$4\,PH_3(g) \rightarrow P_4(g) + 6\,H_2(g),$$

with the simple first-order irreversible reaction rate law:

$$\dot{r}'''_{PH_3} = -m_{PH_3} \cdot k\{T\} \cdot n_{PH_3}\ \frac{\text{kg}}{\text{m}^3\text{-s}},$$

where the reaction rate "constant" has been experimentally found to be:

$$k \cong 1.35 \times 10^{12}\, T^2 \cdot \exp\left(-\frac{43{,}660}{T}\right)\ \text{s}^{-1}$$

(where T is in Kelvins and n_{PH_3} is the phosphine number density (kg-mole/m^3)).

Complete the following preliminary design of a steady-flow reactor vessel that maintains the gas mixture temperature at 953 K (upper limit set by readily available materials of construction) and at about 1 atm pressure, but with negligible mixing in the streamwise direction.

a. How large (total volume) a reactor vessel would be required to continuously decompose 68 percent of a feed-flow rate of 16 kg/hr of pure phosphine (assume the validity of the ideal gas mixture equation-of-state under these proposed operating conditions)?

b. Using the thermochemical data assembled below for the ideal gases PH_3, P_4, and H_2, predict the rate, $\dot{Q}$, at which energy would have to be added (removed?) to maintain the PH_3-decomposition reactor isothermal at 953 K (express your result in kW).

Table 2.14E

Species	M	$\Delta H_f(298)$	$H(1000) - H(298)$	$C_p(1000)$
$PH_3(g)$	33.999	5.470	8.744	15.368
$P_4(g)$	123.90	30.771	13.033	19.445
$H_2(g)$	2.016	0.000	4.944	7.219
Units:	$\frac{\text{g}}{\text{g-mole}}$	$\frac{\text{k cal}}{\text{g-mole}}$	$\frac{\text{k cal}}{\text{g-mole}}$	$\frac{\text{cal}}{\text{g-mole K}}$

Ans. For any fraction decomposed, f, and operating conditions (T, p), we find:

$$V = \frac{\dot{m}}{k(T)} \cdot \left(\frac{p m_{PH_3}}{RT}\right)^{-1} \cdot \left\{\frac{7}{4}\ln\left(\frac{1}{1-f}\right) - \frac{3}{4}f\right\}.$$

For $\dot{m} = 16$ kg/h, $p = 1$ atm, $T = 953$ K, $f = 0.68$, we find $V = 0.98\ \text{m}^3$.

(Adapted from Exercise 3.7 of Denbigh and Turner (1971).)

2.15 A cylindrical storage tank of 2 m diam. (vertical axis, and open to the atmosphere) initially contains 10 m^3 of an organic reagent at 300 K, having a density of 920 kg/m^3 and a viscosity of 100 cP (0.10 Pa-s). The tank has a 1-m long cylindrical discharge "nozzle" of inside diameter 25.4 mm, and is closed *via* a ball valve at its exit.

a. Estimate the time that should be allowed (at 300 K) to drain (*via* gravity flow) 98% of the initial contents of this tank into a second vessel at atmospheric pressure.

b. Estimate the time that would be required (at 300 K) to drain 98% of the initial contents of this tank *if the 1-m long "nozzle" were replaced by a negligibly short 25.4-mm diam. section containing only the ball valve.*

c. Alternatively, estimate the *temperature* to which this reagent could be heated so that the efflux time *with* the 1-m long nozzle in place is within

10% of the 300 K efflux time computed in part (b). (Assume that the activation energy for fluidity of the reagent is about 5 kcal/mole.)

d. Alternatively, to what *pressure* should the 300 K reagent be raised to reduce the efflux time to that calculated in part (b)?

Itemize and discuss each important assumption you make to arrive at these quantitative estimates. (*Hints*: Neglect the fluid kinetic energy in the storage tank, and assume that the steady-flow form of the macroscopic "mechanical"-energy equation is valid at every instant. For part (a), use the fact that the viscous (friction) loss factor for a straight pipe is $4(L/d) \cdot f\{\mathrm{Re}, \ldots\}$, where f is the friction factor (Section 4.5.1). Then use the unsteady macroscopic *mass* balance equation, and obtain the 98% efflux time by integration.)

2.16 Consider the following preliminary design calculation for a "natural-draft" (chimney-produced) stationary combustor (furnace or power-plant) system: A 50-m-high cement-lined chimney, with mean diameter of 3 m and mean roughness height of 1 mm, is expected to contain combustion product gases with a mean temperature of 550 K when the ambient temperature is 306 K and the ambient pressure is 1.00 atm. (See Figure 2.16E.)

a. Estimate the maximum possible "natural-draft" gas-flow rate (kg/s) that can be achieved under these conditions (i.e., what would be the gravity-induced gas-flow rate if the combustor itself had negligible fluid dynamical "resistance"?). (*Hint*: Apply the macroscopic mechanical-energy balance equation (generalized Bernoulli equation), and evaluate the friction-loss factor for the stack in terms of experimentally available friction-factor data for a rough[32] circular pipe.)

b. If the ambient temperature was only 273 K and the corresponding mean combustion gas temperature in the chimney was 520 K, does the "natural draft" increase or decrease?

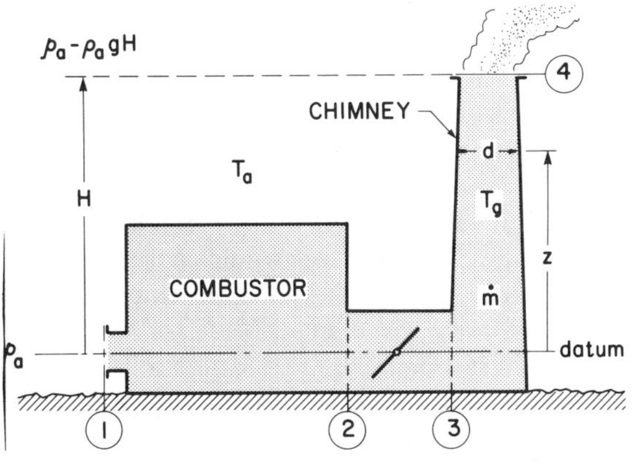

Figure 2.16E

[32] If Re > 5 × 10^6, experiments reveal that C_f asymptotes at 3.75 × 10^{-3} for this "relative roughness."

Numerically complete parts (a) and (b), then answer:

c. If there were a high wind (tangential to the combustor-chimney system), which of your assumptions would have to be modified? Qualitatively, how would you expect the "natural draft" to be influenced by a tangential wind? a local down-draft? How would you propose estimating these effects?

d. If (e.g., using turning vanes) the "entry-elbow" loss to the vertical chimney were dramatically reduced, we come to the conclusion that the induced-"draft," $\dot{m}$, becomes insensitive to chimney height, H. Does this seem reasonable (verify, using the momentum equation applied to the chimney itself)? What, then, would govern the most economic choice of chimney *height* in any new design? Allowing for the maximum "entry-elbow" loss, estimate the fractional change of $\dot{m}$ with a fractional change in chimney height—e.g., estimate the "logarithmic derivative":

$$\left(\frac{H}{\dot{m}}\cdot\frac{d\dot{m}}{dH}\right) = \frac{d\ln\dot{m}}{d\ln H} \qquad \text{near the present design "point."}$$

(Use either a numerical method or differential calculus to evaluate this derivative.)

e. If, instead of relying on the "natural draft" associated with the density difference $(\rho_a - \rho_g)$ in a gravity field, a "mechanical"-draft system were adopted, estimate the *pumping power* (e.g., in kW) that would be required to drive the same mass flow rate, $\dot{m}$, through the combustor-chimney system. Would you install such a system upstream or downstream of the combustor? Why?

f. Note that $(\pi d_c^2/4)\cdot H/(\dot{m}/\rho_c)$ is the mean residence time, t_{flow}, for the combustion products in the chimney. Calculate this time (sec) in the present case. If the ambient temperature changed continuously, but on a time scale large compared to t_{flow}, could you estimate the "response" of the natural draft using the steady-flow $\dot{m}$-formula but evaluated using instantaneous conditions? Critically discuss this commonly used "quasi-steady" approximation.

g. In the (unlikely) event that there were no heat losses from the chimney, the prevailing *viscous dissipation rate* would cause an *increase* in combustion gas temperature, $T_{④} - T_{③}$. Estimate this temperature rise (°K), using c_p- and γ-values for air given in Table 8.1-1. Also estimate the actual pressures (p_a) at stations ④ and ③. Combining the results, what is your estimate of the corresponding relative change in gas *density* between stations ③ and ④? Does this *a posteriori* result justify your earlier use of the "incompressible fluid" form of the mechanical energy equation for this problem (despite the fact that we are dealing with "compressible" *gaseous* combustion products)?

h. In a "second-generation" design, you want to investigate the natural-draft consequences of a "tapered" stack, which, for structural reasons, has a base diameter of 3.5 m and a discharge diameter of 2.5 m. Outline a rational, practical procedure by which the flow-rate consequences of this variable-diameter stack (with a constant mean roughness height of 1 mm) could be calculated, with only a modest additional complication over the procedure used in Part (a).

i. To optimize the overall economics of this particular combustor-chimney natural-draft system, we should consider the consequences of adding an "energy recuperator" (say, tube-bundle) heat-exchanger near the chimney base. If the combustion products ("flue gases") would thereby be cooled from 550 K to 450 K, *at most*, how much energy can be made available to preheat the fuel and/or air required for combustion? Would this heat-exchanger also introduce additional mechanical-energy "losses" within the natural-draft chimney system? Might the adverse consequences of energy recovery offset the more obvious gains in such a "natural-draft" system? "Trade-offs" of this sort must be carefully considered in designing *all* engineering systems. *Remark*: This interesting ME/ChE problem could also be used to discuss *heat* transfer (Ch. 5) and species *mass* transfer (Ch. 6); e.g., the ability to rationally estimate chimney-heat losses, as well as possible deposition (acid condensation, fly ash) on the inner walls of the chimney.

2.17 Consider the ideal, steady-flow physical *separator* shown (schematically) below, which produces N pure product streams from a single N-component feed stream of (molar) composition $y_1, y_2, \ldots, y_N$ at station ①.

a. By systematically using all macroscopic balance principles, Clausius' inequality, and assuming that the mixture is ideal, derive an expression for required *separation work* per mole of processed feed, $-\dot{W}_{sep}/\dot{F}$, (where $\dot{F} \equiv$ molar feed mixture flow rate) if the product streams are at the same pressure and temperature as the feed stream. List and discuss each of your essential assumptions (see Appendix 8.1).

b. For 298 K, 1-atm air treated as a binary mixture of $O_2(g)$ $(y_1 = 0.21)$ and $N_2(g)$ $(y_2 = 0.79)$, evaluate $-\dot{W}_{sep}/\dot{F}$ (in kW-hr/kg-mole).

c. At what mole fraction would the (minimum) required separation work maximize (minimize?) for a *binary* mixture?

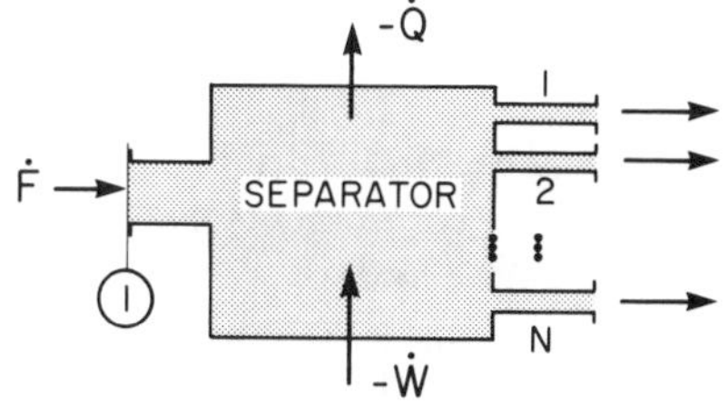

Figure 2.17E

d. In a real separator, there would be a nonzero rate of *entropy production* (associated, say, with steady-state momentum and species *diffusion*). Would this increase or decrease the required separation work?

2.18 a. Show that, for *axisymmetric* fluid motion relative to an inertial coordinate frame, the axial component of the moment (torque) exerted on the fluid can be re-expressed in terms of products of the *cylindrical radius r* and the *tangential* ("*whirl*") component v_θ of the local fluid motion at all inlet and outlet stations;

b. If a steady mass flow rate $\dot{m}$ of fluid passes through a turbomachine at nearly constant radius R_m and experiences a tangential velocity change Δv_θ, show that the magnitude of the torque on the fluid is $\dot{m}R_m\,\Delta v_\theta$, when shear-stress and body-force terms are comparatively negligible (as is often the case);

c. If $\Delta v_\theta = 10^3$ ft/s, $R_m = 0.75$ ft, $\Omega = 2\pi(6000)$rad/min and $\dot{m} = 200$ lb_m/s, calculate the "horsepower" developed (1 hp $\equiv$ 550 ft-lb_f/s), re-express your result in kilowatts.

2.19 Write a valid vector equation expressing *linear* momentum conservation for a *macroscopic* control volume using an accelerating coordinate frame (ACF) (cf. Eq. (2.2-3)). Could this vector equation be used to predict compressible reacting fluid flows in turbomachinery *rotor* passages? Note that Eq. (2.2-3) applies *as is* to a stationary passage ("stator" stage).

2.20 Write a valid vector equation expressing angular (moment of) momentum conservation for a *macroscopic* control volume using an *accelerating coordinate frame* (ACF) (cf. Eq. (2.2-10)). Could this vector equation be used to predict compressible reacting fluid flows in turbomachinery *rotor* passages? Note that Eq. (2.2-10) applies *as is* to a stationary passage ("stator" stage).

2.21 Apply the principle of angular-momentum conservation (Section 2.2.2) to a differential control volume in any particular orthogonal inertial coordinate system and show that in the limit $\Delta V \to 0$ the material derivative:

$$\left|\frac{D}{Dt}\int_V ((\mathbf{x}) \times \rho\mathbf{v})\,dV\right|$$

would be unbounded unless the contact ("extra") stress operator $\mathbf{T}$ were *symmetric* (that is, $\tau_{ij} = \tau_{ji}$ for $i \neq j$ in any orthogonal system) (see also Ex. 3.L_2 of Bird, *et al.* (1960)).

2.22 For any particular physical *vector* quantity (such as the velocity, $\mathbf{v}$, or energy flux vector $\dot{\mathbf{q}}''$), its individual components will, of course, depend upon the particular choice of coordinate system. What characteristic(s) of the vector will be the same in *any* coordinate system? (i.e., "invariant" with respect to coordinate system.) Reconsider this question for a nine- "component" quantity such as $\rho\mathbf{vv}$ or $\mathbf{\Pi}$.

2.23 The vertical distribution of atmospheric density may often be approximated by the relation:

$$\rho = \rho_0 \exp\{-z/H\}.$$

The parameter H is a characteristic scale height, z is the height measured above sea level, and ρ_0 is the sea-level density.

a. Derive an expression for the *spatial gradient* of the local density and evaluate its magnitude and direction for $H = 22{,}000$ ft. and $z = 100{,}000$ ft. (Does the result depend on ρ_0?)

b. Using this $\rho\{z\}$ law, derive a general expression for the rate at which the density changes with respect to a body falling vertically at the local velocity v.

c. Evaluate $(d\rho/dt)$ for $v = 20{,}000$ fps at 100,000 ft.

d. Suppose H were a function of time. Would this influence $d\rho/dt$ as measured with respect to the falling body? If so, why and how?

e. Suppose H were a function of location (longitude, latitude). How would this influence the direction of the vector **grad** ρ? (after Welty *et al.* (1976); Ex. 9.8).

2.24 a. Why is it that local thermodynamic relations, such as:

$$de = T\,ds - p d(1/\rho) + \sum_{i=1}^{N} \hat{\mu}_i\, d\omega_i$$

can be applied to *material* fluid parcels, in the form:

$$\frac{De}{Dt} = T\frac{Ds}{Dt} - p\frac{D}{Dt}\left(\frac{1}{\rho}\right) + \sum_{i=1}^{N} \hat{\mu}_i \frac{D\omega_i}{Dt},$$

but cannot be applied in the *spatial gradient* form:

$$\mathbf{grad}\, e = T\,\mathbf{grad}\, s - p\,\mathbf{grad}\,(1/\rho) + \sum_{i=1}^{N} \hat{\mu}_i\, \mathbf{grad}\, \omega_i\ ?$$

b. Is it justifiable to even invoke such thermochemical state variable interrelationships (e.g., between Dh, Ds, $D(1/\rho)$, $D\omega_i, \ldots$) to describe the behavior of a fluid parcel which may be chemically out of equilibrium, *and* the site of nonzero diffusion fluxes (momentum, energy, and/or species mass)? Defend your position on this important question.

2.25 a. In any particular physicochemical problem, what considerations dictate the most convenient choice of *dependent* variable? Thus, in the "energy equation," should we introduce:

$e \equiv$ specific internal energy,
$h \equiv$ enthalpy $(e + (p/\rho))$,
$h_0 \equiv$ "stagnation" enthalpy $(h + \mathbf{v} \cdot \mathbf{v}/2)$,
$T \equiv$ temperature
$\theta \equiv$ potential temperature, $T(p_{ref}/p)^{(\gamma - 1)/\gamma}$ etc.?

b. In any particular physicochemical problem, what considerations dictate the most convenient choice of *independent* variables (coordinate system)? Thus, should we introduce:

curvilinear orthogonal coordinates?

noninertial coordinates?

nonorthogonal coordinates?

streamline coordinates? (Boussinesq, Von Mises)

coordinates "stretched" using dependent variables? (Howarth) etc.?

c. If you explicitly know the local-level conservation PDEs in one (say, cartesian, inertial) coordinate system, how can you obtain the corresponding PDEs in *another* coordinate system that may be more convenient for a particular (class of) problem(s) you have been asked to solve? Consider, respectively, the treatment of position (Eulerian), fluid velocity, and fluid acceleration (cf. Section 2.6.2).

2.26 Using the notion that the "divergence" can be regarded as the local net outflow rate per unit volume, compute the local divergence of a vector in a *cylindrical* coordinate system by calculating each contribution to the surface integral for an "elemental" volume of "dimensions" $\Delta r, \Delta z, r\,\Delta\theta$, and then passing to the limit $\Delta r \to 0$, $\Delta z \to 0$, $\Delta\theta \to 0$.

2.27 A formal mathematical procedure for deriving the appropriate PDEs for any curvilinear coordinate system is to start with the valid PDE in the *cartesian* system (say) and then transform (*via* the principles of multi-variable calculus) to the coordinate system of interest. For example, to transform from cartesian to *spherical* polar coordinates, we will clearly need to express the "new" (r, θ, ϕ) Eulerian coordinates in terms of the "old" (x, y, z); and the x-, y-, and z-components of any vector, say $\mathbf{v}$, in terms of the r-, θ-, and ϕ-components of that same vector. Derive these required expressions, including expressions for $\partial r/\partial x, \partial\theta/\partial x, \partial\phi/\partial x$, etc., in terms of r, θ, ϕ itself.

2.28 Consider the steady *axisymmetric* flow (in the z-direction) past a motionless, isolated sphere located at the origin ($r = 0$) of a spherical polar coordinate system.

a. Using Eq. (2.5-32) and the scale factor assignments in row 3 of Table 2.5-1, explicitly express the PDE governing *conservation of mass* ("continuity") in this special case.

b. By considering the geometry of the fully incremental control volume "trapped" between the coordinate surfaces r and $r + \Delta r$, θ and $\theta + \Delta\theta$, and ϕ and $\phi + \Delta\phi$, demonstrate the validity of these scale-factor assignments, and show that $\Delta V = r^2 \sin\theta \, \Delta r \, \Delta\theta \, \Delta\phi$.

c. Show that the PDE obtained in Part (a) could equally well be derived by applying the principle of conservation of mass to the "toroidal" (semi-incremental) control volume (with $\Delta V = 2\pi r^2 \sin\theta \, \Delta r \, \Delta\theta$) obtained rotating the CV of Part (b) through an angle, ϕ, of 2π radians.

2.29 "Direct" derivations of the PDEs expressing local conservation (of mass, momentum, energy, etc.) in a *curvilinear* coordinate system require considerable care. As an example, consider the calculation of the radial component of div$\{\rho\mathbf{vv}\}$ occurring in the momentum equation for a fluid moving relative to an *inertial, cylindrical polar* coordinate system. The correct result is:

$$\frac{1}{r}\frac{\partial}{\partial r}(r\rho v_r v_r) + \left\{\frac{1}{r}\frac{\partial}{\partial\theta}(\rho v_\theta v_r) - \frac{\rho v_\theta v_\theta}{r}\right\} + \frac{\partial}{\partial z}(\rho v_z v_r),$$

representing the net rate of outflow of *radial momentum* (by convection) per unit volume. What is the physical origin of the remarkable $-\rho v_\theta v_\theta / r$ term appearing as a result of fluid motion across $\theta =$ constant surfaces? (Note that there will also be a contribution $-\tau_{\theta\theta}/r$ to the net *radial* force per unit volume due to the prevailing viscous stress system. Physically, how can *normal* stresses on $\theta =$ constant surfaces contribute to the *radial* momentum balance?)

REFERENCES

Beek, W.J., and K.M.K. Muttzall, *Transport Phenomena.* London: Wiley Interscience (1975); *see also,* Freeman, J.R., *Flow of Water in Pipes and Fittings.* New York: ASME (1941).

Brenner, H., and J. Haber, *J. Colloid Interface Sci.* 97, 496–514 (1984).

Denbigh, K.G., and J.C.R. Turner, *Chemical Reactor Theory—An Introduction.* (Second Ed.). Cambridge, UK: Cambridge Univ. Press (1971).

Favre, A., "Statistical Equations of Turbulent Gases," in *Problems of Hydrodynamics and Continuum Mechanics.* SIAM, pp. 231–266, Philadelphia, PA (1969); *see also,* Favre, A., *et al., La Turbulence en Mécanique des Fluides.* Gauthier-Villars, Paris (1976).

Geankoplis, C.J., *Transport Processes and Unit Operations.* Boston, MA: Allyn and Bacon (1978).

Hayes, W.D., in Vol. 3, Section D, *Princeton Series in High-Speed Aerodynamics and Jet Propulsion.* Princeton, NJ: Princeton Univ. Press, pp. 416–481 (1958).

King, C.J., *Separation Processes.* New York: McGraw-Hill (1971).

Stull, D.R., and H. Prophet, *JANAF Thermochemical Tables,* Nat. Standard Ref. Data System (NSRDS), NSRDS–NBS 37, Second Ed. (June, 1971).

BIBLIOGRAPHY: CONSERVATION PRINCIPLES

Elementary

Sandler, S.I., *Chemical and Engineering Thermodynamics.* New York: J. Wiley (1977) (Appendix 2.1).

Welty, J.R., C.E. Wicks, and R.E. Wilson, *Fundamentals of Momentum, Heat, and Mass Transfer* (Third Ed.). New York: J. Wiley (1981).

Intermediate

Bird, R.B., W.E. Stewart, and E. Lightfoot, *Transport Phenomena.* New York: J. Wiley (1960). Sons (1960).

Kuethe, A.M., and C-Y. Chow, *Foundations of Aerodynamics: Bases of Aerodynamic Design* (Third Ed.). New York: J. Wiley (1976).

Prigogine, I., *Introduction to Thermodynamics of Irreversible Processes.* New York: Wiley Interscience (Second Revised Ed.) (1961).

Rosner, D.E., "Energy, Mass, and Momentum Transport—The Treatment of Jump Conditions at Phase Boundaries and Fluid-Dynamic Discontinuities" in *Chemical Engineering Education X*, No. 4, 190–194 (Fall 1976).

Thompson, P.A., *Compressible Fluid Dynamics.* New York: McGraw-Hill (1972).

Advanced

Aris, R. *Vectors, Tensors and the Basic Equations of Fluid Mechanics.* Englewood Cliffs, NJ: Prentice Hall (1962).

Boure, J.A., and M. Delhaye, "General Equations and Two-Phase Flow Modeling," Section 1.2 of *Handbook of Multiphase Systems* (G. Hetsroni, ed.). New York Hemisphere-McGraw-Hill, (1982), pp. 1–36 to 1–95; see, also, Abriola, L.M., and N.G. Gray, *Int. J. Multiphase Flow.* New York: Pergamon Press Vol 11, No. 6, 837–852 (1985).

DeGroot, S.R., and P. Mazur, *Nonequilibrium Thermodynamics.* North Holland (1962).

Hirschfelder, J.O., C.F. Curtiss, and R.B. Bird, *Molecular Theory of Gases and Liquids.* New York: J. Wiley (1954).

Moon, P.H., and D.E. Spencer, *Field Theory for Engineers.* Princeton, NJ: Van Nostrand (1961).

Morse, P.M., and H. Feshback, *Methods of Theoretical Physics.* New York: McGraw-Hill (1953).

Newman, J.S., *Electrochemical Systems.* Englewood Cliffs, NJ: Prentice-Hall (1973), Chapter 16.

Schechter, R.S., *The Variational Method in Engineering.* New York: McGraw-Hill (1967); Chapters 4, 5.

Sivashinsky, G.I., "On a Distorted Flame Front as a Hydrodynamic Discontinuity," *Acta Astronantica* Vol. 3, 889–918 (1976).

Slattery, J.C., *Momentum, Energy and Mass Transfer in Continua.* New York: McGraw-Hill (1972).

Tsien, H.S., "The Equations of Gas Dynamics," Section A in *Fundamentals of Gas Dynamics* (H. Emmons, ed.), Vol. 3 of *High-Speed Aerodynamics and Jet Propulsion.* Princeton, NJ: Princeton Univ. Press, pp. 3–63 (1958).

Vavra, M.H., *Aerothermodynamics and Flow in Turbomachines.* New York: J. Wiley (1960).

3

Constitutive Laws: The Diffusion Flux Laws and Their Coefficients

3.1 CLOSURE *via* CONSTITUTIVE LAWS/COEFFICIENTS

We see from the discussion of Chapter 2 that the flow of a multicomponent reacting fluid is described by many field densities that are closely coupled *via* conservation constraints. While these equations are immediately useful for some purposes (Section 1.6), they are *not sufficient* for most predictive purposes in the sense that they contain many thermodynamic functions of local state, the local diffusion fluxes $(-\boldsymbol{\Pi}, \mathbf{j}_i'', \dot{\mathbf{q}}'')$, and reaction-rate laws $\dot{r}_i'''$, all of which must be explicitly related to the remaining field densities. The information needed to "close" the predictive problem is outlined below.

3.1.1 Equations of State

The local mixture is assumed to be describable in terms of the usual thermodynamic "state" variables, such as p, T, composition $(\rho_1, \rho_2, \rho_3, \ldots, \rho_N)$, e, $h[\equiv e + (p/\rho)]$, s, $f(\equiv h - Ts)$, etc.; however, the nature of the interrelations between these variables may differ from fluid to fluid (perfect gases, liquid solutions, dense vapors, etc.). Accordingly, laws governing these interrelations must be supplied—appropriate to the fluid mixture under consideration. This information comes from equilibrium chemical thermodynamics.

3.1.2 Chemical Kinetics

The individual net chemical species source strengths $\dot{r}_i'''$ (and $\dot{r}_i''$ for the relevant boundaries) must be explicitly related to the state variables (p, T, composition, etc.) defining the local reacting mixture. This information comes from the science of *chemical kinetics*, which provides either comprehensive expressions based on all of the relevant (molecular-level) *elementary steps*, or "global" expressions empirically describing the net rate of production of species i in terms of fewer field densities. In either extreme these "source-strength functions" are considered among the "knowns" in quantitative treatments of the design (e.g., sizing) of chemical reactors (see, e.g., Exercise 2.14, Section 6.1.3, and Froment and Bischoff (1979), Part Two), but are sought in laboratory applications of, say, flow reactors to the experimental inference of chemical kinetic laws. Reaction rate laws can be algebraically quite simple or quite complicated (see, e.g., Smith (1970), Boudart (1968), and Glassman (1977)). They need satisfy only the following general constraints:

- No net mass production: $\sum_{i=1}^{N} \dot{r}_i''' = 0 \quad (i = 1, 2, \ldots, N)$ (corollary of conservation of each chemical element: $\dot{r}_{(k)}''' = 0$ for $k = 1, 2, \ldots, N_{\text{elem}}$),
- No net charge production: $\sum_{i=1}^{N} z_i F(\dot{r}_i'''/m_i) = 0,$

and vanishing of the net production rates of each chemical species *at* ("dynamical") *local thermochemical equilibrium* (LTCE); that is,

- $\dot{r}_i'''\{T, p; \rho_{1\,\text{LTCE}}, \rho_{2\,\text{LTCE}}, \ldots, \rho_{N\,\text{LTCE}}\} = 0,$

where the $\rho_{i\,\text{LTCE}} (i = 1, 2, \ldots, N)$ are calculated (say, *via* a Gibbs free-energy minimization procedure) at the prevailing local mixture temperature, pressure, and chemical-element ratios. Analogous statements apply to the corresponding source-strength functions at interfaces separating phase boundaries (the function $\dot{r}_i''\{\ldots\}$ appearing in Eq. (2.6-5)); but, in general, these functions can depend upon composition variables evaluated on *both* sides of the relevant phase boundary (see, e.g., Rosner (1972, 1976)).

3.1.3 Diffusion Flux-Driving Force Laws/Coefficients

The conservation equations presented in Chapter 2 are seen to involve "*diffusive*" *fluxes* of linear momentum ($-\mathbf{\Pi}$), species mass ($\mathbf{j}_i''$), and energy ($\dot{\mathbf{q}}''$), which must be explicitly related to the other field densities. These relations, however, are also particular to certain classes of fluids and are, hence, called "constitutive." In the remainder of this chapter, we consider the simplest set of such laws, compatible with the Second Law of Thermodynamics, in which the fluxes are considered to be linearly proportional to their "driving forces," i.e., appropriate local spatial gradients of the field densities.[1] As discussed further in Section 3.5, such laws are valid for chemically reacting gas mixtures provided state variables do not undergo an appreciable fractional change in:

a. a spatial region of the dimension *ca.* one molecular mean-free-path;
b. a time interval of the order of the mean time between molecular collisions.

An interesting and very important corollary of the use of such laws is the following: Since the conservation laws have been shown to involve spatial derivatives of the fluxes (*via* div( )) and the diffusion fluxes themselves will be seen to involve spatial derivatives of the field densities, *in the presence of diffusion* the resulting field equations (PDEs) will of necessity be mathematically of *second order* (i.e., involve *second* space derivatives of the field densities).

3.1.4 General Constraints on the Diffusion-Flux Laws

Several rather general constraints guide the choice of constitutive laws governing the relationship between the diffusion fluxes of momentum, energy, and species mass and the coefficients that appear therein. Briefly, these are:

1. *Positive entropy production.* The laws chosen must lead to positive entropy production (Clausius) in the presence of diffusion, irrespective of the direction of the diffusion fluxes.
2. *Material Frame Invariance.* The resulting conservation laws must retain their same *form* irrespective of certain changes in vantage point; i.e., they must remain invariant with respect to translations, rotations, reflections, and shifts in time-origin if the fluid is isotropic.
3. *Local Action* (Space and Time).[2] Of all otherwise possible constitutive laws, we select those that relate the *local, instantaneous* fluxes to the present (cf. past or future) local fields.

Moreover, for simplicity, we restrict ourselves to laws exhibiting:

4. *Isotropy.* Most (but by no means all) fluids of interest, including those comprised of nonspherical molecules, can be considered isotropic—i.e., their

[1] Despite the suggestive nomenclature, it is usually *not* fruitful to regard the "driving forces" as *causing* the corresponding fluxes, any more than the fluxes can be regarded as *causing* the spatial gradients of the field densities. The most generally useful viewpoint is that these fluxes and (appropriate) spatial gradients of field densities simply "go together"—i.e., the existence of either implies the other, rather than a cause-and-effect relationship.

[2] As discussed in Section 3.5, this rules out "action-at-a-distance" phenomena (e.g., radiation transfer) as well as fluids "with a memory."

transport properties (the coefficients discussed below) are *not direction-dependent*, but rather functions of the local state variable alone.

5. *Linearity*. From among all laws satisfying the above constraints, we select those that are explicitly linear in the local field variables and/or their spatial gradients.

3.2 LINEAR-MOMENTUM DIFFUSION (CONTACT STRESS) *VS.* RATE OF FLUID-PARCEL DEFORMATION

3.2.1 The "Extra" Stress Operator and Its Components

Let us consider that portion of the total local stress $\mathbf{\Pi}$ not accounted for by the ordinary thermodynamic (normal) scalar pressure p (which exists even for an equilibrium fluid in the absence of any spatial nonuniformities). We focus, then, on the "extra" stress $\mathbf{T}$ associated with fluid motion, where:

$$\mathbf{T} \equiv \mathbf{\Pi} - (-p\mathbf{I}), \tag{3.2-1}$$

sometimes called the "viscous" stress. Here

$$\mathbf{I} \equiv \begin{pmatrix} 1 & 0 & 0 \\ 0 & 1 & 0 \\ 0 & 0 & 1 \end{pmatrix}$$

is the so-called "unit tensor," introduced to make this a meaningful (homogeneous) tensor equation—that is, equivalent to six independent scalar equations, one for each component of these *symmetrical* stress tensors (cf. Section 2.2.2).

The components of the extra (or viscous) stress, $\mathbf{T}$, have a simple physical interpretation, which also brings out clearly why the local state of stress cannot be defined by only one or three numbers. For example, consider the cartesian coordinate system (x, y, z), which requires the specifications of no less than nine components:

$$\mathbf{T} = \begin{pmatrix} \tau_{xx} & \tau_{xy} & \tau_{xz} \\ \tau_{yx} & \tau_{yy} & \tau_{yz} \\ \tau_{zx} & \tau_{zy} & \tau_{zz} \end{pmatrix}, \tag{3.2-2}$$

"only" six of which are independent because of symmetry ($\tau_{xy} = \tau_{yx}$, etc.). This is readily appreciated since each component is physically a *force per unit area* and:

a. the first subscript is chosen to denote the *kind of coordinate surface* on which the force acts (e.g., x denotes the surfaces $x =$ constant);

b. the second subscript is chosen to denote the *direction* of that force.

Thus, τ_{xx} is a *normal* (tensile) stress and τ_{xy} is a shear stress (force/area in y direction on x = constant surface), etc.

3.2.2 Stokes' Extra Stress *vs.* Rate of Deformation Relation

Now $-\mathbf{T}$ can be considered the rate of *linear momentum diffusion* and, for gases, J. C. Maxwell showed that in the presence of *fluid velocity* (linear momentum) *spatial gradients*, there is inevitably a corresponding flux of linear momentum (due to random molecular "traffic" between adjacent fluid layers). Indeed, Maxwell estimated the magnitude of the proportionality constant (between local shear stress and velocity gradient)—the so-called "viscosity coefficient"—and was able to relate it to molecular (mass, size) and state (T) properties (see Section 3.2.5). Thus, for a low-density gas in simple shear flow ($v_x(y)$ alone), Maxwell showed that

$$\tau_{yx} = \mu(T) \cdot \frac{\partial v_x}{\partial y}. \qquad (3.2\text{-}3)$$

For more general flows, G. Stokes showed that the appropriate generalization of Eq. (3.2-3) is to consider that $\mathbf{T}$ is linearly proportional to the local *rate of deformation of a fluid parcel*, the latter being comprised of two contributions (Figure 3.2-1):

a. rate of *angular* deformation, readily shown to be, in the x-y plane:

$$\left(\frac{\partial v_x}{\partial y} + \frac{\partial v_y}{\partial x}\right);$$

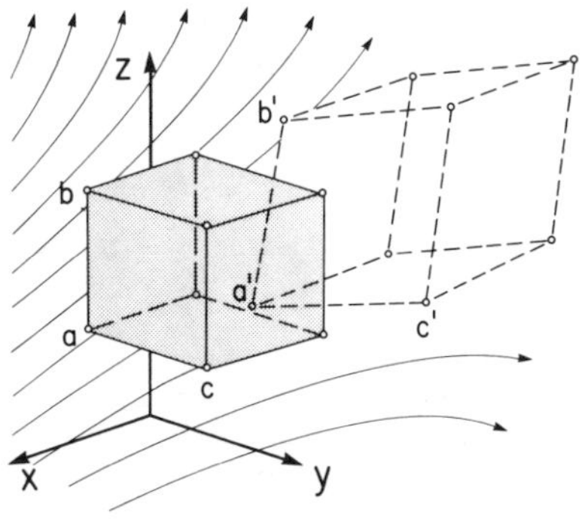

Figure 3.2-1 Subsequent shape of an initially cubical fluid parcel exhibiting both angular and volumetric deformation.

b. rate of *volumetric* deformation, readily shown to be:

$$\frac{\partial v_x}{\partial x} + \frac{\partial v_y}{\partial y} + \frac{\partial v_z}{\partial z}$$

in cartesian coordinates (or div{**v**} in any coordinates (see Eq. (2.5-6a)).

Accordingly, Stokes proposed that the *constitutive law for the local extra* (*viscous*) *stress* should be written:

$$\mathbf{T} = 2\mu \cdot \mathbf{Def\,v} + \left(\kappa - \frac{2}{3}\mu\right) \cdot \mathrm{div}\{\mathbf{v}\}\mathbf{I}, \tag{3.2-4}$$

where, in general,[3]

$$\mathbf{Def\,v} \equiv \frac{1}{2}\left[(\mathbf{grad\,v}) + (\mathbf{grad\,v})^{\dagger}\right],$$

which defines *two* scalar viscosity coefficients, μ and κ, assumed to be functions of local state, determinable *via* independent experiments (i.e., "phenomenological"). The first of these coefficients, μ, is called the *dynamic viscosity*, with units such that $\mu/\rho \equiv \nu$ is a "diffusivity" (cm^2/s). The second, called the "bulk" viscosity, is sometimes neglected for simple fluids (e.g., monatomic gases).[4] By convention, a fluid that is well-described by Eq. (3.2-4) is called "Newtonian," after I. Newton [who explored the consequences of simple linear laws (Eq. (3.2-3)) even before 1700].

It can be verified that this momentum-flux driving-force constitutive law (Eq. (3.2-4)) satisfies each of the constraints given in Section 3.1.4 (see Exercise 1). When this law is inserted into the PDE governing linear momentum conservation (Eq. (2.5-9)), the resulting equation (equivalent to three scalar PDEs) is called the *Navier–Stokes equation*, and forms the basis for most analyses of viscous fluid mechanics.

The six independent components of **Def v** and hence **T** can be calculated in any desired coordinate system. For example, in cylindrical polar coordinates, it is found that:

$$\tau_{rz}(=\tau_{zr}) = \mu\left(\frac{\partial v_z}{\partial r} + \frac{\partial v_r}{\partial z}\right), \tag{3.2-5a}$$

[3] $(\mathbf{grad\,v})^{\dagger}$ means the "transpose" of **grad v**, that is, the result of simply interchanging all off-diagonal components of **grad v** in any particular orthogonal coordinate system.

[4] This coefficient has been chosen in such a way (Eq. (3.2-4)) that even if div{**v**} $\neq 0$, a deforming fluid with $\kappa = 0$ will exhibit a local *mean normal stress* (e.g., $(1/3)(\pi_{xx} + \pi_{yy} + \pi_{zz})$ in cartesian (x, y, z) coordinates) equal to $-p$. However, while in such cases the *mean* normal "extra" stress vanishes (e.g., $(1/3)(\tau_{xx} + \tau_{yy} + \tau_{zz}) = 0$), this does *not* imply that *each* of the *normal* components of the "extra" stress vanishes. For these reasons, it is a misnomer to call **T** the *shear* stress operator. For compressible fluids with nonzero κ, we note that $(1/3)(\tau_{xx} + \tau_{yy} + \tau_{zz}) = \kappa\,\mathrm{div}\,\mathbf{v}$.

whereas

$$\tau_{r\theta}(=\tau_{\theta r}) = \mu\left(\frac{1}{r}\frac{\partial v_r}{\partial \theta} + r\frac{\partial}{\partial r}\left(\frac{v_\theta}{r}\right)\right). \quad (3.2\text{-}5b)$$

It is interesting to explicitly note here that the local, instantaneous viscous stresses have nothing to do with the local ("solid-body"-like) *rotational* motion of a fluid parcel; they relate only to its *deformation* rate. Thus, we can always formally decompose **grad v** into a symmetrical and antisymmetrical part:

$$\mathbf{grad\ v} = \underbrace{\frac{1}{2}\left[(\mathbf{grad\ v}) + (\mathbf{grad\ v})^{\dagger}\right]}_{\mathbf{Def\ v}} + \underbrace{\frac{1}{2}\left[(\mathbf{grad\ v}) - (\mathbf{grad\ v})^{\dagger}\right]}_{\mathbf{Rot\ v}}.$$

While the first (symmetrical) portion is associated with the local *deformation* rate of the fluid parcel at **x**, t, the second (antisymmetric)[5] portion, called here **Rot v** (the "spin rate"), completely defines the local *rotational* motion of a fluid parcel. By its definition **Rot v** has only three nonzero components (since it is antisymmetric *and* its diagonal components vanish), which can be identified with the three independent components of a *vector* $\boldsymbol{\omega}$ describing the *local fluid parcel rotation rate* (including direction). Fluid mechanicists call $2\boldsymbol{\omega}$ the local *vorticity*, $\boldsymbol{\zeta}$, of the fluid parcel. While, as remarked above, the local viscous stress is independent of the local fluid parcel *vorticity*, it will be shown in Chapter 4 that, in the presence of *local viscous stresses*, fluid vorticity will necessarily "diffuse," in much the same way that energy and/or species mass diffuse!

3.2.3 Energy Equation in Terms of the Work Done by the Fluid against the Extra Stress

Perhaps the most useful form of the total *energy* equation [cf. Eq. (2.3-3)] can be written in terms of the "extra" (viscous) stress **T**, lumping the work associated with $-p\mathbf{I}$ (loosely called the "flow-work") into the *convective* term on the LHS. Thus,

[5] A square matrix is called "antisymmetric" if its corresponding off-diagonal components are equal in magnitude but *opposite in sign.* Note that the scalar quantity $\mathbf{T}:\mathbf{grad\ v}$ appearing in our expressions for the viscous dissipation ratio (see, e.g., Eq. (2.5-20)) and its associated rate of entropy production (Eq. (2.5-28)) can clearly be written $\mathbf{T}:\{\mathbf{Def\ v} + \mathbf{Rot\ v}\}$. But $\mathbf{T}:\mathbf{Rot\ v} = 0$ (in view of the symmetry of **T** and antisymmetry of **Rot v**); therefore $\mathbf{T}:\mathbf{grad\ v} = \mathbf{T}:\mathbf{Def\ v}$. For this reason (cf., the Newton–Stokes' relation; Eq. (3.2-4)), the local entropy production rate associated with momentum diffusion in a Newtonian fluid flow of arbitrary complexity will necessarily be positive.

Eq. (2.3-3) may now be rewritten:

$$\frac{\partial}{\partial t}\int_V \rho\left(e+\frac{v^2}{2}\right)dV + \int_S \rho\left(h+\frac{v^2}{2}\right)\mathbf{v}\cdot\mathbf{n}\,dA$$

$$= -\int_S \dot{\mathbf{q}}''\cdot\mathbf{n}\,dA + \int_V \dot{q}'''\,dV + \int_S \mathbf{v}\cdot(\mathbf{T}\cdot\mathbf{n}\,dA) + \sum_{i=1}^{N}\int_V \dot{\mathbf{m}}_i''\cdot\mathbf{g}_i\,dV, \qquad (3.2\text{-}6a)$$

where we have introduced the familiar state function, *specific enthalpy*, h, defined by $h \equiv e + (p/\rho)$. Note that whereas h appears in the net convective outflow term, e still appears in the accumulation (transient) term. Thus, only for the analysis of *steady-flow* problems is the *enthalpy* variable particularly advantageous.

It is interesting to compare this equation with its "mechanical-energy" counterpart (e.g., Eq. (2.5-23)) for the special case of incompressible flow with all $\mathbf{g}_i = \mathbf{g} = -\mathbf{grad}\,\phi(\mathbf{x})$. Whereas the surface stress work[6] $\int_s \mathbf{v}\cdot(\mathbf{T}\cdot\mathbf{n}\,dA)$ is common to both equations, only the mechanical-energy equation contains the viscous-dissipation rate (sink) term $-\int_V \mathbf{T}:\mathbf{grad}\,\mathbf{v}\,dV$.

Indeed, subtracting Eq. (2.5-23) from Eq. (3.2-6a) leads to a very instructive result, *viz.*:

$$\frac{\partial}{\partial t}\int_V \rho e\,dV + \int_S \rho e\mathbf{v}\cdot\mathbf{n}\,dA = \left(-\int_S \dot{\mathbf{q}}''\cdot\mathbf{n}\,dA + \int_V \dot{q}'''\,dV\right) + \int_V \mathbf{T}:\mathbf{grad}\,\mathbf{v}\,dV, \qquad (3.2\text{-}6b)$$

demonstrating that, in general, both the rate of heat addition (term on RHS in parentheses) and the rate of viscous dissipation contribute to the accumulation rate and/or net outflow rate of thermodynamic internal energy of/from the macroscopic incompressible fluid system considered. Note that this scalar equation is free of any explicit reference to the rate at which work is done by the surface stresses and/or the body forces. As two important special cases, note further that, in the absence of energy addition (removal), to compensate for the rate of viscous dissipation within the fluid system, there must be:

a. an accumulation rate of internal energy for a no-net-flow system (e.g., fluid contained inside a "journal" bearing), and
b. a net outflow rate of internal energy in a steady-flow system.

---

[6] This term is usually negligible for control volumes whose boundary surfaces are either inlets, outlets, or fixed solid surfaces (as opposed to rotating shafts). At inlets and outlets, $\mathbf{v}$ is appreciable but the appropriate (normal) components of $\mathbf{T}$ are negligible. Conversely, at the solid walls, where the shear components of $\mathbf{T}$ are appreciable, the local tangential velocity vanishes (no slip condition).

These are important *thermal* consequences of viscous dissipation in macroscopic fluid systems. Equation (3.2-6b) will be recognized as the macroscopic form of the PDE Eq. (2.5-17) for the case of an incompressible fluid subject to the exclusive body-force gravity.

3.2.4 Viscous Dissipation and Its Consequences

Having defined:

a. the "extra" stress operator **T** such that $\mathbf{T}\cdot\mathbf{n}\,dA$ is the surface force on the differential area $\mathbf{n}\,dA$, associated with all contact stresses *other* than the local thermodynamic pressure;
b. div **T** such that it is, physically, the local net (resultant) contact force per unit volume in the limit of vanishing volume,

there is sufficient information to compute all contributions to the local scalar

$$\mathrm{div}(\mathbf{T}\cdot\mathbf{v}) - \mathbf{v}\cdot(\mathrm{div}\,\mathbf{T}) \equiv \mathbf{T}:\mathbf{grad}\,\mathbf{v} \tag{3.2-7}$$

in any particular coordinate system. For example, in cartesian (x, y, z) coordinates one finds that $\mathbf{T}:\mathbf{grad}\,\mathbf{v}$ is the sum of the following nine terms:

$$\mathbf{T}:\mathbf{grad}\,\mathbf{v} = \begin{cases} \tau_{xx}\dfrac{\partial v_x}{\partial x} + \tau_{yx}\dfrac{\partial v_y}{\partial x} + \tau_{zx}\dfrac{\partial v_z}{\partial x} \\[2ex] \tau_{xy}\dfrac{\partial v_x}{\partial y} + \tau_{yy}\dfrac{\partial v_y}{\partial y} + \tau_{zy}\dfrac{\partial v_z}{\partial y} \\[2ex] \tau_{xz}\dfrac{\partial v_x}{\partial z} + \tau_{yz}\dfrac{\partial v_y}{\partial z} + \tau_{zz}\dfrac{\partial v_z}{\partial z} \end{cases} \quad \text{(Local viscous dissipation rate per unit volume.)} \tag{3.2-8}$$

In fact, for any two tensors **A**, **B**, with components (in any particular orthogonal coordinate system) A_{ij} and $B_{ij}(i = 1,2,3; j = 1,2,3)$,

$$\mathbf{A}:\mathbf{B} = \sum_{i=1}^{3}\sum_{j=1}^{3} A_{ij}B_{ij}.$$

Since angular-momentum conservation at the local level leads to the conclusion that, for simple ("nonpolar") fluids, **T** is "symmetric" (e.g., in the cartesian system $\tau_{xy} = \tau_{yx}$, $\tau_{xz} = \tau_{zx}$, $\tau_{yz} = \tau_{zy}$), for most fluids the viscous dissipation rate term $\mathbf{T}:\mathbf{grad}\,\mathbf{v}$ can equally well be written $\mathbf{T}:\mathbf{Def}\,\mathbf{v}$, where $\mathbf{Def}\,\mathbf{v}$ completely describes the local fluid parcel deformation rate:

$$\mathbf{Def\,v} = \frac{1}{2}[(\mathbf{grad\,v}) + (\mathbf{grad\,v})^{\dagger}] \tag{3.2-9}$$

and is also symmetric.[7]

Since the rate of entropy production due to linear-momentum diffusion (Eq. (2.5-28)) is therefore proportional to $\mathbf{T}:\mathbf{Def\,v}$, taking $\mathbf{T}$ *itself* to be linearly proportional to $\mathbf{Def\,v}$ (Stokes' constitutive law for Newtonian fluids, Eq. (3.2-4a), ensures positive entropy production for arbitrary angular motions of a fluid with positive viscosity coefficient (cf. Section 3.1.4).)

Consider, for definiteness, a simple shear flow ($v_\theta(r)$, $v_z = 0$, $v_r = 0$) in cylindrical coordinates—such as in the annular space between a rotating shaft and its journal bearing. Then, since (Eq. (3.2-5b))

$$\tau_{r\theta} = \mu\left[r\frac{\partial}{\partial r}\left(\frac{v_\theta}{r}\right)\right], \tag{3.2-5c}$$

we find that:

$$(\mathbf{T}:\mathbf{Def\,v}) = \mu\left[r\frac{\partial}{\partial r}\left(\frac{v_\theta}{r}\right)\right]^2 = \mu\left[\frac{\partial v_\theta}{\partial r} - \frac{v_\theta}{r}\right]^2, \tag{3.2-10}$$

which is seen to be quadratic in $\partial v_\theta/\partial r$, contributing an additional nonlinearity to the energy conservation PDEs.

For turbulent flows, in which all velocities are fluctuating ($\mathbf{v} = \bar{\mathbf{v}} + \mathbf{v}'$), such nonlinearities contribute additive "correlation terms" to the time-averaged energy equations (cf. Section 2.6.2). For example, consider the time-average value of $\mathbf{T}:\mathbf{Def\,v}$ in an incompressible turbulent flow of a constant-property Newtonian fluid. We see that:

$$\frac{\overline{\mathbf{T}:\mathbf{Def\,v}}}{\rho} = 2\frac{\mu}{\rho}\mathbf{Def\,\bar{v}}:\mathbf{Def\,\bar{v}} + 2\frac{\mu}{\rho}\overline{\mathbf{Def\,v}':\mathbf{Def\,v}'}, \tag{3.2-11}$$

with the second set of terms comprising the *viscous dissipation rate per unit mass* associated with the turbulent (as opposed to time-averaged) motions of the fluid.[8] Thus, for "steady" turbulent flows through straight ducts, enlargements, contractions, valves, elbows, etc., viscous dissipation associated with *both* the time-mean and fluctuating velocity fields contributes to:

---

[7] $(\mathbf{grad\,v})^{\dagger}$ means the "transpose" of **grad v**, that is, the result of simply interchanging all off-diagonal components of **grad v**.

[8] The quantity is of particular importance in simple ("two-PDE") computational models of turbulent flow (cf. Section 5.7.3).

a. the net inflow rate (negative outflow rate) of $(p/\rho) + (v^2/2) + \phi$ per unit mass flow[9] (cf. Eq. (2.5-23)), and to
b. a corresponding rise in the internal energy per unit mass of fluid (cf. Eq. (3.2-6b)).

The heating associated with local viscous dissipation can strongly modify the local temperature field and hence all temperature-dependent properties (such as the viscosity coefficient itself, especially for liquids (Section 3.2.5), chemical rate "constants," etc.) as well as the Fourier heat flow itself. This will be seen to be particularly important in high-Mach-number viscous flows (e.g., rocket exhausts, hypersonic vehicle viscous boundary layers) as well as certain low-Reynolds'-, low-Mach-number flows in restricted passages (e.g., journal bearings, packed beds in high-pressure liquid chromatographic columns, etc.).

3.2.5 The Dynamic Viscosity Coefficient of Gases and Liquids[10]—Real and Effective

Experimentally, viscosities are obtained by establishing a simple flow (e.g., steady laminar flow in a capillary tube) and fitting the observables (e.g., pressure drop for a given mass-flow rate) to the predictions of mass and linear-momentum conservation for that constitutive law and configuration (see Section 4.5). Figure 3.2-2 displays the sensitive viscosity–temperature relation for a number of important Newtonian *liquids* (see below). The ordinate is expressed in centi-poise, where one Poise (expressed in g/cm · s) is the cgs unit of dynamic viscosity (after Poiseuille; see Section 4.5.1). We will frequently require the *ratio* of μ to the fluid *density* ρ, a ratio that is called the momentum "diffusivity," or "kinematic viscosity," with the units cm^2/s ("Stoke"), or ft^2/s.

For low-density *gases*, μ is found to be independent of pressure and proportional to a power of the local temperature, T, between 1/2 ("hard-spheres") and near unity. These dependencies, as well as the effect of molecular properties (molecular mass, m, size, σ, and interaction energy parameter, ε) are well understood in terms of the kinetic theory of gases (nonequilibrium statistical mechanics), which provides the following explicit expression for the dynamic viscosity of a pure gas (Enskog, Chapman):

$$\mu = \frac{5}{16} \cdot \frac{(\pi m k_B T)^{1/2}}{(\pi\sigma^2) \cdot \Omega_\mu \{k_B T/\varepsilon\}}. \qquad (3.2\text{-}12)$$

[9] Note that $\overline{v^2}/2$ will also be comprised of $(\bar{\mathbf{v}} \cdot \bar{\mathbf{v}}/2) + (\overline{\mathbf{v}' \cdot \mathbf{v}'}/2)$, that is, a mean-flow KE/mass and a turbulent KE/mass. A convective diffusion equation for the latter quantity is frequently used as an adjunct to the "prediction" of turbulent flows (see Section 5.7.3).

[10] Noncrystalline "solids" (called "glasses") are really highly viscous fluids. For example, the dynamic viscosity of noncrystalline silica (SiO_2) at room temperature is *ca.* $\mu_{SiO_2}(300\ K) \cong 10^{20}$ Poise. Even a polycrystalline solid "flows" (continuously deforms) under applied stress—a phenomenon termed "creep."

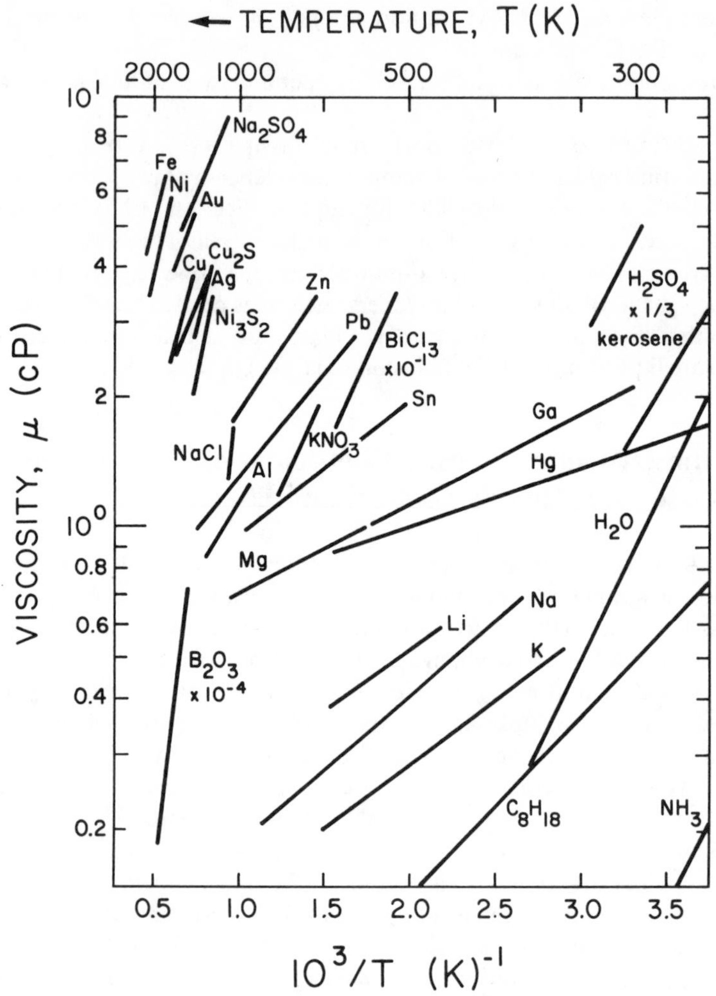

Figure 3.2-2 Temperature dependence of the dynamic viscosity, μ, of selected liquids (near atmospheric pressure).

Here $\Omega_\mu\{k_B T/\varepsilon\}$ is a calculable function of the dimensionless temperature $T/(\varepsilon/k_B)$ once the nature of the intermolecular potential function is specified.[11] The parameter ε is the depth of the potential energy "well" and σ the intermolecular spacing at which the potential crosses zero (see Figure 7.1-3). By matching the

---

[11] $\Omega_\mu\{k_B T/\varepsilon\}$ is defined in such a way that $\Omega_\mu = 1$ for a nonattracting hard sphere ("billiard ball") interaction; for the power-law attraction and repulsion (Lennard–Jones), this function is well tabulated (see Hirschfelder, Curtiss, and Bird (1954)) and approximately given by $1.22(k_B T/\varepsilon)^{-0.16}$ for $3 \leqslant k_B T/\varepsilon \leqslant 200$. Only "quasi-spherical molecules" are considered here; that is, the potential is regarded as spherically symmetrical. Ω_μ-values for other spherically symmetric potentials (e.g., exponential repulsion, power-law attraction) are also available (*loc. cit.*).

predictions of this theory with experimental viscosity data σ- and ε-values for many vapors have been determined (see Table 3.2-1). Since these parameters also correlate with more accessible quantities (e.g., condensed-phase molar volumes, and melting, boiling, or critical temperatures, respectively), reasonable estimates can be made even for cases where direct viscosity data are not available. For example, in the absence of more direct information for any pure substance, the following estimates

Table 3.2-1 Lennard-Jones Potential Parameters (after Svehla (1962))

Substance		σ, Å	ε/k_B, K
Ar	Argon	3.542	93.3
He	Helium	2.551	10.22
Kr	Krypton	3.655	178.9
Ne	Neon	2.820	32.8
Xe	Xenon	4.047	231.0
Air	Air	3.711	78.6
AsH_3	Arsine	4.145	259.8
BCl_3	Boron chloride	5.127	337.7
BF_3	Boron fluoride	4.198	186.3
$B(OCH_3)_3$	Methyl borate	5.503	396.7
Br_2	Bromine	4.296	507.9
CCl_4	Carbon tetrachloride	5.947	322.7
CF_4	Carbon tetrafluoride	4.662	134.0
$CHCl_3$	Chloroform	5.389	340.2
CH_2Cl_2	Methylene chloride	4.898	356.3
CH_3Br	Methyl bromide	4.118	449.2
CH_3Cl	Methyl chloride	4.182	350
CH_3OH	Methanol	3.626	481.8
CH_4	Methane	3.758	148.6
CO	Carbon monoxide	3.690	91.7
COS	Carbonyl sulfide	4.130	336.0
CO_2	Carbon dioxide	3.941	195.2
CS_2	Carbon disulfide	4.483	467
C_2H_2	Acetylene	4.033	231.8
C_2H_4	Ethylene	4.163	224.7
C_2H_6	Ethane	4.443	215.7
C_2H_5Cl	Ethyl chloride	4.898	300
C_2H_5OH	Ethanol	4.530	362.6
C_2N_2	Cyanogen	4.361	348.6
CH_3OCH_3	Methyl ether	4.307	395.0
CH_2CHCH_3	Propylene	4.678	298.9
CH_3CCH	Methylacetylene	4.761	251.8
C_3H_6	Cyclopropane	4.807	248.9
C_3H_8	Propane	5.118	237.1

(*Table 3.2-1 continued*)

Substance		σ, Å	ε/k_B, K
n-C_3H_7OH	n-Propyl alcohol	4.549	576.7
CH_3COCH_5	Acetone	4.600	560.2
CH_3COOCH_5	Methyl acetate	4.936	469.8
n-C_4H_{10}	n-Butane	4.687	531.4
iso-C_4H_{10}	Isobutane	5.278	330.1
$C_2H_5OC_2H_5$	Ethyl ether	5.678	313.8
$CH_3COOC_2H_5$	Ethyl acetate	5.205	521.3
n-C_5H_{12}	n-Pentane	5.784	341.1
$C(CH_3)_4$	2,2-Dimethylpropane	6.464	193.4
C_6H_6	Benzene	5.349	412.3
C_6H_{12}	Cyclohexane	6.182	297.1
n-C_6H_{14}	n-Hexane	5.949	399.3
Cl_2	Chlorine	4.217	316.0
F_2	Fluorine	3.357	112.6
HBr	Hydrogen bromide	3.353	449
HCN	Hydrogen cyanide	3.630	569.1
HCl	Hydrogen chloride	3.339	344.7
HF	Hydrogen fluoride	3.148	330
HI	Hydrogen iodide	4.211	288.7
H_2	Hydrogen	2.827	59.7
H_2O	Water	2.641	809.1
H_2O_2	Hydrogen peroxide	4.196	289.3
H_2S	Hydrogen sulfide	3.623	301.1
Hg	Mercury	2.969	750
$HgBr_2$	Mercuric bromide	5.080	686.2
$HgCl_2$	Mercuric chloride	4.550	750
HgI_2	Mercuric iodide	5.625	695.6
I_2	Iodine	5.160	474.2
NH_3	Ammonia	2.900	558.3
NO	Nitric oxide	3.492	116.7
NOCl	Nitrosyl chloride	4.112	395.3
N_2	Nitrogen	3.798	71.4
N_2O	Nitrous oxide	3.828	232.4
O_2	Oxygen	3.467	106.7
PH_3	Phosphine	3.981	251.5
SF_6	Sulfur hexafluoride	5.128	222.1
SO_2	Sulfur dioxide	4.112	335.4
SiF_4	Silicon tetrafluoride	4.880	171.9
SiH_4	Silicon hydride	4.084	207.6
$SnBr_4$	Stannic bromide	6.388	563.7
UF_6	Uranium hexafluoride	5.967	236.8

of σ (Å) and ε/k_B (K) are often acceptable for engineering calculations:

$$\sigma \approx 1.16\, V_{\text{nbp}}^{1/3} \approx 0.841\, V_c^{1/3} \approx 2.44\left(\frac{T_c}{p_c}\right)^{1/3} \approx 1.22\, V_{\text{s,mp}}^{1/3}$$

and

$$\frac{\varepsilon}{k_B} \approx 1.18 T_{\text{nbp}} \approx 0.77 T_c \approx 1.92 T_{\text{mp}}$$

(where the molar volumes are expressed in $\text{cm}^3/\text{g-mole}$, temperatures are in Kelvins, and the critical pressure in atm) (cf. Table 3.2-2).

The dependence of viscosity on mixture composition can be complex (cf. Figure 3.7E, but is usually adequately described by semi-empirical "mixing rules" relating the mixture μ to the individual μ_i's and the prevailing compositions (e.g., mole fractions, y_i). While rigorous "mixing laws" are available (Hirschfelder, Curtiss, and Bird (1954)), most engineering calculations are done using acceptably accurate simpler relations, such as the "square-root rule":

$$\mu_{\text{mix}} \approx \frac{\sum_{i=1}^{N} M_i^{1/2} y_i \mu_i}{\sum_{i=1}^{N} M_i^{1/2} y_i}. \tag{3.2-13}$$

A major qualitative difference between liquid and gas viscosity behavior is in the temperature dependence. In contrast to the above-mentioned behavior for gases, the viscosity of liquids *decreases* with increased temperature, usually following the (Andrade–Eyring) two-parameter law:

$$\mu = \mu_\infty \cdot \exp\left(\frac{E_\mu}{RT}\right). \tag{3.2-14}$$

Table 3.2-2 Correlation between Lennard-Jones Parameters and Accessible Macroscopic Parameters.

Quantity \ *Characteristic State*	nbp	c	mp[a]
$\sigma/V_{\text{char}}^{1/3}$	1.16	0.841	1.22
$(\varepsilon/k_B)/T_{\text{char}}$	1.18	0.77	1.92

Units: σ(Å), T(K), $V(\text{cm}^3/\text{g-mole})$

Nomenclature: c ≡ critical, char ≡ characteristic
[a] Solid

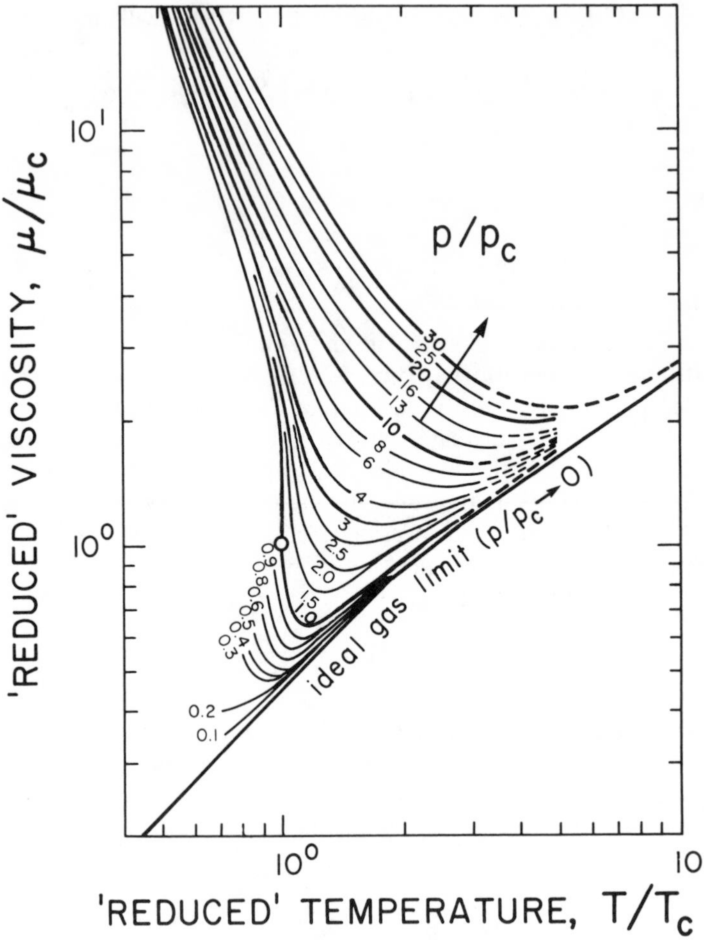

Figure 3.2-3 Corresponding states correlation for the viscosity of simple fluids; based primarily on data for group VIIIA elements (after Bird, *et al.* (1960)).

Here $E_\mu (> 0)$ is the so-called "activation energy for fluidity" (inverse viscosity), R the universal gas constant, and the pre-exponential factor μ_∞ the (hypothetical) dynamic viscosity at infinite temperature (see Figure 3.2-2, with coordinates chosen such that Eq. (3.2-14) plots as a straight line). For example, near room temperature (*ca.* 273–310 K) the viscosity of certain diesel fuels can be well described by Eq. (3.2-14) with $\mu_\infty \approx 3.3 \times 10^{-5}$ g/cm · s ("poise") and $E_\mu \approx 4.1$ kcal/mole.[12]

[12] For many substances E_μ (cal/g-mole) is of the order of 7.6 times the normal boiling point temperature (K).

The nature of the dramatic transition from "gas behavior" to "liquid behavior" is clearly displayed in Figure 3.2-3, constructed primarily from viscosity data on the elements of group VIIIA (He, Ne, Ar, Kr). This is a so-called "corresponding-states" transport property correlation (cf. Section 7.1.3.1) because it compares μ, T, and p values to their corresponding values μ_c, T_c, and p_c at the thermodynamic critical state for that same substance—a representation which tends to make all such substances "look alike." Note that at any given T/T_c, low-density gas behavior is approached at sufficiently low p/p_c values. However, especially at subcritical temperatures ($T/T_c < 1$), the above-mentioned liquid-like behavior is approached at high p/p_c values. This is due to an increasing contribution of direct intermolecular forces to the momentum transfer between adjacent fluid layers (cf. momentum transfer by interlayer molecular "traffic," Section 3.2.2).

There are no simple, yet reasonably general relations for the viscosity of *liquid* solutions as a function of composition (x_i) and the constituent viscosities (μ_i). For this purpose, empirical relationships, specific to certain classes of mixture (e.g., molten oxide "slags," glass formulations, etc.) are employed.

Gases and liquids in *turbulent* motion (Sections 2.6.2, 2.7.6) will be seen to behave as though they had "augmented" viscosities; that is,

$$\mu_{\text{eff}} = \mu + \mu_t,$$

where μ is the intrinsic viscosity of the (nonturbulent) fluid and the turbulent contribution μ_t is far more dependent on the local condition of turbulence than on the particular fluid in question.[13] Semi-empirical methods for estimating μ_t will be discussed in Chapter 4.

3.3 ENERGY DIFFUSION FLUX *VS.* SPATIAL GRADIENTS OF TEMPERATURE AND SPECIES CONCENTRATION

3.3.1 Fourier's Heat-Flux Law

On experimental grounds, for energy diffusion ("conduction") in pure isotropic solids, J. Fourier [1808] proposed that $\dot{\mathbf{q}}''$ is linearly proportional to the *spatial gradient* of the local temperature, and the direction of energy flow is "down" the temperature gradient. Thus, Fourier took:

$$\dot{\mathbf{q}}'' = -k\,\mathbf{grad}\,T. \tag{3.3-1}$$

This is a *vector* equation, equivalent to the following *three* scalar (component)

[13] Indeed, μ_t itself usually depends on the time-averaged deformation rate $\mathbf{Def}\,\bar{\mathbf{v}}$, giving the fluid an apparently non-Newtonian character.

equations in, say, cylindrical polar coordinates:

$$\dot{q}''_r = -k\left(\frac{\partial T}{\partial r}\right),$$

$$\dot{q}''_\theta = -k\frac{1}{r}\frac{\partial T}{\partial \theta}, \qquad (3.3\text{-}2a,b,c)$$

$$\dot{q}''_z = -k\left(\frac{\partial T}{\partial z}\right).$$

Here k is the local *thermal conductivity*, a measurable scalar function of the field variables defining the local material state.

Of course, many important engineering materials are *not isotropic*. Such cases, in which the energy flux, $\dot{\mathbf{q}}''$, and $-\mathbf{grad}\,T$ need not be colinear, can be quantitatively described by a generalization of Fourier's law involving a thermal conductivity *tensor* (vector operator, rather than scalar). For heat flow in, say, a cylindrically "layered" material (like wood), the situation is actually not much more complicated than that described above, provided the coordinate system for analyzing the energy transport is "lined up" with the intrinsic directions of the material. Thus, if we distinguish between the thermal conductivities in the *radial* direction, k_{rr}, *azimuthal* direction, $k_{\theta\theta}$, and *axial* direction, k_{zz}, Eqs. (3.3-2a,b,c) simply generalize to:

$$\dot{q}''_r = -k_{rr}\left(\frac{\partial T}{\partial r}\right),$$

$$\dot{q}''_\theta = -k_{\theta\theta}\left(\frac{1}{r}\frac{\partial T}{\partial \theta}\right), \qquad (3.3\text{-}2d,e,f)$$

$$\dot{q}''_z = -k_{zz}\left(\frac{\partial T}{\partial z}\right).$$

Note that, because of the presumed alignment between the material and the coordinate system, there are above no "off-diagonal" contributions to the transport—e.g., no radial heat flow produced by an azimuthal temperature gradient, etc. *Anisotropic* materials are seen to require the specification of three independent scalar conductivities, defining the heat-flow properties of the material in its three intrinsic spatial directions.

3.3.2 Species Diffusion Contribution to Energy Flux

For multicomponent systems (e.g., reacting gas mixtures), it is not sufficient to merely generalize Eq. (3.3-1) by considering k as a function of local composition. This is because each diffusing species also transports energy in accordance with its own specific enthalpy, h_i. For this reason, a sufficient generalization of Eq. (3.3-1) for multicomponent systems in the absence of long-range transport is usually:

$$\dot{\mathbf{q}}'' = -k\,\mathbf{grad}\,T + \sum_{i=1}^{N} \mathbf{j}_i'' h_i \tag{3.3-3}$$

(which, for positive k, is consistent with the requirement of locally positive entropy production irrespective of the sign of **grad** T (Eq. (2.5-28)), as well as the remaining constraints outlined in Section 3.1.4 (Exercise 3.1). Radiation energy transport, which involves "action at a distance" (that is, nonlocal behavior) ordinarily cannot be treated as a diffusion process and must be dealt with separately (Section 5.9.4).

3.3.3 Entropy Production and Diffusion Associated with "Fourier" Energy Diffusion

It is interesting to note that the additive contribution of energy diffusion ("conduction") to the local rate of entropy production per unit volume [cf. Eq. (2.5-28)], is now seen to be:

$$(\dot{s}''')_{\substack{\text{energy transport}\\ \text{by diffusion}}} = \frac{k(\mathbf{grad}\,T)^2}{T^2}. \tag{3.3-4}$$

As in the case of all the remaining nonequilibrium flux contributions (momentum, mass), it is quadratic in the gradient of the relevant local field density (hence, positive for any flux direction). Likewise, in the absence of multicomponent (species) diffusion, the *entropy diffusion flux vector* is simply:

$$\mathbf{j}_s'' = \frac{-k\,\mathbf{grad}\,T}{T} = -k\,\mathbf{grad}\{\ln T\}. \tag{3.3-5}$$

Apparent "violations" of the Second Law of Thermodynamics (Clausius' inequality in the form of Eq. (2.4-8)) are often "rescued" by merely noting the importance of entropy flows by *diffusion* as well as *convection.*

3.3.4 The Thermal Conductivity Coefficient of Gases, Liquids, and Solids—Real and Effective

Experimentally, thermal conductivities are usually determined by matching the results of steady-state or transient heat-diffusion experiments with the predictions of the relevant conservation and constitutive laws, usually in the absence of convection[14] or other complicating phenomena.

[14] See Section 5.4.2 concerning criteria for the avoidance of convection.

Figure 3.3-1, which covers over four decades, displays the results of such measurements for a number of technologically important, relatively simple substances in the nominal temperature range 200–1300 K. In most of the cases shown, the temperature dependence of k is modest; however, in many technological applications, temperature differences are large enough to necessitate allowance for these variations.

Again, for low-density gases and their mixtures, a reliable theoretical formalism exists (Enskog, Chapman, Hirschfelder). For pure gases, this "kinetic theory" leads to:[15]

$$k \approx \frac{15}{4}\frac{R}{M}\mu \cdot \left[1 + \frac{4}{15}\left(\frac{C_p}{R} - \frac{5}{2}\right)\right], \tag{3.3-6}$$

where μ is the corresponding viscosity (Eq. (3.2-12)), C_p is the molar specific heat, and R the universal gas constant.

This theory also provides reliable but somewhat cumbersome "mixing laws." For engineering estimates it is often sufficient to use simpler "rules," such as the cube-root "law":

$$k_{\text{mix}} \approx \frac{\sum_{i=1}^{N} M_i^{1/3} y_i k_i}{\sum_{i=1}^{N} M_i^{1/3} y_i}. \tag{3.3-7}$$

For mixtures of condensed phases, simple yet reasonably general mixing laws are not available, and a greater dependence on direct experimental data is necessary.

The transition from gas-like to liquid-like thermal conductivity (energy diffusion) behavior for simple substances is qualitatively similar to that already discussed for viscosity (momentum diffusion, Section 3.2.5) but somewhat less dramatic (cf. the "corresponding states" correlation for k/k_c (Figure 3.3-2) with that for μ/μ_c (Figure 3.2-3)). For the "saturated liquid" we are observing the intermolecular ("phonon") contribution to the condensed-phase thermal conductivity—not additional contributions (e.g., electron, photon) that can be important for other substances and/or conditions (molten metals, and high-temperature inorganic melts, respectively).

Gases and liquids in *turbulent* motion (Sections 2.6.2, 2.7.6) will be seen to behave as though they had "augmented" thermal conductivities; that is,

$$k_{\text{eff}} = k + k_t, \tag{3.3-8}$$

---

[15] The bracketed (Eucken) correction factor (= 1 for a monatomic gas without electronic excitation) corrects the "translational" thermal conductivity for the contribution of "internal" energy transfer. It can be derived by adopting the "LTCE-mixture" approximation discussed in Section 5.8 and Ex. 3.14, where in this case each (quantized) energy state of the molecular species is considered to be a distinct "chemical species." Then "internal" energy transfer is associated with the sum of the $\mathbf{j}_i''h_i$ products appearing in Eq. (3.3-3) (see Hirschfelder (1957)).

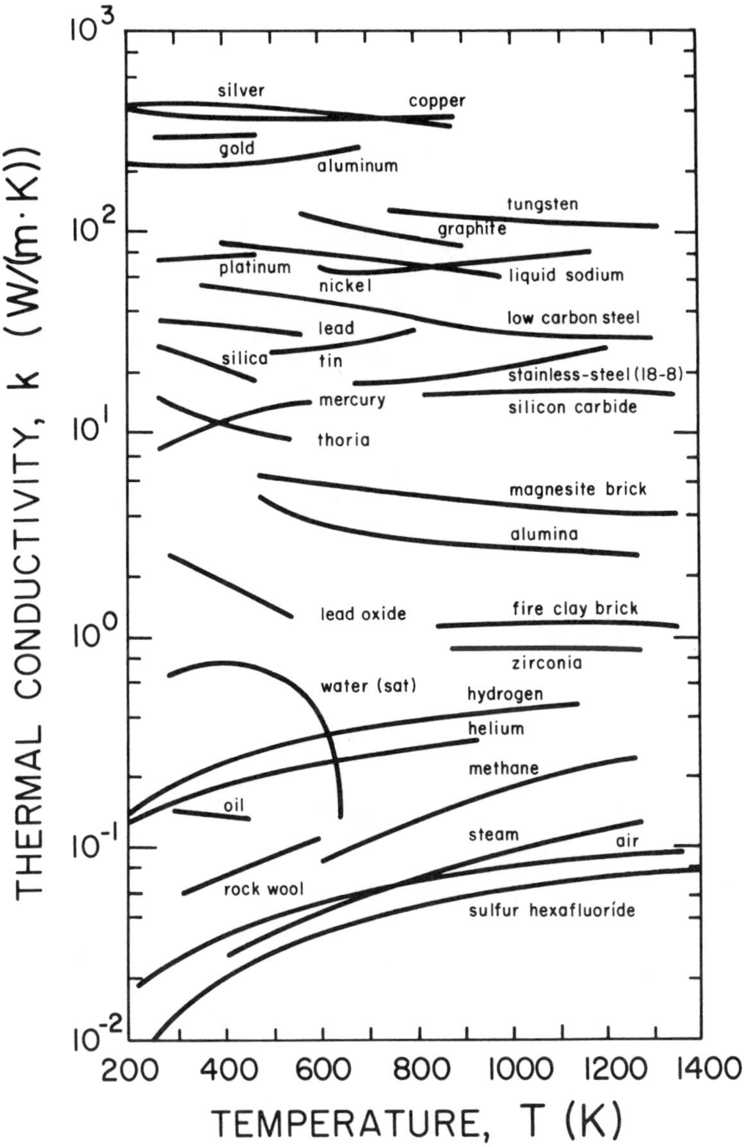

Figure 3.3-1 Thermal conductivities of various substances (adapted from Rohsenow and Choi (1961)).

where k is the intrinsic thermal conductivity of the (quiescent) fluid and the turbulent contribution k_t is far more dependent on the local conditions of turbulence than on the particular fluid in question. Several methods for estimating k_t will be discussed in Chapter 5; however, the simplest of these takes $k_t/(\rho c_p)$ to be the same as μ_t/ρ (Section 3.2.5).

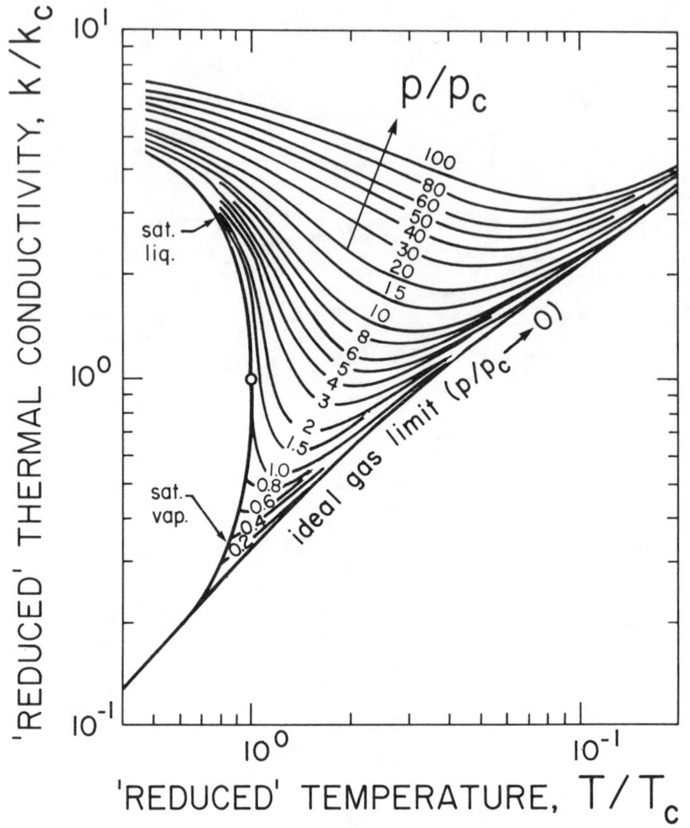

Figure 3.3-2 Corresponding-states correlation for the thermal conductivity of simple fluids; based primarily on data for group VIIIA elements (after Bird, *et al.* (1960)).

Similarly, chemically reacting mixtures in local thermochemical equilibrium (LTCE) may exhibit effective thermal conductivities far greater than those estimated from Eq. (3.3-7). This phenomenon, due to the additional terms $\Sigma \mathbf{j}_i'' h_i$ (Eq. (3.3-3)), is illustrated in Ex. 3.14, outlined in Section 5.8, and discussed more fully in the references cited therein.

3.4 MASS DIFFUSION FLUX *VS.* SPATIAL GRADIENTS OF COMPOSITION

3.4.1 Fick's Diffusion-Flux Law for Chemical Species

On experimental grounds, A. Fick [1855] proposed that species mass diffusion in pure, isothermal, isotropic materials is linearly proportional to the local concentration gradient, and directed "down" the gradient. One of the simplest such laws for a multicomponent system would be:

$$\mathbf{j}_i'' = -D_i\rho \cdot \mathbf{grad}\, \omega_i, \tag{3.4-1}$$

where

$\omega_i \equiv \rho_i/\rho$ is the local *mass fraction* of species i

and

$D_i \equiv$ Fick diffusion coefficient (scalar "diffusivity") for species i transport in the prevailing mixture.

Actually, while Eq. (3.4-1) is often an adequate approximation, the laws governing *multicomponent* diffusion (see Eq. (3.4-6) below) are somewhat more complicated because:

a. other "forces" (besides $-\mathbf{grad}\, \omega_i$) can "drive" species i diffusion (e.g., $-\mathbf{grad}\{\ln T\}$, $-\mathbf{grad}\, p$, etc.);
b. due to interspecies "drag" or "coupling," the flux of species i may be influenced by the fluxes (and hence composition gradients) of the *other* species ($-\mathbf{grad}\, \omega_j$, where $j \neq i$).

For the present purposes, however, Eq. (3.4-1) will ordinarily[16] prove adequate, especially when species i is present as a "trace" constituent ($\omega_i \ll 1$), or the mixture only has two components ($N = 2$).

3.4.2 Corresponding Chemical-Element Diffusion Fluxes

Once expressions are available for the local *species* diffusion fluxes, the corresponding expression for each *element* flux follows immediately. Thus, if the symbol $\omega_{k/i}$ denotes the mass fraction of the kth chemical element in species i, clearly

$$\mathbf{j}_{(k)}'' = \sum_{i=1}^{N} \omega_{k/i}\mathbf{j}_i'' \qquad (k = 1, 2, \ldots, N_{\text{elem}}). \tag{3.4-2}$$

As an example of Eq. (3.4-2), consider the local diffusional flux of the element oxygen

[16] A notable exception, important when large molecular-weight disparities and large temperature gradients co-exist, is "thermal" (Soret) diffusion (Rosner (1980)) (cf. Eq. (3.4-6)) or particle "thermophoresis." (See Sections 6.2.5.3 and 8.2.1.)

in a reacting multicomponent combustion gas mixture. We can evidently write:

$$\mathbf{j}''_{(O)} = \frac{32}{32}\mathbf{j}''_{O_2} + \frac{16}{18}\mathbf{j}''_{H_2O} + \frac{32}{44}\mathbf{j}''_{CO_2} + \frac{16}{28}\mathbf{j}''_{CO} + \frac{16}{17}\mathbf{j}''_{OH} + \cdots \text{(etc.)}, \tag{3.4-3}$$

where each species diffusion flux $\mathbf{j}''_i$ will be given, approximately, by Eq. (3.4-1).

3.4.3 Multicomponent Diffusion Flux Law: Entropy Production and Entropy Diffusion Associated with Chemical Species Diffusion

A more generally applicable primary choice of driving force for chemical species diffusion, compatible with the constraint of positive entropy production (Eq. (2.5-28))[17] is the combination:

$$\mathbf{X}_i = \underbrace{-\mathbf{grad}_{T,p}\hat{\mu}_i}_{\substack{\text{Concentration}\\ \text{diffusion}}} + \underbrace{(v - \bar{v}_i)\mathbf{grad}\, p}_{\substack{\text{Pressure}\\ \text{diffusion}}} + \underbrace{\left(\mathbf{g}_i - \sum_{j=1}^{N} \omega_j \mathbf{g}_j\right)}_{\substack{\text{Body force}\\ \text{"diffusion"}}}, \tag{3.4-4}$$

where, in general, the chemical potential $\hat{\mu}_i$ is dependent on mixture composition *via* the "activity" a_i, appearing in:

$$\hat{\mu}_i = \overbrace{\hat{\mu}_i\{T; p, x_i = 1\}}^{f_i\{p,\, T\}} - \frac{RT}{M_i} \cdot \ln \frac{1}{a_i\{x_1, x_2, \ldots\}}, \tag{3.4-5}$$

and the notation $\mathbf{grad}_{T,p}$ means: spatial gradient holding T and p constant. Moreover, certain well-known "coupling" phenomena exist, such as the effect of heat diffusion on species mass diffusion (Soret effect). These considerations, and the constraints of frame invariance, linearity, and isotropy (cf. Section 3.1.4) lead to a general species mass flux law of the form:

$$\mathbf{j}''_i = \sum_{j=1}^{N} \alpha_{ij}\mathbf{X}_j + \alpha_i\left(-\frac{\mathbf{grad}\, T}{T}\right) \qquad (i = 1, 2, \ldots, N-1), \tag{3.4-6}$$

where the α's are scalar "phenomenological" coefficients, that is, capable of "direct"

---

[17] Note that, since $\sum_i \mathbf{j}''_i = 0$, any constant vector can be added to $-\mathbf{grad}_T\hat{\mu}_i + \mathbf{g}_i$ without altering the condition of positive entropy production. The driving force (Eq. 3.4-4), chosen such that $\sum \omega_j \mathbf{X}_j = 0$, clearly exhibits distinct contributions due to concentration gradients, pressure gradients, and "differential" body force, all of which, however, will appear with the same diffusion coefficient.

measurement for each local set of state variables. Clearly, Eq. (3.4-1) represents a highly degenerate uncoupled form of this type of flux law.

The number of "phenomenological" coefficients that actually have to be measured to describe a system with many components is approximately halved by the existence of necessary ("reciprocal") relations between them, as proved by L. Onsager [1933] on the basis of statistical mechanical arguments beyond the reach of *macroscopic* irreversible thermodynamics. Thus, in Eqs. (3.4-6) the matrix α_{ij} is symmetric ($\alpha_{ij} = \alpha_{ji}$), and the same coefficients α_i appearing in the $\mathbf{j}_i''$ expressions (Eq. (3.4-6)) also appear in a more general expression for the energy flux vector $\dot{\mathbf{q}}''$ (cf. Eq. (3.3-3)).

While not discussed further here, or in Chapter 6 (Mass Transport), it is of theoretical interest to mention that the "matrix structure" of the multicomponent species diffusion flux laws (Eq. (3.4-6)) and the resulting set of PDEs governing the fully coupled species concentration fields can be exploited in simple cases (nearly constant thermophysical properties, no chemical reaction or first-order homogeneous chemical reaction) which yield to an "uncoupling transformation." In such cases the elegant methods of linear matrix algebra can be used to find linear combinations of concentrations for which the transformed α_{ij}-diffusivity matrix is *diagonal* (see, e.g., Toor (1964), and Stewart and Prober (1964)). Often these "pseudo-concentrations" satisfy uncoupled PDE boundary-value problems whose solutions are already known. Then the solution to the originally posed coupled-multicomponent diffusion problem can be constructed from available solutions to simpler uncoupled (pseudo-binary) diffusion problems. This strategy has also been extended to simplify iterative numerical computations of chemically reacting multicomponent nonisothermal gases (see, e.g., Tambour and Gal-Or (1976)).

If we turn our attention to the *diffusional flux of entropy*, we see that Eq. (2.5-27) implies that, in addition to the "Fourier contribution" $-k(\mathbf{grad}\, T)/T$ (Eq. (3.3-4)), we should add:

$$(\mathbf{j}_s'')_{\substack{\text{species}\\ \text{diffusion}}} = \sum_{i=1}^{N} \mathbf{j}_i'' \left[s_i(T, p, x_i = 1) + \frac{R}{M_i} \ln \frac{1}{x_i} \right] \tag{3.4-7}$$

for the case of multicomponent species diffusion in a thermodynamically ideal solution ($a_i = x_i$). Note that each bracketed quantity is simply $\bar{s}_i$, the *partial specific entropy* of chemical species i ($i = 1, 2, \ldots, N$) in the prevailing mixture, defined such that $s_{\text{mixture}} = \sum \omega_i \bar{s}_i$.

The same quantity enters the convective flux of entropy, given by

$$\left(\sum_i \rho_i \bar{s}_i\right)\mathbf{v} = \rho\mathbf{v}\left(\sum_i \omega_i \bar{s}_i\right).$$

The "mixing entropy" contributions (the $(R/M_i)\ln(1/x_i)$ terms) are the origin of the minimum *work* required to separate mixtures into their pure constituents (see Exercise 2.17).

3.4.4 Solute Diffusivities in Gases, Liquids, and Solids—Real and Effective

The quantity $D_{i,\text{eff}}$, defined by Eq. (3.4-1), is called the effective "mass diffusivity" of species i in the prevailing (homogeneous or pseudo-homogeneous) medium. There are important situations in which it is *not* a scalar (single number), but rather a tensor defined by nine (or six independent) local numbers, because diffusion may be easier in some directions than others. This is true for solute diffusion in

a. anisotropic solids (e.g., single crystals or layered materials),
b. anisotropic fluids (e.g., turbulent shear flow).

For such cases, diffusion is not "down the concentration gradient," but, rather, skewed with respect to $-\mathbf{grad}\,\omega_i$. Fortunately, however, $D_{i,\text{eff}}$ often can be treated as a single *scalar* coefficient, applying equally to mass diffusion in any direction (e.g., in a fine-grained polycrystalline metal or quiescent fluid (gas or liquid)).

Values of the diffusivities D_i are obtainable from experiments, usually done in simple geometries with no flow (see Section 6.4.1) or an ultra-simple flow. However, as in the case of μ- and k-measurements, the task of covering all mixture compositions, pressures, and temperatures would be never-ending, so the guidance of molecular theory is essential.

Fortunately, reasoning on a molecular level, it is easy to estimate the magnitude of D_i and its dependence on molecular and environmental parameters.

For dilute solute i diffusion in low-density gases, $D_{i,\text{eff}}$ can be calculated from a knowledge of all of the relevant *binary* diffusion coefficients D_{ij} governing encounters with species $j\,(j \neq i)$, where:[18]

$$D_{ij} = \frac{3k_B T}{8p} \cdot \left[\frac{k_B T}{2\pi}\left(\frac{m_i + m_j}{m_i m_j}\right)\right]^{1/2} \cdot \frac{1}{\sigma_{ij}{}^2 \Omega_D\{k_B T/\varepsilon_{ij}\}} \tag{3.4-8}$$

and

$$D_{i\text{-mix}} \approx (1 - y_i) \cdot \left[\sum_{\substack{j=1,\\ j\neq i}}^{N} \frac{y_j}{D_{ij}}\right]^{-1} \qquad \text{for } N \geqslant 3. \tag{3.4-9}$$

Here, y_j is the mole fraction of species j and $y_i \ll 1$. This transport can be interpreted as the net *molecular flux* (Section 6.2.3) across each gap of thickness equal to the

[18] Here σ_{ij} is the intermolecular separation at which the interaction potential between molecules i and j vanishes and ε_{ij} is the depth of the energy "well" (see Figure 7.1-3). The function Ω_D of the "reduced" temperature $T/(\varepsilon/k_B)$, which accounts for departures from "hard-sphere" behavior, has been computed for interaction potentials of various functional forms (see Hirschfelder, Curtiss, and Bird (1954)). For the Lennard–Jones (power-law attraction and repulsion) potential $\Omega_D \approx 1.12\,(k_B T/\varepsilon)^{-0.17}$ for $3 \leqslant k_B T/\varepsilon \leqslant 200$.

molecular mean-free-path at the point in question. Compared to the "activated" process of atom migration in a solid (Eq. (3.4-15) below), Eq. (3.4-8) predicts D_i values that are much less sensitive to temperature (usually behaving like T^n, where $n \geqslant 3/2$. Indeed, the curve-fits collected for 39 binary mixtures in Table 3.4-1 reveal values of n between 1.597 and 2.072, with $n \approx 1.8$ being a representative value). The tabulated a-values (Column 2) are the corresponding D_{ij}-values (cm^2/s) at 1 atm, 300 K.

Comparison of these (mostly experimental) D_{ij} values (Table 3.4-1) with the predictions of Eq. (3.4-8) reveals that the rules:

$$\sigma_{ij} \approx \frac{1}{2}(\sigma_{ii} + \sigma_{jj}), \tag{3.4-10}$$

$$\frac{\varepsilon_{ij}}{k_B} \approx \left(\frac{\varepsilon_{ii}}{k_B} \cdot \frac{\varepsilon_{jj}}{k_B}\right)^{1/2}, \tag{3.4-11}$$

together with the σ_{ii} and ε_{ii}/k_B values collected in Table 3.4-1, would provide reasonable engineering estimates in the absence of direct data on binary diffusivities, or binary interaction parameters. For disparate-size systems ($\sigma_{ii} \gg \sigma_{jj}$), an improved procedure is suggested in Exercise 3.11.

It is interesting and important to note that, for mixtures of similar *gases*, the species mass diffusivity D_i is always of the same order of magnitude as the *momentum diffusivity* $\mu/\rho \equiv \nu$ ("kinematic viscosity") and the *energy diffusivity* $k/(\rho c_p) \equiv \alpha$. This is because, for gases, the *mechanisms* of mass, momentum, and energy transfer are identical (*viz.*, random molecular motion between adjacent fluid layers). Thus, the following important dimensionless ratios are found to be near unity for such mixtures:[19]

$$\mathrm{Sc}_i \equiv \frac{\nu}{D_{i\text{-mix}}} \quad \text{(Schmidt number)}, \tag{3.4-12}$$

$$\mathrm{Le}_i \equiv \frac{D_{i\text{-mix}}}{\alpha} \quad \text{(Lewis number)}. \tag{3.4-13}$$

This near equality of the various diffusivities is *not* the case for solutes in *liquids* (e.g., anions in aqueous solution), for which Sc_i-values can exceed 10^3. Moreover, if the molecular-weight disparity is sufficiently great (as in the case of particles suspended in a gas (an "aerosol")), we also anticipate $\mathrm{Sc}_i \gg 1$ (see Figure 3.4-1).

---

[19] As a corollary, the same may be said for the Prandtl number, $\mathrm{Pr} \equiv \nu/\alpha$. Indeed, for a monatomic gas (without electronic excitation), kinetic theory predicts $\mathrm{Pr} = 2/3$. More generally, Eq. (3.3-6) is equivalent to

$$\mathrm{Pr} = \left[1 + \frac{5}{4} \cdot \frac{\gamma - 1}{\gamma}\right]^{-1},$$

where $\gamma \equiv C_p/C_v$.

Table 3.4-1 Power-Law Curve-Fit[a] to Available $D_{ij}(T)$ Data for Some Low-Density Binary Gas Mixtures; Parameters a and n in $D_{ij}\,(\text{cm}^2/\text{s}) = a/p(\text{atm}) \cdot (T(K)/300)^n$

Gas Pair (*System*)	*a*	*n*	*T range* (K)	*Gas Pair* (*System*)	*a*	*n*	*T range* (K)
He–CH_4	0.677	1.750	298–10^4	N_2–H_2O	0.254	2.072	282–373
He–O_2	0.752	1.710	244–10^4	CO–O_2	0.211	1.724	285–10^4
He–air	0.725	1.729	244–10^4	CO–air	0.216	1.730	285–10^4
He–CO_2	0.603	1.720	200–530	CO–CO_2	0.169	1.803	282–473
He–SF_6	0.415	1.627	290–10^4	O_2–H_2O	0.256	2.072	282–450
Ne–H_2	1.15	1.731	90–10^4	O_2–H_2O	0.307	1.632	450–1070
Ne–N_2	0.330	1.743	293–10^4	air–H_2O	0.254	2.072	282–450
Ne–CO_2	0.268	1.776	195–625	air–H_2O	0.303	1.632	450–1070
Ar–CH_4	0.207	1.785	307–10^4	CO_2–N_2[b]	0.177	1.755	300–1100
Ar–N_2	0.198	1.752	244–10^4	CO_2–N_2	0.118	1.866	195–550
Ar–CO	0.198	1.752	244–10^4	CO_2–O_2[b]	0.165	1.800	300–1100
Ar–O_2	0.195	1.736	243–10^4	CO_2–C_3H_8	0.0880	1.896	298–550
Ar–air	0.197	1.749	244–10^4	CO_2–SF_6	0.0658	1.886	328–472
Kr–N_2	0.155	1.766	248–10^4	H–He	2.77	1.732	275–10^4
Kr–CO	0.155	1.766	248–10^4	H–Ar	0.131	1.597	275–10^4
Xe–N_2	0.127	1.789	242–10^4	H–H_2	2.16	1.728	190–10^4
H_2–CH_4	0.737	1.765	293–10^4	N–N_2	0.327	1.774	280–10^4
H_2–O_2	0.814	1.732	252–10^4	O–He	1.01	1.749	280–10^4
H_2–air	0.787	1.750	252–10^4	O–Ar	0.273	1.841	280–10^4
CH_4–N_2	0.216	1.750	298–10^4	O–N_2	0.327	1.774	280–10^4
CH_4–air	0.219	1.747	298–10^4	O–O_2	0.327	1.774	280–10^4
N_2–O_2	0.211	1.724	285–10^4				

[a] Source (unless otherwise specified): Marrero and Mason (1972).
[b] Source: Westenberg (1966).

For *dilute solutes in liquids or dense vapors*, D_i can be estimated using a fluid-dynamics approach in which each solute molecule is viewed as drifting in the host viscous fluid in response to a net force associated with the gradient in its partial (osmotic) pressure (Einstein [1905]). This argument leads to the well-known Stokes–Einstein equation:

$$D_i = \frac{k_B T}{3\pi\mu\sigma_{i,\text{eff}}}, \tag{3.4-14}$$

where $\sigma_{i,\text{eff}}$ is the effective molecular diameter of solute molecule i and μ is the Newtonian viscosity of the host "solvent." Equation (3.4-14)[20] also applies to the

[20] The "no-slip" assumption underlies Eq. (3.4-14); however, this is questionable when applied to small solute molecules. Often Eq. (3.4-14) is modified by a (Stokes–Cunningham) slip-correction factor.

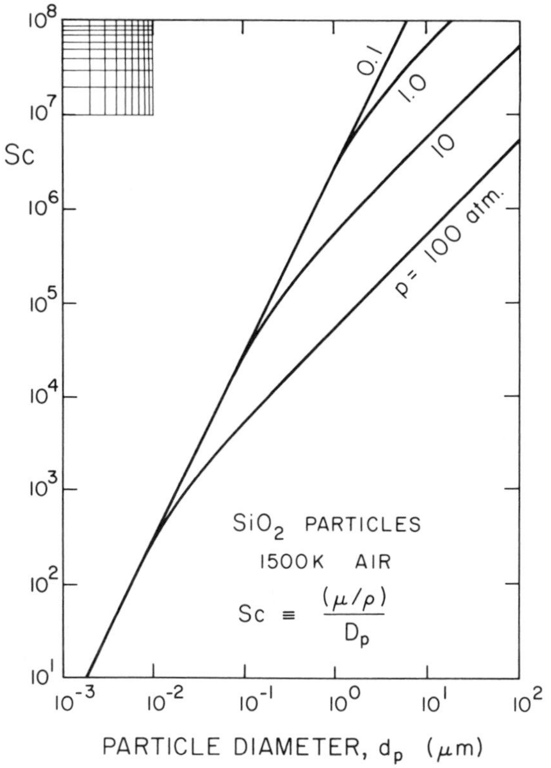

Figure 3.4-1 Predicted small-particle Schmidt number (ν/D_p) for SiO_2 particles in 1500 K air; transition between free-molecule (Eq. (3.4-8)) and continuum (Eq. (3.4-14)) behavior (Rosner and Fernandez de la Mora (1982)).

Brownian diffusion of particles in a gas, provided the particle diameter is large compared to the prevailing mean-free-path ($d_p > 10^{-1}$ μm, say, at STP) (see Figure 3.4-1).

For *solute diffusion in ordered solids*, D_i can be calculated from the net flux of solute atoms jumping between interstitial positions (sites) in the lattice. This always leads to an expression of the form:

$$D_i = \begin{pmatrix}\text{Geometric}\\\text{factor}\end{pmatrix} \cdot \begin{pmatrix}\text{Lattice}\\\text{dimension}\end{pmatrix}^2 \cdot \begin{pmatrix}\text{Attempted jump}\\\text{frequency}\end{pmatrix} \cdot \exp\left(-\frac{\varepsilon_D}{k_B T}\right), \quad (3.4\text{-}15)$$

where ε_D is the energy barrier encountered in moving an atom of solute i from one interstitial site ("cage") to another.

In more complex situations the "effective" diffusivity may, however, be governed by quite different considerations, as illustrated below. We first consider the important example of solute diffusion through the fluid contained in the interconnected pores of a solid porous structure when the solid itself is impervious to the solute. If the solute mean-free-path is small compared to the mean pore diameter, then the effect of the porous structure is to reduce $D_{i,\text{eff}}$ below $D_{i\text{-fluid}}$ by an amount dependent on the interconnected pore volume fraction, ε. In general, this effect is described by a relation of the form:

$$D_{i,\text{eff}} = \frac{\varepsilon}{\tau_{\text{pore}}\{\varepsilon\}} \cdot D_{i\text{-fluid}}, \tag{3.4-16}$$

where the reduction factor ε in the numerator is obvious in the trivial limiting case of straight pores in the direction of diffusion. In the absence of surface diffusion (i.e., adsorbed solute migration *along* the pore walls), the function $\tau_{\text{pore}}\{\varepsilon\}$ (>1, in the denominator) is therefore a correction for the "tortuosity" of the actual passages (i.e., their variable direction and variable effective diameter). While usually determined experimentally, $\tau_{\text{pore}}\{\varepsilon\}$ can be computed theoretically for "model" porous materials comprised of simple periodic arrays of "obstacles." (For arrays of impermeable spheres, $\tau_{\text{pore}}\{\varepsilon\}$ is only $1 + \frac{1}{2}(1 - \varepsilon)$, but for most actual "random" porous solids $\tau_{\text{pore}}\{\varepsilon\}$ exceeds this by a considerable amount, often being nearer to ε^{-1}.)

When the solute mean-free-path is *larger* than the mean pore diameter, as is often the case for gas diffusion through microporous solid media at atmospheric pressure, then the solute molecule "rattles" down each pore by successive collisions with the pore walls (rather than with "solvent" molecules). For a single straight cylindrical pore, M. Knudsen [1909] showed that in this limit:

$$D_{i\text{-Knudsen}} = \frac{1}{3} \cdot \begin{pmatrix} \text{Mean thermal speed} \\ \text{of molecule } i \end{pmatrix} \cdot (\text{Pore diameter}), \tag{3.4-17}$$

where, from the viewpoint of gas kinetic theory, the pore diameter is now seen to play the role of the solute mean-free-path. For Knudsen diffusion in a porous solid, we then expect $D_{i,\text{eff}}$ to be considerably less than the value $D_{i\text{-Knudsen}}$ calculated from Eq. (3.4-17) using the *mean* pore diameter; that is,

$$D_{i,\text{eff}} = \frac{\varepsilon}{\tau_{\text{pore}}\{\varepsilon\}} \cdot D_{i\text{-Knudsen}}, \tag{3.4-18}$$

which (in contrast to Eq. (3.4-16)) when the fluid is, say, an ideal gas, is independent of pressure. A simple interpolation formula, actually rigorous for a dilute gaseous species at any combination of gas mean-free-path and pore size, is:

$$D_{i,\text{eff}} = \frac{\varepsilon}{\tau_{\text{pore}}\{\varepsilon\}} \cdot \frac{1}{(D_{i\text{-fluid}})^{-1} + (D_{i\text{-Knudsen}})^{-1}}, \tag{3.4-19}$$

where $D_{i\text{-fluid}}$ is calculated from Eq. (3.4-9) and $D_{i\text{-Knudsen}}$ is calculated from Eq. (3.4-17) using the mean pore size. This relation for the "effective" diffusion coefficient of a dilute species i is widely used in the description of gas diffusion through porous solids (e.g., catalyst "support" materials (Section 6.4), coal char, natural adsorbents, etc.).

Similarly, a solute species i in *turbulent fluid flow* will have an effective diffusivity, $D_{i,\mathrm{t}}$, practically unrelated to its *molecular* diffusivity, but closely related to the prevailing effective *momentum* diffusivity ν_{t} in the local *flow*. Indeed, when the solute can "follow" the turbulence (cf. Section 6.6.3), it is usually sufficient to assume:

$$\frac{\nu_{\mathrm{t}}}{D_{i,\mathrm{t}}} \equiv \mathrm{Sc}_{i,\mathrm{t}} = \text{pure number near unity} \qquad \text{("turbulent Schmidt number").}$$

For example, tracer dispersion measurements made near the centerline of ducts containing a Newtonian fluid in *turbulent* flow ($\mathrm{Re}(\equiv Ud/\nu) > 2 \times 10^3$) reveal that

$$D_{i,\mathrm{t}} = \frac{1}{\mathrm{Pe}_{\mathrm{eff}}\{\mathrm{Re}\}} \cdot Ud_{\mathrm{w}}, \tag{3.4-20}$$

where $\mathrm{Pe}_{\mathrm{eff}}\{\mathrm{Re}\}$, a weak function of Re, takes on values between about 250 and 1000 [see, e.g., Sherwood, Pigford, and Wilke (1975)]. Thus, in such flows $D_{i,\mathrm{t}}$ is almost proportional to the product Ud_{w}, with the multiplier being a pure number of order 10^{-3}. (This should be compared with the *molecular* diffusivity which, for gases, is proportional to the product of the mean molecular thermal speed $\bar{c}$ and the molecular mean-free-path ℓ, with the multiplier being 1/3.) From this viewpoint the Reynolds' number Ud/ν can itself be regarded as the ratio of the effective ("eddy") momentum diffusivity to the molecular-momentum diffusivity (magnified by a factor of about 10^3).

With respect to the *effective diffusion* of solutes in high-Reynolds'-number fluid flow through a fixed bed of granular material (with characteristic (particle) diameter d_{p}), the situation is much like *turbulent* flow in a homogeneous medium in the sense that D_{eff} is almost proportional to the product of the average interstitial fluid velocity, u_i, and the *particle size*, d_{p}. For example, in a packed cylindrical duct with $\mathrm{Re}_{\mathrm{bed}} > 10^2$, it is found that:

$$(D_{i,\mathrm{eff}})_r = \frac{1}{\mathrm{Pe}_{\mathrm{eff},r}\{\mathrm{Re}_{\mathrm{bed}}\}} \cdot u_i d_{\mathrm{p}} \tag{3.4-21}$$

and

$$(D_{i,\mathrm{eff}})_z = \frac{1}{\mathrm{Pe}_{\mathrm{eff},z}\{\mathrm{Re}_{\mathrm{bed}}\}} \cdot u_i d_{\mathrm{p}}, \tag{3.4-22}$$

where both dimensionless functions (called effective Peclet numbers) are weakly dependent on the bed Reynolds' number (see Sections 4.7, 5.5.5, and 6.5.1), being near the values 10 and 2, respectively (Wilhelm (1962)). Thus, time-averaged solute mixing in such situations, which appears to be anisotropic, is much more rapid than would be expected based on *molecular* motions alone (especially for liquids). Analogous remarks can be made for the fluid dynamical contribution to momentum and energy "diffusivities" (ν_{eff} and α_{eff}, respectively) in fixed beds.

3.5 LIMITATIONS OF LINEAR LOCAL FLUX *VS.* LOCAL DRIVING FORCE CONSTITUTIVE LAWS

It is prudent to briefly restate the principal limitations underlying the use of the above-mentioned diffusion flux (momentum, energy, mass)-driving force laws (cf. Section 3.1.4), and indicate those situations of engineering importance requiring special treatment.

3.5.1 "Nonlinear" Fluids

Many fluids of engineering significance (especially high-mass loaded "slurries," such as clay in water, etc.) are termed "non-Newtonian" since the dynamic viscosity appearing in Eq. (3.2-4) is itself noticeably[21] dependent on (an invariant of) the deformation rate or extra stress (as well as local state variables like T, p, x_i). The *dynamics* of such fluids,[22] exhibiting many "unusual" phenomena, is the province of *rheology*. Analogous nonlinearities are possible in the domain of *heat* and *mass* transfer, including interesting couplings (e.g., solute diffusion through human blood is augmented in the presence of a deformation rate). Apart from *turbulence*, which makes Newtonian fluids appear non-Newtonian, this important branch of transport phenomena will not be considered further here.

3.5.2 Nonlocal Spatial Behavior

The phrase "diffusion" implies spatially *local* behavior; however, the mechanisms of momentum, energy, and mass transfer are such that the "diffusion" picture may break down in important circumstances. This occurs when the characteristic transport length (e.g., "mean-free-path") is no longer negligible compared to the overall length over which the transport process occurs. Well-studied examples include transport across very-low-density gases in ducts, electronic conduction in microcrystallites, and radiation transport through nearly transparent media.

---

[21] Of course, at sufficiently small driving forces, almost all fluids will appear to be linear. The fluids here considered "nonlinear" exhibit changes in their coefficients at driving forces that are ordinarily encountered, and for which simpler fluids behave linearly.

[22] Solids that "creep" (continuously deform) under an applied stress behave like non-Newtonian fluids.

3.5.3 Nonlocal Temporal Behavior—Fluids with Memory

Non-Newtonian behavior may also be due to the fluid responding not only to the *present* applied stresses, but also to *previous* stresses—that is, the so-called stress "history." For example, "visco-elastic" fluids (such as many "gels") will partially return to their original state when an applied stress is interrupted. Analogous behavior is possible for energy and mass transfer; for example, time lags between the imposition of spatial gradients in field quantities and their associated fluxes, as well as effects associated with the finiteness of thermal and concentration wave speeds. However, in what follows we confine our attention to "time-independent" fluids, in which all local instantaneous fluxes and reaction rates can be uniquely related to the local instantaneous fields with reasonable accuracy.

3.5.4 Multiphase Effects: Nonlinear Species "Drag" Laws

The division between a single-phase *mixture* containing species of unequal molecular weights and a "multiphase" system (e.g., droplet-laden gas) is not a sharp one and relates to the vectorial differences between the resulting drift velocities $\mathbf{v}_i - \mathbf{v}_j$, as well as the magnitudes of the individual "diffusion" velocities $\mathbf{v}_i - \mathbf{v}$. Linear momentum and mass "diffusion" laws can be expected to apply only when such velocity differences are acceptably small, just as energy transfer can be considered a single-phase diffusion process if the characteristic temperatures of all local "species" are not too disparate.

Quantitative criteria for the validity of a linear "diffusion" treatment of momentum and mass transfer will be presented and discussed in Sections 6.6 and 8.3.

SUMMARY

- The *general* conservation laws must be supplemented by laws that describe the behavior of certain classes of matter. These "constitutive" laws govern:
  a. thermodynamic interrelations between local state variables, e.g., $p\{T, \rho_1, \rho_2, \rho_3, \ldots, \rho_N\}$, $e\{T, \rho_1, \rho_2, \rho_3, \ldots, \rho_N\}$;
  b. local relations between the diffusion fluxes of species mass, linear momentum and energy, and spatial gradients of appropriate field densities;
  c. local "kinetic" relations between the individual net chemical production rates and local state variables; e.g., $\dot{r}_i'''\{T, \rho_1, \rho_2, \ldots, \rho_N\}$.
- The diffusion flux-driving force laws are chosen to satisfy several important constraints; *viz.*, that only the local, present fields are relevant, that the form of the law should be unchanged relative to translating and rotating observers, that the presence of nonzero fluxes and their associated driving forces must correspond to positive entropy production; moreover, we focus our attention on

isotropic media for which the transport coefficients (viscosity, thermal conductivity, diffusion coefficients) appearing in these laws are at most scalar functions of the local fluid state variables.

- "Newtonian" fluids are those for which the stresses associated with fluid motion are linearly proportional to the local, instantaneous *rates of fluid deformation*—both angular and volumetric. The motions of solids differ fundamentally from those of fluids in that, while solids can exhibit spatial gradients of velocity (e.g., in solid body rotation) only compressible fluids *continuously deform* (angle and volumetric changes).
- While, in principle, the coefficients in the flux-driving force laws are measurable ("phenomenological"), in practice, molecular theory is essential to provide enough insight into the dependence of viscosities, thermal conductivities, and diffusivities on both state variables and molecular parameters (molecular weight, molecular size, etc.) to permit the rational interpolation and extrapolation of available data.

TRUE/FALSE QUESTIONS

3.1 T F J. C. Maxwell showed that the *dynamic viscosity* of a perfect gas depends inversely on the pressure level and directly on the square root of the molecular mass.

3.2 T F The near-equality of the molecular diffusivities for momentum, heat, and mass in low-density gases indicates that the microscopic *mechanisms* of momentum, heat, and mass are the same in such "fluids."

3.3 T F Information on the nature of individual molecular encounters can be derived from careful measurements of the temperature dependence of gas viscosity.

3.4 T F The presence of velocity gradients ensures the simultaneous presence of momentum diffusion (i.e., shear stresses) in a continuum.

3.5 T F According to Stokes, the local viscous stress in a Newtonian fluid is proportional to the local *deformation* of the continuum (both angular and volumetric).

3.6 T F All liquids have Prandtl numbers not very different from unity.

3.7 T F Electrons play an important role in liquid-metal *heat* diffusion, but not in liquid-metal *momentum* diffusion.

3.8 T F Radiation makes no contribution to the apparent thermal conductivity of partially transparent materials.

3.9 T F All substances of engineering interest are "isotropic" in their thermal properties.

3.10 T F Polyatomic gases with low (near unity) values of the equilibrium-specific heat ratio, $\gamma(\equiv c_p/c_v)$, usually have higher Prandtl numbers than those characterizing monatomic gases (such as He, Ar, etc.).

EXERCISES

3.1 Consider the Fourier energy flux law [1808]:

$$\dot{\mathbf{q}}'' = -k\,\mathbf{grad}\,T$$

(where the thermal conductivity k is a scalar function of local state variables) from the points of view of the constraints of Section 3.1.4, i.e., nonnegative entropy production, determinism (local action temporally), local action (spatially), frame indifference, isotropy, simplicity (linearity).

a. *Must* the $\dot{\mathbf{q}}''$-law satisfy these conditions? Explain.
b. *Does* this particular law "satisfy" each of these conditions? Demonstrate.
c. Does the addition of $\sum \mathbf{j}_i'' h_i$ in multicomponent ideal mixtures (Section 3.3.2) cause any of these conditions to be violated?
d. Under what conditions would the Fourier heat-flux law break down? (Consider both rapid transients and domains of small dimension; cf. Section 3.5.)
e. For what sort of materials must k be considered a tensor? What are some important implications of tensorial conductivity?
f. When k is a function of T alone, the quantity

$$\phi_T \equiv \int_{T_{\text{ref}}}^{T} k\{T\} \cdot dT$$

is sometimes called a "heat-flux potential." Is this viewpoint justified? Would $\phi_T\{T\}$ be a useful dependent variable in the energy equation?
g. Under what conditions could radiation (photon) energy transfer be included in k? (Consider the magnitude of the "photon mean-free-path.")

3.2 Do the local diffusion flux-driving force laws

$$\mathbf{T} = 2\mu\,\mathbf{Def}\,\mathbf{v} + \left(\kappa - \frac{2}{3}\mu\right)(\text{div}\,\mathbf{v})\mathbf{I} \qquad \text{and} \qquad \dot{\mathbf{q}}'' = -k\,\mathbf{grad}\,T + \sum_i \mathbf{j}_i'' h_i$$

imply that the scalar coefficients μ, κ, k, D_i must be constants with respect to space and/or time? Discuss.

3.3 Consider the following possibilities in the light of the *multicomponent* species diffusion-flux law, Eq. (3.4-6):

a. Species i diffuses even though $-\mathbf{grad}\,\omega_i$ locally vanishes;
b. Species i does not diffuse even though $-\mathbf{grad}\,\omega_i$ is nonzero;
c. Species i diffuses "up" the local concentration gradient, in the direction of $\mathbf{grad}\,\omega_i$ (not $-\mathbf{grad}\,\omega_i$).

Do these situations, which have been observed experimentally, violate the principle of a positive entropy production? What condition *does* the Second Law of Thermodynamics impose on the NXN *matrix* α_{ij}?

3.4 Since the diffusion fluxes (species mass, energy, and linear momentum) are often themselves proportional to spatial gradients (e.g., $-\mathbf{grad}\,\omega_i$, $-\mathbf{grad}\,T$, and $\mathbf{Def}\,\mathbf{v}$, respectively) and the conservation equations involve the local divergence of each flux, to express each conservation equation in any particular coordinate system we will clearly need an expression for the operator div **grad** (the so-called Laplacian) in that system.

a. Using the results given in Section 2.5.7 (for any orthogonal coordinate system ξ_1, ξ_2, ξ_3, and its associated scale factors h_1, h_2, h_3), show that:

$$\text{div}\,\mathbf{grad} = \frac{1}{h_1 h_2 h_3} \cdot \left\{ \begin{aligned} &\frac{\partial}{\partial \xi_1}\left(\frac{h_2 h_3}{h_1} \cdot \frac{\partial}{\partial \xi_1}\right) \\ &+ \frac{\partial}{\partial \xi_2}\left(\frac{h_1 h_3}{h_2} \cdot \frac{\partial}{\partial \xi_2}\right) \\ &+ \frac{\partial}{\partial \xi_3}\left(\frac{h_1 h_2}{h_3} \cdot \frac{\partial}{\partial \xi_3}\right) \end{aligned} \right\}.$$

b. Using row 2 of Table 2.5-1, show that, in a cylindrical polar coordinate system (r, θ, z) (Figure 2-2), the result of part (a) is equivalent to:

$$\text{div}\,\mathbf{grad} = \frac{1}{r}\frac{\partial}{\partial r}\left(r\frac{\partial}{\partial r}\right) + \frac{1}{r^2}\frac{\partial^2}{\partial \theta^2} + \frac{\partial^2}{\partial z^2}.$$

c. Observe that the results obtained in parts (a) and (b) contain no "mixed" (cross) second derivatives. What property of these curvilinear coordinate systems is responsible for this important simplification?

3.5 In Section 3.2.1 it is stated that the rate of *angular* deformation of a fluid parcel in the x-y plane, and the r-θ plane are, respectively:

$$\left(\frac{\partial v_x}{\partial y} + \frac{\partial v_y}{\partial x}\right) \quad \text{and} \quad \left(r\frac{\partial}{\partial r}\left(\frac{v_\theta}{r}\right) + \frac{1}{r}\frac{\partial v_r}{\partial \theta}\right).$$

a. Derive the first of these two expressions by calculating the change in *included angle* of a small fluid parcel in the x-y plane between times $t + \Delta t$ and t, and then dividing by Δt and passing to the limit $\Delta t \to 0$.

b. In the expression for $(\mathbf{Def}\,\mathbf{v})_{r\theta}$ what is the origin of the term $-(v_\theta/r)$, which arises after radial differentiation of (v_θ/r) above?

c. Using a method analogous to that in Part (a), prove that, at least in cartesian coordinates, the rate of *volumetric* deformation of a fluid parcel is given by (div $\mathbf{v}$). Since div $\mathbf{v} = (Dv/Dt)/v$ (by mass conservation, where $v \equiv 1/\rho$), does this result apply to all coordinate systems?

3.6 "Low-density" gas-kinetic theories yield accurate expressions for the *transport* properties only when the prevailing mean-free-path, ℓ, is large compared to the molecular diameter σ. Using Maxwell's hard-elastic-sphere estimate:

$$\ell = \frac{1}{\sqrt{2}\,\pi n \sigma^2},$$

derive an expression for the "limiting" *number* density $n_{\lim}$, and associated *mass* density $\rho_{\lim} = mn_{\lim}$ at which the mean-free-path, ℓ, would be, say, only four times σ. Evaluate this mass density for *water* (use effective hard-sphere diameter of H_2O of 4.2 Å) and compare your result to the well-known density of *liquid* water. [For "dense" vapors the law of "corresponding states" is usually extended (into the domain of *transport* properties) in order to generalize transport-property data and make engineering predictions at high pressures and/or low temperatures (see Sections 3.2.5 and 7.1.3.1). Also, Enskog developed formulae that describe small *corrections* to "low-density" gas-kinetic predictions.]

3.7 Figure 3.7E displays the experimentally observed composition dependence of the viscosity μ of a mixture of ammonia and hydrogen at 306 K, 1 atm. N.B.: The units used for μ in Figure 3.7E are μP (micro-poise) where 1 poise $\equiv$ 1 g/cm · s. The MKS unit of viscosity is 1 kg/m · s or 1 Pascal-second.

a. Using Chapman–Enskog (CE) Theory and the Lennard–Jones (LJ) potential parameters in our table, examine the agreement for the two endpoints, i.e., the viscosities of *pure* NH_3 and *pure* H_2 at 306 K, 1 atm.
b. Using the "square-root rule," what would you have predicted for the composition dependence of the mixture viscosity in this case? What conclusions do you draw from this comparison?
c. How would you expect the $NH_3 + H_2$ curve to shift if the temperature

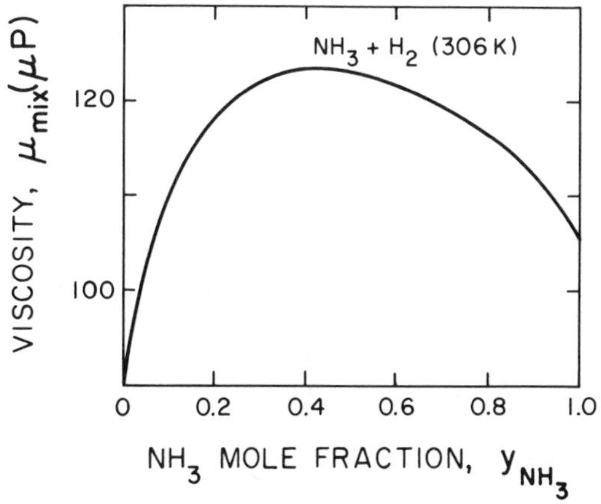

Figure 3.7E

were increased to 1000 K and the pressure were increased ten-fold? (Discuss the *basis* for your expectation.)

d. Estimate the binary diffusion coefficient $D_{NH_3-H_2}$ (cm^2/s) from Chapman–Enskog theory. (*Ans.* 0.789 cm^2/s.)

e. Calculate the Schmidt number for an equimolar mixture of $H_2(g)$ and $NH_3(g)$. (*Ans.* 0.411.)

3.8 Consider the properties of a "binary gas" mixture comprised of 0.2μm-diameter $SiO_2(c)$ particles[23] present in mass fraction $\omega_1 = 0.33$ with combustion products (with properties not very different from $N_2(g)$) present in mass fraction $\omega_2 = 0.67$. Treating the silica "aerosol" as a high-molecular-weight "vapor," and at $T = 1600$ K, $p = 1$ atm:

a. Calculate the mean molecular weight, M, of this binary mixture.

b. Calculate the *specific heat*, c_p, of this binary mixture.

c. Estimate the effective *dynamic viscosity*, μ, of this mixture, using both the "square-root rule" (Eq. (3.2-13)) and the semi-theoretical (Wilke) equation:

$$\mu_{mix} \approx \sum_{i=1}^{N} \frac{y_i \mu_i}{\sum_{j=1}^{N} y_j \Phi_{ij}}$$

where

$$\Phi_{ij} = 8^{-1/2} \cdot \left(1 + \frac{M_i}{M_j}\right)^{-1/2} \cdot \left[1 + \left(\frac{\mu_i}{\mu_j}\right)^{1/2} \cdot \left(\frac{M_j}{M_i}\right)^{1/4}\right]^2. \quad \text{(N.B.: } \Phi_{ii} = 1.\text{)}$$

d. Estimate the effective *thermal conductivity*, k, of this mixture, using both the "cube-root rule" (Eq. (3.3-7)) and the semi-theoretical (Mason–Saxena) equation:

$$k \approx \sum_{i=1}^{N} \frac{y_i k_i}{\sum_{j=1}^{N} y_j \Phi_{ij}}$$

identical in structure to that given above (in the viscosity formula).

e. Estimate the binary Fick diffusivity, D_{12}, under these conditions (using an equivalent "hard-sphere" diameter for each N_2 molecule).

f. Predict the effective *Prandtl number* characterizing this pseudo-single-phase *mixture.* How does it compare to that of pure $N_2(g)$?

g. Predict the *Schmidt number* characterizing this pseudo-single-phase *mixture.*

---

[23] Consider the intrinsic density of each SiO_2 microparticle to be 2.6 g/cm^3. $SiO_2(c)$ has a heat capacity at 1600 K of about 0.30 cal/g · K.

h. Calculate the ratio of the $N_2(g)$ mean-free-path to the $SiO_2(c)$ particle diameter. Is this ("particle Knudsen number," Kn_p) ratio large enough to validate your treatment of this mixture as a pseudo-single-phase "gas mixture"? If at $T = 1600$ K the pressure level was 20 atm (rather than 1 atm), would this approach remain equally valid?

i. Despite the fact that the silica fume is present with an appreciable *mass fraction* (0.33), it is a "trace species" on a *number* (mole-fraction) basis. Study your results for the predicted effect of the SiO_2-fume on each of the *mixture* properties (M, c_p, μ, k) and discuss the possible significance/implications of your conclusions.

3.9 For the preliminary design of a chemical reactor to carry out the removal of trace sulfur compounds (e.g., C_4H_4S) from petroleum naptha vapors (predominantly heptane, C_7H_{16}) (see Exercise 6.10), it is necessary to estimate the Newtonian viscosity and corresponding momentum diffusivity of the vapor mixture at 660 K, 30 atm; these are rather extreme conditions for which direct measurements are not available.

a. Using the vapor composition, and the assembled data displayed in the table below, together with selected results from the kinetic theory of ideal vapors and dense vapors, what are your best estimates for μ_{mix} and ν_{mix}?

b. If this vapor mixture is to be passed at the rate of 2 g/s through each 2.54 cm diameter circular tube (in a parallel array), calculate the corresponding Reynolds' number, $Re \equiv U\, d_w/\nu_{mix}$ within each tube (where U is the average vapor-mixture velocity). This dimensionless ratio will be seen to be required to estimate the mechanical energy required to pump the vapor through such tubes (see Section 4.5.1).

c. Estimate the effective (pseudo-binary) Fick diffusion coefficient for thiophene (C_4H_4S) migration through this mixture, and the corresponding diffusivity *ratio*: $Sc \equiv \nu_{mix}/D_{3-mix}$.

Ans. a) $\mu_{mix} \approx 1.21 \times 10^{-4}$ Poise; $\nu_{mix} \approx 1.16 \times 10^{-2}$ cm^2/s; b) $Re \approx 3.26 \times 10^3$.

Table 3.9E Composition and Thermophysical Data for Hydrodesulfurization Reactor Example

i	*Species*	y_i	m_i	σ_i	ε_{ii}/k_B	p_c	T_c
1	$H_2(g)$	0.828	2.016	2.827	59.7		
2	$C_7H_{16}(g)$	0.172	100.128	8.88	282	26.8	540
3	$C_4H_4S(g)$	trace	84.13	5.3	410		
Units:			g/g-mole	Å	K	atm	K

3.10 $SF_6(g)$ is used in gas-cooled electrical equipment because of its high-electrical-breakdown resistance. Whether the equipment is cooled by forced or natural convection, it is necessary to estimate the transport properties μ and k near 1 atm over a temperature range of perhaps 300-600 K.

a. Given the data below, use Chapman-Enskog kinetic theory to estimate μ, k, and Pr at 300, 400, 500, and 600 K:

Table 3.10E Properties of Sulfur Hexafluoride Vapor

Molecular weight	146.07 g/g-mole
Critical Pressure	37.1 atm
Critical Temperature	318.7 K
Critical volume	198 cm^3/g-mole
Molar heat capacity	23.28 @ 300 K
(cal/g-mole K)	27.83 @ 400 K
	30.70 @ 500 K
	32.54 @ 600 K

b. Justify the quantitative use of Chapman–Enskog kinetic theory under these conditions.

c. For natural convection engineering calculations, we will also need the coefficient of thermal expansion, that is,

$$\beta_T \equiv \frac{1}{v}\left(\frac{\partial v}{\partial T}\right)_p = -\frac{1}{\rho}\left(\frac{\partial \rho}{\partial T}\right)_p.$$

Estimate this coefficient at 500 K.

d. Compare your Pr-values with the corresponding Prandtl numbers for air (tabulated, e.g., in Table 8.1-1). Discuss any significant similarities (or differences).

e. To justify the continuum approximation in calculating heat transfer to SF_6(g), estimate its molecular mean-free-path at 1 atm, 600 K.

f. Estimate the viscosity of SF_6 at its thermodynamic critical state, that is, $\mu\{T_c, p_c\} \equiv \mu_c$ using your previous "ideal vapor" estimate of μ and assuming that the viscosity behavior of SF_6 is an accordance with the principle of corresponding states (cf. Figure 3.2-3).

3.11 For gas mixtures containing molecules of rather disparate size (for example, $\sigma_{ii} \gg \sigma_{jj}$) which interact according to a Lennard–Jones-type potential, there is more theoretical justification for the unlike-molecule energy-well relation:

$$\varepsilon_{ij}\sigma_{ij}{}^6 \cong (\varepsilon_{ii}\sigma_{ii}^6\varepsilon_{jj}\sigma_{jj}{}^6)^{1/2}$$

than there is for the more familiar approximation (Eq. (3.4-11)):

$$\varepsilon_{ij} \cong (\varepsilon_{ii}\varepsilon_{jj})^{1/2}$$

(see, e.g., Ferziger and Kaper (1972)). If one retains, for the unlike-molecule interaction size-parameter estimate,

$$\sigma_{ij} \cong \frac{1}{2}(\sigma_{ii} + \sigma_{jj}),$$

a. Show that the above-mentioned suggestion is equivalent to modifying Eq. (3.4-11) by the correction factor:

$$(\sigma_{ii}{}^3 \sigma_{jj}{}^3) \Big/ \left(\frac{\sigma_{ii} + \sigma_{jj}}{2}\right)^6 .$$

b. Test the success of this suggestion by selecting two disparate molecular-size systems from Table 3.2-2 and compare predicted values of $D_{ij}(300\,\text{K}, 1\text{ atm})$ using both Eq. (3.4-11) and its "corrected" counterpart. What is your tentative conclusion, and what steps do you propose to strengthen it?

3.12 Porous solids are used to "support" catalyst microcrystals in fixed-bed catalytic reactors. To estimate the rates of reactant diffusion to the interior of such porous solids (and rates of product escape), it is necessary to estimate the *effective diffusion coefficient*, $D_{\text{A,eff}}$, for concentration diffusion of reactant "A" through the gas-mixture-filled "tortuous" pores. Consider a fixed-bed reactor for catalytically oxidizing CO to CO_2 (automotive catalytic "after-burner") under the following operating conditions (see, also, Exercise 6.6):

$T = 800\,\text{K}$	void fraction $\equiv \varepsilon = 0.45$
$p = 1.05$ atm	tortuosity $\equiv \tau_{\text{pore}} \cong 2$
CO dilute in "air"	mean pore diam. $\equiv \bar{d}_{\text{pore}} = 100\text{Å} = 10^{-2}\,\mu\text{m}$

a. What is $D_{\text{CO-air}}(800\,\text{K}, 1.05\text{ atm})$ under these conditions? (See Table 3.4-1.)

b. If the pore diameter was much larger than the gas mean-free-path, what would $D_{\text{CO-eff}}$ be in the *porous medium*?

c. For the actual mean pore diameter, what would be the value of $D_{\text{CO-eff}}$ (Knudsen) if the mean-free-path of gas in the pores was much larger than $\bar{d}_{\text{pore}}$?

d. Consider both pore wall and background gas "resistances" (and neglecting "surface-diffusion"), under actual operating conditions, what is your estimate of $(D_{\text{CO}})_{\text{eff}}$ within the porous catalyst support?

e. Under actual environmental conditions, at what mean pore diameter would the gas resistance and pore-wall diffusional resistance become equal?

3.13 Accurate values of the Fick diffusion coefficients of individual cations and anions at "infinite dilution" in the polar solvent water have been inferred *via* electrical conductivity measurements.

a. Complete the table below, applicable to water at $T = 25°C$.

Table 3.13E Ion Diffusion Coefficients in 25°C Water

Ion i	*Charge* z_i	$10^5 D_i(\text{cm}^2/\text{s})$	$\sigma_i \left(\begin{matrix}\textit{effective Stokes'}\\ \textit{diameter}\end{matrix}\right)$ Å
H^+	1	9.312	
Na^+	1	1.334	
K^+	1	1.957	
OH^-	−1	5.260	
$SO_4^=$	−2	1.065	

b. Consider (separately) an electrically neutral dilute aqueous solution of each of the salts Na_2SO_4 and K_2SO_4. What would be the effective (ambipolar) diffusion coefficient of each "salt" at 25°C? Exploit the fact that, for a binary electrolyte, individual ion migration is electrostatically coupled, with (Nernst–Planck):

$$(D_{\text{salt}})_{\text{eff}} = \frac{(p+q)D_+ D_-}{pD_+ + qD_-},$$

where p and q are, respectively, the *numbers* of charges on each cation and anion involved (that is, $p = z_i$ and $q = -z_i$; cf. Section 2.1.2).

c. Assuming Stokes–Einstein behavior for each of the ions involved, and neglecting the (weak) temperature dependence of the effective Stokes' diameters, predict the Na_2SO_4 and K_2SO_4-salt (ambipolar) diffusion coefficients for each salt in water at 5°C.

d. Based on the assumptions above, what is the apparent activation energy for diffusion, E_{diff} (kcal/mole) in such dilute aqueous solutions, and what is its relation to the activation energy for the solvent "fluidity" (reciprocal viscosity)? How do each of these values compare to the heat of vaporization of the solvent?

e. Calculate the expected diffusion coefficient of KOH in 25°C water. Why is the result so much larger than that of K_2SO_4?

Note: The Newtonian viscosities of pure water at 25°C and 5°C are 0.4987 cP and 1.5188 cP, respectively.

3.14 As stated in Section 3.3.4, a chemically reacting mixture in LTCE exhibits an effective thermal conductivity which can be written in the additive form:

$$k_{\text{LTCE}} = (k_{\text{mix}})_{\text{CF}} + (k_{\text{chem}})_{\text{LTCE}},$$

where $(k_{\text{mix}})_{\text{CF}}$ is the prevailing-mixture thermal conductivity if the composition did not change ("chemically frozen") and $(k_{\text{chem}})_{\text{LTCE}}$ is the so-called "chemical contribution," associated with the terms

$$\sum_{i=1}^{N} \mathbf{j}_i'' h_i$$

in Eq. (3.3-3). Consider a thermodynamically ideal binary-gas mixture comprised of the molecular (or atomic) species "A," and its "dimer" A_2, in LTCE at constant pressure, that is,

$$A_2(g) \leftrightarrows 2A(g) \begin{cases} \Delta H\{T\} \equiv 2H_1\{T\} - H_2\{T\} \\ K_p\{T\} = p_1^2/p_2 = p y_1^2/(1 - y_1) \end{cases}$$

a. Assuming that Fick diffusion completely dominates Soret (thermal) diffusion, derive an expression for the chemical contribution $(k_{\text{chem}})_{\text{LTCE}}$ to the effective thermal conductivity, k_{LTCE} (cf., also, Eq. (5.8-2)) as a function of the local temperature T and pressure level p (note that even though this is an ideal gas mixture, the dissociation–association equilibrium introduces a *pressure* dependence to the effective thermal conductivity). In evaluating the rate of change of composition with respect to temperature recall that: $d \ln K_p/d(1/T) = -\Delta H/R$ (Van't Hoff).
b. Derive a relation between the LTCE "thermal diffusivity" associated with $(k_{\text{chem}})_{\text{LTCE}}$, that is,

$$[k_{\text{chem}}/(\rho c_{p,\text{chem}})]_{\text{LTCE}}$$

and the Fick binary diffusivity, $D_{12}\{T; p\}$. Here, $(c_{p,\text{chem}})_{\text{LTCE}}$ is the chemical contribution, $(c_{p,\text{mix}})_{\text{LTCE}} - (c_{p,\text{mix}})_{\text{CF}}$, to the effective heat capacity per unit mass of mixture, $[(\partial h/\partial T)_p]_{\text{LTCE}}$.
c. Using the numerical data in Table 3.14E for the important particular case $H_2 \rightleftarrows 2H$, evaluate $(k_{\text{chem}})_{\text{LTCE}}$ at $T = 3000$ K, $p = 1$ atm, and compare your result to $k_2(3000\text{ K})$ and $k_1\{3000\text{ K}\}$.
d. Evaluate k_{LTCE} for the prevailing LTCE binary "hydrogen" gas mixture at 3000 K, 1 atm.

Ans. (Part a):

$$(k_{\text{chem}})_{\text{LTCE}} = \frac{D_{12}p}{T} \cdot \left(\frac{\Delta H}{RT}\right)^2 \cdot \left[\frac{1}{2}\omega_1(1 - \omega_1)\right],$$

where

$$\omega_1\{T; p\} = \left\{\frac{[K_p\{T\}/4p]}{1 + [K_p\{T\}/4p]}\right\}^{1/2}.$$

Note that this result will also apply to a system in which "A" is itself polyatomic, i.e., $N_2O_4 \rightleftarrows 2NO_2$. For applications of these results, see Exercises 5.19 and 6.13.

Table 3.14E Thermodynamic[a] and Estimated Transport[b] Data: $H_2 \rightleftarrows 2H$ System

Quantity	*Symbol*	*Value*	*Units*
Dissociation energy	$\Delta H(3000\,\text{K})$	109.84	kcal/g-mole
Equilibrium "constant"	$K_p(3000\,\text{K})$	2.477×10^{-2}	atm
Fick binary diffusion coefficient	$D_{12}(3000\,\text{K}, 1\,\text{atm})$	115.5	cm^2/s

Species	*i*	M_i	σ_{ii}	ε_{ii}/k_B	$C_{pi}(3000\,\text{K})$
H	1	1.008	2.708	37.0	4.968
H_2	2	2.016	2.827	59.7	8.859
Units		g/g-mole	Å	K	cal/g-mole K

[a] JANAF Thermochemical Data.
[b] Table 3.2–2; Chapter 3; and Svehla, R.A., NASA Report R-132 (1962).

3.15 Examine the success of the Lennard–Jones potential parameter estimate: $\varepsilon/k_B \approx 0.77\ T_c$ by selecting any ten substances from Table 3.2-2 and then constructing a plot of ε/k_B (Table 3.2-2) *vs.* T_c for that substance. Determine the slope of the best straight line through your "data," and compare this (dimensionless) slope to the value 0.77 recommended in Section 3.2.5. Use your correlation to estimate ε/k_B for an important substance not already in Table 3.2-2. Using the rough estimate: $\sigma \approx 2.44\ (T_c/p_c)^{1/3}$, also include the likely value for the Lennard–Jones *size parameter* (Å) of that additional substance. What further information would you now need to estimate the *thermal conductivity* of an ideal vapor comprised of this substance (cf. Eq. 3.3-6)?

REFERENCES

Boudart, M., *Kinetics of Chemical Processes.* Englewood Cliffs, NJ: Prentice-Hall (1968).

Froment, G.F., and K.B. Bischoff, *Chemical Reactor Analysis and Design.* New York: J. Wiley (1979).

Glassman, I., *Combustion.* New York: Academic Press (1977).

Hirschfelder, J. O., *J. Chem. Phys. 26*, 282–285 (1957).

Rohsenow, W.M., and H.Y. Choi, *Heat, Mass, and Momentum Transfer.* Englewood Cliffs, NJ: Prentice-Hall (1961).

Rosner, D.E., and J. Fernandez de la Mora, *ASME Trans.–J. Engrg. for Power* **104**, pp. 885–894 (1982).

Rosner, D.E., *J. PhysicoChem. Hydrodynamics* (London: Pergamon Press) **1**, pp. 159–185 (1980).
Smith, J.M., *Chemical Engineering Kinetics* (Third Ed.). New York: McGraw-Hill (1981).
Stewart, W.E., and R. Prober, *Ind. Engrg. Fundam.* **3**, 224 (1964).
Svehla, R.A., *NASA TR R-132* (1962).
Tambour, Y., and B. Gal-Or, *Phys. Fluids* **19**, No. 2, 219 (1976).
Toor, H.L., *AIChE J.* **10**, 448 (1964).
Wilhelm, R.H., *Pure Appl. Chem.* **5**, pp. 403–421 (1962).

BIBLIOGRAPHY: CONSTITUTIVE LAWS

Elementary

Cowling, T.G., *Molecules in Motion.* New York: Harper (TB 516) (1960).
Feynman, R.P., R.B. Leighton, and M. Sands, "Diffusion," Chapter 43 in *The Feynman Lectures on Physics*, Vol. 1. Reading, MA: Addison-Wesley (1963); pp. 43–1 to 43–10.
Girifalco, L.A., *Atomic Migration in Crystals.* New York: Blaisdell (1964).
Lavenda, B.H., "Brownian Motion," *Scientific American* **252**, No. 2, 70–85 (1985).

Intermediate

Bird, R.B., W. Stewart, and E. Lightfoot, *Transport Phenomena.* New York: J. Wiley (1960).
Cussler, E.L., *Diffusion; Mass Transfer in Fluid Systems.* Cambridge: Cambridge University Press (1984).
Denbigh, K.G., *The Thermodynamics of the Steady State.* Methuen: Monograph (1958).
Jost, W., *Diffusion in Solids, Liquids, and Gases.* New York: Academic Press (1960).
Launder, B., and D.B. Spalding, *Mathematical Models of Turbulence.* New York: Academic Press (1972).
Lightfoot, E.N., and E.L. Cussler, *Diffusion in Liquids. ChE Progress Symp. Series* **61** (58) (1965).
Marrero, T.R., and E.A. Mason, "Gaseous Diffusion Coefficients," *J. Phys. Chem. Ref. Data* **1**, No. 1, 3–118 (1972).
Turkdogan, E.T., *Physical Chemistry of High-Temperature Technology.* New York: Academic Press (1980).
Reid, R.C., and T.K. Sherwood, *The Properties of Gases and Liquids* (Third edition). New York: McGraw-Hill (1976).
Satterfield, C.N., *Diffusion in Heterogeneous Catalysis.* Cambridge, MA: MIT Press (1970).
Shewmon, P.G., *Diffusion in Solids.* New York: McGraw-Hill (1963).
Westenberg, A.A., "A Critical Survey of the Major Methods for Measuring and Calculating Dilute Gas Transport Properties," in Vol. 3, *Adv. Heat Transfer.* New York: Academic Press (1966).

Advanced

Chapman, S., and T.G. Cowling, *Mathematical Theory of Nonuniform Gases.* Cambridge, England: Cambridge Univ. Press (1960).

Clarke, J.F., and M. McChesney, *The Dynamics of Real Gases*. London: Butterworths (1964).

Cussler, E.L., *Multicomponent Diffusion*. Amsterdam: Elsevier (1976).

deGroot, S.R., and P. Mazur, *Nonequilibrium Thermodynamics*. Amsterdam: Elsevier (1960).

Dixon-Lewis, G., "Computer Modeling of Combustion Reactions in Flowing Systems with Transport," in *Combustion Chemistry*, ed. W.C. Gardiner, Jr. New York: Springer-Verlag (1984), pp. 21–125.

Ferziger, J.H., and H.G. Kaper, *Mathematical Theory of Transport Processes in Gases*. Amsterdam: North-Holland (1972).

Fitts, D.D., *Nonequilibrium Thermodynamics*. New York: McGraw-Hill (1962).

Hirschfelder, J.O., C.F. Curtiss, and R.B. Bird, *Molecular Theory of Gases and Liquids*. New York: J. Wiley (1954).

Hirschfelder, J.O., R.B. Bird, C.F. Curtiss, and E.L. Spotz, "The Transport Properties of Gases and Gaseous Mixtures," Section D in *Thermodynamics and Physics of Matter* (Vol. 1, High-Speed Aerodynamics and Jet Propulsion). Princeton, NJ: Princeton Univ. Press (1955); pp. 339–418.

Krishna, R., and R. Taylor, "Multicomponent Mass Transfer—Theory and Applications," in Chapter 7 *Handbook of Heat and Mass-Transfer Operations* (N.P. Cheremisinoff, ed.). Gulf Publishing Corp. (1985).

Slattery, J.C.: *Momentum Energy and Mass Transfer in Continua*. New York: McGraw-Hill (1972).

Woods, L.C., *The Thermodynamics of Fluid Systems*, Vol. 2 Oxford Engineering Series. Oxford: Clarendon Press (Paperback) (1985).

4

Momentum Transport Mechanisms, Rates, and Coefficients

4.1 RELEVANCE OF FLUID DYNAMICS AND THE CLASSIFICATION OF FLUID FLOW SYSTEMS

4.1.1 Role of Fluid Mechanics in the Analysis/Design of Chemical Reactors, Separators, etc.

At first glance, especially when one considers very idealized configurations, it appears that fluid dynamics governs only such overall characteristics as pressure drop across the device and, hence, fluid pumping costs. However, a closer look always reveals that the performance of the device *as a chemical reactor*, *separator*, etc., is intimately connected with:

a. reagent contacting patterns (e.g., where fuel meets oxidizer within the device; extent of recirculation or backmixing, etc.), and
b. local turbulence levels, which influence species mass and energy transport rates, and even time-averaged homogeneous reaction rates.

Thus, the performance of a reactor, separator, etc. is inevitably sensitive to the laws governing *momentum* transfer discussed in subsequent sections.

4.1.2 Criteria for Quiescence

In most engineering devices, *convection* of momentum, mass, and energy accompanies *diffusion*; indeed, convection usually dominates diffusion in the streamwise direction. This convection may be the result of an *imposed*- or "*forced*"-flow (e.g., that produced by a moving piston, or an upstream or downstream fan, or motion of the entire device (e.g., aircraft) through otherwise "still" fluid) or the result of so-called "buoyancy forces"[1] when chemical reaction- or transport-induced local density differences are subjected to a body force field. In the latter case, the convection is termed "natural" convection or "free" convection. Criteria for the negligibility of convection compared to diffusion are discussed in Sections 5.4.2 and 6.4.2. Such criteria are important, say, to experimentalists interested in determining transport coefficients (such as k, $D_{i-\text{mix}}$) in truly "stagnant" fluids. However, except in fixed solids or very viscous fluids, convection is actually difficult to avoid, and, not surprisingly, ubiquitous.

4.1.3 Further Classification of Continuum Fluid Flows

In addition to the categories discussed in Sections 2.7 and 4.1.2 above, it is also useful to distinguish between:

4.1.3.1 Internal vs. External Flows

Confined flows, such as those in ducts (of any cross section) are termed "internal," as opposed to flows such as that over an airplane, or past a falling droplet, which are called "external." In some cases, this division is not sharp—e.g., the flow over an airfoil mounted in a wind-tunnel test section; however, the distinction is useful because quantities that are known in one class of problem (e.g., pressures or total flow rates, etc.) may have to be found in the other class of problem, and *vice versa*.

4.1.3.2 Constant-Property vs. Variable-Property Flows

Chemically reacting multicomponent flows usually exhibit large spatial and temporal variations not only in the fields of primary interest (velocity, temperature, and species composition), but also in such fluid thermophysical properties as total density, transport coefficients (μ, k, D_i), chemical rate constants, etc. In simple cases, e.g., low-Mach-number gas or liquid flows subjected to modest overall temperature and concentration nonuniformities, these latter properties can often be assumed constant, greatly simplifying the problem of determining the primary field variables.

4.1.3.3 Single Phase vs. Multiphase Flows

When the region of interest strictly contains only a single phase (e.g., gas mixture or liquid solution), or there is a distinct (e.g., dispersed) phase whose motion is "tightly coupled" to that of the host phase (see Section 6.6.1), then a single-phase continuum formulation is appropriate for making performance predictions. Commonly,

[1] The "buoyancy force" per unit volume is not an additional force of nature—it is simply the combination $-\mathbf{grad}\, p + \rho\mathbf{g}$ which is locally nonzero in variable density situations.

however, explicit transport laws have to be written for each participating phase in a *multiphase fluid* flow (e.g., droplets in gas, bubbles in liquid), as briefly outlined in Section 2.6.3.

4.1.4 Interactive Role of Experiment and Theory

The geometric complexity of most engineering devices ensures the need for experimental measurements of flow patterns, local pressures, turbulence levels, etc. However, the cost of such experimentation can be minimized, if not eliminated, by the ingenious use of the above-mentioned conservation and constitutive laws. This can be the result of:

a. substituting accessible measurements for relatively difficult ones leading to the quantity of interest (QOI);
b. reducing the number of "independent" flow measurements that have to be made in any full-scale configuration;
c. using these principles to design and interpret relevant small-scale model experiments (Chapter 7), often without simultaneous chemical reaction (e.g., "cold-flow" experiments);
d. making analytical predictions or, more often, numerical predictions of the effects of varying parameters (flow rates, shapes, etc.) away from those cases for which direct experimental data are available.

An understanding of the principles of momentum, energy, and mass transport will allow these techniques to be pushed to their limits, and define what these limits are, thereby minimizing costly "surprises." In some cases, optimizations (e.g., selecting the best configuration and operating conditions from among many possible choices) may be possible based on "computer models," previously validated by selected comparisons with experiment.

4.1.5 Overall vs. Local Momentum and Mass Balances

As an example of point (a) above, suppose you were interested in *total drag force* on an object of prescribed shape in a uniform fluid stream. One experimental approach would be to carefully measure the *distribution* of shear stress and normal pressure on the object and vectorially sum all of the resulting forces to calculate the resultant force in the direction of the approach flow. A simpler approach might be to directly measure this net force by an appropriate model mounting and load cell. Still a simpler approach might be to survey the momentum flow rates and pressures in the wake far behind the object and exploit the macroscopic momentum balance (Eq. (2.2-3)) in the approach-stream direction.

But the overall force may not be sufficient to answer other important questions, or suggest clever ways to reduce this drag. For this purpose, knowledge of the spatial distribution of surface stresses, as well as the nature of the fluid flow in the vicinity of the object may be essential. Consider, for example, the flame "holding" device sketched in Figure 1.2-4, the performance of which is intimately connected to the behavior of the recirculating wake flow ("near-wake"). In such cases, overall macroscopic CV balance equations are *not* sufficient, and we must invoke conservation equations *at the local level*—i.e., the PDEs of Chapter 2 governing mass momentum and energy conservation locally. In the remainder of this chapter we illustrate the use of these principles for instructive prototypical situations involving mainly momentum *convection* (e.g., inviscid duct flow of a gas (Section 4.3)), as well as situations involving momentum *diffusion* (due to intrinsic and/or "eddy" viscosity) as well as convection (Sections 4.5 and 4.6).

4.2 MECHANISMS OF MOMENTUM TRANSPORT, THEIR ASSOCIATED TRANSPORT PROPERTIES, AND ANALOGIES TO ENERGY AND MASS TRANSPORT

The two *mechanisms* of momentum *transport* with which we deal are convection and diffusion.

4.2.1 Momentum Convection

We have already emphasized that the fluid velocity vector $\mathbf{v}$ can be fruitfully viewed as the *specific linear momentum* (per unit mass of fluid mixture); hence, the product[2] $\dot{\mathbf{m}}''\mathbf{v}$ or $\rho\mathbf{v}\mathbf{v}$ (where $\dot{\mathbf{m}}'' = \rho\mathbf{v}$ is the local convective *mass flux vector*) completely defines the local flux of linear momentum by the mechanism of *convection* ("bodily" transport). Thus, for a uniform flow of velocity U in, say, the x-direction, the x-momentum convective flow rate per unit area is ρU^2—that is, the product of the mass flux ρU and the linear (x-)momentum U per unit mass of mixture.

4.2.2 Momentum Diffusion

Comparing the linear momentum conservation PDE to those for species mass, energy, and entropy reveals that $-\mathbf{\Pi}$ (where $\mathbf{\Pi}$ is the contact stress operator) can be viewed as the linear momentum flux by diffusion, with $-\mathbf{T}$ being that part of the linear momentum diffusion associated with the fluid deformation rate (i.e., not present in a quiescent fluid, or one that is undergoing a "solid body" motion (uniform translation, solid body rotation, etc.)).

In turbulent flow (Section 2.6.2) the time-average effect of momentum *convection* is to produce an *apparent* (augmented) *momentum diffusion* (cf. the so-called Reynolds' stress, Eq. (2.6-9)).

---

[2] In any coordinate system, this product has nine components, which transform in a very particular way from one system to another. Hence, this object is called the momentum flux *tensor*.

The science of fluid dynamics can be viewed as comprising the consequences of one or both of these momentum transport mechanisms, subject to the necessary constraints of momentum and mass *conservation.*

4.2.3 Real and Effective Fluid Viscosities

Associated with momentum transport by diffusion is the phenomenological coefficient *viscosity* μ (see Section 3.2) appearing in the Stokes' extra-stress law Eq. (3.2-4). For laminar flows μ is a function of local fluid-state variables (Section 3.2.5), whereas the effective viscosity $\mu + \mu_t$ in a turbulent flow is itself dependent on the time-mean fluid deformation rate—usually making even a Newtonian fluid appear to be non-Newtonian.

4.2.4 Analogies to Energy and Mass Transport, and Their Uses

Since:

a. all of the conservation (balance) statements are of the same *form*—involving the net outflow rate by convection and the net inflow rate by diffusion, and
b. momentum, energy, and species mass are each transported by both convection and diffusion,

it should not be surprising that the *consequences* of these underlying laws are often remarkably similar, allowing quantitative information on one type of transfer to be used to predict another. These "analogies" or "co-relations" will be indicated and exploited in Chapters 5, 6, and 7. Again, an understanding of these quantitative analogies can be used to safely replace a difficult measurement by a simpler one, e.g., to allow a prediction of (say) shear stress distribution to be made from, say, accurate electrochemical measurements of large Sc convective *mass* transfer coefficients. However, most commonly these analogies are used quantitatively in the reverse direction; i.e., momentum transfer measurements are used to predict what mass or energy transfer would take place in the same system if (sufficiently small) concentration or temperature nonuniformities were introduced (see Section 5.7.2).

4.3 CONVECTIVE MOMENTUM TRANSPORT IN GLOBALLY INVISCID FLOW

Often the effects of momentum *diffusion* are confined to a negligibly small fraction of the volume of a device (reactor, separator, duct, etc.). For example, momentum *diffusion* may be important compared to *convection* only in:

a. the immediate vicinity of solid walls—where the fluid is brought locally "to rest" (Section 4.5.2);

b. within abrupt fluid dynamic transitions—such as shock waves or detonation waves, which appear to be discontinuities in an otherwise (piecewise) continuous inviscid[3] flow field (see Section 4.3.3).

In such cases, the laws of momentum transport simplify considerably, as will be discussed below for the steady-state cases of quasi-one-dimensional compressible fluid flow, and multi-dimensional incompressible fluid flow. In each case we must include consideration of energy transfer and the species equations to illustrate the effect of *coupling* between the various conservation equations.

4.3.1 Steady One-Dimensional Compressible Fluid Flow (cf., e.g., Crocco (1958))

It is instructive to first consider the implications of the basic mass, momentum, and energy conservation laws for the predominantly one-dimensional steady flow of a gas mixture in a duct of (slowly) varying area $A\{\xi\}$ and perimeter $P\{\xi\}$, where ξ is the length coordinate measured along the duct axis.

It is simplest to start with the conservation laws for an arbitrary but Eulerian macroscopic control volume and apply them to the control volume shown in Figure 4.3-1, *viz.*, one of streamwise *length* $\Delta\xi$ with inlet and outlet areas of $A\{\xi\}$ and $A\{\xi + \Delta\xi\}$, respectively. We will then divide each conservation statement term-by-term by the volume $\Delta V = A\{\xi\}\Delta\xi + \ldots$ and pass to the limit $\Delta\xi \to 0$ to obtain ordinary differential equations.

4.3.1.1 Total Mass

If we drop the transient (accumulation) term in Eq. (2.1-2) and evaluate the net convective outflow term, we obtain:

$$\lim_{\Delta\xi \to 0} \frac{1}{A\,\Delta\xi} \cdot [(\rho u A)|_{\xi+\Delta\xi} - (\rho u A)|_{\xi}] = 0 \qquad (4.3\text{-}1)$$

or

$$\frac{1}{A}\frac{d}{d\xi}(\rho u A) = 0, \qquad (4.3\text{-}2)$$

which implies:

$$\rho u A = \text{constant} = \dot{m}. \qquad (4.3\text{-}3)$$

---

[3] Frequently, because the diffusivities μ/ρ, $k/(\rho c_p)$, $D_{i-\text{mix}}$ or μ_t/ρ, $k_t/(\rho c_p)$, $D_{i,t}$ are of comparable magnitude, the same localization of the effects of diffusion is true for *energy* and species *mass* transport as well.

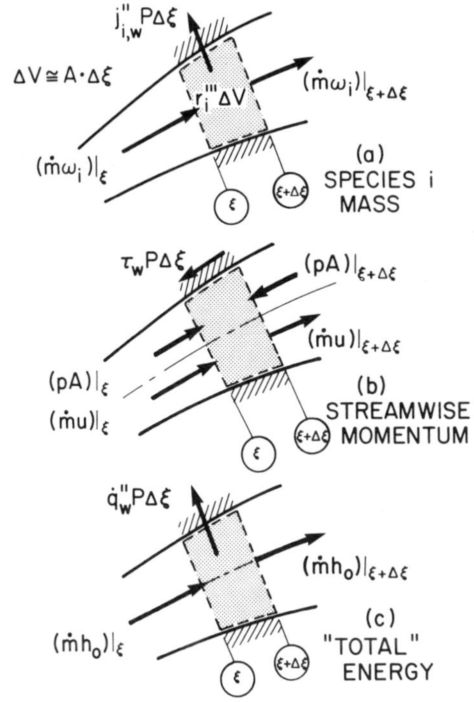

Figure 4.3-1 Application of conservation principles to a semi-differential (incremental) Eulerian control volume; (a) chemical species mass; (b) streamwise linear momentum; (c) energy.

Thus, the density, velocity, and area may each be varying in the streamwise direction, but in steady flow their product (the total mass flow rate $\dot{m}$) must be the same everywhere.

4.3.1.2 *Species Mass*[4] ($i = 1, 2, \ldots, N$)

If we drop the accumulation rate term of Eq. (2.1-7a) and evaluate the remaining terms we obtain:

$$\lim_{\Delta\xi\to 0} \frac{1}{A\,\Delta\xi}\{(\dot{m}\omega_i)|_{\xi+\Delta\xi} - (\dot{m}\omega_i)|_{\xi} = -j''_{i,w}P(\xi)\,\Delta\xi + \dot{r}'''_i A(\xi)\,\Delta\xi\}, \qquad (4.3\text{-}4)$$

where we have neglected streamwise diffusion compared to the (perimeter-averaged)

[4] While we will shortly focus on the *momentum* transfer implications, we include the chemical species mass-balance here since changes in mixture composition will influence the thermodynamic properties of the local mixture through the appropriate equations of state.

transverse diffusion flux $j''_{i,w}$. Utilizing (4.3-3), Eq. (4.3-4) immediately leads to

$$\rho u \frac{d\omega_i}{d\xi} = -j''_{i,w} \cdot \frac{P\{\xi\}}{A\{\xi\}} + \dot{r}'''_i, \tag{4.3-5}$$

only $N - 1$ of which are independent of Eq. (4.3-2).

4.3.1.3 Streamwise Momentum (of Mixture)

If we drop the accumulation rate term of Eq. (2.2-3) and assume that:

1. streamwise momentum convection dominates streamwise momentum diffusion;
2. each species is acted upon by the same streamwise body force g_ξ per unit mass (e.g., gravity); and
3. the shear force exerted by the duct *wall* on the fluid is nearly in the upstream $(-\xi)$ direction,

then

$$\lim_{\Delta\xi \to 0} \frac{1}{A\,\Delta\xi} \{(\dot{m}u)|_{\xi+\Delta\xi} - (\dot{m}u)|_\xi$$
$$= (pA)|_\xi - (pA)|_{\xi+\Delta\xi} + p_{\xi+\Delta\xi}\Delta A - \tau_w P\{\xi\}\Delta\xi + \rho g_\xi A\{\xi\}\Delta\xi\}. \tag{4.3-6}$$

Carrying out the indicated operations[5] gives the single ODE:

$$\rho u \frac{du}{d\xi} = -\frac{dp}{d\xi} - \tau_w \frac{P}{A} + \rho g_\xi. \tag{4.3-7}$$

4.3.1.4 Energy

If we drop the accumulation rate term of Eq. (2.5-5) and assume that:

1. streamwise energy *convection* dominates streamwise energy diffusion,
2. the fluid does not do work at the duct surface in overcoming the wall friction because the streamwise velocity is brought to zero there, and
3. each species is acted upon by the same streamwise body force g_ξ per unit mass (e.g., gravity), having a gravitational potential ϕ,

[5] In fact, u(avg) can be *defined* as the actual streamwise momentum flow rate divided by the actual mass flow rate. An analogous remark can be made for the cross-sectional-averaged enthalpy appearing in Eq. (4.3-9) and entropy appearing in Eq. (4.3-12).

then we can write

$$\lim_{\Delta\xi \to 0} \frac{1}{A\,\Delta\xi} \cdot \{(\dot{m}(h_0 + \phi))|_{\xi+\Delta\xi} - (\dot{m}(h_0 + \phi))|_{\xi} = -\dot{q}''_{\mathrm{w}} P\{\xi\}\,\Delta\xi + \dot{q}'''A\,\Delta\xi\}, \tag{4.3-8}$$

so that:

$$\rho u \frac{d}{d\xi}(h_0 + \phi) = -\dot{q}''_{\mathrm{w}} \frac{P\{\xi\}}{A\{\xi\}} + \dot{q}''', \tag{4.3-9}$$

where

$$h_0 \equiv e + \frac{p}{\rho} + \frac{u^2}{2} \tag{4.3-10}$$

is the so-called local "stagnation" (or "total") enthalpy.

4.3.1.5 Entropy

If we drop the accumulation term of Eq. (2.4-2) and additionally assume that:

a. entropy production due to streamwise momentum diffusion, energy diffusion, and species *i* mass diffusion are negligible in the bulk (core) of the fluid, and
b. the gas mixture is thermodynamically ideal,

then we can write:

$$\lim_{\Delta\xi \to 0} \frac{1}{A\,\Delta\xi} \{(\dot{m}s)|_{\xi+\Delta\xi} - (\dot{m}s)|_{\xi} = -j''_{s,\mathrm{w}} P\{\xi\}\,\Delta\xi + \dot{s}'''A\,\Delta\xi\}. \tag{4.3-11}$$

Thus, the streamwise change in the mixture-specific entropy is governed by the ODE:

$$\rho u \frac{ds}{d\xi} = -j''_{s,\mathrm{w}} \cdot \frac{P\{\xi\}}{A\{\xi\}} + \dot{s}'''. \tag{4.3-12}$$

In the absence of volumetric energy addition ($\dot{q}'''$)—e.g., *via* radiation absorption—the wall entropy-diffusion flux is given by Eq. (2.5-26):

$$-j''_{s,\mathrm{w}} = \frac{-(\dot{q}''_{\mathrm{w}} - \sum_{i=1}^{N} j''_{i,\mathrm{w}} h_i)}{T_{\mathrm{w}}} + \sum_{i=1}^{N} (-j''_{i,\mathrm{w}}) \left(s_i + \frac{R}{M_i} \ln \frac{1}{y_i}\right)_{\mathrm{w}}, \tag{4.3-13}$$

and the volumetric entropy production rate is due to finite-rate *homogeneous* chemical reactions, that is,[6]

$$\dot{s}''' = -\frac{1}{T}\sum_{i=1}^{N} \dot{r}_i''' \cdot \left[f_i - \frac{RT}{M_i} \cdot \ln\frac{1}{y_i} \right]. \tag{4.3-14}$$

Equations (4.3-2, -5, -7, -9, and -12) will form the basis of the following discussion of steady *nonreacting*[7] gas flow in ducts of

a. variable area, but without friction and heat addition;
b. constant area with heat addition; and
c. constant area with friction (momentum "withdrawal").

4.3.1.6 Steady, Frictionless Flow of a Nonreacting Gas Mixture in a Duct of Variable Area ("Nozzle") without Heat Addition

In this case, the conservation equations take the following simple forms:

$$\dot{m} = \text{constant} \qquad \text{(total mass)}, \tag{4.3-15}$$

$$\omega_i = \text{constant} \qquad \text{(species } i \text{ mass)} \; (i = 1, 2, \ldots, N-1), \tag{4.3-16}$$

$$\rho u \frac{du}{d\xi} = -\frac{dp}{d\xi} \qquad \text{(linear momentum)}, \tag{4.3-17}$$

$$h_0 = \text{constant} \qquad \text{(energy)}, \tag{4.3-18}$$

$$s = \text{constant} \qquad \text{(entropy)}. \tag{4.3-19}$$

Thus, for such flows the mass flow-rate, mixture composition, and the total enthalpy and entropy all remain constant, and velocity changes are related to pressure changes *via* Eq. (4.3-17) (Euler equation). It is convenient to imagine that the flow is initiated from a large area "reservoir," where the velocity is necessarily very small. There:

$$\omega_i = \omega_i \,\text{(reservoir)}, \tag{4.3-20}$$

$$h_0 = h_0(\text{reservoir}) = h(\text{reservoir}) = h\{T_0\}, \tag{4.3-21}$$

$$s = s(\text{reservoir}) \equiv s_0. \tag{4.3-22}$$

Alternatively, these qualities can be regarded as the *hypothetical* values of ω_i, h, and s obtained if the prevailing mixture were isentropically decelerated to rest—i.e., attaining their so-called "local stagnation" values. Moreover, if we suppress $d\xi$,

---

[6] Here f_i is the specific Gibbs free energy $h - Ts$ for pure substance i at the same p, T, and y_i is the species i *mole* fraction.

[7] Thus, in what follows we neglect volume or duct wall surface reaction. However, we note that these same equations are the basis of the analysis of so-called "plug-flow" chemical reactors (see, also, Section 6.7.1).

Eqs. (4.3-15–19) are equivalent to the following interrelations between differentials:

$$d(\rho u A) = 0 \qquad \text{(mass)}, \tag{4.3-23}$$

$$\rho u\, du = -dp \qquad \text{(momentum)}, \tag{4.3-24}$$

$$c_p\, dT + u\, du = 0 \qquad \text{(energy)}, \tag{4.3-25}$$

$$c_p\, dT - \frac{dp}{\rho} = 0 \qquad \text{(entropy)},^{8} \tag{4.3-26}$$

where, for a perfect gas mixture of molecular weight M:

$$\rho = \frac{pM}{RT} \qquad \text{(state)} \tag{4.3-27}$$

and[9]

$$c_p = \frac{\gamma}{\gamma - 1} \cdot \frac{R}{M} \qquad \text{(corollary of EOS)}. \tag{4.3-28}$$

Note that Eq. (4.3-23) relates thermodynamic state, velocity, and local area changes, Eq. (4.3-24) relates velocity and state variables, as does Eq. (4.3-25), whereas the entropy condition Eq. (4.3-26) provides an interrelation only between state variables. This relationship is especially simple if the heat capacities (hence γ) are themselves constant,[10] in which case:

$$\frac{p}{p_0} = \left(\frac{T}{T_0}\right)^{\gamma/(\gamma-1)}. \tag{4.3-29}$$

---

[8] See the Gibbs relation (Eq. (2.5-24b)) for the case $ds = 0$, $d\omega_i = 0$.

[9] Here $\gamma\{T\} \equiv c_p\{T\}/c_v\{T\}$ is the *specific heat ratio* of the ideal gas mixture.

[10] Such a gas is sometimes called "calorically perfect." It is noteworthy that, since the *enthalpy* of a perfect gas mixture is given by the mass-fraction weighted sum of its constituent specific enthalpies, $\sum \omega_i h_i\{T\}$, the effective local *heat capacity*, $(\partial h/\partial T)_p$, can change dramatically due to changes in *chemical composition* (as, e.g., for a combustion gas mixture in LTCE). In such cases Eq. (4.3-28) remains valid for each constituent, but in general:

$$(c_p)_{\text{mix,eff}} \neq \sum_{i=1}^{N} \omega_i \left(\frac{\gamma_i}{\gamma_i - 1}\right) \frac{R}{M_i},$$

and similarly:

$$\gamma \neq \left(\sum_{i=1}^{N} y_i \left(\frac{\gamma_i}{\gamma_i - 1}\right)\right) \Bigg/ \left(\sum_{i=1}^{N} y_i \left(\frac{1}{\gamma_i - 1}\right)\right).$$

Only for the limiting case of a gas mixture of *constant composition* do these inequalities become equalities. (For this reason the RHS of the first inequality is sometimes called the "chemically frozen" heat capacity of a prevailing gas mixture.)

At this point we introduce the local speed of sound ("acoustic" speed) a, in the prevailing gas mixture, shown in Section 4.3.2, to satisfy:

$$a^2 = \left(\frac{\partial p}{\partial \rho}\right)_{s=\text{const}}. \quad (4.3\text{-}30)$$

Accordingly, in this isentropic flow, locally:

$$dp = a^2\, d\rho, \quad (4.3\text{-}31)$$

so that changes in temperature (Eq. 4.3-26), pressure, and density are all of the same sign. Combining this relation with the momentum conservation (Euler) constraint then gives the following interesting relationship:

$$\frac{dG}{du} = \rho[1 - (\text{Ma})^2], \quad (4.3\text{-}32)$$

where G is the *mass flux*:

$$G \equiv \rho u, \quad (4.3\text{-}33)$$

and Ma the local *Mach number*:

$$\text{Ma} \equiv \frac{u}{a}. \quad (4.3\text{-}34)$$

We see from Eq. (4.3-32) that G increases with increasing u for *subsonic* flow, *decreases* with u for *supersonic* flow, and has its maximum value G_* where Ma = 1 (where $u = a = a_*$). It follows that for these reservoir conditions the maximum mass flow that can be passed through the duct is:

$$\dot{m}_{\text{max}} = G_* A_{\text{min}} = \rho_* a_* A_{\text{min}}, \quad (4.3\text{-}35)$$

where A_{min} is the minimum (or "throat") area, which can also be written A_*. But since h_0 = constant,

$$c_p(T_0 - T_*) = \frac{u_*^2}{2} = \frac{a_*^2}{2}. \quad (4.3\text{-}36)$$

Moreover, for a perfect gas, Eq. (4.3-30) gives

$$a_*^2 = \frac{\gamma R T_*}{M}. \quad (4.3\text{-}37)$$

Therefore one finds:

$$\frac{T_*}{T_0} = \frac{2}{\gamma + 1}, \tag{4.3-38}$$

and from Eq. (4.3-29),

$$\frac{p_*}{p_0} = \left(\frac{2}{\gamma + 1}\right)^{\gamma/(\gamma - 1)}. \tag{4.3-39}$$

Inserting those results in Eq. (4.3-35), one obtains:

$$\dot{m}_{max} = \left(\frac{p_0 M}{R T_0}\right)\left(\frac{\gamma R T_0}{M}\right)^{1/2} \cdot \left(\frac{2}{\gamma + 1}\right)^{(\gamma + 1)/[2(\gamma - 1)]} \cdot A_{min}. \tag{4.3-40}$$

This is a valuable result for the mass flow achieved through the duct when sonic conditions have been achieved at the "throat." This occurs for all downstream ("back") pressures below the critical value[11] given by Eq. (4.3-39). Equation (4.3-40) provides the basis for the use of so-called "critical orifice flowmeters," which have the useful features that:

a. provided the downstream pressure is *less* than p_*, the mass flow rate is given by Eq. (4.3-40), irrespective of the actual value of the *downstream* pressure;
b. the mass flow rate is linearly proportional to the *upstream* (stagnation) pressure.

It is clear from the preceding equations that for γ = constant isentropic gas flow, all local state properties can be uniquely related to their corresponding "stagnation" values and the *local Mach number* Ma. For future reference, these relations are given in the reciprocal form:[12]

$$\frac{T_0}{T} = 1 + \frac{\gamma - 1}{2}(\text{Ma})^2, \tag{4.3-41}$$

$$\frac{p_0}{p} = \left[1 + \frac{\gamma - 1}{2}(\text{Ma})^2\right]^{\gamma/(\gamma - 1)} \tag{4.3-42}$$

$$\frac{\rho_0}{\rho} = \frac{p_0}{p} \cdot \frac{T}{T_0} = \left[1 + \frac{\gamma - 1}{2}(\text{Ma})^2\right]^{1/(\gamma - 1)}, \tag{4.3-43}$$

from which we can calculate any other function of local thermodynamic state.

[11] 0.5283 p_0 for $\gamma = 1.4$ (e.g. air near 300 K), and 0.5457 p_0 for $\gamma = 1.3$ (high temperature air), etc.
[12] Note that, when Ma = 1, we recover Eqs. (4.3-38 and 39).

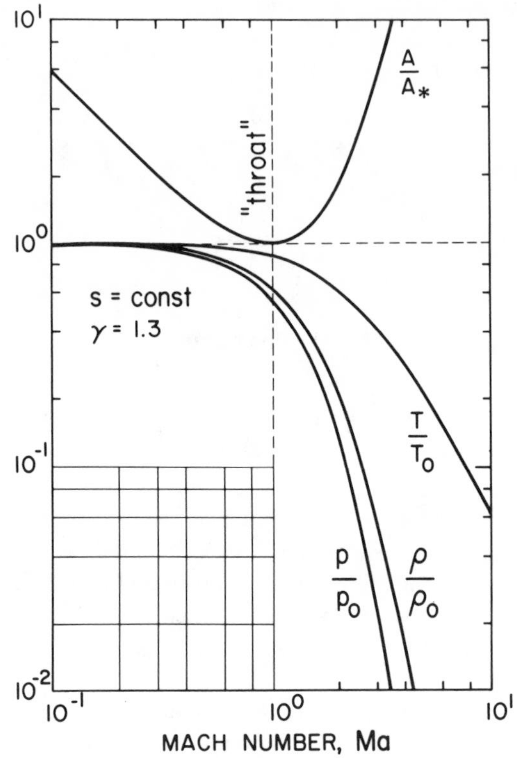

Figure 4.3-2 Steady one-dimensional isentropic flow of a perfect gas with $\gamma = 1.3$ (adapted from Shapiro (1953)).

Moreover, since $\dot{m} = \dot{m}_* =$ constant, we find that the Mach number at any area must satisfy:

$$\frac{A}{A_*} = \frac{G_*}{G} = \frac{\rho_*}{\rho}\frac{u_*}{u} = \frac{1}{\mathrm{Ma}} \cdot \left\{\frac{2}{\gamma + 1}\left[1 + \frac{\gamma - 1}{2}(\mathrm{Ma})^2\right]\right\}^{(\gamma + 1)/[2(\gamma - 1)]} . \quad (4.3\text{-}44)$$

Inspection of the A/A_* vs. Ma relation (Figure 4.3-2) shows that to achieve *supersonic* velocities in the duct, the area of the duct downstream of the throat must *increase.*[13] Such converging–diverging nozzles, called "DeLaval" nozzles, are widely used in stream turbines, gas turbines, and rocket engines. This remarkable property (that the velocity continues to increase despite area *increases*) is a direct consequence of the accompanying drop in gas density (cf. Eq. (4.3-23)).

---

[13] Conversely, in *supersonic* flow, to decrease the velocity and *increase* the pressure one must *constrict* the area. Thus, in contrast to the familiar subsonic case, supersonic "diffusers" are *converging* ducts.

4.3.1.7 Steady Compressible Flow of a Nonreacting Gas Mixture in a Constant-Area Duct with Heat Addition, but without Friction[14]

In this particular case, our basic equations become:

$$G = \text{const} \qquad \text{(mass)},$$

$$\rho u \frac{du}{d\xi} = -\frac{dp}{d\xi} \qquad \text{(momentum)}, \tag{4.3-45}$$

$$\rho u \frac{dh_0}{d\xi} = \dot{q}''' \qquad \text{(energy)}, \tag{4.3-46}$$

where we have again neglected the body force (gravity) terms. Thus:

$$G\,du = -dp$$

and

$$Gc_p dT_0 = \frac{(\dot{q}'''\,A\,d\xi)}{A} \equiv G\,dq. \tag{4.3-47}$$

According to the mass and momentum conservation equations above:

$$p + Gu = \text{const} = p_0 \tag{4.3-48}$$

or, since $u = G/\rho = Gv$,

$$p + G^2 v = p_0. \tag{4.3-49}$$

This locus on the p-v (or the corresponding T-s) thermodynamic state variable plane is called a Rayleigh "line" (locus), so that the extent of heat addition (or removal) will determine which "couple" p, v one reaches at the prevailing G-value. If we define the local "stagnation (or total) temperature":

$$T_0 \equiv T + \frac{u^2}{2c_p}, \tag{4.3-50}$$

then for a constant c_p-gas mixture between any two duct sections 1 and 2, the change in T_0 is dictated by the heat addition per unit mass:

$$q_{12} = c_p(T_{0,②} - T_{0,①}). \tag{4.3-51}$$

[14] In a gas, it is impossible to have *wall* heat transfer without friction (analogy between momentum and energy transfer). Therefore, we consider here the presence of a volumetric energy additive term $\dot{q}'''$. In fuel-lean systems, the heat effects of homogeneous combustion can often be approximated in this way.

Moreover, since

$$Tds = c_p dT - v\,dp \qquad \text{(Gibbs)}, \tag{4.3-52}$$

one can also calculate the corresponding entropy change, as well as the Mach number, Ma, at each point along the Rayleigh locus. In this way, it is found that heat addition to a *subsonic* duct flow causes the pressure to drop and both the temperature and the Mach number to *increase* up to the point at which Ma = 1 is attained ("thermal choking").[15] Further heat addition at this same mass flux is not possible, since it would correspond to a specific entropy decrease, violating Clausius' inequality (Eq. (2.4-8)). Figure 4.3-3 shows the interrelation between state variables, stagnation temperature, and Mach number along the Rayleigh locus, for the case $\gamma = 1.3$.

4.3.1.8 Steady Compressible Flow of a Nonreacting Gas Mixture in a Constant-Area Duct with Friction but no Heat Addition

In this case, our basic conservation equations become:

$$G = \text{constant} \qquad \text{(mass)}, \tag{4.3-53}$$

$$G\frac{du}{d\xi} = -\frac{dp}{d\xi} - \tau_w \frac{P}{A} \qquad \text{(momentum)}, \tag{4.3-54}$$

$$h_0 = \text{constant} \qquad \text{(energy)}, \tag{4.3-55}$$

and the entropy changes can be computed from the corresponding changes in state variables (e.g., T, p). Combining Eqs. (4.3-53–55) we obtain the equation of the so-called *Fanno locus, viz.*:

$$c_p(T_0 - T) = \frac{1}{2}\left(\frac{G}{\rho}\right)^2 = \frac{1}{2}(Gv)^2, \tag{4.3-56}$$

which, with the help of the perfect gas equation of state (and Gibbs equation) can also be represented on the thermodynamic p-v or T-s planes. One finds that, for a subsonic adiabatic flow in a duct of constant area, friction *increases* the Mach number, velocity, and entropy, but *decreases* the temperature. Friction cannot, however, cause the flow to accelerate into the supersonic branch of the Fanno locus for this same G-value, since this would correspond to *decreasing* exit gas entropy, again violating Clausius' inequality. The duct length required to achieve such "frictional choking" must be found by numerical integration of Eq. (4.3-54), using a

---

[15] In contrast, heat addition to a *supersonic* stream causes *deceleration* of the gases (e.g., combustion heat addition in a supersonic ramjet outlet), also driving the exit Mach number toward unity.

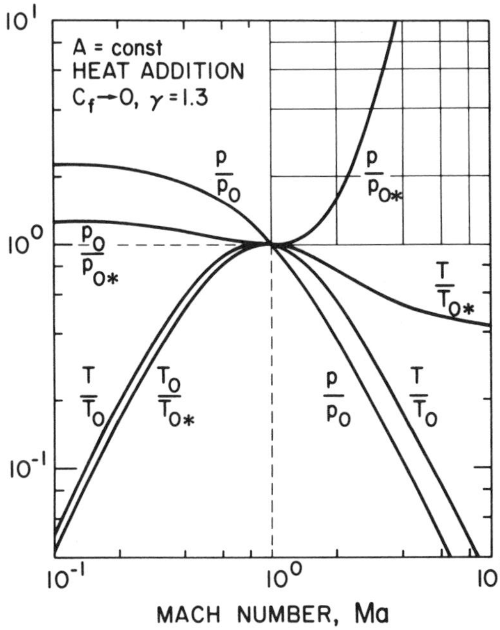

Figure 4.3-3 Steady one-dimensional flow of a perfect gas (with $\gamma = 1.3$) in a constant-area duct; frictionless flow with heat addition (adapted from Shapiro (1953)).

reasonable expression of the local "skin friction coefficient"[16] C_f, defined by:

$$C_f \equiv \frac{\tau_w}{\frac{1}{2}\rho u^2} \tag{4.3-57}$$

(cf. Section 4.5). Representative results for the (usually unrealistic) case of *constant*[17] C_f and $\gamma = 1.3$ are shown in Figure 4.3-4.

4.3.2 "Shock" Waves, Sound Waves, Detonation Waves, and "Deflagration" Waves

There are a number of important situations, not directly involving the presence of solid surfaces, in which diffusion can be neglected upstream and downstream of an abrupt fluid dynamic transformation, which then appears to be a discontinuity separating two adjacent continua. For simplicity, consider here steady-flow

[16] This wall momentum-transfer coefficient must be obtained from either experimental data, or *via* a detailed analysis of the PDEs governing tangential momentum transfer in the immediate vicinity of the wall—i.e., the wall "viscous boundary layer" (see Sections 4.4, 4.5, and 4.6).

[17] This would be unreasonable for laminar duct flow, and a crude approximation for turbulent duct flow (see Figure 4.5-3).

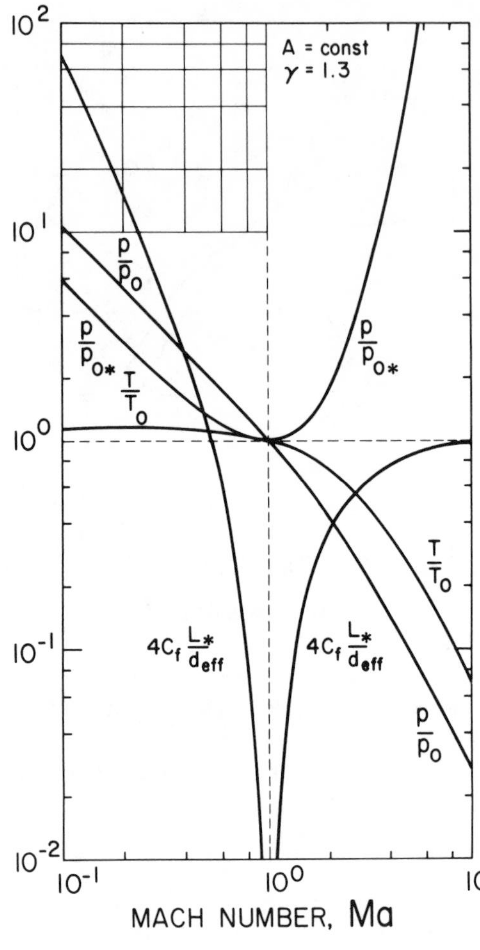

Figure 4.3-4 Steady one-dimensional flow of a perfect gas (with $\gamma = 1.3$) in a constant-area duct; adiabatic flow with friction (adapted from Shapiro (1953)).

situations in which the same equation of state is valid on both sides of the transformation[18]—e.g., that governing a mixture of perfect gases. We ask: *What constraints do the above-mentioned macroscopic* CV *conservation laws impose on the field variables on both sides of such discontinuities*?

Without loss in generality, we imagine that the discontinuity is (locally) planar, fixed in space, and "fed" by a gas stream with known velocity $u_{①}$ normal to the discontinuity[19] and known thermodynamic state properties (see Figure 4.3-5). The

[18] This, for example, rules out transformations in *liquid* or *solid* monopropellants in which high-temperature *gases* can be produced as the result of a detonation wave. Typically, such waves have speeds in the range 5 to 9 km/s!

[19] Oblique waves are governed by the same laws if expressed in terms of the *normal* components of the velocities. Moving discontinuities can be treated by adopting a coordinate system fixed relative to the discontinuity.

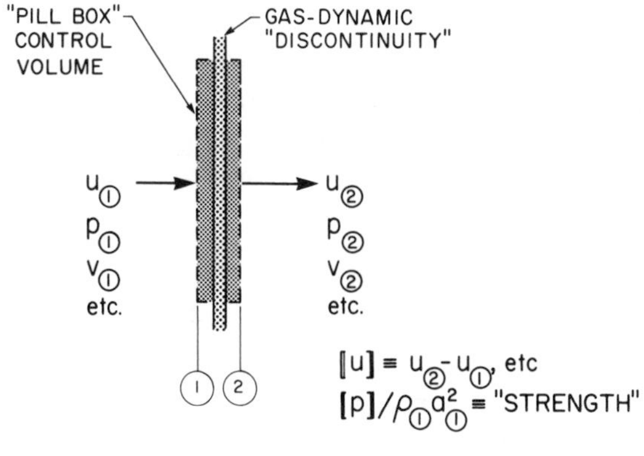

Figure 4.3-5 Control volume and station nomenclature for applying conservation principles across a gas-dynamic discontinuity separating two regions of flow in which diffusion processes can be neglected.

basic conservation laws (Chapter 2) are applied by considering a macroscopic control volume that is shrunk down to a "pillbox" of unit area straddling the wave (Figure 2.1-3). Since most of the relationships we will write involve the "jump" in field quantities between stations two (downstream) and one (upstream), we introduce the "jump operator" notation:

$$[\ \ ] \equiv (\ \ )_{②} - (\ \ )_{①}. \tag{4.3-58}$$

4.3.2.1 Shock Waves and Sound Waves

In these terms the conservation equations for a transformation *without* chemical reaction take the simple form:

$$[\rho u] = 0 \quad \text{(total mass)}, \tag{4.3-59a}$$

$$[\rho u \omega_i] = 0 \quad \text{(species mass)}, \tag{4.3-60a}$$

$$[\rho u u] = -[p] \quad \text{(momentum)}, \tag{4.3-61a}$$

$$[\rho u h_0] = 0 \quad \text{(energy)}, \tag{4.3-62a}$$

$$[\rho u s] \geqslant 0 \quad \text{(entropy)}. \tag{4.3-63a}$$

Introducing the mass flux $G \equiv \rho u$, these conditions can obviously be rewritten:

$$[G] = 0 \quad \text{(total mass)}, \tag{4.3-59b}$$

$$[\omega_i] = 0 \quad \text{(species mass)}, \tag{4.3-60b}$$

$$G[u] = -[p] \quad \text{(normal momentum)}, \tag{4.3-61b}$$

$$[h_0] = 0 \quad \text{(energy)}, \tag{4.3-62b}$$

$$[s] \geqslant 0 \quad \text{(entropy)}, \tag{4.3-63b}$$

showing that, in general, G, ω_i, and h_0 are *continuous* (undergo no jump) in crossing the discontinuity, but u, p, T, $v(\equiv 1/\rho)$, and s will experience jumps that must be compatible with these conservation equations, together with the relevant equations of state. Combining the total mass and normal momentum conditions provides the interesting interrelation:

$$u_{①}u_{②} = \frac{[p]}{[\rho]}. \tag{4.3-64}$$

We show below that when the discontinuity becomes a sufficiently weak compression wave, the positive entropy jump becomes negligible; hence both $u_{①}$ and $u_{②}$ approach the thermodynamic derivative:

$$a = \left(\frac{\partial p}{\partial \rho}\right)^{1/2}_{s=\text{const}}. \tag{4.3-65}$$

Evaluating this derivative for a perfect gas gives the well-known equation for the acoustic (sound) speed.[20]

$$a = \left(\frac{\partial p}{\partial \rho}\right)^{1/2}_{s} = \left(\frac{\gamma RT}{M}\right)^{1/2} \tag{4.3-66}$$

(utilized in Section 4.3.1, where we introduced the local *Mach number* $\text{Ma} \equiv u/a$).

Returning to the case of a transformation of arbitrary "strength" (e.g., measured, say, in terms of $[p]/(\rho_{①}a_{①}{}^2)$), note that Eqs. (4.3-59–63) also indicate that the final state must lie on the *intersection* of the Fanno and Rayleigh loci passing through point ① on the *T-s* diagram, and corresponding to the common mass flux G (Figure 4.3-6). This is because the Rayleigh locus links all states with the same $p + Gu$ irrespective of heat addition, and the Fanno locus links all states with the same stagnation enthalpy (or stagnation temperature, for the nonreactive cases), irrespective of viscous dissipation (cf. Sections 4.3.1.7, 4.3.1.8).

Another important state variable interrelation, due to Rankine [1876] and Hugoniot [1887], (RH), can be derived from Eqs. (4.3-62) and (4.3-64), *viz.*:

$$[h] = \frac{1}{2}[p]\cdot(v_{②} + v_{①}). \tag{4.3-67}$$

Since at constant composition the enthalpy is uniquely defined by the two state variables p and v, the RH relation defines a locus (called the "shock adiabat") on the p-v plane along which state point ② must lie (Figure 4.3-6). For a perfect gas this

[20] For example, if we insert the properties of air at 300 K (Table 8.1-1 gives $\gamma = 1.400$, $M = 28.964$), we obtain $a = 0.3472$ km/s.

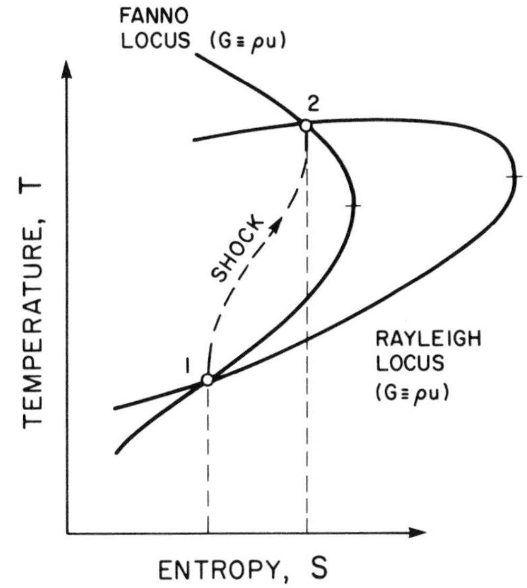

Figure 4.3-6 Fanno and Rayleigh loci for the same mass flux G, displayed on the T-s plane. The normal shock transition goes from the supersonic intersection to the subsonic intersection.

relation is:

$$\frac{p_{②}}{p_{①}} = \frac{\left(\frac{\gamma+1}{\gamma-1}\right) - \frac{v_{②}}{v_{①}}}{\left(\frac{\gamma+1}{\gamma-1}\right)\cdot\frac{v_{②}}{v_{①}} - 1}, \tag{4.3-68}$$

which, for sufficiently "strong" waves, differs appreciably from the isentropic relation $p_{②}/p_{①} = (v_{①}/v_{②})^{\gamma}$. Equation (4.3-64) also reveals that, on the p-v plane, the chord joining points ① and ② has the slope $-G^2$, which approaches $-\rho_{①}{}^2 a_{①}{}^2$ as point ② approaches point ①. This implies that for strong compression waves (called "shock waves") the *upstream* flow is *supersonic*, and the relative normal *downstream* flow is *subsonic*.

We now note that for a perfect gas one can show from the $s(p, T)$ relation that:

$$\exp\frac{[s]}{R/M} = \frac{p_{0,①}}{p_{0,②}} = \left(\frac{2\gamma}{\gamma+1}\cdot \mathrm{Ma}_{①}{}^2 - \frac{\gamma-1}{\gamma+1}\right)^{1/(\gamma-1)} \cdot \left[\frac{2}{(\gamma+1)\mathrm{Ma}_{①}{}^2} + \frac{\gamma-1}{\gamma+1}\right]^{\gamma/(\gamma-1)}, \tag{4.3-69}$$

which, combined with Clausius' inequality, rules out the possibility of rarefaction ($\mathrm{Ma}_{①} < 1$) shocks (see Figure 4.3-7).

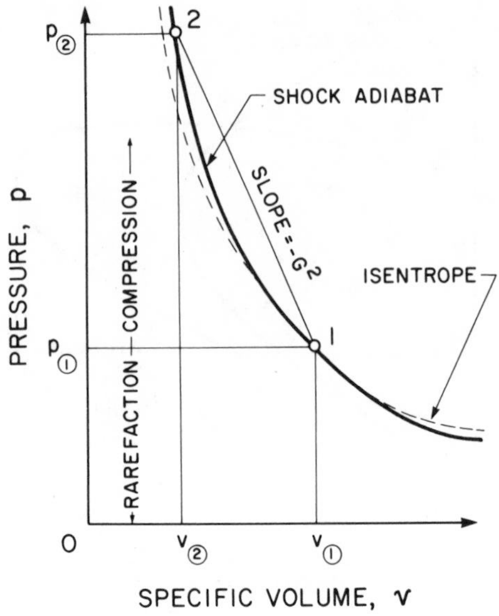

Figure 4.3-7 Rankine–Hugoniot "shock adiabat" on the *p*-*v* plane.

Finally, in terms of the upstream (normal) Mach number, we find (cf. Figure 4.3-8 for $\gamma = 1.3$)

$$\frac{p_{②}}{p_{①}} = 1 + \frac{2\gamma}{\gamma+1}(\mathrm{Ma}_{①}^2 - 1), \tag{4.3-70}$$

$$\frac{T_{②}}{T_{①}} = 2\left[1 + \frac{\gamma-1}{2}\mathrm{Ma}_{①}^2\right]\frac{2\gamma\mathrm{Ma}_{①}^2 - (\gamma-1)}{(\gamma+1)^2\mathrm{Ma}_{①}^2}, \tag{4.3-71}$$

and

$$\frac{\rho_{②}}{\rho_{①}} = \frac{(\gamma+1)\mathrm{Ma}_{①}^2}{(\gamma-1)\mathrm{Ma}_{①}^2 + 2}. \tag{4.3-72}$$

It is interesting to note that, whereas both $p_{②}/p_{①}$ and $T_{②}/T_{①} \to \infty$ when $\mathrm{Ma}_{①} \to \infty$ (hypersonic limit), the density ratio $\rho_{②}/\rho_{①}$ instead approaches the *finite* limit $(\gamma + 1)/(\gamma - 1)$. (See Figure 4.3-8.)[21] This feature plays an important role in approximate theories of "hypersonic" ($(\mathrm{Ma})^2 \gg 1$) flow (Hayes and Probstein (1966)).

[21] Note also that when $\mathrm{Ma}_{①} \to \infty$ (hypersonic limit), $\mathrm{Ma}_{②}$ approaches the *finite* limit $[(\gamma - 1)/2\gamma]^{1/2}$.

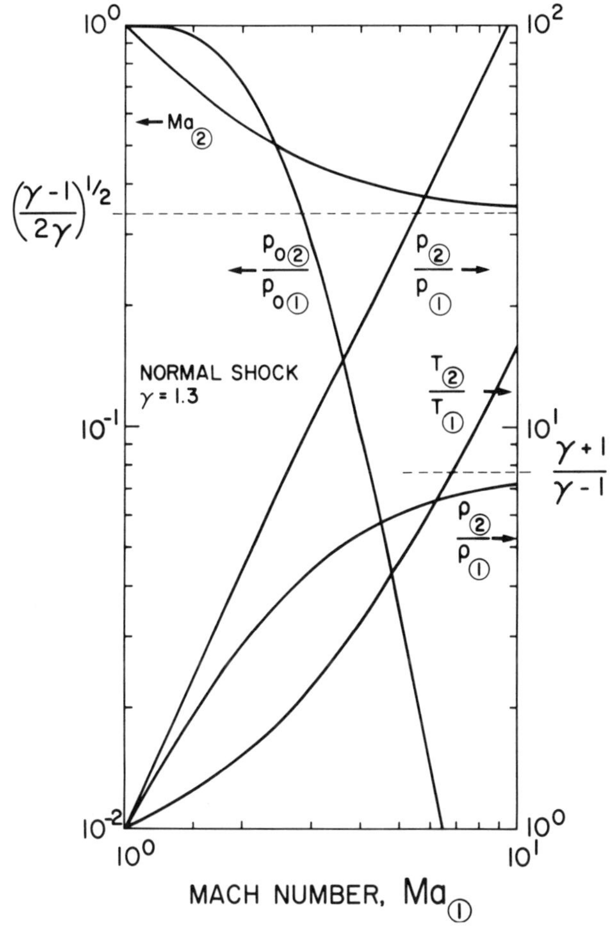

Figure 4.3-8 Normal shock property ratios as a function of upstream (normal) Mach number Ma (for $\gamma = 1.3$).

4.3.2.2 Detonation Waves and Deflagration Waves

Apart from the condition $[\omega_i] = 0$, Eqs. (4.3-59, -61, -62, -63) apply equally well to abrupt transitions *accompanied by chemical reactions*—an example being a shock-wave–reaction-zone combination, called a "detonation." However, in such a case it must be realized that the specific enthalpy function, h, must include *chemical* contributions, and with reaction there is a change in chemical composition in going from ① to ②. Alternatively, we may regard the reaction as merely adding an amount of heat q per unit mass to a perfect gas mixture of constant specific heat γ—a picture which adequately describes many fuel-lean/air mixtures. Then the generalized RH conditions become:

$$[G] = 0, \tag{4.3-73}$$

$$G[u] = -[p], \tag{4.3-74}$$

$$c_p[T_0] = q, \tag{4.3-75}$$

and

$$[s] \geqslant \frac{q}{T_{②}}. \tag{4.3-76}$$

The Rankine–Hugoniot Eq. (4.3-67) still applies, but now the *detonation adiabat* will be above the *shock adiabat* (cf. Eq. (4.3-68)) by an amount depending upon the heat release q (see Figure 4.3-8). Detonations are observed to propagate with an end state at or near the so-called Chapman–Jouguet point CJ (Figure 4.3-9), which is the singular point at which:

a. the combustion products have the minimum possible entropy compatible with Eqs. (4.3-73–76), and
b. the normal velocity, $u_{②}$, of the products (with respect to the detonation wave) is exactly *sonic* (that is, $Ma_{②} = 1$).

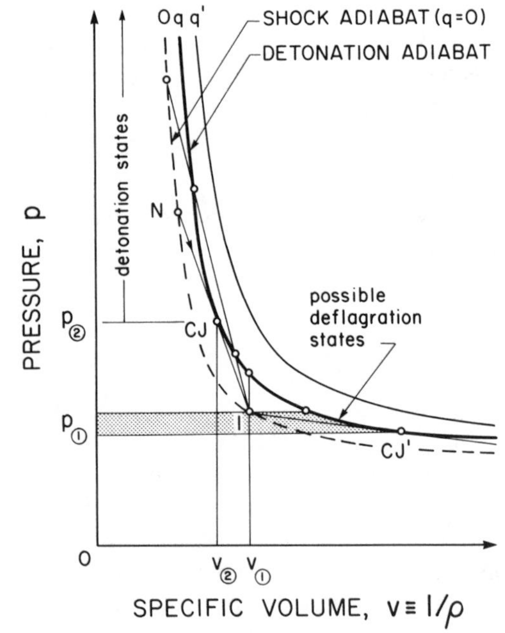

Figure 4.3-9 Rankine-Hugoniot "detonation adiabats" on the p-v plane.

Imposing the condition $\mathrm{Ma}_{②} = 1$ gives the following result for eligible discontinuities with heat release, but constant γ (Thompson (1972)):

$$\mathrm{Ma}_{①} = (1 + \mathscr{H})^{1/2} \pm \mathscr{H}^{1/2} \tag{4.3-77}$$

where

$$\mathscr{H} \equiv \frac{\gamma^2 - 1}{2\gamma} \cdot \frac{Mq}{RT_{①}}. \tag{4.3-78}$$

The + sign in Eq. (4.3-77) gives the upstream Mach number for a CJ-*detonation* (compression wave), whereas the − sign gives the upstream Mach number for a CJ-"deflagration." Whereas the former waves *are* observed, subsonic combustion waves ("deflagrations") are observed to propagate much more slowly than predicted by Eq. (4.3-77), owing to transport and kinetic limitations *within* the wave structure (see Section 7.2.2.2).

4.3.3 Remarks on Multidimensional Inviscid Steady Flow

If we neglect the effects of diffusion and nonequilibrium chemical reaction (i.e., consider them confined to boundary layers and shock waves), then the equations expressing local conservation of mass, momentum, and energy for steady-flows of a perfect gas simplify considerably, but nevertheless remain *partial* differential equations interrelating many dependent field variables (the three components of $\mathbf{v}$, ρ, T, and p). Predictions of the flow field past a body of prescribed shape require that these coupled PDEs be solved subject to conditions imposed on the pressure, temperature, and velocity fields "at infinity," and the velocity (normal component) along the effective surface[22] of the body.

For such flows a simpler procedure is available, based on the possibility of reducing the above-mentioned conservation equations to *one* (higher-order) PDE involving only *one* unknown—the so-called "velocity potential" $\phi_v(\mathbf{x})$, having the property:

$$\mathbf{v} = -\mathbf{grad}\, \phi_v. \tag{4.3-79}$$

It can be proved[23] that all inviscid compressible flows admit such a "potential,"

[22] Actually, the outer edge of the thin viscous boundary layer that exists along the true solid surface of the body (see Section 4.5).

[23] The condition of vanishing vorticity (Section 3.2.2) ensures the existence of a path-independent fluid velocity potential (cf., e.g., Milne–Thompson (1955)). It is interesting to note that, in some coupled two-phase flows (e.g., suspended particles in a "dusty" gas; cf. Chapter 8), when the interphase coupling force is proportional to $|\mathbf{v}_p - \mathbf{v}|$, *both* phase velocities $\mathbf{v}(\mathbf{x})$ and $\mathbf{v}_p(\mathbf{x})$ admit potentials (Robinson (1956), and Fernandez de la Mora and Rosner (1981)).

which in the present case would have a constant gradient far from the body, and satisfy $\partial\phi_v/\partial n = 0$ everywhere along the effective body surface of prescribed shape. This single scalar function $\phi_v(\mathbf{x})$ must satisfy the nonlinear second-order PDE (see, e.g., Milne–Thompson (1955))

$$a^2 \mathrm{div}(\mathbf{grad}\,\phi_v) = \mathbf{grad}\,\phi_v \cdot \mathbf{grad}\left(\frac{1}{2}(\mathbf{grad}\,\phi_v)^2\right), \tag{4.3-80}$$

where

$$a^2 = \frac{\gamma R T_0}{M} - \frac{\gamma - 1}{2}(\mathbf{grad}\,\phi_v)^2 \tag{4.3-81}$$

(the square of the *local* acoustic speed). An interesting and widely exploited property of this equation is the fact that, when $a^2 \to \infty$ (as in the case of an incompressible liquid), ϕ_v merely has to satisfy the well-known linear PDE (Laplace's Equation):

$$\mathrm{div}(\mathbf{grad}\,\phi_v) = 0. \tag{4.3-82}$$

Thus, many simple inviscid flows can be constructed using the classical *analytical* methods of "potential theory," e.g., the *superposition* of fundamental (source, sink) solutions or *numerical* methods based on Eq. (4.3-80). However, in general, the following two features of such problems must be dealt with:

a. The mathematical character of Eq. (4.3-80) changes discontinuously, depending upon whether:

$$(\mathbf{grad}\,\phi_v)^2 \overset{?}{\gtrless} a^2$$

that is, depending upon whether the *local* flows are supersonic or subsonic.

b. For supersonic or transonic conditions at upstream "infinity," abrupt gas-dynamic discontinuities (called "shock" waves, cf. Section 4.3.2) appear *within* the flow field—which then becomes only piecewise (zonally) continuous. These discontinuities provide a means by which energy can be dissipated even though the fluid is considered inviscid and nonheat-conducting. Not surprisingly, momentum and heat diffusion play an important role *within* these "discontinuities"—i.e., the interplay of *diffusion* and convection determines the "structure" of such apparent discontinuities.

It will be seen that, in very simple situations (heat or mass diffusion through a constant-property stagnant fluid or solid), the temperature field $T(\mathbf{x})$ or the concentration field(s) $\omega_i(\mathbf{x})$ will also satisfy Laplace's equation[24] (4.3-82) (see Chapters 5 and 6).

[24] Actually obtained first by L. Euler in the present (incompressible, inviscid fluid flow) context.

4.4 VELOCITY FIELDS AND CORRESPONDING SURFACE MOMENTUM-TRANSFER COEFFICIENTS

4.4.1 Relation between Local Velocity Fields, Wall Momentum-Transfer Rates, and Wall Coefficients

In the design of chemical reactors [including furnaces (Figure 1.1-2) and combustion chambers (Figures 1.1-1 and 1.2-8)], as well as separators and other equipment, we are usually interested in predicting the flow fields within them under operating conditions—and their dependence on such parameters as inlet momentum fluxes, gravitational (body) forces, and momentum diffusion to the containment-walls and internal bodies, for various possible geometries. Similarly, in the design of automobiles and, especially, aircraft we are interested in the flow fields about them and their relation to the momentum exchange between the vehicle and the surrounding fluid—which manifests itself as a net *force* ("drag" in the streamwise direction, "lift" in the direction perpendicular to the motion and resulting moments (torques)). Often this information can be obtained by numerically solving the relevant momentum- and mass-conservation equations, subject to (a) appropriate conditions imposed upstream and along the solid surfaces, and (b) a knowledge of the fluid properties (density, viscosity, etc.). More often, however, this information is obtained by a combination of laborious probing of the momentum density field, and pressure and force measurements, either on full-scale equipment, or using smaller "scale models" (see Chapter 7).

Figure 4.4-1 shows a small representative portion of either a fixed confining wall, or an immersed body, including the steady[25] velocity (specific linear momentum) profiles in its vicinity. Suppose the approach stream direction is X, distance *along the surface* is specified by the running coordinate x, and distance *normal* to the surface by n. We consider here the relationship between the local pressure $p\{x,0\}$, the velocity field $v_x\{x,n\}$, and the momentum exchanged locally between the fluid and surface. This *momentum exchange manifests itself as a force on the fluid and an equal and opposite force on the solid.*

If the solid surface is motionless, then its impermeability ensures the kinematic condition $v_n\{x,0\} = 0$, however, of even greater importance with respect to momentum exchange with the surface is the value of the local *tangential* (x-direction) velocity $v_x\{x,0\}$.

Whereas a nonzero "slip" velocity $v_x\{x,0\}$ would *not* violate any conservation law, it is *experimentally* found that when the continuum viewpoint is valid (gases at reasonable pressure levels, or liquids), the local tangential velocity of the fluid can be taken as equal to that of the solid—in the present case, zero. This "no-slip"

[25] Alternatively, in a transient flow we can interpret all quantities appearing in the following discussion as those prevailing at the instant in question.

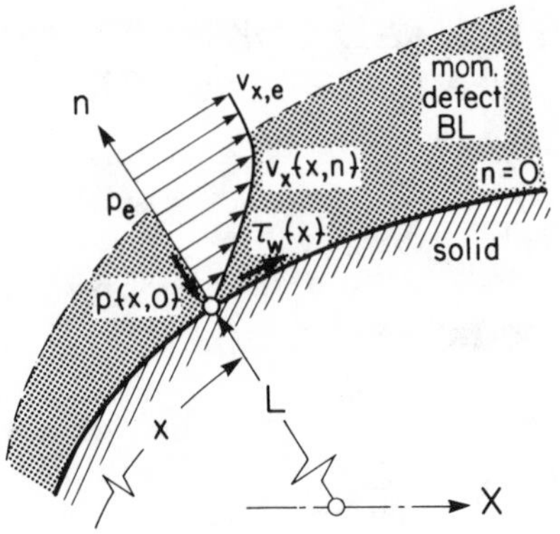

Figure 4.4-1 Momentum exchange between the moving fluid and a representative segment of a solid surface (confining wall or immersed body).

condition is central to the dynamics of real (viscous) fluids, and, in particular, the following arguments.

In a situation such as that sketched in Figure 4.4-1, the local momentum-exchange rate per unit surface area is dominated by the following two terms: the local *pressure* $p\{x, 0\}$, and the local wall *shear stress* τ_{nx} (evaluated at x and $n = 0$), hereafter written $\tau_w\{x\}$. Each of these forces (per unit area) will have a component in the X-direction, contributing to the total *drag* on the body (or a "drain" of X-momentum from the fluid), and a perpendicular component contributing to the "lift" on the body.

The relevant local fluid-deformation rate which determines the stress component $\tau_{nx}\{x, 0\}$ is $(\partial v_x/\partial n)_{n=0}$; hence, if the local fluid viscosity is μ_w, we can write:

$$\tau_w\{x\} = \mu_w\left(\frac{\partial v_x}{\partial n}\right)_{n=0}. \tag{4.4-1}$$

Equation (4.4-1) relates the local tangential stress to the local normal gradient of the tangential fluid velocity. Accordingly, if v_x measurements could be made near enough to the solid surface $n = 0$ to accurately determine this local gradient, and the prevailing fluid *viscosity* were known, in principle[26] this information would be sufficient to determine $\tau_w\{x\}$ and, hence, the components of this force (per unit area).

Suppose experimental values of the *dimensional* pressure and shear stress *distributions* were obtained, that is, $p\{x, 0\}$ and $\tau_w\{x\}$. These could, of course, be reported directly, for each set of experimental conditions—i.e., for each upstream fluid density ρ_∞, X-velocity U, mainstream turbulence intensity, etc. However, we show in Chapter 7 that such data are actually much more useful if reported in terms

[26] In practice, $\tau_w\{x\}$-values are rarely obtained in this way; more often they are obtained "indirectly" *via* tangential momentum surveys and the use of the linear-momentum balance equation [Eq. (2.2-3)].

of the following *dimensionless local momentum transfer coefficients*:

$$C_p\{x\} \equiv \frac{p\{x,0\} - p_\infty}{\frac{1}{2}\rho_\infty U^2} \tag{4.4-2}$$

and

$$C_f\{x\} \equiv \frac{\tau_w\{x\}}{\frac{1}{2}\rho_\infty U^2}. \tag{4.4-3}$$

The first of these is called the local "*pressure coefficient*,"[27] and the second is called the local "*skin-friction coefficient*." Conversely, in the absence of such measurements, there is considerable interest in predicting these local *dimensionless momentum transfer coefficients* everywhere along the specified solid surfaces.

In Section 4.5.2 we show that at sufficiently large convective momentum fluxes (i.e., at sufficiently large Reynolds' numbers), tangential-velocity profiles in the vicinity of the surface $n = 0$ have the qualitative character sketched in Figure 4.4-1; i.e., they approach an "asymptotic" value $v_{x,e}\{x\}$ ($\neq U$) in a normal distance small compared to the characteristic body dimension L or x itself. Moreover, the local pressure there, written $p_e\{x\}$, is approximately equal to $p\{x, 0\}$. In such cases an alternate skin-friction coefficient can be defined in terms of properties at the outer "edge" of this "tangential-momentum (defect) boundary layer"—*viz.*:

$$c_f\{x\} \equiv \frac{\tau_w\{x\}}{\frac{1}{2}\rho_e v_{x,e}^2\{x\}}. \tag{4.4-4}$$

This coefficient (hereafter written with a lower-case C) is of even greater theoretical interest than C_f, as will become clear in Section 5.5.2.

When the fluid is incompressible, $\rho_\infty = \rho_e$, there is a simple interrelation between these coefficients,[28] since a "Bernoulli" equation relates the difference between $v_{x,e}^2$ and U^2 to the difference between $p_e\{x\}$ and p_∞. Thus, in the absence of compressibility and gravitational body-force effects,

$$p_0 = p_\infty + \frac{1}{2}\rho U^2 = p_e\{x\} + \frac{1}{2}\rho v_{x,e}^2, \tag{4.4-5}$$

provided there are no appreciable fluid velocities in the direction perpendicular to the plane of Figure 4.4-1. This equation (which follows from the negligibility of viscous dissipation outside of the tangential-momentum defect "boundary layer")

---

[27] Not to be confused with the molar heat capacity at constant pressure!

[28] Using the results of Section 4.3.1.1, it is not difficult to generalize Eqs. (4.4-5 and -6) to the case of isentropic *compressible* flow, when γ = constant.

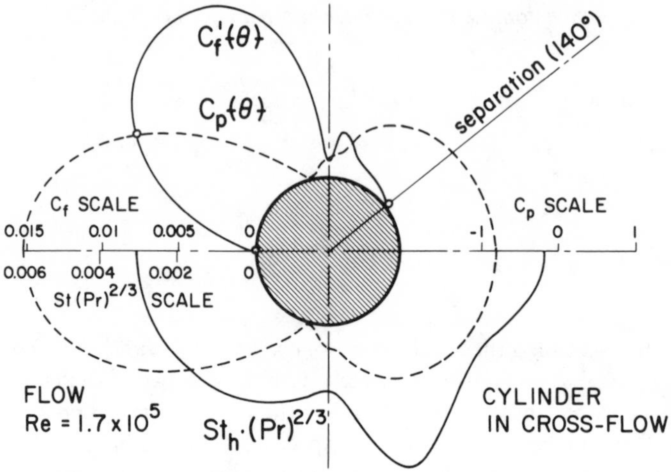

Figure 4.4-2 Experimentally determined angular dependence of the skin-friction—and pressure—coefficients around a circular cylinder in cross-flow at Re = 1.7×10^5.

implies that:

$$\frac{v_{x,e}\{x\}}{U} = (1 - C_p)^{1/2}. \tag{4.4-6}$$

Comparing the defining equations for C_f and c_f (Eqs. (4.4-3 and -4) we see that:

$$c_f\{x\} = \frac{C_f\{x\}}{1 - C_p\{x\}} \tag{4.4-7}$$

Figure 4.4-2 displays the results of pressure and shear stress distributions *measured* on a circular cylinder at a crossflow Reynolds' number of 1.7×10^5 and for $Ma_\infty^2 \ll 1$. Plotted, in polar form,[29] are the local pressure coefficients, C_p, and the local skin-friction coefficients, C_f.

It is noteworthy that:

a. The pressure coefficient C_p is unity at $\theta = 0$, but at this Reynolds' number it drops to negative values over most of the cylinder surface. Thus, beyond about $\pm 30°$ the local pressure drops *below* p_∞, and remains below p_∞ even near the rear "stagnation" region.
b. C_f (the skin-friction coefficient) drops to zero near $\pm 140°$; that is, the local skin friction τ_w vanishes there.[30]

---

[29] The angle θ is measured from the forward "stagnation" point, where $v_{x,e} \to 0$. Thus $\theta = x/(d_w/2)$, where d_w is the circular cylinder diameter.

[30] Since $v_{x,e} \to 0$ at $\theta = 0$ and $\theta = 180°$, it can be shown that C_f is also zero at the forward and rear stagnation points. Between $\theta = 140°$ and $180°$ there is *reverse* flow and τ_w actually changes sign.

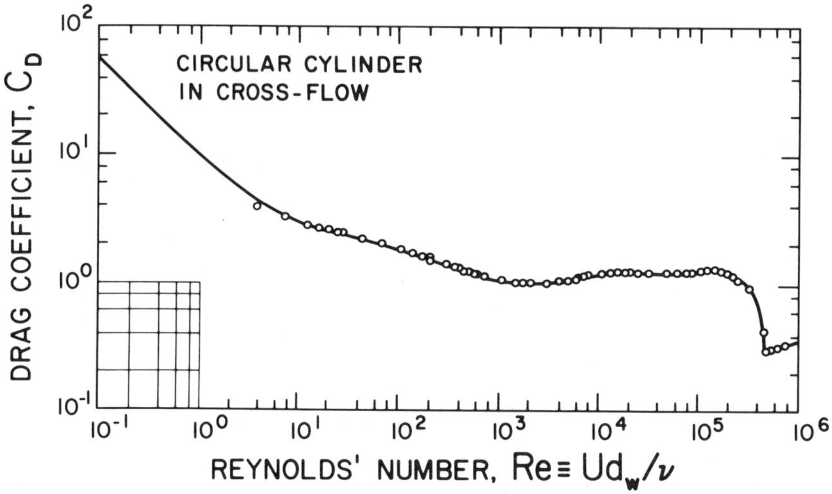

Figure 4.4-3 Experimental values for the overall drag coefficient (dimensionless total drag) for a cylinder (in cross-flow), over the Reynolds' number range $10^{-1} \leqslant \mathrm{Re} < 10^6$ (adapted from Schlichting (1979)).

Suppose we are interested in the *total* (net) *drag force* D' acting per unit length of cylinder.[31] It is customary to compare this force to a reference force:

$$D'_{\mathrm{ref}} \equiv \frac{1}{2}\rho_\infty U^2 A'_{\mathrm{proj}}, \tag{4.4-8}$$

where

$$(A'_{\mathrm{proj}})_{\mathrm{cylinder}} = d_{\mathrm{w}} \tag{4.4-9}$$

is the *projected* ("frontal") area of the cylinder (per unit length). The ratio D'/D'_{ref} defines the dimensionless *drag coefficient* of the cylinder (Figure 4.4-3), *viz.*:

$$(C_D)_{\mathrm{cylinder}} \equiv \frac{D'}{\frac{1}{2}\rho_\infty U^2 d_{\mathrm{w}}}, \tag{4.4-10}$$

which can obviously be calculated in terms of the above-mentioned $C_{\mathrm{p}}(\theta)$ and $C_{\mathrm{f}}(\theta)$ data.[32] If we project the pressure and shear forces in the direction of the approach flow and sum them up, we obtain:

$$D' = 2\left\{\int_0^\pi [p(\theta)\cos\theta + \tau_{\mathrm{w}}(\theta)\sin\theta]\frac{d_{\mathrm{w}}}{2}\,d\theta\right\}, \tag{4.4-11}$$

[31] Since the C_{p} and C_{f} distributions on this (nonrotating) cylinder exhibit $\pm\,\theta$ symmetry, there is clearly no *lift* force, or net moment (about the cylinder axis).

[32] Assuming we had access to $C_{\mathrm{f}}(\theta)$ data between 140° and 180°. Actually, this (thrust!) contribution is found to be negligible.

so that

$$(C_D)_{\text{cylinder}} = \int_0^{\pi} \left\{ C_p\{\theta\} \cos\theta + [C_f\{\theta\}] \sin\theta \right\} d\theta, \tag{4.4-12}$$

where the polar angle is now expressed in radians rather than degrees (π radians = 180°).

Note that the cos θ term arises from the (locally normal) pressure force, whereas the sin θ term arises from the (locally tangential) aerodynamic *shear* force. Thus, Eq. (4.4-12) can be split into its two separately calculable contributions, *viz*.:

$$(C_D)_{\text{cylinder}} = (C_D)_{\text{form}} + (C_D)_{\text{friction}}, \tag{4.4-13}$$

where the "*form*" *drag* is due to the *pressure* forces, and the *friction drag* due to the aerodynamic *shear* forces.[33]

If this were the only way to determine the total drag coefficient, it would not be known for many flow conditions (e.g., as a function of Reynolds' and Mach number) or body geometries. Fortunately, it is much simpler to directly determine the net force D acting per unit axial length over a range of flow conditions. One can then report the function $C_D\{\text{Re}\}$ for a cylinder in crossflow, with typical experimental results shown in Figure 4.4-3. Corresponding results for an isolated *sphere* are given in Figure 4.4-4.[34]

If the *total drag* were the only quantity of interest, such data would be sufficient; however, to gain an *understanding* of the various regimes ("creeping" flow for Re < 1, constancy of C_D for $10^3 \leqslant \text{Re} \leqslant 10^5$, the cause of the abrupt drop in C_D near Re = 2×10^5, etc.), the detailed *distributions* of $p\{x, 0\}$, $\tau_w\{x\}$, and $v_x\{x, n\}$ are of great interest. Moreover, it will be shown in Chapters 5 and 6 that the distribution of shear stress can be used to predict the corresponding *distribution* of diffusional *heat* or *mass* transfer around a solid body. Unfortunately, even at this date, *predicting* these pressure and shear-stress distributions is possible only over limited ranges of the Reynolds' number, and often (at high Reynolds' numbers) only over a portion of the cylinder.[35] For this purpose *asymptotic* theories (valid for Re ≪ 1 or Re ≫ 1) and numerical methods are widely exploited. In what follows (Section 4.5.2) we outline this approach for the case of greatest engineering interest, *viz*.: Re ≫ 1.[36]

---

[33] The word "form" here is not especially appropriate since the "form" of a body also determines its friction drag. In the present case, it can be shown that $(C_D)_{\text{form}} \gg (C_D)_{\text{friction}}$.

[34] Note that just beyond Re = 1.7×10^5 (cf. Figure 4.4-3), the cylinder-drag coefficient undergoes an abrupt drop, corresponding to abruptly larger pressures in the rear of the cylinder. This has been shown to be associated with turbulence in the momentum-defect boundary layer, delaying the point where $\tau_w = 0$. This corresponds to a smaller region of recirculation and greater pressure recovery.

[35] This is largely due to fluid dynamic instabilities which cause motion within the momentum-defect boundary layers to become *turbulent*. This greatly complicates the prediction of *time-average shear-stress distributions*.

[36] Note (Eq. (4.4-10)) that only when C_D = const is the drag force proportional to the *square* of the relative velocity. Figure 4.4-3 also shows that at sufficiently *low* Re-values, $C_D \sim \text{Re}^{-1}$, corresponding to a drag force *linear* in the relative velocity. In particular, for a solid *sphere* at Re < 1, $C_D \approx 24/\text{Re}$, a well-known viscous-flow drag result first obtained by G. Stokes.

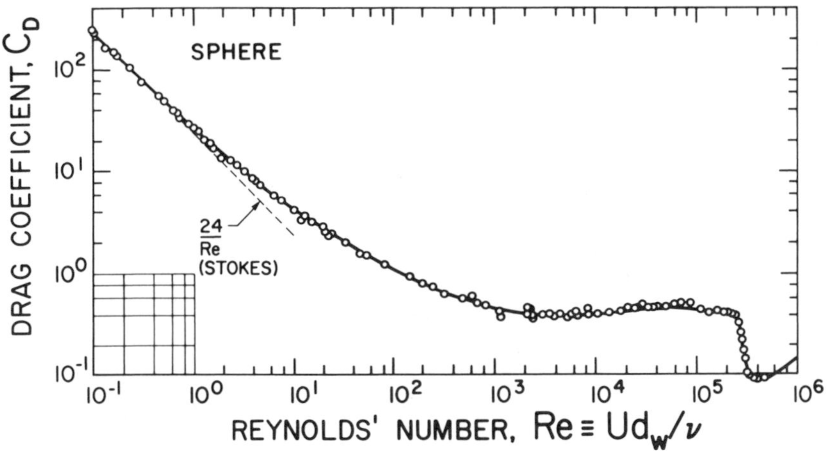

Figure 4.4-4 Experimental values for the overall drag coefficient (dimensionless total drag) for a sphere over the Reynolds' number range $10^{-1} \leqslant \text{Re} \leqslant 10^6$ (adapted from Schlichting (1979)).

We conclude this section with the remark that chemical reactions can alter the results discussed above. For example, if the cylinder is the flame "stabilizer" in a fuel vapor-air stream (cf. Figure 1.2-4), then heat release in the cylinder wake would significantly alter $p(x)$-values there, and, hence, modify, C_D at the same Reynolds' number. In principle, with sufficient "base-region" heat release one can, in fact, produce a net thrust! This is the basis of several schemes for *drag reduction or propulsion.*

4.4.2 Conservation Equations Governing the Velocity and Pressure Fields

Most flow structure problems in "pure" Newtonian fluid mechanics are discussed in terms of the so-called Navier–Stokes (*linear momentum conservation*) laws,[37] obtained by combining Eqs. (2.5-9), (3.2-1), and (3.2-4):

$$\rho\left[\frac{\partial \mathbf{v}}{\partial t} + \mathbf{v} \cdot \mathbf{grad}\,\mathbf{v}\right] = -\mathbf{grad}\,p + \text{div}\left[2\mu\,\mathbf{Def}\,\mathbf{v} - \frac{2}{3}\mu(\text{div}\,\mathbf{v})\,\mathbf{I}\right] + \rho\mathbf{g}, \quad (4.4\text{-}14)$$

[37] This is a nonlinear vector PDE, equivalent to three independent-scalar, second-order PDEs in any particular (inertial) coordinate system. Equation (4.4-14) includes "Stokes' postulate" (that the bulk viscosity κ can be neglected). Many authors call only the incompressible, constant μ form of this vector equation the N-S equations.

together with the *total mass conservation* ("continuity") equation:

$$\frac{\partial \rho}{\partial t} + \mathbf{v} \cdot \mathbf{grad}\, \rho = -\rho \operatorname{div} \mathbf{v}. \tag{4.4-15}$$

However, this provides only four PDEs for the five fields: $\mathbf{v}$ (three scalar fields), p, and ρ. As is also clear from our discussion of one-dimensional steady-flow gas dynamics (Section 4.3), it is, therefore, necessary to specify an equation of state, unless $\rho =$ constant ("incompressible" flow, for which Eq. (4.4-15) also simplifies to $\operatorname{div} \mathbf{v} = 0$). In general, one cannot escape simultaneous consideration of *energy* conservation and a "caloric" equation of state (e.g., h as a function of T and p for a pure substance) in addition to the usual equation of state (in the form $p\{\rho, T\}$ for a pure substance). Moreover, one must specify how the *dynamic viscosity* μ depends on local state variables[38] (see Section 3.2.5).

For turbulent flows Eqs. (4.4-14 and -15) are "time-averaged" (Section 2.6.2) with the result that the time-averaged fields of velocity, pressure, and density satisfy equations of the same *form*, but with the intrinsic viscosity μ replaced by an effective viscosity $\mu + \mu_t$, where μ_t accounts for the *Reynolds' stress* (Section 2.6.3). Clearly, in such cases, μ_t must also be related to the above-mentioned time-averaged fields to "close" the problem of computing them.

4.4.3 Typical Boundary Conditions

We have already encountered one set of "boundary" conditions (BCs) in the form of the Rankine–Hugoniot "jump" constraints of Section 4.3.2. There, we dealt with single-phase fluid dynamic (nonmaterial) discontinuities *through which there was linear momentum and mass transfer*. However, as was remarked, the "upstream" phase could equally well have been a chemically unstable solid material (such as TNT), in which case we would be dealing with a solid/gas interface through which there was momentum and mass transfer (as well as energy transfer). Moreover, in general we would have to include *diffusive* fluxes of mass, momentum energy, and entropy across such interfaces, as well as their convective counterparts.

In dealing with chemically reacting flow systems, this viewpoint concerning "boundaries" is actually more fruitful than the usual *ad hoc* viewpoint presented in treatises of classical fluid mechanics. This is because the boundaries of interest may indeed be the site of chemical reaction, phase change, etc., and may indeed be moving (e.g., receding) with respect to a fixed (Eulerian) observer. Moreover, the notion that useful necessary boundary conditions result from the systematic application of our battle-tested conservation laws to a "pillbox" control volume straddling the interface (Section 2.6.1) is a generally useful one (Rosner (1976)).

---

[38] The dependency also involves an invariant of **Def v** itself for nonlinear (non-Newtonian) fluids (see Section 3.5.1).

Relative to a moving interface and the normal component of the mass flux can be written G_n; hence the mass and momentum balances usually simplify to:

$$[G_n] = 0 \qquad \text{(cf. Eq. (2.6-4))}, \tag{4.4-16}$$

$$G_n[v_n] = -[p] + [\tau_{nn}] \qquad \text{(cf. Eq. (2.6-5))}. \tag{4.4-17}$$

If we here[39] use t to denote quantities in the *tangent* plane, then conservation of tangential linear momentum implies:

$$G_n[v_t] = [\tau_{nt}] \tag{4.4-18}$$

(equivalent to two independent equations (component directions in the tangent plane)). However, even these momentum conditions must be generalized if the interface exhibits an appreciable surface tension.[40]

It is significant that:

a. Conservation Condition (4.4-16) does *not* preclude a discontinuity in the normal component of velocity.
b. Conservation Condition (4.4-17) does *not* preclude a discontinuity in the pressure across the interface.
c. Conservation Condition (4.4-18) does *not* preclude a discontinuity in the tangential velocity (i.e., "slip") across the interface.

With this perspective one appreciates that the "boundary" conditions,

$$[v_n] = 0, \qquad [v_t] = 0, \qquad [p] = 0, \qquad [\tau_{nt}] = 0, \tag{4.4-19(a–d)}$$

found in treatises on classical fluid mechanics, are only "sometimes" true.[41]

Turning our attention briefly to *time* as the remaining independent variable in Eqs. (4.4-14 and -15), one might consider the specification of "initial conditions" (i.e., all independent field variables at $t = 0$) and hereafter abbreviated ICs, as a remaining (nonphysical) "boundary" condition. Thus, $t = 0$ could be considered to be:

a. the time of "start-up" of a chemical reactor, separator, etc.;
b. the time when the piston is at "top dead center" in Figure 1.1-1;
c. the present, if we are interested in predicting future weather and/or climate.

---

[39] As distinct from the alternate meaning of t in connection with descriptions of *turbulence*.

[40] In that case (a) the normal component of the linear-momentum balance must be augmented by a term involving $\sigma(R_1^{-1} + R_2^{-1})$, where σ is the local *surface tension*, and R_1, R_2 the principal radii of curvature; and (b) the tangential component of the linear momentum balance must be augmented by a term involving the gradient of σ in the tangent plane (Levich (1969)).

[41] A similar remark can be made concerning commonly found statements about $[T]$, $[\dot{q}''_n]$, $[j''_{i,n}]$, etc. (see Chapters 5, 6).

In this respect it is significant that the governing conservation equations are *first-order* in time,[42] and, hence, invariant with respect to a shift in the origin (zero point) chosen for time. Moreover, in accord with the principle of "local" action (in time)—sometimes called "determinacy"—the "future" has no way of influencing the "present!"[43] This is *not* the case in the *spatial* domain, only in the time domain.

4.4.4 Outline of Solution Methods

At the outset it should be pointed out that the coupled PDEs governing mass and linear-momentum conservation at the local level (i.e., Eqs. (4.4-14 and -15)), combined with appropriate BCs and ICs need not always be "solved" to extract valuable information from them. This is demonstrated in Chapter 7 in the context of establishing similarity conditions essential to the experimental use of "scale models" in engineering investigations.

Moreover, only relatively simple fluid-dynamic problems must be solved in order to interpret the readings of *instruments* (e.g., a capillary flowmeter (Section 4.5.1), an "impact" pressure probe for inferring flow velocities in a locally supersonic flow, etc.). These instruments can then be used to define a very complex flow (either at full scale or on a scale model), in which case we, in effect, leave to "Mother Nature" the otherwise arduous task of finding those fields that simultaneously satisfy all conservation constraints!

However, for reasons of economy (intellectual, as well as financial), *mathematical* solutions of the governing PDEs (+BCs and ICs) *are* of considerable interest, and increasingly possible with the help of large-memory, ultra-high-speed electronic digital[44] computers. These have become widely available in the last 30 years, only about ten percent of the (*ca.* 300 year) period during which the science of chemically reacting fluids was developed. Thus, especially from an *engineering* viewpoint, there is good reason to be optimistic about the future implications of "computational fluid mechanics."

For problems that are sufficiently simple, both geometrically and thermodynamically, or that, while complex, can be "modularized" into simpler subregions, the methods of classical mathematical analysis (i.e., those developed for the solution of PDEs and ODEs) can be invoked, sometimes leading to explicit results in terms of well-known special functions (e.g., sines, cosines, Bessel functions, Legendre polynomials, etc.). Several of these methods will be illustrated in Chapters 5 and 6. More

---

[42] It should also be remarked that in *transient* problems one should consider the possibility of nonnegligible instantaneous *accumulation* rates (of mass, momentum, etc.) at the "boundaries" under consideration, representing another additive contribution (to the LHS) of Eqs. (4.4-16, 17, 18).

[43] However, geologists are routinely interested in "predicting" the past from the present—i.e., reconstructing events, climates, etc. from quantitative observations made in the present.

[44] The digital nature of these computers implies that to exploit them we must convert the problem of approximately solving the governing PDEs (subject to BCs and ICs) to an *equivalent algebraic problem*. This is accomplished by the methods of finite differences (MFD; see Appendix 8.2) and the closely related method of finite elements (MFE; see Appendix 8.3).

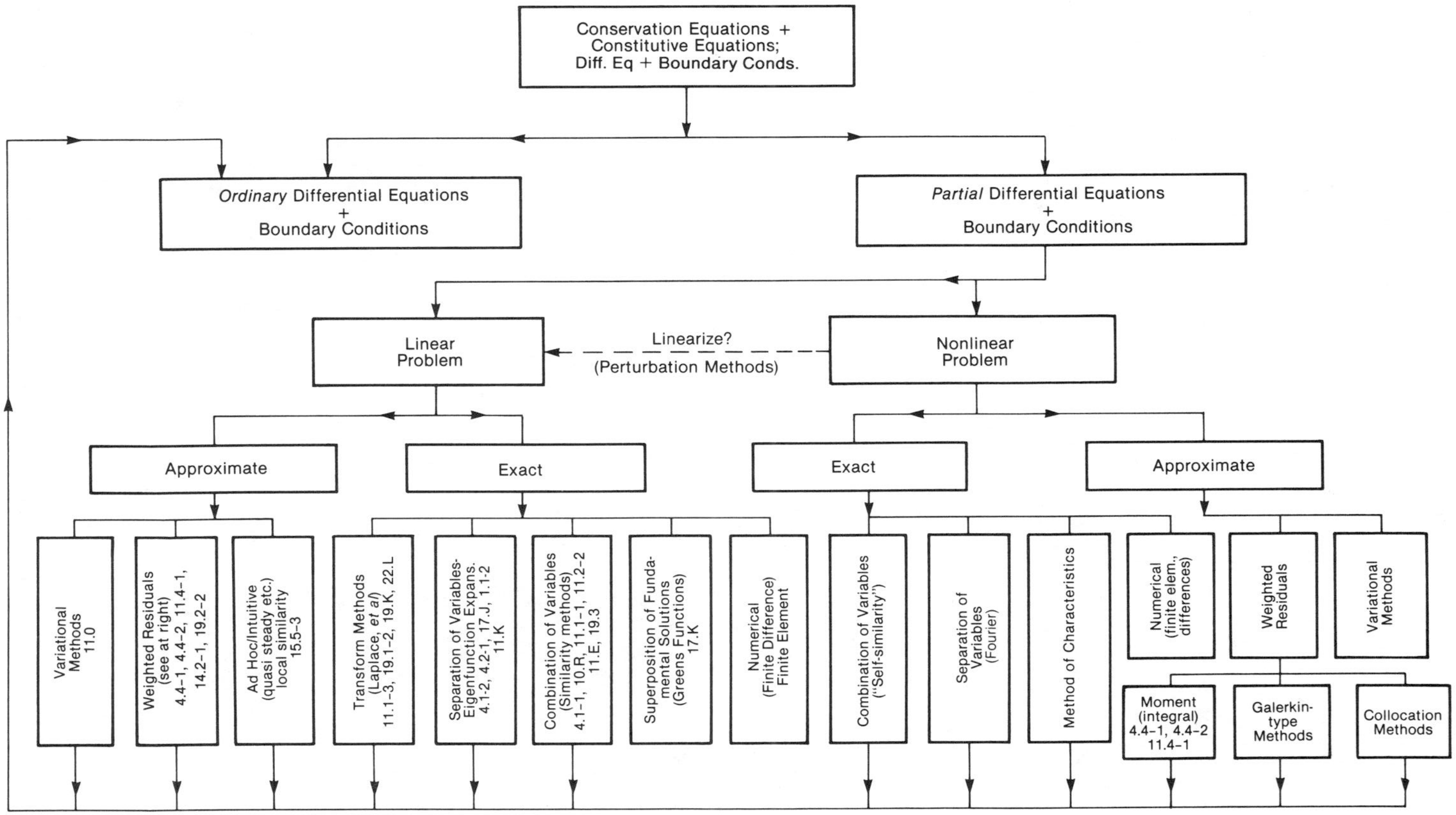

Figure 4.4-5 "Road map" of common methods of solution to problems in transport (convection/diffusion) theory.

often, the task of obtaining the solution of the governing PDEs (+ BCs and ICs) is first "reduced" to the solution of one (or more) nonlinear ODEs, which must then be solved *numerically* (see, e.g., Section 4.5.2 and Appendix 8.2).

Figure 4.4-5 provides an outline of the principal solution methods exploited in the field of mass, momentum, and energy transport theory.[45] In the selected examples that follow (see Section 4.5), the reader should "pinpoint" which method has, in fact, been utilized, and consider whether alternate methods would have been more, or less, expedient. Needless to say, the *results* (but not the effort required) should be independent of the *method* chosen!

Finally, it is important that the engineer confronted with complex problems should not be "impatient" with the extreme idealizations that have sometimes been made in order to apply the methods of classical analysis. These "prototypical" problems and their solutions often play an essential *conceptual role* in the subject since they illustrate most (if not all) of the important *qualitative features* of more complex problems. Hence they pay handsome dividends in the *insight* they provide. Section 4.5.2 (dealing with high Re steady flow past a flat plate) provides one such example.

The above-mentioned exact solutions (to highly idealized problems) are also widely used for *checking* more complex numerical codes (in suitable limiting cases) and for "starting" numerical calculations near singular points, regions, etc. (see Appendixes 8.2 and 8.3).

4.5 VELOCITY FIELDS AND SURFACE MOMENTUM-TRANSPORT COEFFICIENTS: STEADY LAMINAR FLOW OF AN INCOMPRESSIBLE NEWTONIAN FLUID

For the conditions described in the title, we have the following simplified forms of the PDEs governing the steady velocity and pressure fields:

$$\mathbf{v} \cdot \mathbf{grad}\, \mathbf{v} = -\frac{1}{\rho}\mathbf{grad}\, p + \nu\, \mathrm{div}(\mathbf{grad}\, \mathbf{v}) + \mathbf{g} \qquad \text{(Navier–Stokes)} \qquad (4.5\text{-}1)$$

and

$$\mathrm{div}\, \mathbf{v} = 0 \qquad \text{(Mass Conservation);} \qquad (4.5\text{-}2)$$

cf. Eqs. (4.4-14 and -15). Moreover, we adopt the "no-slip" condition (empirical) at stationary solid boundaries—so that:

$$\mathbf{v} = 0 \qquad \text{at fixed solid boundaries.} \qquad (4.5\text{-}3)$$

---

[45] The examples explicitly identified in Figure 4.4-5 pertain to those in a well-known text on transport phenomena, *viz.*, that of Bird, *et al.* (1960).

We now consider the following two simple, but rather important, special cases:

a. fully developed steady axial flow in a straight duct of constant, circular cross section[46] (Poiseuille); and
b. two-dimensional steady flow at high Re-number past a thin flat plate aligned with the approach stream (Prandtl, Blasius).

In each case we seek, and find, a useful expression for the *momentum transfer coefficient*.

4.5.1 Momentum Transfer from the Fluid Flowing in a Straight Duct of Circular Cross Section

The cylindrical coordinate system (and nomenclature) chosen to describe this simple viscous fluid flow is sketched in Figure 4.5-1. Cylindrical polar coordinates (see Figure 4.5-1) are, of course, expedient here since the wall BCs (Eq. 4.5-3) are imposed on a simple "coordinate surface": r = constant = a_w (the duct radius). By "fully developed," it is implied that we are sufficiently far downstream of the duct inlet that the fluid velocity field is no longer a function of the axial coordinate z.[47] From the symmetry of the problem (none of the fields will depend upon θ) and because of the absence of swirl, we also conclude:

$$v_\theta = 0 \text{ everywhere (not just at } r = 0 \text{ and } r = a_w). \tag{4.5-4}$$

Moreover, the equation of *conservation of mass* takes the form (cf. Eq. 2.1-4), with ρ = const):

$$\frac{1}{r}\frac{\partial}{\partial r}(r v_r) + \frac{1}{r}\frac{\partial}{\partial \theta}(v_\theta) + \frac{\partial}{\partial z}(v_z) = 0, \tag{4.5-5}$$

so that if v_z is independent of the axial coordinate z we conclude that the first term vanishes; that is,

$$v_r = 0 \quad \text{(everywhere, not just at } r = 0 \text{ and } a_w). \tag{4.5-6}$$

Consequently, we require PDEs to find the remaining two fields: $v_z(r)$ and $p(r, z)$. These are provided by the radial and axial components of the linear-momentum

[46] It is straightforward to generalize this treatment to include (i) ducts of noncircular (but constant) cross section (see Table 5.5-1 for C_f. Re-results (fourth column)), and (ii) ducts of slowly varying cross section (using O. Reynolds' "lubrication theory" approximations (see, e.g., Denn (1980)).

[47] For "developing" flow in a duct, see Exercises 4.7 and 4.8.

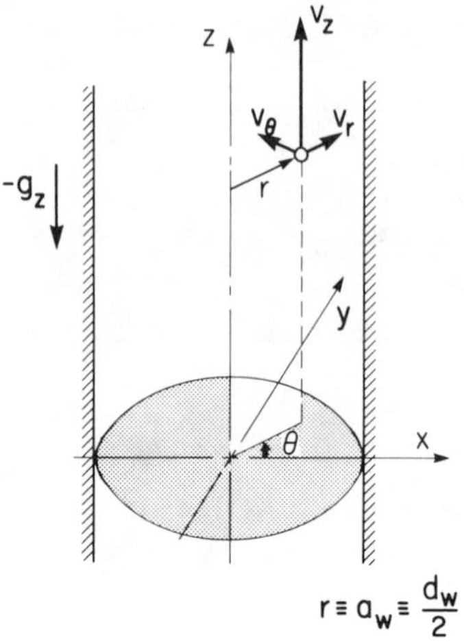

Figure 4.5-1 Cylindrical polar-coordinate system for the analysis of viscous flow in a straight circular duct of constant cross section.

conservation (N-S) equations (4.5-1), which simplify, respectively, to:

$$0 = -\frac{\partial p}{\partial r} \qquad (4.5\text{-}7)$$

and

$$0 = -\frac{1}{\rho}\frac{\partial p}{\partial z} + \nu\left(\frac{1}{r}\frac{\partial}{\partial r}\left(r\frac{\partial v_z}{\partial r}\right)\right) + g_z. \qquad (4.5\text{-}8a)$$

From the first of these we conclude that the pressure is, at most, a function of z alone. From the second we note that if $p = p(z)$ and $v_z = v_z(r)$, then the rearranged Eq. (4.5-8),

$$\frac{1}{\rho}\cdot\frac{\partial p}{\partial z} - g_z = \nu\left(\frac{1}{r}\cdot\frac{\partial}{\partial r}\left(r\frac{\partial v_z}{\partial r}\right)\right), \qquad (4.5\text{-}8b)$$

implies that a function of z alone (LHS) must equal a function of r alone (RHS). This is possible only if the LHS and RHS of Eq. (4.5-8b) are each equal to the same *constant*,[48] say C_1. Thus, we can rewrite Eq. (4.5-8b) as two separate ODEs, *viz.*:

$$\frac{1}{\rho}\cdot\frac{dp}{dz} - g_z = C_1 \qquad (4.5\text{-}9)$$

[48] Shown below to be negative.

and

$$\nu \frac{1}{r} \cdot \frac{d}{dr}\left(r \frac{dv_z}{dr}\right) = C_1. \qquad (4.5\text{-}10)$$

Note that, if we define a new pressure variable $\mathscr{P}$ such that:

$$\mathscr{P} \equiv p - \rho g_z z \qquad (4.5\text{-}11)$$

(i.e., by subtracting the prevailing "hydrostatic" contribution $\rho g_z z$), then Eq. (4.5-11) implies that the constant C_1 can be regarded as $(d\mathscr{P}/dz)/\rho$ and, hence, $\mathscr{P}$ varies linearly with distance z:

$$\mathscr{P}(z + \Delta z) = \mathscr{P}(z) + C\rho\,\Delta z. \qquad (4.5\text{-}12)$$

Turning to Eq. (4.5-10) we note that it can be formally integrated twice to give:

$$v_z = \frac{C_1 r^2}{4\nu} + C_2 + C_3 \cdot \ln r. \qquad (4.5\text{-}13)$$

However, since:

i v_z is finite when $r = 0$; therefore $C_3 = 0$.
ii v_z is zero when $r = a_w$; therefore $C_2 = -C_1 a_w{}^2/4\nu$.

We conclude that the shape of the velocity profile is *parabolic*, that is,

$$v_z(r) = -\frac{C_1 a_w{}^2}{4\nu} \cdot \left[1 - \left(\frac{r}{a_w}\right)^2\right]. \qquad (4.5\text{-}14)$$

Since $r \leqslant a_w$ and $v_z > 0$, we also see that C_1 is itself a *negative* constant—that is, the nonhydrostatic pressure *drops* linearly along the duct. Thus, if we consider any finite length, Δz, of duct, we have:

$$-C_1 = \frac{1}{\rho} \cdot \left(\frac{-\Delta\mathscr{P}}{\Delta z}\right), \qquad (4.5\text{-}15)$$

and Eq. (4.5-14) can be written in its final form:

$$v_z(r) = \frac{1}{4} \cdot \left(\frac{-\Delta\mathscr{P}}{\Delta z}\right) \cdot \frac{a_w{}^2}{\mu} \cdot \left[1 - \left(\frac{r}{a_w}\right)^2\right]. \qquad (4.5\text{-}16)$$

Having the detailed fluid-velocity profile, we are now in a position to calculate two quantities of immediate interest, the total flow rate and the nondimensional wall-friction coefficient.

4.5.1.1 *Total Flow Rate*

Clearly the total mass flow rate will be the sum of all the contributions $\rho v_z 2\pi r\, dr$ through annular rings each of area $2\pi r\, dr$. Thus:

$$\dot{m} = \int_0^{a_w} \rho v_z(r) 2\pi r\, dr. \tag{4.5-17}$$

Inserting Eq. (4.5-16) and carrying out the indicated steps yields the well-known Hagen [1839]–Poiseuille [1840] law:

$$\dot{m} = \frac{\pi}{8} \frac{a_w^{\ 4}}{\nu} \cdot \left(-\frac{\Delta \mathscr{P}}{\Delta z}\right), \tag{4.5-18}$$

which interrelates the axial pressure drop and the mass flow rate. For fluids of known Newtonian viscosity, this law forms the basis of the so-called "capillary-tube flowmeter"; i.e., one measures the pressure drop and computes the corresponding mass flow rate (from the length Δz between pressure taps, the bore ($2a_w$), and the fluid kinematic viscosity ν. Conversely, Eq. (4.5-18) is used to experimentally determine fluid viscosities.

It is interesting to note that if we define an *average* velocity, U, by the equation:

$$\rho U(\pi a_w^{\ 2}) \equiv \dot{m}. \tag{4.5-19}$$

Then we find:

$$U = \frac{(\dot{m}/\rho)}{\pi a_w^{\ 2}} = \frac{1}{2} \cdot v_z(0) = \frac{1}{8} \cdot \left(\frac{-\Delta \mathscr{P}}{\Delta z}\right) \cdot \frac{a_w^{\ 2}}{\mu}; \tag{4.5-20}$$

that is, the maximum (centerline) velocity is just *twice* the average value.[49] This allows Eq. (4.5-16) to be written in the equivalent form:

$$v_z(r) = 2U\left[1 - \left(\frac{r}{a_w}\right)^2\right], \tag{4.5-21}$$

which will prove useful in discussing heat or mass transfer to/from fluids flowing within circular ducts (Section 5.5.4).

[49] For planar (slot) viscous fluid flow, the velocity profile is also found to be parabolic, but the center-plane velocity exceeds U by the factor 3/2.

4.5.1.2 Nondimensional Wall-Friction (Tangential-Momentum Transfer-) Coefficient

The wall shear stress, τ_w, and its dimensionless counterpart C_f, defined by:[50]

$$C_f \equiv \frac{\tau_w}{\frac{1}{2}\rho U^2}, \tag{4.5-22}$$

can now be calculated in either one of two ways.

The "direct" method is to note that:

$$\tau_{rz} = \mu\left(\frac{\partial v_z}{\partial r} + \frac{\partial v_r}{\partial z}\right). \tag{3.2-5a}$$

Therefore:

$$\tau_w \equiv \tau_{rz}\Big|_{r=a_w} = \mu\left\{\frac{d}{dr}\left[2U\left(1-\left(\frac{r}{a_w}\right)^2\right)\right]\right\}\Big|_{r=a_w}. \tag{4.5-23}$$

Carrying out the indicated operations, we readily find:

$$\tau_w = \frac{8\mu U}{d_w}, \tag{4.5-24}$$

a simple result equivalent to the *dimensionless momentum transfer coefficient*:

$$C_f = 16/\text{Re}, \tag{4.5-25}$$

which, accordingly, holds for *all* Newtonian fluids (gases and all Newtonian liquids). Here $\text{Re} \equiv Ud_w/\nu$ and such flows are experimentally found to be stable only up to Reynolds' numbers of about 2100 (see Figure 4.5-3).

This important result can also be found from an *overall linear-momentum balance* (Eq. 2.2-3) on the macroscopic control volume $A\,\Delta z$ shown in Figure 4.5-2, that is,

$$\left\{\begin{matrix}\text{Net outflow rate of} \\ \text{axial momentum}\end{matrix}\right\} = \{\text{Net force on fluid}\}. \tag{4.5-26}$$

Since the flow is "fully developed" axial velocities do not change with z and this reduces to the axial *force* balance:

$$0 = (pA)|_z - (pA)|_{z+\Delta z} - \rho g_z A\,\Delta z - \tau_w 2\pi a_w\,\Delta z. \tag{4.5-27}$$

---

[50] Also denoted f in many treatments (Fanning friction factor). For fully developed flow in a straight circular duct of length L the friction-loss-factor K (Section 2.5.5) can be shown to be $4(L/d_w)\cdot C_f$.

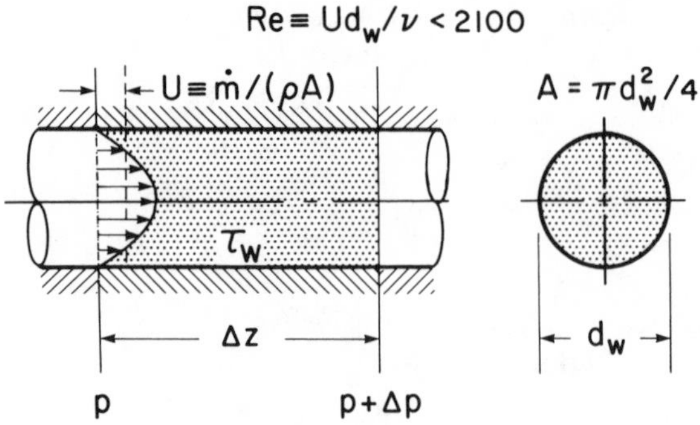

Figure 4.5-2 Configuration and notation: steady flow of an incompressible Newtonian fluid in a straight circular duct of constant cross section.

Solving this for τ_w and introducing the definition of $\mathscr{P}$ (Eq. (4.5-11) gives:

$$\tau_w = \frac{a_w}{2} \cdot \left(\frac{-\Delta \mathscr{P}}{\Delta z} \right), \tag{4.5-28}$$

which, in view of Eq. (4.5-18), is identical to Eq. (4.5-24)—found *via* the tangential velocity gradient at the duct wall. Note that by satisfying axial-momentum conservation at the *local* level, we have, *ipso facto*, satisfied it on the *macroscopic* level. This also indicates how the macroscopic conservation principles can be used to check analyses based on local (detailed behavior) *within* the chosen CV.

As is evident in Figure 4.5-3, *above* Re $= 2100$ *experimentally* observed friction coefficients are much higher than those predicted by the present laminar-flow analysis (over an order-of-magnitude difference for Re $\geqslant 2 \times 10^4$). Measurements and subsequent stability analyses (initiated by O. Reynolds) reveal that this is due to "transition" to turbulent flow within the duct, which (a) causes the Newtonian fluid to behave as though it were non-Newtonian, and (b) augments the transport of axial momentum to the duct wall. In this fully developed turbulent regime (Blasius): (a) C_f is found to vary approximately like Re$^{-1/4}$ for a duct with smooth walls (corresponding to τ_w and $-\Delta p$ proportional to $U^{7/4}$, rather than U^1 (cf. Eq. (4.5-24)) and (b) C_f becomes sensitive to the *roughness* of the inner wall. In fact, for sufficiently rough walls, C_f becomes nearly independent of Re, corresponding to τ_w and $-\Delta p$ proportional to U^2.

Time-averaged velocity profile *measurements*, combined with these C_f measurements, have been used to discover semi-empirical rules governing the effective eddy momentum diffusivity, $\nu_t \equiv \mu_t/\rho$, in such turbulent flows. This is discussed further in Sections 5.7.2 and 6.5.2 where we deal with methods capable of using C_f-data to predict corresponding *heat* and *mass* transfer coefficients.

It is interesting to note that, when the duct flow is fully *turbulent*, the perimeter-average skin friction and pressure drop can be reasonably well estimated, even for ducts of *noncircular cross-sectional area*, by using Eq. (4.5-13) and treating the actual

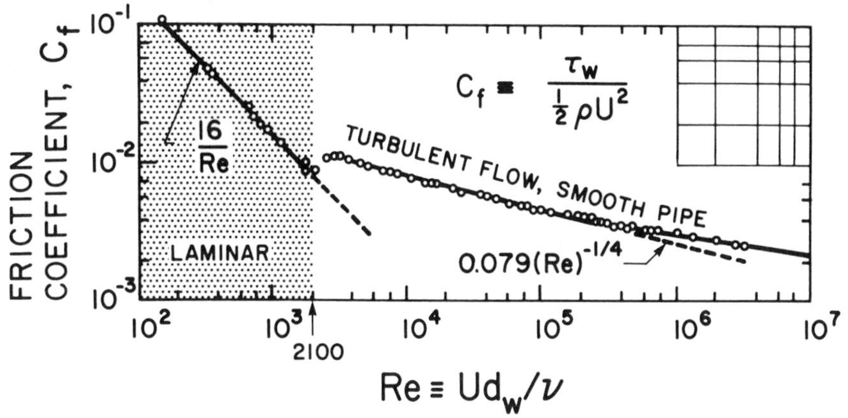

Figure 4.5-3 Experimental and theoretical friction coefficients for incompressible Newtonian fluid flow in a straight smooth-walled circular duct of constant cross section (after Denn (1980)).

duct as a circular duct with the "effective diameter"

$$d_{\text{eff}} \equiv \frac{4A}{P}, \tag{4.5-29}$$

where P is the "wetted perimeter." However, reference to the $C_f \cdot \mathrm{Re}$-values collected in column 4 of Table 5.5-1 shows that this is *not* an accurate (or necessary) approximation for *laminar* duct flow.[51]

4.5.2 Momentum Diffusion Boundary-Layer Theory: Example of Laminar Flow Past a Flat Plate at Zero Incidence

4.5.2.1 The Momentum Diffusion "Boundary-Layer" Concept

In 1904, L. Prandtl introduced a decisive simplification applicable to the prediction of momentum transport to immersed solids (hence, their friction drag) in the important limit of *very large but finite Reynolds' numbers.* It was well known that, for a hypothetical "ideal" inviscid fluid ($\mathrm{Re} = \infty$) only $v_n = 0$ at the fluid/solid interface and the fluid would "slip" along the surface, producing no friction drag. However, for a *real* fluid flow, with large but finite Re, *both* v_n and v_x vanish at the solid surface. Therefore, Prandtl reasoned that, in this $\mathrm{Re} \gg 1$ asymptotic limit, there must be a

[51] Note also that in laminar flow a duct of *square* cross section exhibits a $C_f \cdot \mathrm{Re}$-value about 11 percent *lower* than that of a circular duct. For a two-dimensional "slot," $C_f \cdot \mathrm{Re} = 24$ (Table 5.5-1).

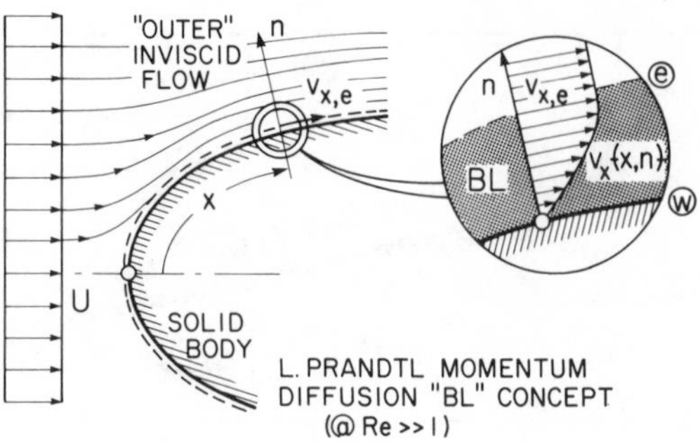

Figure 4.5-4 Division of flow field at $Re^{1/2} \gg 1$ into an inviscid "outer" region and a thin tangential momentum diffusion boundary layer (BL) (after L. Prandtl).

very thin transition layer near the surface across which the tangential fluid velocity, v_x, rather abruptly drops to zero. Inside this "boundary layer" the velocity gradients $\partial v_x/\partial n$ are large enough to make momentum *diffusion* important, even though the viscosity coefficient is small. Outside of this boundary layer the effects of momentum diffusion can be neglected. Thus (cf. Figure 4.5-4), Prandtl suggested that a flow field at $Re \gg 1$ can fruitfully be divided into two simpler domains: an "exterior" inviscid region (governed by the solution of, say, Eq. (4.3-80)),[52] matched to a very thin *momentum diffusion boundary layer* (BL) "hugging" the fluid/solid interface. Moreover, Prandtl showed that at sufficiently large Reynolds' numbers, this BL is so thin that:

a. within the BL, $v_n \ll v_x$ (*cf*. Figure 4.5-4);
b. within the BL, x-momentum *diffusion* is important, but only in the normal direction (equivalently, $\tau_{nx} \gg \tau_{xx}$);
c. *across* the BL the pressure at any streamwise station x is nearly constant—that is, $p \cong p_e(x)$.

Thus, *within* the BL, when written in a "body-oriented" (x, n)-coordinate system, the linear-momentum conservation PDEs simplify considerably. Prandtl reasoned that, if solutions to these simpler BL equations could be found which properly matched "inner behavior" of the "exterior" *inviscid flow*, then the resulting flow field would be a self-consistent approximation, valid at large Reynolds' numbers. The approach is a very general one and continues to be widely exploited in the area of high-Re fluid mechanics. Moreover, it will be seen that the same approach can be used to deal with the corresponding energy and species mass diffusion "boundary layers" (Chapters 5 and 6). We illustrate its implementation and results below in the simplest fluid-dynamic situation; *viz*., planar (two-dimensional) steady flow of an incompressible

[52] Based on a superposition method, efficient computer codes are now available to predict the near-wall behavior of the steady inviscid flow about any solid body whose shape can be reasonably approximated by a very large number of small flat "panels" (Hess, J.L., and A.M.O. Smith (1967)).

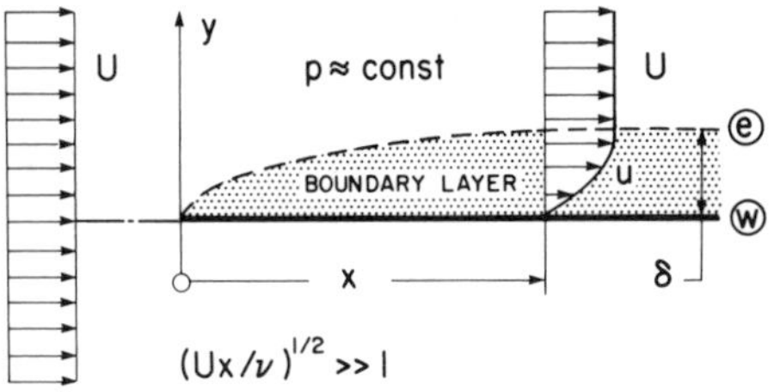

Figure 4.5-5 Newtonian incompressible fluid flow past a flat plate; configuration, nomenclature, and coordinate system.

constant-property Newtonian fluid past a semi-infinite flat plate at zero incidence (see Figure 4.5-5).

4.5.2.2 Laminar Boundary Layer on a Flat Plate (Blasius [1908])

Subject to the Prandtl BL assumptions discussed above, for a thin flat plate the pressure will be constant everywhere. This eliminates one field and the need for the y-momentum equation, leaving us with two scalar PDEs governing the two components of the fluid velocity field $v_x \equiv u\{x, y\}$ and $v_y \equiv v\{x, y\}$, *viz.*:

$$\frac{\partial u}{\partial x} + \frac{\partial v}{\partial y} = 0 \qquad \text{(mass)}, \tag{4.5-30}$$

$$u\frac{\partial u}{\partial x} + v\frac{\partial u}{\partial y} = \nu\frac{\partial^2 u}{\partial y^2} \qquad (x\text{-momentum}). \tag{4.5-31}$$

This coupled nonlinear system of PDEs, subject to the boundary conditions:

$$u\{-\infty, y\} = U, \tag{4.5-32a}$$

$$u\{x, \infty\} = U, \tag{4.5-32b}$$

$$u\{x, 0\} = 0, \tag{4.5-32c}$$

$$v\{x, 0\} = 0, \tag{4.5-32d}$$

was solved by Blasius (1908) using a powerful (when applicable) method called the method of "combination of variables."[53] It consists of seeking and finding a *combination* of the independent variables (and parameters) such that the field variables (including their boundary values) only depend upon this combination, and hence are governed by an *ordinary* (rather than partial) differential equation. Thus,

[53] Also called the method of finding "self-similar" solutions (see, also, Section 4.6.1).

Blasius showed that in the present case:

$$\frac{u}{U} = \mathrm{fct}_1\left\{\frac{1}{2}\frac{y}{x}\cdot\left(\frac{Ux}{\nu}\right)^{1/2}\right\} \equiv \mathrm{fct}_1\{\eta\} \tag{4.5-33}$$

and

$$\frac{v}{U}\left(\frac{Ux}{\nu}\right)^{1/2} = \mathrm{fct}_2\left\{\frac{1}{2}\frac{y}{x}\cdot\left(\frac{Ux}{\nu}\right)^{1/2}\right\} \equiv \mathrm{fct}_2\{\eta\}. \tag{4.5-34}$$

Moreover, he derived and numerically solved the nonlinear ODE governing[54]

$$f\{\eta\} \equiv \int_0^{\eta}\frac{u\{\eta\}}{U}\,d\eta,$$

from which the results for the normalized tangential fluid-velocity profiles are constructed in Figure 4.5-6 (cf. the experimental data at $\mathrm{Re}_x \equiv Ux/\nu$-values between 1.08×10^5 and 7.28×10^5).

Note that if we define a local *BL thickness* $\delta\{x\}$ by the y-location at which $u/U = 0.99$, say, then, this occurs at $\eta \simeq 5$. Therefore, we conclude:

$$\delta \simeq 5x\left(\frac{Ux}{\nu}\right)^{-1/2}. \tag{4.5-35}$$

Note that the BL thickness grows as the square root of the distance x from the leading edge of the plate.

From the tangential velocity profile one can calculate the *wall shear stress*, which, in the present case, is found to be:

$$\tau_w\{x\} = 0.332\,\rho U^2\cdot\left(\frac{Ux}{\nu}\right)^{-1/2}. \tag{4.5-36}$$

This corresponds to a *local dimensionless skin-friction (momentum-transfer) coefficient* c_f given by:

$$\frac{\tau_w}{\frac{1}{2}\rho U^2} \equiv c_f\{x\} = 0.664\left(\frac{Ux}{\nu}\right)^{-1/2}. \tag{4.5-37}$$

Equation (4.5-36) indicates that the wall shear stress falls off like $x^{-1/2}$ as one proceeds down the plate; however, the singularity at $x = 0$ (where $\tau_w \to \infty$) is an

---

[54] Blasius first introduced a dimensionless "stream function" $f\{\eta\}$ such that $u/U = df/d\eta$, $(v/U)\cdot(Ux/\nu)^{1/2} = [\eta(df/d\eta) - f]$ and the *mass* conservation equation (4.5-30) is automatically satisfied. Then, from the x-momentum PDE (4.5-31) he showed that $f\{\eta\}$ must satisfy the following "two-point" boundary-value problem (BVP): $d^3f/d\eta^3 + (f/2)(d^2f/d\eta^2) = 0$ subject to: $f\{0\} = 0$, $(df/d\eta)_0 = 0$, and $\lim_{\eta\to\infty}(df/d\eta) = 1$. It can be shown that every two-dimensional (planar or axisymmetric) viscous flow possesses a "stream function" from which the fluid-velocity components can be recovered by appropriate partial differentiation.

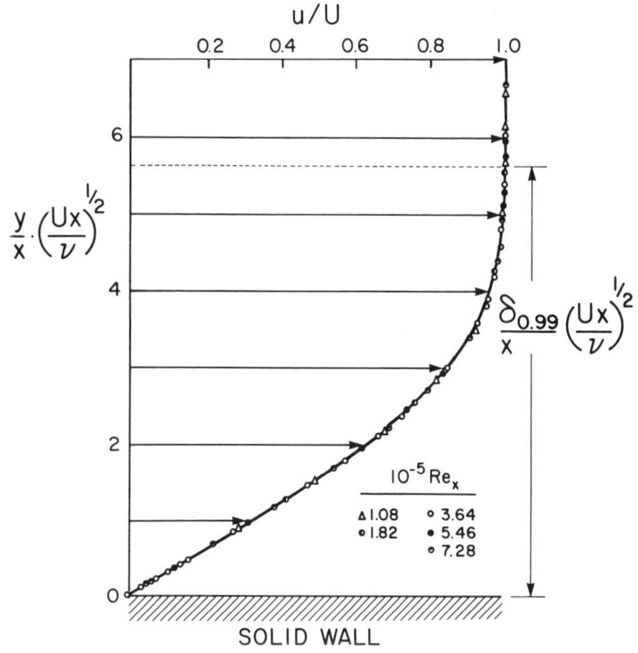

Figure 4.5-6 Tangential velocity profile within the laminar BL on a flat plate at zero incidence (after Blasius [1908] and Schlichting (1979)).

artifact of BL theory—*viz.*, the underlying assumption $(Ux/\nu) \gg 1$ obviously breaks down at sufficiently small downstream distances.

Despite this singularity at $x = 0$, it is possible to sum up all of the local drag contributions (on both sides of the plate) *up to the streamwise station x*. This leads to the total *friction drag coefficient*.[55]

$$\bar{c}_f \equiv \frac{D(\text{both sides})}{\frac{1}{2}\rho U^2 A_w} = \frac{1.328}{(Ux/\nu)^{1/2}}. \tag{4.5-38}$$

Moreover, since *streamwise* diffusion of momentum is negligible, if the plate is truncated at $x = L$ (that is, we now consider a *finite* length plate) then to a first approximation Eq. (4.5-38) is expected to apply, with x set equal to L.

It is obvious from the comparison with experiment shown in Figure 4.5-6 that the asymptotic theory is highly accurate for Re_x values of the order of 10^5. However, for Re_x values above about 3×10^6 it is found to be impossible to maintain a stable *laminar* BL—that is, turbulence sets in and the friction drag law is modified (cf. our analogous discussion of transition to turbulence in duct flow, Section 4.5.1).

Many of these features are exhibited by BLs that develop along bodies (e.g.,

[55] Here A_w is the "wetted" (*not* frontal) area of the plate up to station x.

wings, gas turbine blades, fuselages) of more complex geometry.[56] Indeed, the *procedure* outlined in the introduction to this section is now routinely carried out for bodies of prescribed geometry, using readily available digital computer programs. It is also possible to systematically calculate correction terms that become important at *smaller* Re-values (Van Dyke (1969)), using the method of matched asymptotic expansions.

While not pursued here quantitatively, because of its importance in our later discussions of *energy* and *mass* transport, it should be commented here that local (and downstream) momentum transport coefficients (e.g., c_f-values) can be modified by the injection ("blowing") or "suction" of fluid through the "solid" (porous) surface. Thus, "blowing" can be used to produce a significant reduction in skin friction drag, which is readily measurable or calculable using the methods of BL-theory. Such results are usually expressed in the form:

$$c_f\{x/L, \mathrm{Re}, \ldots\} = [c_f\{x/L, \mathrm{Re}, \ldots\}]_0 \cdot F\{\text{blowing}\}, \tag{4.5-39}$$

where $(c_f)_0$ is the corresponding no-blowing momentum-transfer coefficient and $F\{\text{blowing}\}$ is primarily a function of the dimensionless variable:

$$\frac{\rho_w v_w}{\rho_e u_e c_{f,0}}. \tag{4.5-40}$$

Here $\rho_w v_w = \dot{m}''_w$ is the local mass injection rate (cf., e.g., Schlichting (1979), and Cebeci and Smith (1974)).

4.5.2.3 Vorticity Diffusion Equation

For a planar (two-dimensional), constant-property (composition, density), viscous fluid flow it can be shown (see, e.g., P.A. Thompson (1972), Section 2.4) that the *vorticity vector*, $\boldsymbol{\zeta}$ (Section 3.2.2) actually satisfies a much simpler PDE than the specific momentum (velocity) vector $\mathbf{v}\{\mathbf{x}, t\}$ itself, *viz.*:

$$\frac{D\zeta}{Dt} = \nu \operatorname{div}\{\mathbf{grad}\, \zeta\} \equiv \nu \nabla^2 \zeta \tag{4.5-41}$$

where $\nu \equiv \mu/\rho$ is the momentum ("kinematic") diffusivity. Thus, for such a viscous fluid flow in the x-y plane, the quantity:

$$\zeta_z = \frac{\partial v_y}{\partial x} - \frac{\partial v_x}{\partial y} \tag{4.5-42}$$

satisfies a simple *convective-diffusion PDE, irrespective of the presence of pressure gradients and/or (conservative) body forces.* It will be seen below that in simple cases the fluid *temperature* field $T\{\mathbf{x}, t\}$, and species (or chemical element) mass fraction fields $\omega_i\{\mathbf{x}, t\}$, satisfy convective diffusion PDEs of this same *form*, the only difference being the relevant *diffusivity* (see Sections 5.3.2 and 6.3.2).

---

[56] However, locally symmetrical bodies with blunt noses have *vanishing* shear stress and nonzero BL thickness at the forward stagnation point.

4.6 MOMENTUM TRANSFER IN "STEADY" TURBULENT FLOWS: ENTRAINMENT BY JETS

4.6.1 Introduction

It is instructive to consider the role of experiment in the understanding and "prediction" of *turbulent* fluid flows, and, for this purpose, we discuss the important case of a circular jet discharging into an otherwise quiescent fluid. Remarkably enough,[57] sufficiently far from the jet orifice a "fully" turbulent round jet is found to have all of the properties of a *laminar* round jet, except the effective kinematic viscosity, ν_t, is much greater than the intrinsic kinematic viscosity, ν, of the fluid, being given by:

$$\nu_t = 0.0161\left(\frac{\dot{J}}{\rho}\right)^{1/2}. \tag{4.6-1}$$

Here $\dot{J}$ is the jet axial-momentum flow rate—a quantity which (by linear-momentum conservation) remains constant across any plane $z = \text{constant}$ perpendicular to the jet axis.

4.6.2 Laminar Round Jet of an Incompressible Newtonian Fluid: Far-Field

The far-field properties of steady-flow *laminar* round jets of an incompressible, Newtonian fluid are well known in the boundary-layer approximation[58] (Schlichting [1933]). In that case the PDEs governing mass and axial momentum conservation in r, θ, z coordinates also admit an exact solution by the method of "combination of variables," *viz.*, the dependent variables,

$$\tilde{u} \equiv v_z \cdot \left[\frac{(\dot{J}/\rho)}{\nu z}\right]^{-1} \quad \text{and} \quad \tilde{v} \equiv v_r \cdot \left[\frac{(\dot{J}/\rho)^{1/2}}{z}\right]^{-1}, \tag{4.6-2a,b}$$

are uniquely determined by the *single* independent variable:

$$\xi(r, z) \equiv \left(\frac{3}{16\pi}\right)^{1/2} \cdot \frac{(\dot{J}/\rho)^{1/2}}{\nu} \cdot \frac{r}{z}. \tag{4.6-2c}$$

[57] More generally, turbulent flows exhibit apparent kinematic viscosities, ν_t, which are *not* uniform throughout the flow—a simple example being fully developed pipe flow, discussed in Section 5.7.2. Thus, while the far field of a round turbulent jet is similar to that of a Newtonian fluid with a higher viscosity, in most cases turbulent flows "appear" to be the result of a non-Newtonian-stress law. Incidentally, using Eq. (4.6-1) the basic self-consistency condition, $\nu_t \gg \nu$ is seen to be equivalent to the condition: $\mathrm{Re}_j \gg 7$. In practice, such jets are found to be turbulent above about $\mathrm{Re}_j \simeq 3 \times 10^4$.

[58] The "BL approximation" (neglecting momentum *diffusion* in the streamwise direction) is seen to be meaningful even for flows *without solid boundaries*!

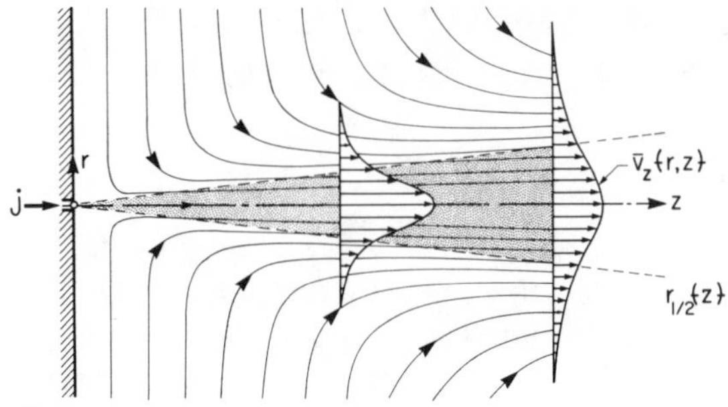

Figure 4.6-1 Streamline pattern and axial velocity profiles in the far-field of a *laminar* (Newtonian) or fully turbulent unconfined round jet (adapted from Schlichting (1968)).

The streamlines are plotted in Figure 4.6-1.[59] Of special interest is the $\tilde{u}\{\xi\{r, z\}\}$ profile (shown at two stations), simply given by:

$$\tilde{u} = \frac{3}{8\pi}\left(1 + \frac{\xi^2}{4}\right)^{-2}. \tag{4.6-3}$$

If one calculates the total mass-flow rate past any station z far from the jet mouth, that is,

$$\dot{m}\{z\} = \int_0^{\infty} \rho v_z\{r, z\} \cdot (2\pi r\, dr), \tag{4.6-4}$$

one finds the remarkable result:

$$\frac{\dot{m}}{\rho} = 8\pi\nu z. \tag{4.6-5}$$

That is, the mass flow "in the jet" *increases* with downstream distance! This does *not* violate mass conservation (indeed *local* mass conservation was one of the starting points). Rather, it indicates that such a jet *entrains the ambient fluid* in the process of being decelerated (*via* the radial diffusion of its initial axial momentum).

4.6.3 Fully Turbulent Round Jet: Far-Field

All of these results carry over to the fully turbulent round jet with ν_t (Eq. (4.6-1)) replacing ν. Thus, the *entrainment rate*, $d\dot{m}/dz$, is found to be constant at the value:

[59] These profiles exhibit the so-called "self-similarity" property (see, also, the Blasius flat-plate tangential-velocity profiles, Section 4.5). This means that, if we know the profile at one station z_1, say, the profile at any *other* station z_2 can be obtained from it. (In the present case the product $v_z z$ remains constant along any co-axial cone r/z = constant.) For this reason the method of combination of variables is sometimes called the "self-similarity" method.

$$\left(\frac{d\dot{m}}{dz}\right)_{\text{turb}} = 8\pi\rho\left[0.0161\left(\frac{\dot{J}}{\rho}\right)^{1/2}\right]. \tag{4.6-6}$$

Further consequences of this phenomenon of *jet entrainment* will be taken up in Section 4.6.5.

4.6.4 Generalization: Turbulence "Modeling"

That ν_t, above, is proportional to the square root of the jet "kinematic momentum flow rate," $\dot{J}/\rho$, can be deduced purely from *dimensional* arguments (see Chapter 7), but the constant 0.0161 appearing in Eq. (4.6-1) can be obtained only by comparison with time-averaged velocity profile *measurements* made in such turbulent jet flows. However, when we have done this, it is natural to ask if this constant can somehow be used to predict *related* turbulent jet flows—e.g., the properties of the analogous *two-dimensional* turbulent jet, or the rate of turbulent mixing between two adjacent streams of initially unequal axial velocity (momentum/mass). In fact, this is the role played by *theory* in the case of *turbulent* momentum (and mass- and energy-) transfer. It serves to relate more complex turbulent flows to the known (measured) properties of "simple" turbulent flows.

To illustrate how this might be done, one could re-express ν_t in terms of two "local"[60] *flow* properties, such that their product also has the units of a length × velocity—e.g., the product of a characteristic *jet width* and the centerline *velocity* $v_z(0, z)$. In the present (round jet) case, a convenient choice of characteristic transverse dimension is the radius $r_{1/2}$, at which the local axial velocity is half that on the centerline. When this dimension is used, it can be shown that Eq. (4.6-1) is fully equivalent to:

$$\nu_t = 0.0256[r_{1/2}(z)] \cdot [v_z(0, z)] \tag{4.6-7}$$

(where 0.0256 is sometimes called the Reichardt constant). Now suppose it is desired to calculate turbulent diffusion of momentum (entrainment rates, etc.) in more complex "BL" cases—for example, with a co-flowing outer ducted stream and axial pressure gradients. A reasonable postulate would be to replace Eq. (4.6-7) with:

$$\nu_t(z) = 0.0256 r_{1/2}(z) \cdot \left| v_z(0, z) - v_z(\text{“}\infty\text{”}, z) \right|, \tag{4.6-8}$$

where $r_{1/2}(z)$ is now defined as the radius at which:

$$\frac{v_z(r, z) - v_z(\text{“}\infty\text{”}, z)}{v_z(0, z) - v_z(\text{“}\infty\text{”}, z)} = \frac{1}{2}, \tag{4.6-9}$$

[60] The reader will notice that the choices made here are not quite local—i.e., they are characteristic of the entire jet at station z.

and include the pressure gradient term $\partial p/\partial z$ in the axial-momentum conservation PDE. This simple procedure (which does not require the solution of any additional PDEs) is reasonably successful; however,

a. It is found that one cannot use the same constant to treat analogous planar (two-dimensional "BL") jet mixing problems with reasonable accuracy;
b. It is not clear how to generalize such results to include cases where the density of the outer stream and jet are significantly different, etc.

For such reasons, more elaborate turbulence models have been developed (see, e.g., Launder and Spalding (1972)) to predict turbulent reacting flows in the presence of swirl, density gradients, etc. (These are very briefly outlined in Section 5.7.3.) However, even the simplest of these require the solution of additional PDEs to keep track of the accumulation, convection, diffusion, and (net) production of those local turbulence quantities presumed to determine the local value of ν_t (and its counterparts α_t and D_t for energy and mass transfer, respectively). Of course, the goal is a turbulence model which will be reasonably universal, without incurring too great an additional computational cost (Bradshaw, Cebeci, and Whitelaw (1981)).

4.6.5 Entrainment Limitations and Recirculation in Confined Ducts

The entrainment phenomenon discussed previously is obviously essential to provide for mutual reactant access when the jet is comprised of one reagent (e.g., fuel vapor) and the surrounding fluid contains the second reagent (e.g., the oxidizer). We mention here an interesting and important phenomenon that occurs in the case of *ducted* coaxial jets—a common configuration in engineering equipment. If the entrainment "demands" of the central jet can be satisfied by the co-flowing fluid, then the resulting time-averaged flow pattern is sketched in Figure 4.6-2, Panel (a). However, suppose that, because of duct confinement and inadequate axial momentum flow in the co-flowing stream, the entrainment demands of the central jet can*not* be met without "self-entrainment"; that is, a "recirculation" flow develops (Panel (b)), with a portion of the primary jet fluid cycling *back* near the duct walls only to re-enter the primary jet upstream! In a still more extreme case,[61] shown in

[61] A dimensionless parameter called the Craya–Curtet number (cf. Figure 4.6-2) correlates this turbulent self-entrainment behavior rather well. It is constructed from the relevant primary (jet) and secondary fluid stream momentum fluxes ($\dot{J}_j$ and $\dot{J}_s$, respectively) and jet-to-duct area ratio (see, e.g., Beer and Chigier (1972)). However, a simpler alternative is the dimensionless grouping:

$$\left(\frac{\dot{m}_j + \dot{m}_s}{\dot{m}_j}\right)\cdot\frac{d_j}{d_w}\cdot\left(\frac{\rho_j}{\rho_s}\right)^{1/2},$$

proposed by Thring and Newby (1953) on physical grounds, and found to dictate the fraction of the total mass which recirculates, the downstream position of the recirculation zone (measured in duct diameters), etc. Recirculation is experimentally observed for values of the Thring–Newby parameter *below* about 0.6.

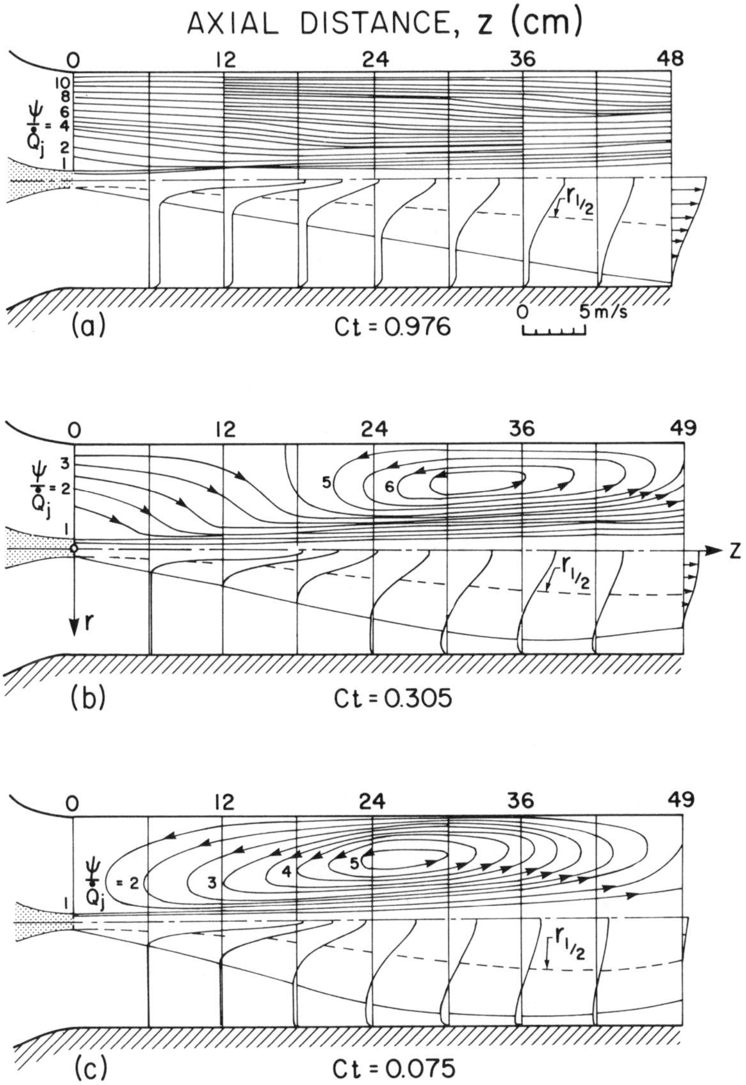

Figure 4.6-2 Experimentally observed streamline pattern and axial velocity profiles for turbulent jet mixing of co-flowing streams in a constant area duct (after Barchilon and Curtet (1964)).

Figure 4.6-2, Panel (c), there is no co-flowing surrounding fluid—that is, the entire entrainment demand is satisfied by the mechanism of "self-entrainment" (recirculation). This recirculation flow dramatically alters the fluid-parcel residence-time distribution in the device (see Section 6.7) and may provide a useful "feedback" mechanism for energy and reaction intermediates to help stabilize a reaction zone, especially in a duct without recirculation patterns produced by *other* (geometric) means (e.g., immersed bluff bodies (Figure 1.2-4) or sudden enlargements in duct area (Figure 1.2-7).

4.6.6 Concluding Remarks on Turbulent Jet Mixing

In closing this discussion of the principle of momentum conservation applied to fluids, several observations should be made about our "self-similar" round-jet turbulent mixing example:

a. *Near-Field Behavior*. No matter what the detailed *shape* of the jet orifice (square, elliptical, circular, etc.), *sufficiently far from the jet mouth* all such jets will be governed by the round-jet behavior discussed in Sections 4.6.1 and 4.6.3, with $(\dot{J}/\rho)^{1/2}$ being the crucial parameter. However, these laws clearly do not apply in the immediate vicinity (say, $z/d_{\mathrm{j,eff}} \leqslant 10$) of the jet mouth; indeed, when $z \to 0$ Eq. (4.6-3) with $\nu \to \nu_{\mathrm{t}}$ formally predicts the nonphysical result: $v_z(0, z) \to \infty$. The so-called "near-field" is the region where the detailed nozzle shape *is* important, and over which $v_z(r, z)$ makes its transition between the actual jet exit velocity profile $v_z(r, 0)$ (for a round axially symmetrical jet), and $v_z(r, z)$ obtained from Eq. (4.6-3) with $\nu \to \nu_{\mathrm{t}}$. This type of transition, shown in Figure 5.7-1 (for the case of a secondary-to-primary jet velocity ratio of 1/4 and a secondary-to-primary jet *density* ratio of 1.09) reveals the presence of a so-called "potential core," within which jet profiles are essentially unaltered by the peripheral (and, of course, downstream) momentum diffusion process. Thus, along the centerline, $v_z(0, z)$ is found to retain its (initially uniform) value U_j until about 4.4 jet diameters downstream (in the absence of a secondary jet), whereupon $v_z(0, z)$ makes a relatively abrupt transition to a slightly shifted[62] $(z - z_0)^{-1}$ decay law (cf. Eq. (4.6-3), with $\nu \to \nu_{\mathrm{t}}$).

b. *Swirling Jets*. Momentum diffusion and entrainment rates can be dramatically altered in the simultaneous presence of tangential "swirl" in the primary and/or secondary jets. Indeed, "predicting" the structure of such flows, of enormous technological significance, constitutes a stringent test of the utility of any postulated turbulence model.

   For the case of swirl only in the primary jet, a significant additional parameter governing the structure of such jets is the dimensionless "swirl number" (cf., for example, Beer and Chigier (1972)):

$$\frac{(\dot{J}_\star/\rho)}{d_{\mathrm{j}}(\dot{J}/\rho)}.$$

Here $\dot{J}_\star$ is the flow rate of *angular* momentum at the jet mouth, i.e., for

---

[62] Matching of the near- and far-field introduces a "virtual origin" of the jet, z_0.

an axisymmetric jet:

$$\dot{J}_{\star} \equiv \int_0^{d_j/2} [rv_\theta(r, 0)] \cdot [\rho v_z(r, 0) \cdot 2\pi r\, dr],$$

where $rv_\theta(r, 0)$ is the angular momentum (moment of momentum relative to the axis $r = 0$) per unit mass of fluid.

c. *Additional Parameters.* The diversity of additional parameters that can influence momentum diffusion and entrainment rates in turbulent jets[63] (e.g., initially nonuniform density, viscous dissipation, and chemical heat release, the presence of a second (dispersed) phase, etc.) is readily visualized, along with the associated need for ongoing experimentation and ancillary-turbulence modeling techniques able to cope with this level of complexity.

An interesting example of how physical insight can be used to relate such situations to the simpler situations treated above pertains to the effects (on the far-field) of the jet fluid having a different density than that of the secondary (surrounding) fluid. Since (as we have seen) the far-field turbulent motion is dictated by the z-momentum flux, and is insensitive to jet-nozzle shape, Thring and Newby (1953) proposed and verified that the far-field behavior of such jets is the same as that for the case $\rho_j = \rho_s$, provided one introduces the *effective jet diameter*:

$$d_{j,\,\mathrm{eff}} = d_j \cdot \left(\frac{\rho_j}{\rho_s}\right)^{1/2}.$$

This is nothing more than the nozzle diameter required for the jet fluid to flow with the prevailing flow rates of mass and z-momentum *if the jet fluid had the density* ρ_s *rather than* ρ_j.

4.7 MOMENTUM TRANSFER FOR FLUID FLOW IN POROUS MEDIA OR PACKED BEDS

In many types of engineering equipment, as well as natural environments, we encounter the flow of a viscous fluid through the void spaces between many solid objects, each of dimension small on the scale of our overall interest. Thus, the volume of interest may contain a "solids fraction" ϕ made up of:

granular particles (sand, pebbles)
"wool" ("steel wool," fiberglass, etc.)
gauzes, screens (e.g., woven metals)
porous pellets (absorbent, catalyst support)

---

[63] Analogous remarks can be made about turbulent "wakes"—that is, regions of axial momentum "defect," such as the wake behind a bluff-body flame-holder (see Figure 1.2-4).

or commercially available packing shapes (rings, saddles, *et al.*), leaving a "void" fraction $\varepsilon(=1-\phi)$ for the fluid flow.[64] Usually the geometry is "random," although occasionally a regular packing is used (or, in effect, the geometry constitutes a regular "packing" (e.g., a dense bundle of regularly spaced tubes in cross-flow)).

Especially if the void fraction were large (say, $\varepsilon > 0.8$), it would be natural to think of the flow about each object as a "perturbed" *external* flow, with each wetted object contributing its own drag (momentum withdrawal) in accord with some effective approach flow. But, more often, the void fraction is less than $\frac{1}{2}$, in which case it is perhaps more natural to view the situation as an *internal* flow through the tortuous "ducts" comprising the spaces between the particles. In what follows (and Sections 5.5.5 and 6.4.4) we adopt the latter viewpoint in order to introduce an appropriate dimensionless transfer coefficient and Reynolds' number. We then turn to *experiments* to establish the quantitative interrelation between them.

In the case of flow through a *straight* duct of noncircular cross section we already introduced the notion of an effective duct diameter, *viz.*:

$$d_{\text{eff}} \equiv \frac{4(\text{area available for fluid})}{(\text{wetted perimeter})}. \tag{4.5-29}$$

In a similar manner, for the flow through a packed bed we can introduce an effective average *interstitial* ("duct") diameter $d_{\text{i, eff}}$ defined by:

$$d_{\text{i, eff}} \equiv \frac{4(\text{volume available for flow/total volume})}{(\text{wetted } area\text{/total volume})} = \frac{4\varepsilon}{a'''}, \tag{4.7-1}$$

where:

$$\begin{aligned} a''' &\equiv \left(\frac{\text{particle surface area}}{\text{particle volume}}\right)\left(\frac{\text{particle volume}}{\text{total volume}}\right) = \left(\frac{A_{\text{p}}}{V_{\text{p}}}\right)\cdot(1-\varepsilon) \\ &= \frac{6}{d_{\text{p, eff}}}\cdot(1-\varepsilon). \end{aligned} \tag{4.7-2}$$

Moreover, since each particle of the bed can be viewed as having an effective diameter[65]:

$$d_{\text{p, eff}} \equiv 6\left(\frac{V_{\text{p}}}{A_{\text{p}}}\right). \tag{4.7-3}$$

[64] In some cases, of course, not all of the void volume will be accessible to the fluid.

[65] When the packing is composed of a distribution of particle sizes $d_{\text{p,j}}$, each of volume fraction ϕ_{j} it is often sufficient to treat this as a packing with effective (uniform) particle size:

$$d_{\text{p, eff}} = \left(\sum \frac{\phi_{\text{j}}}{d_{\text{p,j}}}\right)^{-1}.$$

Equation (4.7-1) can, therefore, be rewritten:

$$d_{\mathrm{i,\,eff}} = \frac{2}{3} \cdot \left(\frac{\varepsilon}{1-\varepsilon}\right) \cdot d_{\mathrm{p,\,eff}}. \tag{4.7-4}$$

An appropriate Reynolds' number to use to characterize this "internal" flow would evidently be:

$$\mathrm{Re}_{\mathrm{bed,\,eff}} = \mathrm{const} \cdot \frac{\rho v_{\mathrm{i}} d_{\mathrm{i,\,eff}}}{\mu} = \mathrm{const} \cdot \frac{G_{\mathrm{i}} d_{\mathrm{i,\,eff}}}{\mu}, \tag{4.7-5}$$

where $\rho v_{\mathrm{i}} \equiv G_{\mathrm{i}}$ is the *interstitial mass velocity* related to the hypothetical "empty duct" (called "superficial"[66]) mass velocity $\rho v_0 \equiv G_0 = \dot{m}/A_0$ by:

$$G_{\mathrm{i}} = \frac{G_0}{\varepsilon}. \tag{4.7-6}$$

Combining Eqs. (4.7-4) and (4.7-5) (and selecting the constant in Eq. (4.7-5) to be 3/2) leads to the choice:

$$\mathrm{Re}_{\mathrm{bed,\,eff}} \equiv \frac{G_0 d_{\mathrm{p,\,eff}}}{\mu(1-\varepsilon)}. \tag{4.7-7}$$

It remains to define an appropriate dimensionless *momentum transport coefficient*, f_{bed}, and then resort to experiments to define the quantitative $f_{\mathrm{bed}}(\mathrm{Re})$ relationship. The same strategy will be adopted and pursued (in Sections 5.5.5 and 6.4.4) for *energy* and *mass* transport, respectively.

Recall that, in the case of a short length of a single straight duct, we introduced the dimensionless momentum-transfer coefficient:

$$C_{\mathrm{f}} \equiv \frac{\tau_{\mathrm{w}}}{\frac{1}{2}\rho U^2} = \frac{\frac{d_{\mathrm{eff}}}{4}\left[-\left(\frac{d\mathscr{P}}{dz}\right)\right]}{\frac{1}{2}\rho U^2}, \tag{4.7-8}$$

where $\mathscr{P} \equiv p + \rho |g_z| z$. This suggests that we now introduce:

$$f_{\mathrm{bed}} \equiv \mathrm{const} \cdot \frac{\frac{1}{2} d_{\mathrm{i,\,eff}}(-d\mathscr{P}/dz)}{\frac{1}{2}\rho v_{\mathrm{i}}^2}. \tag{4.7-9}$$

[66] This term is (unhappily) in widespread use.

A simple choice consistent with this requirement is:

$$f_{\text{bed}} \equiv \frac{\left(\dfrac{\varepsilon^3}{1-\varepsilon}\right) d_{\text{p, eff}} \left(-\dfrac{d\mathscr{P}}{dz}\right)}{G_0{}^2/\rho}. \tag{4.7-10}$$

Available experimental data (cf. Figure 4.7-1) correlate well on this basis, and, over the entire Re-range of interest, f_{bed} can be approximated by (Ergun (1952)):

$$f_{\text{bed}} \simeq \frac{150}{\text{Re}_{\text{bed}}} + 1.75. \tag{4.7-11}$$

The range $\text{Re}_{\text{bed}} < 10$, within which $f_{\text{bed}} \simeq 150/\text{Re}_{\text{bed}}$, is the laminar flow region, analogous to the branch 16/Re for a single, straight, circular-cross-section duct (Figure 4.5-3). The linear relationship between G_0 and $(-d\mathscr{P}/dz)$ equivalent to $f_{\text{bed}} = 150/\text{Re}_{\text{bed}}$ is a form of Darcy's law which yields the effective local *permeability*, $G_0\nu/(-d\mathscr{P}/dz)$, of such a bed (here assumed to be much greater than the intrinsic permeability of *each* particle). Because of the averaging effect of turbulent transition in a random distribution of "ducts" there is no abrupt single transition Reynolds' number for the *bed*, but there is a fully turbulent asymptote $f_{\text{bed}} \simeq 1.75$ (Burke–Plummer) for $\text{Re}_{\text{bed}} > 10^3$. Equations (4.7-11, -7, -10) form the basis of most engineering pressure-drop calculations in quasi-one-dimensional

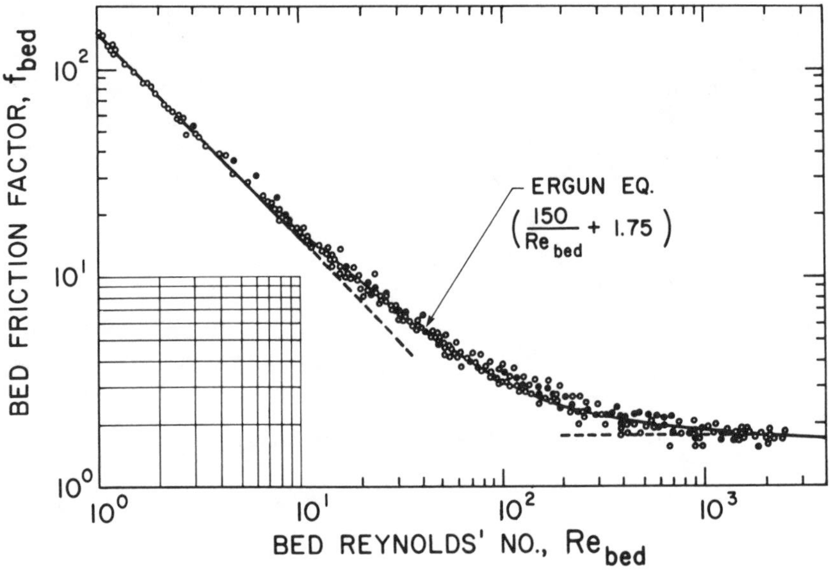

Figure 4.7-1 Experimentally determined dependence of fixed-bed friction factor f_{bed} on the bed Reynolds' number (adapted from Ergun (1952)).

packed ducts. Moreover, by equating $-\Delta p$ to the bed weight per unit area, one can estimate the ("incipient fluidization") velocity at which an unrestrained bed will be "levitated," achieving what is called a "fluidized" bed.

Still another generalization of practical interest is the calculation of the momentum exchange experienced by a viscous fluid forced through a fixed-bed when a second fluid (immiscible in the first, and which wets the packing) is also passing through the same bed (often falling under the action of gravity, as in a so-called irrigated "fixed bed"). When the dynamical interaction between the two fluids is small, in effect this calculation can be made using the equations of this section, by simply increasing $d_{p,\text{eff}}$ to allow for the viscous film present on and between each packing element. In counterflow situations the upward flowing fluid may transmit enough momentum to the falling second liquid to prevent its motion through the packed column. This sets a practical upper limit (called the "flooding limit") to the flow rate of the upwardly flowing fluid. For engineering correlations allowing the estimation of these flow-rate limitations, see Lobo, *et al.* (1945).

These fixed-bed momentum transfer results can not only be generalized to embrace *multidimensional* flows through isotropic fixed beds, but can also be used to estimate the interphase forces associated with the relative motion between a dense cloud (e.g., spray of viscous droplets) and its "host" (or "carrier") fluid.

Further details on the momentum-transfer situations we have explicitly discussed here, and many others, will be found in the Reference list. However, with the quantitative understanding of momentum convection and diffusion in both "external" and "internal" situations gained from this chapter, we are now in a position to quantitatively take up *energy* (Chapter 5) and *mass* (Chapter 6) *transport processes* in these same physical situations, even allowing for the effects of chemical reaction (Sections 5.8, 6.5.3, 6.5.4, and 6.5.5) on energy and/or mass-transfer rates. Of course, energy and/or species mass can *diffuse* in a completely *quiescent* medium; i.e., one free of any *momentum* convection or diffusion. For completeness, several simple, but technologically important, examples of energy and mass diffusion in quiescent media will be treated early in Chapters 5 and 6.

SUMMARY

- Variable density in a fluid flow produces surprising effects, such as the result that the flow velocity can *increase* in a diverging section of duct (in steady supersonic flow) without violating the conservation of mass constraint.
- The local speed of sound (and, hence, Mach number) plays an important role in compressible fluid flow, *via* the mechanism of communicating pressure information. Once sonic speed is reached locally, downstream pressure changes leave the upstream flow unaltered, leading to so-called "choking" phenomena in ducts.
- Gas-dynamic "discontinuities," called "shock" waves, can appear in nonreactive compressible fluid flows. These are abrupt transitions, fully consistent with the conservation laws, "available" to the fluid to meet what would have been

incompatible boundary conditions in their absence. In chemically *reactive* systems, deflagration waves (subsonic) and detonation waves (supersonic) play the same role.

- The important parameter governing the momentum-transfer behavior of viscous fluids is the Reynolds' number, Re:

$$\mathrm{Re} \equiv \frac{\rho U U}{\mu U/L} = \frac{\text{convective momentum flux}}{\text{diffusive momentum flux}}.$$

- When $\mathrm{Re}^{1/2} \ll 1$, viscous effects prevail everywhere in the flow field. When $\mathrm{Re}^{1/2} \gg 1$, viscous effects are confined to thin boundary layers (BLs) near (or produced by) solid walls, simplifying the analysis of such flows (Prandtl BL theory).
- In many cases, the quantity of interest is the *friction drag*. The distribution of shear stress along a surface can often be predicted using simplified forms of the momentum-balance equation valid within thin boundary layers. The relevant dimensionless momentum-transfer coefficient is the *skin-friction coefficient*:

$$C_{\mathrm{f}} \equiv \frac{\tau_{\mathrm{w}}}{\frac{1}{2}\rho U^2},$$

which will be Re-dependent at each streamwise position x/L. Half the local skin-friction coefficient ($c_{\mathrm{f}}/2$) plays the same role (in momentum transfer) as the *Stanton numbers* play in forced convective *energy* and *mass* transfer.

- However, *friction* is not the only contributor to the drag; for "bluff" bodies the more important contributor is the streamwise component of the net *pressure* forces, associated with the so-called "*form drag*."
- Turbulence greatly augments *momentum* transfer rates. This is often a disadvantage (increased drag, pressure drop). Most *turbulent* c_{f}-values are either measured, or "predicted" with the help of turbulent data on a simpler configuration (e.g., fully developed turbulent flow in a straight pipe of circular cross section). In *jets* and *wakes* turbulence also dramatically increases momentum diffusion rates, and associated entrainment rates.

TRUE/FALSE QUESTIONS

4.1 T F The local fluid velocity vector, **v**, can also be viewed as the local linear momentum per unit mass of fluid.

4.2 T F The "no-slip" condition is an immediate consequence of the principle of conservation of tangential momentum at fluid/solid interfaces.

4.3 T F The velocity profile for steady viscous flow of a non-Newtonian fluid in a straight circular pipe is parabolic, provided the flow is laminar.

4.4 T F "Reynolds' analogy" is a quantitative relation between the dimensionless skin-friction coefficient and the corresponding heat or mass-transfer coefficients for the same fluid/flow conditions.

4.5 T F The surface roughness of a conduit has a greater effect on fluid motion in the turbulent regime than in the laminar flow regime.

4.6 T F The terminal settling velocity of an isolated solid sphere of known mass can be used to infer the fluid viscosity, provided the Reynolds' number is small enough to be in the Stokes' regime.

4.7 T F Chemically reacting gas flows can always be treated as incompressible provided the square of the prevailing Mach number is sufficiently small.

4.8 T F Linear momentum is not conserved in swirling flows; only *angular momentum* is conserved in such flows.

4.9 T F The friction factor for viscous fluid flow through a fixed granular bed undergoes an abrupt increase at a "transition" bed Reynolds' number of the order of 2000.

4.10 T F The "buoyancy force" per unit volume is not an additional fundamental force of nature. It is simply the combination $-\mathbf{grad}\, p + \rho\mathbf{g}$, which is locally nonzero in variable-density situations accompanied by a body force $\mathbf{g}$ per unit mass.

4.11 T F Natural ("free"-) convection flows do not exhibit transition to turbulence.

4.12 T F Poiseuille's law is the basis of both the capillary viscosimeter and the capillary flowmeter.

4.13 T F If the drag coefficient for an object is independent of Reynolds' number, then in this range the actual drag force will increase as the cube of the velocity.

4.14 T F The drag force on an object is necessarily directly proportional to its "frontal area."

4.15 T F In a duct of increasing cross-sectional area the velocity of a steady-flow must necessarily decrease (as a consequence of total mass conservation).

EXERCISES

4.1 It is often stated that at solid/"real" fluid interfaces the fluid velocity equals the velocity of the surface itself. Moreover, Batchelor (1967) states that "unless rupture occurs" at the interface between two contacting media, the *normal* component of fluid velocity must be continuous across the interface. Suppose, however, that the solid is subliming into a gas (e.g., napthalene in air). Would the above statements be true? Discuss.

4.2 It is often stated that the shear stress is negligible at gas/liquid interfaces (e.g., along a falling liquid film). However, when wind drives a film of rainwater up your car windshield against the pull of gravity, would this assumption be justified? Discuss.

4.3 Can the following interfaces or "fronts" be treated in accordance with the principles discussed in Sections 2.6.1 and 4.4.3?

a. a 1 mm thick "laminar" flame?
b. a 1 cm thick "turbulent" flame?
c. a 1 km thick meteorological "cold front"?
d. a 1 light-year "front" of ion number density in an intergalactic gas cloud?

4.4 For a combustion turbine materials-test program it is desired to expose specimens to a *sonic* but atmospheric pressure jet of combustion-heated air with a post-combustion (stagnation) chamber temperature of 2000 K.

a. What should the *pressure* in the combustion chamber (upstream of the nozzle) be?
b. What should the *shape* of the nozzle be?
c. How much air flow (g/s) must be handled if the exit jet must be 2.5 cm. in diameter?
d. What will the exit (jet) *velocity* be?
e. What will be the flow rate of axial momentum at the nozzle exit?
f. By what factor will the gas *density* change in going from the nozzle inlet to the nozzle outlet?

4.5 Consider a dilute acetylene–air mixture at atmospheric pressure and 300 K with a C_2H_2 *mass* fraction of 4.5%. If the heat of reaction of acetylene is 11.52 kcal/gm C_2H_2, estimate:

a. the speed with which a Chapman–Jouguet *detonation* would propagate through this mixture (km/s);
b. the stagnation pressure (atm) immediately behind such a detonation wave;
c. the corresponding Chapman–Jouguet *deflagration* speed (km/s).

(It will be interesting to compare this to the *measured* "laminar" flame speed at this composition.)

4.6 Consider the steady axisymmetric flow of hot air in a straight circular tube of radius a_w and cross-sectional area A.

Conditions (at exit):

$$p = 1 \text{ atm (uniform)}$$
$$T = 1500 \text{ K (uniform)}$$
$$a_w = 5 \text{ mm}$$

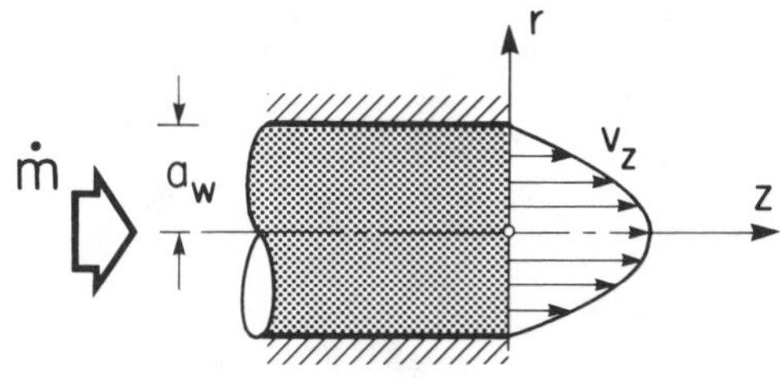

Figure 4.6-1E

Suppose it has been observed that the *axial-velocity profile* $v_z\{r, z(\text{exit})\}$ is, in this case, well described by the simple equation:

$$v_z = 2U\left[1 - \left(\frac{r}{a_w}\right)^2\right],$$

where $U = 10^3$ cm/s. Using this observation, the conditions above, and Table 8.1-1, answer the following questions:

a. If the molecular mean-free-path in air is approximately given by the equation:

$$\ell(\text{air}) \approx 0.065\left(\frac{T}{300}\right)^{1.2}\frac{1}{p} \qquad (\text{in } \mu\text{m}),$$

estimate the prevailing mean-free-path ℓ and the ratio of ℓ to the duct diameter—i.e., the relevant *Knudsen number* for the gas flow:

$$\text{Kn} \equiv \frac{\ell}{2a_w}.$$

What conclusions can you now draw concerning the validity of the *continuum* approach in this case?

b. Calculate the convective mass flow rate $\dot{m}$ (expressed in g s^{-1}) through the entire exit section. For this purpose assume the approximate validity of the "perfect" gas law, *viz.*:

$$\rho = \frac{pM}{RT} \qquad (\text{g/cm}^3).$$

Here p is the pressure (expressed in atm), M is the molecular weight (g/g-mole) (28.97 for air), $R = 82.06$ (univ gas const), and T is the absolute temperature (expressed in kelvins). Also note that for this axisymmetric flow a convenient area element is the annular ring sketched in Figure 4.6-2E, i.e.,

$$d\mathbf{A} = (2\pi r \cdot dr)\mathbf{e}_z$$

(where $\mathbf{e}_z$ is the unit vector in the z-direction).

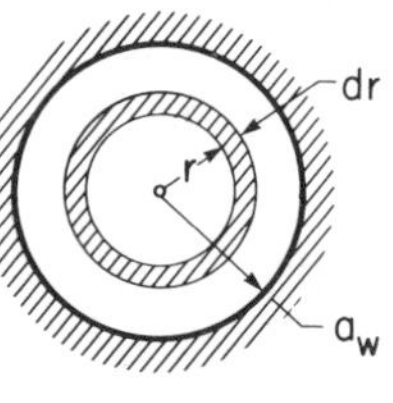

Figure 4.6-2E

c. Also calculate the *average* gas velocity $U_{avg} \equiv \dot{m}/(\rho A)$ at the exit section and the corresponding Reynolds' number:

$$\mathrm{Re} = \frac{\rho U_{avg} U_{avg}}{\left[\frac{\mu U_{avg}}{2a_w}\right]} = \frac{\rho U_{avg}(2a_w)}{\mu}.$$

d. Calculate the *convective axial-momentum flow rate* (expressed in $g \cdot cm/s^2$) through the exit section. Is your result equivalent to $\dot{m}U_{avg}$? Why or why not?

e. Calculate the *convective kinetic-energy flow rate* (expressed in $g \cdot cm^2/s^3$). Is your result equivalent to $\dot{m}(U_{avg}^2/2)$? Why or why not?

f. If, in addition to the axial component of the velocity v_z, the air in the duct also has a swirl component $v_\theta(r, z)$, how would this influence your previous estimates (of *mass* flow rate, momentum flow rate, kinetic energy flow rate)? Briefly discuss.

g. If the local shear stress τ_{rz} is given by the following degenerate form of Newton's law:

$$\tau_{rz} = \mu \frac{\partial v_z}{\partial r},$$

at what radial location does τ_{rz} maximize? Calculate the maximum value of τ_{rz} and express your result in dyne cm^{-2} and Newton m^{-2}. Calculate the *skin-friction coefficient*, c_f (dimensionless), at the duct exit. At what radius does τ_{rz} take on its *minimum* value? Can $-\tau_{rz}$ be regarded as the radial diffusion flux of axial momentum? Why or why not? Does the rate at which work is done by the stress τ_{rz} maximize at either of the two locations found above? Why or why not?

h. Characterize this flow in terms of the descriptors of Section 2.7, and defend your choices.

4.7 A Newtonian fluid of constant density ρ enters a horizontal straight duct of radius a_w with the radially uniform velocity $v_{z①} = U(= \text{const})$ and steadily emerges from the duct with the radially *nonuniform* velocity profile:

$$v_{z②}(r) = 2U \cdot [1 - (r/a_w)^2].$$

a. Calculate the net outflow rate of z-momentum for this (hydrodynamically "developing") flow.

b. What is the relationship between your answer to Part (a) and the two surface *forces*:

$$-(p_② - p_①) \cdot \pi a_w^2 \qquad \text{and} \qquad 2\pi a_w \int_{z①}^{z②} \tau_w(z)\, dz,$$

where $\tau_w\{z\} = \mu(\partial v_z/\partial r)_{r=a_w}$ is the local shear stress τ_{rz} along the duct surface $r = a_w$.

c. If at a downstream station ③ the velocity distribution is the same as at station ②, then what is the relation between the same surface forces mentioned in Part (b), but evaluated using conditions at stations ③ and ②? (Derive a useful expression which relates the wall shear stress τ_w to the readily measured pressure drop $-(p_{③} - p_{②})$.)

4.8 The streamwise average skin friction factor for Newtonian laminar flow in a straight duct of circular cross section has been shown to approach 16/Re at values of $\xi_{mom} \equiv \mathrm{Re}^{-1}(z/d_w)$ large enough for the flow to be "fully developed." Near the duct inlet, however, where the velocity profile more closely approximates "plug-flow" ($v_z = U = \text{const}$), values of this momentum transfer coefficient C_f are much larger, and may be described *via* a dimensionless momentum entrance function; $F\{\xi_{mom}\}$, defined by:

$$\bar{C}_f \equiv \frac{1}{\frac{1}{2}\rho U^2} \cdot \frac{1}{z} \int_0^z \tau_w\{z\}\, dz \equiv F\{\xi_{mom}\} \cdot \frac{16}{\mathrm{Re}},$$

where $F\{\infty\} = 1$ (see, also, Part (c)).

a. Using Blasius' result for the laminar boundary layer skin-friction distribution on a *flat plate* (Eq. (4.5-37)), show that $F\{\xi_{mom}\}$ defined above must have the "initial" (small ξ_{mom}-) behavior:

$$F\{\xi_{mom}\} \sim \frac{2(0.6641)}{16} \cdot \xi_{mom}^{-1/2}.$$

b. If the entire function $F\{\xi_{mom}\}$ is approximated by a simple two-parameter (A, m) curve-fit of the form:

$$F\{\xi_{mom}\} \cong [1 + (A\xi_{mom})^{-m/2}]^{1/m} \qquad (0 \leqslant \xi_{mom} \leqslant \infty),$$

determine the appropriate value of A (*via* Part (a) above), and outline a possible method for estimating the "best" value of the remaining parameter m.

c. On the basis of the results of Part (b), estimate the value of ξ_{mom} at which $F\{\xi_{mom}\} = 1.02$ (i.e., is within 2% of $F\{\infty\}$).

4.9 Using Figures 4.4-2 and -3 and Eq. (4.5-38), estimate the drag force (Newtons)[67] per meter of length for each of the following long objects

[67] Also express your results in terms of the force exerted by the earth's gravity on a 1 kg mass.

of transverse dimension 5 cm if placed in a heated air stream with the following properties:[68] $T_\infty = 1200$ K, $p_\infty = 1$ atm, and $U = 10$ m/s.

a. A circular cylinder.
b. A thin "plate" perpendicular to the stream (i.e., at 90° incidence).
c. A thin plate aligned with the stream (i.e., at 0° incidence).

In each case qualitatively discuss how the drag is apportioned between "form" (pressure difference) drag and "friction" drag (see Section 4.4.1).

For Part (c), is the application of *laminar* boundary-layer theory (Section 4.5.2) likely to be valid? (Briefly discuss your reasoning.) If so, what would be the estimated BL thickness, δ, at the trailing edge of the plate, i.e., at $x = L$? Suppose *two* such adjacent plates were separated by a distance much greater than $\delta(L)$—would they strongly "interact" with respect to momentum transfer?

d. Justify the use of an *incompressible* Newtonian fluid C_D(Re, shape) curve to solve Part (a) (involving the *gas* air) by showing (using Eq. (4.3-43)) that $\frac{1}{2}(\mathrm{Ma})^2$ is small enough under these conditions to neglect $(\Delta\rho/\rho)_{\mathrm{air}}$.

4.10 A domestic automobile manufacturer recently announced that it has developed a passenger car with a drag coefficient, C_D, (at cruise conditions) of only 0.15. This is to be compared with the values 0.30 to 0.35 typical of most production passenger cars.

a. At a cruise speed of 60 mph, compare the power (in kW and HP) needed to overcome the aerodynamic drag for both a new and conventional car with a frontal area of 3.75 m^2.
b. What is the Reynolds' number (per meter) at cruise speed when $p = 1$ atm, $T = 300$ K?
c. How do the aforementioned C_D-values compare to the drag coefficient of a sphere in the same Reynolds'-number range?
d. What would be the corresponding power and drag coefficient for a 2 m high by 6 m long thin plate *aligned* with a 60 mph flow?

4.11 Reconsider the sonic jet test facility specified in Exercise 4.4, from the viewpoint of turbulent jet momentum exchange (mixing) with the surrounding atmosphere, and the entrainment of that atmosphere (Section 4.6).

a. Calculate the Reynolds' number $\mathrm{Re}_j \equiv U d_j/\nu_j$ at the nozzle exit and compare it to the value 3×10^4 above which such jets almost certainly lead to *turbulent* mixing with the surrounding atmosphere.
b. *Estimate* the appropriate value of ν_t by assuming that the relevant density in Eq. (4.6-1) is about the arithmetic mean between ρ_j and ρ_∞. How much larger is the effective turbulent momentum diffusivity, ν_t, than the intrinsic momentum diffusivity of the jet fluid ($\nu_j \equiv \mu_j/\rho_j$)?
c. *Estimate* the downstream distance at which the time-averaged velocity $\bar{v}_z$ (axial momentum per unit mass) along the jet centerline will be reduced to 10% of the initial jet velocity U_j (axial momentum per unit mass) as a

[68] See Table 8.1-1 for properties of air under these conditions.

result of momentum diffusion. Compare this to the result that would have been obtained had the jet remained laminar (with kinematic viscosity ν_j).

d. At this location what is the approximate ratio between the entrained (laboratory air) mass flow and the "primary" (combustion-heated air) jet? (See Eq. (4.6-5) with $\nu \rightarrow \nu_t$.)

4.12 To predict the distribution of skin-friction coefficient, $c_f(x)$ (Section 4.5.2) and heat-transfer coefficient $St_h(x)$ (Chapter 5) along the surface of each blade of an air-breathing axial-flow combustion turbine, it is necessary to calculate the local mass flux ("mass velocity") $G_e \equiv \rho_e(x) \cdot u_e(x)$ along the outer edge of the viscous boundary layer (BL), where $u_e(x)$ is the tangential gas velocity measured *relative* to the blade (which may be a rotor blade or stator ("vane" or "nozzle")). Consider a *rotor* blade stage with the inlet stagnation[69] state conditions: $T_0 = 1400\,\text{K}$, $p_0 = 20\,\text{atm}$, and calculate u_e, ρ_e, G_e, p_e and the local *Mach* number at a position along the blade at which the turbine designer has determined that $u_e/u_* = 0.75$ (where u_* is the sonic (acoustic) speed for an isentropic expansion from the stated stagnation state conditions). Defend your assumptions, method of calculation, and choice of thermophysical parameters (γ, etc.). Compare the actual local pressure $p_e(x)$ to the pressure that would correspond to the same local velocity if the *incompressible* Bernoulli equation were (erroneously) used, assuming the density remained equal to the stagnation density ρ_0. If the relative gas velocity entering (approaching) the rotor stage is $0.413\,u_*$, what is the gas *static* pressure entering the stage?

4.13 For "steady" turbulent fluid flows (i.e., flows for which *time-averaged* dependent variables are themselves independent of time), we have shown that each unit mass of fluid has a *turbulent kinetic energy* κ_t, where (in a Cartesian coordinate system):

$$\kappa_t = \frac{1}{2}(\overline{v'_x v'_x} + \overline{v'_y v'_y} + \overline{v'_z v'_z}).$$

a. For a fixed macroscopic control volume enveloping a "steady" *turbulent* fluid flow, should the mechanical-energy equation (generalized Bernoulli equation) now be written:

$$\int_S \bar{\rho}\left(\frac{\bar{p}}{\bar{\rho}} + \frac{\bar{v}^2}{2} + \phi + \kappa_t\right)\bar{\mathbf{v}} \cdot \mathbf{n}\,dA = -\int_V \left(\frac{\text{visc. dissip.}}{\text{volume}}\right)dV - \bar{\dot{W}}_{\text{shaft}}$$

(where the energy density $\bar{\rho}\kappa_t$ is now included in the net outflow surface integral, and all other variables are *time-averaged*)? Defend your conclusion.

[69] Based on gas velocity measured relative to the (moving) blade.

b. When would it be important to take into account changes in turbulent kinetic energy?

c. How would you go about deriving the *p*artial *d*ifferential *e*quation which κ_t must locally satisfy within such a flow field? Is this equation useful?

4.14 Can a "sharp" front between a nonturbulent region of flow and a turbulent region of flow be treated using jump-conservation conditions? What would the jump conditions across such a "front" look like; *i.e.*, what "new" (explicit) terms would they have to contain?

4.15 For any two-dimensional (planar or axisymmetric) steady (viscous) fluid flow, the incompressible fluid *mass conservation* condition, div $\mathbf{v} = 0$ (Eq. (2.1-3)) ensures the existence of a single *stream function* $\Psi\{\mathbf{x}\}$ from which the two fluid-velocity components can be obtained by appropriate partial differentiation. Rework the classical Poiseuille viscous-flow problem (Section 4.5.1) in terms of a single (Stokes) stream function, Ψ, for which:

$$v_z = -\frac{1}{r}\cdot\frac{\partial\Psi}{\partial r}, \qquad v_r = \frac{1}{r}\cdot\frac{\partial\Psi}{\partial z};$$

that is, what differential equation (DE) and boundary conditions (BC) does Ψ satisfy, and what is your explicit result for $\Psi\{r\}$? Also show that if v_z and v_r are calculated as above, then div $\mathbf{v} = 0$ (Eq. (4.5-5)) is automatically satisfied for this incompressible, steady, axisymmetric flow without swirl. Could the stream-function approach be used for a *variable density*, two-dimensional, steady, chemically reacting flow governed by the total mass conservation PDE div$\{\rho\mathbf{v}\} = 0$? If so, how would v_z and v_r (cf. equations above) be obtained from Ψ? What is the advantage of working with a single stream function instead of the two individual fluid-velocity components?

4.16 If a fluid parcel has a vorticity $\boldsymbol{\zeta}$ (twice the local fluid rotational rate vector; Section 3.2.2) relative to a fixed observer (ICF):

a. What will be its *apparent vorticity*, $\boldsymbol{\zeta}_{\text{app}}$, relative to an observer on a coordinate frame (ACF; Section 2.6.2) accelerating with angular velocity $\boldsymbol{\Omega}$? (*Ans.*: $\boldsymbol{\zeta}_{\text{app}} = \boldsymbol{\zeta} - 2\boldsymbol{\Omega}$.)

b. Will a "solid-body"-type rotational flow therefore appear "irrotational" to an observer on a coordinate frame rotating at an appropriate angular velocity?

4.17 The linear-momentum conservation PDE for a constant-property Newtonian fluid can be rewritten in a form that indicates that fluid rotation (hence "vorticity") "diffuses" (with diffusivity $\nu \equiv \mu/\rho$; cf. Section 4.5.2) into regions of fluid formerly without fluid rotation. Derive the convective–diffusion PDE for vorticity in the presence of nonequilibrium chemical reaction (cf. Eq. [4.5-39]).

4.18 A sample "core" (8 cm long, 2 cm diam.) of a porous rock obtained from an oil reservoir is placed in a cylindrical holder to measure the flow of pressurized water through it. With a pressure drop of 1.0 atm, the water flow through the core at 20.2°C was measured as 2.60 cm^3s^{-1}. What is the *permeability* of this rock (expressed in Darcy), and is this value relevant to the flow of *other* Newtonian viscous fluids through this same material? (N.B.: The "permeability" of a porous material is the product of the liquid viscosity, μ, and the (superficial) fluid velocity produced by a unit pressure gradient; in particular, the Darcy is 1 cm^2 cP s^{-1} atm^{-1}, where 1 cP $\equiv$ 1 centiPoise.)

4.19 To regenerate (clean) a 0.2 m thick sand-bed filter, it is proposed to "fluidize" the sand bed at minimum conditions in a constant-area vessel using water at 24°C. The sand particles comprising the filter are nominally spherical and have a density of 2550 kg/m^3 and an average diameter of 0.40 mm. Assuming the sand filter has the void fractions 0.38 and 0.42 before and during regeneration, respectively,:

a. Predict the height of the "expanded" bed under these minimum (incipient) fluidization conditions.
b. What is the pressure drop under regeneration conditions?
c. Estimate the minimum water velocity required for the incipient fluidization cleaning process.

REFERENCES

Barchilon, M., and R. Curtet, *ASME Trans., J. Basic Engrg.* **86**, pp. 777–787 (1964).

Bird, R.B., W. Stewart, and E.N. Lightfoot, *Transport Phenomena.* New York: J. Wiley (1960).

Blasius, H., *Z. Math. u. Phys.* **56**, 1 (1908); see also *NACA Tech Memo No. 1256* (translation into English).

Ergun, S., *Chem. Engrg. Prog.* (*AIChE*) **48**, p. 89 (1952).

Fernandez de la Mora, J., and D.E. Rosner, *PhysicoChem. Hydrodynamics* **2**, pp. 1–21 (1981).

Hess, J.L., and A.M.O. Smith in *Progress in Aeron. Sci.*, Vol. 8 (Pergamon Press, N.Y), pp. 1–138 (1966). See, also, Hess, J.L., in *Computer Methods in Applied Mechanics and Engineering* 5, pp. 145–196 (1975).

Levich, V.G., and V.S. Krylov, in Vol. 1, *Annual Review of Fluid Mechanics.* Palo Alto, CA: Annual Reviews, Inc., pp. 293–316 (1969).

Lobo, *et al., Trans. AIChE* **41**, 693 (1945).

Robinson, A., *Commun. Pure Appl. Math.* **IX**, pp. 69–84 (1956).

Rodi, W., *AIAA J.*, **20**, pp. 872–879 (1982).

Thring, M.W., and M.P. Newby, in *Fourth Symposium on Combustion.* Baltimore, MD: Williams and Wilkins (1953); pp. 789–796.

Van Dyke, M., in Vol. 1, *Annual Review of Fluid Mechanics.* Palo Alto, CA: Annual Reviews, Inc., pp. 265–292 (1969).

BIBLIOGRAPHY: MOMENTUM TRANSPORT

Elementary

Prandtl, L., and O.G. Tietjens, *Applied Hydro. and Aeromechanics.* New York: Dover, 1957.

Von Karman, T., *Aerodynamics*, New York: McGraw-Hill 1963 (paperback).

Intermediate

Ball, T.F. ed., *Handbook of Supersonic Aerodynamics*, U.S. Navy Dept., Bureau of Ordnance, Washington, DC, NAVORD Report 1488, Vol. 1, Section 4, April 1, 1950.

Batchelor, G.K., *An Introduction to Fluid Dynamics.* Cambridge, U.K.: Cambridge Univ. Press (1967).

Beer, J.M., and N.A. Chigier, *Combustion Aerodynamics.* London: Appl. Sci. Publishers (1972).

Bradshaw, P., *An Introduction to Turbulence and its Measurement.* Elmsford, NY: Pergamon Press (1971).

Churchill, S.W., *The Practical Use of Theory in Fluid Flow.* Etanor Press (1980).

Denn, M., *Process Fluid Mechanics.* Englewood Cliffs, NJ: Prentice-Hall (1980).

Launder, R.E. and D.B. Spalding, *Mathematical Models of Turbulence.* New York: Academic Press (1972).

Liepmann, H.W., and A. Roshko, *Elements of Gas Dynamics.* New York: J. Wiley (1957).

Pai, S.-I., *Modern Fluid Mechanics.* Science Press (Beijing; Van Nostrand Reinhold, distributor, New York) (1981).

Rosenhead, L. (ed.), *Laminar Boundary Layers.* London: Oxford Univ. Press (1963).

Rosner, D.E., *Chemical Engineering Education.* Fall 1976, pp. 192–194.

Shapiro, A., *The Dynamics and Thermodynamics of Compressible Fluid Flow.* New York: Ronald Press (1953).

Thompson, P.A., *Compressible Fluid Dynamics.* New York: McGraw-Hill (1972).

Thwaites, B. (ed.), *Incompressible Aerodynamics.* Oxford: Clarendon (1960).

White, F.M., *Fluid Mechanics.* New York: McGraw-Hill (1979).

Whittaker, S., *Introduction to Fluid Mechanics.* Englewood Cliffs, NJ: Prentice-Hall, Inc. (1968).

Zucrow, M.J., and J.D. Hoffman, *Gas Dynamics* (two volumes). New York: J. Wiley (1976).

Advanced

Bradshaw, P., T. Cebeci, and J.H. Whitelaw, *Engineering Calculation Methods for Turbulent Flow.* New York: Academic Press (1981).

Cebeci, T., and A.M.O. Smith, *Analysis of Turbulent Boundary Layers* (Vol. 15 Appl. Math. Mechanics Monographs (F. Frankiel, G. Temple, eds.)) New York: Academic Press, (1974).

Crocco, L., "One-Dimensional Treatment of Steady Gas Dynamics," in *Fundamentals of Gas Dynamics* (Vol. 3, *High-Speed Aerodynamics and Jet Propulsion*). Princeton, NJ: Princeton Univ. Press (1958).

Emmons, H. (ed.), *Fundamentals of Gas Dynamics* (Vol. 3, *High-Speed Aerodynamics and Jet Propulsion*). Princeton, NJ: Princeton Univ. Press (1958).

Goldstein, S., *Modern Developments in Fluid Dynamics* (Vol. II). London: Oxford (Clarendon) Press (1938); pp. 676–680.

Hayes, W.D., and R.F. Probstein, *Hypersonic Flow Theory* (Vol. 1). New York: Academic Press (1966).

Hinze, J.L., *Turbulence.* New York: McGraw-Hill (1975); second edition.

Lamb, H., *Hydrodynamics.* New York: Dover (1945); sixth edition.

Landau, L.D., and E.M. Lifshitz, *Fluid Mechanics* (Vol. 6, Course in Theoretical Physics). Reading, MA: Addison-Wesley (1959).
Milne-Thompson, L.M., *Theoretical Hydrodynamics.* New York: MacMillan (1955); third edition.
Moore, F.K. (ed.), *Theory of Laminar Flows* (Vol. 4, *High-Speed Aerodynamics and Jet Propulsion*). Princeton, NJ: Princeton Univ. Press (1964).
Schlichting, H., *Boundary-Layer Theory.* New York: McGraw-Hill (1979); seventh edition.
Temam, R., *Navier-Stokes Equations—Theory and Numerical Analysis,* Third (Revised) Edition. Amsterdam: Elsevier/North-Holland (1984).
Van Dyke, M., *Perturbation Methods in Fluid Mechanics.* New York: Academic Press (1969).
Wolfshtein, M., D. Naot, and A. Lin, "Models of Turbulence," Ch. 1 in *Topics in Transport Phenomena* (C. Gutfinger, ed.). Washington, D.C.: Hemisphere Press (1975); pp. 3–45.

Additional Sources of Information

Annual Reviews of Fluid Mechanics, Annual Reviews, Inc. Palo Alto, CA.
Journal of Fluid Mechanics.
Physics of Fluids, American Physical Society.
AIAA J, Amer. Inst. Aeronautics and Astronautics.
Journal of Heat and Fluid Flow (*Butterworth*).
ASME Trans.-J. Fluids Engineering.

5

Energy Transport Mechanisms, Rates, and Coefficients

5.1 RELEVANCE

Again using combustors as a representative example, it is clear that the *rate of energy transfer* strongly influences, if it does not *control*, the rate at which combustion regions can "propagate" into regions containing unburned fuel, especially if oxidizer is already present there.

In the case of condensed fuels, energy transfer is necessary to (1) vaporize the fuel prior to homogeneous reactions (cf. Figure 1.1-3) and (2) heat the resulting vapors to an effective "ignition" temperature above which reaction rates (and associated energy release rates) are adequate to sustain combustion. The latter process is necessary even in the combustion of premixed *vapors* (cf. Figure 1.1-1). It is therefore not surprising that the subject of *energy transport* plays a central role in the subject of combustion science/technology in particular, and chemical reaction engineering, generally. Often the relevant laws of energy transport are no different than those applicable in systems without chemical reaction (Sections 5.2–5.7). However, in some cases, even the basic laws of energy transport must be modified for chemically reactive systems (Section 5.8)—usually due to energy transport as a result of the diffusion of energetic (high-enthalpy) chemical species.[1] Our objective in the present chapter is to present concisely the essential features of energy transport in both quiescent and flowing media. Particular emphasis will be placed on *analogies to momentum transport* (Chapter 4) and topics essential in understanding/predicting important combustion phenomena. For details beyond the scope of this summary, consult the references at the end of the chapter.

5.2 MECHANISMS OF ENERGY TRANSPORT

There are three important *mechanisms of energy transport*; *viz.*:

[1] Radiation transport from chemical species "pumped" into nonequilibrium energetic states by chemical-energy release would be another example. Such radiation, called "chemiluminescence," will not be treated in Section 5.9, but has extremely important applications (e.g., chemical lasers).

1. Convection;
2. Diffusion (also called Heat Conduction); and
3. Radiation;

and these are discussed sequentially below. As will be seen, the first two (*convection* and *diffusion*) are completely analogous to their counterparts for linear-momentum transport in material media (cf. Chapters 2 and 4). However, *radiation*, a form of energy transport that can occur even through a perfect vacuum, is unique, and governed by rather different laws. All three mechanisms participate in most high-temperature, chemically reacting systems. Radiation becomes especially important in high temperature systems that are large (e.g., ignited oil spills or forest fires) and/or at high pressures (e.g., rocket-motor combustion).

5.2.1 Convection

Whether a material medium is deformable or not, if it is in a state of *motion* then it transports energy by *convection*. Thus, if $\rho\mathbf{v}$ is the local mass-flux vector, and h the enthalpy per unit mass, then $\rho\mathbf{v}h$ is the corresponding *local enthalpy flux by convection*. Note that, according to this viewpoint, *moving solids* transport energy by convection.[2] More commonly perhaps, solids are at rest and convection usually occurs in *fluids*. When the fluid motion is mechanically produced and not itself the result of localized energy (or mass) transfer, the convection is said to be "forced." But even in the absence of "forced" motion, energy (or mass) transfer can lead to local density nonuniformities. In such cases each unit volume of fluid in a body force field $\mathbf{g}$ experiences a *local* force

$$-\mathbf{grad}\, p + \rho\mathbf{g} \equiv \text{“Buoyancy” force}, \tag{5.2-1}$$

which, in variable density fluids, produces a convection termed "natural" convection, or "free" convection. In general, both natural and forced convection can operate simultaneously; however, whatever the "cause" of the motion, the local enthalpy flux by *convection*[3] remains $\rho\mathbf{v}h$.

5.2.2 Diffusion (Heat Conduction)

The word diffusion implies "spreading" from one material region to an *adjacent* material region. *Energy* can diffuse in a motionless solid or liquid *via* the propa-

[2] Examples would be energy transfer *via* hot "cannon balls," firebrands, and the like. When a hot torch is played upon a rotating solid part (e.g., on a glass lathe), energy is transported by *convection* associated with solid body rotation. Similarly, when a "traveling" grate loaded with a bed of pellets (e.g., as in zinc sulfide "roasting") moves past a cross-flow of air, energy (and mass) are transported by *solid* convection.
[3] The etymology of the word *convection* (and *advection*, used by meteorologists) is the Latin word for "carry."

gation of molecular ("lattice") vibrations, and through "motionless" gases *via* the motion (convection!) of its constituent molecules. Whatever the microscopic mechanism by which energy can "diffuse" (spread) in macroscopically motionless continua, the process is called *heat diffusion*, heat conduction, molecular conduction, or Fourier conduction. The Fourier heat flux, written:

$$(\dot{\mathbf{q}}'')_{\text{Fourier}} = -k\,\mathbf{grad}\,T, \tag{5.2-2}$$

defines the local *thermal conductivity*, k, of the material (cf. Eqs. (3.3-1 and 3.3-3)).

When the continuum solid or fluid is both in macroscopic *motion and* non-uniform in temperature, then energy transport occurs by *both* diffusion and convection. This is the usual situation, i.e., energy diffusion occurs in moving media, hence *along with* convection.

When the fluid motion is *turbulent*, small parcels of fluid ("eddies") are tossed about, much as individual molecules are in gases. In the presence of a spatial gradient in the time-averaged fluid temperature, the time-averaged effect of this "eddy convection" is a net energy flux, ironically called "eddy diffusion." Such a moving fluid appears to have a local thermal diffusivity, α_{eff}, larger than $k/(\rho c_p)$ by an amount written as α_t. However, whereas $k/(\rho c_p)$ is a local *fluid* property, the turbulent contribution α_t itself depends on the local fluid motion (see Section 5.7.2, and cf. our previous discussion of linear-momentum transport in Section 2.6.2). In so-called "fully" turbulent regions ($\alpha_t \gg (k/\rho c_p)$), however, where turbulence is strongly damped (e.g., near a motionless solid wall), $\alpha_t \to 0$ and $\alpha_{\text{eff}} \to k/(\rho c_p)$.

5.2.3 Radiation

In contrast to convection and diffusion, energy transport by "radiation" occurs through *empty* space. This can be regarded either as the propagation of *electromagnetic waves* or as the flow of "photons," i.e., corpuscles of radiation. The latter view is perhaps most useful in the present context, since there are certain useful analogies between the behavior of photons and material molecules. Indeed, radiation-filled space can be described as containing a "photon gas" or photon "phase," which may interact with the co-existing "material phase."

While photons can pass through empty space, they require matter for their emission, scattering (deflection), and absorption (capture). Two features of radiation important to what follows are:

a. All matter, at any temperature above 0 K, *emits* a distribution of ("thermal") photons. The rate of emission and the distribution of this energy/photon is strongly temperature-dependent (see Section 5.9.1).
b. In contrast to ordinary molecules, in many engineering applications the distance a photon travels before absorption or scattering is *not* small compared to the relevant macroscopic dimensions of the system (cf. Figure 1.2-1). Thus,

energy transfer by radiation *cannot* usually be treated as a "diffusion" process, and Eq. (5.2-2) does not apply.[4]

In a sense (cf. Section 2.7.1), whereas in engineering applications the *material molecules* are well into the continuum flow region, the photons are often in the free-"molecule" flow regime. In the limiting case where the photons do not appreciably interact with the intervening gas, then energy-transfer predictions simplify considerably, since the real-gas motion and photon-gas motion become completely uncoupled and the photons travel in straight lines from boundary surface-to-surface, "rattling around" in the reactor. In that case, radiation energy-transfer appears explicitly in striking an energy balance on the boundary surfaces, but disappears from the energy balance on a representative parcel of intervening gas (e.g., fuel vapor, air, combustion product vapors). (Cf. Sections 5.9.3 and 5.9.4.)

5.3 OBTAINING TEMPERATURE FIELDS AND CORRESPONDING SURFACE ENERGY-TRANSFER RATES AND COEFFICIENTS

5.3.1 Relation between Temperature Field, Wall Heat Transfer Rates, and Heat-Transfer Coefficients

Isotherms or selected temperature profiles can be experimentally mapped by probing a flow field (see, e.g., Figures 1.2-3 and 1.2-4), using either:

a. "immersion" methods (e.g., small thermocouples); or
b. "remote" or "nonintrusive" (spectroscopic) methods.

More often, measurements are made of the corresponding energy *fluxes* $\dot{q}''_w$ only at important *boundary* surfaces (subscript w), using wall-mounted local heat-flux gauges (see, e.g., Thompson (1981)). Simpler still are measurements of the *overall* heat transfer rate $\dot{q}_w$ for a surface of total area A_w, corresponding to an *average* energy (heat) flux $\overline{\dot{q}''_w} \equiv \dot{q}_w/A_w$.

In reporting the results of such experiments on energy transfer from/to surfaces of overall characteristic dimension L, maximum use of such data can be made by comparing the observed average heat flux $\overline{\dot{q}''_w}$ to:

$$\dot{q}''_{\text{ref}} \equiv k\left(\frac{T_w - T_\infty}{L}\right), \tag{5.3-1}$$

[4] Equation (5.2-2) is said to satisfy the principle of "local action"—i.e., the *present* value of the *local* heat flux is displayed as related to the present values of the *local* temperatures about the point **x** in question. This is generally *not* true for the radiation-energy flux vector in gases, but may be approximated in dense, particulate-laden combustion gases (Section 5.9.2) or in high-temperature *packed beds*.

which is the energy (flux) that would diffuse (be *conducted*) across a quiescent fluid gap of dimension L subjected to the same overall temperature "driving force" $(T_w - T_\infty)$. This *ratio, viz.*:

$$\frac{(\dot{q}_w/A_w)}{k\left(\dfrac{T_w - T_\infty}{L}\right)} \equiv \overline{\mathrm{Nu}}_h, \qquad (5.3\text{-}2)$$

will be called the *dimensionless heat-transfer coefficient* or Nusselt number, $\overline{\mathrm{Nu}}_h$, named after the German engineer, W.E.K. Nusselt [1882–1957].

Often engineering results are discussed in terms of the average heat flux achieved per unit temperature difference, called the *dimensional heat-transfer coefficient*, i.e.,

$$\frac{(\dot{q}_w/A_w)}{(T_w - T_\infty)} \equiv \text{dimensional heat-transfer coefficient.} \qquad (5.3\text{-}3)$$

This quantity is sometimes written $\bar{h}$ (not to be confused with the specific enthalpy) and its units must be fully stated, e.g., $\mathrm{cal(s)^{-1}\ cm^{-2}\ (K)^{-1}}$; $\mathrm{kW\ m^{-2}\ K^{-1}}$, or $\mathrm{BTU\ (hr)^{-1}\ (ft)^{-2}\ (^\circ F)^{-1}}$, etc.

If one imagined the actual average energy flux was the result of pure diffusion across a quiescent ("stagnant") fluid film, then the thickness of this "equivalent stagnant film" would have to be $\delta_{\mathrm{eff}} = k_{\mathrm{fluid}}/\bar{h}$. From the definitions above it is therefore clear that $\overline{\mathrm{Nu}}_h$ can also be regarded simply as the ratio of the reference body dimension L to $\bar{\delta}$ or, equivalently, the ratio of (L/k_{fluid}) to $(1/\bar{h})$. As shown in Section 5.4.3, $\overline{\mathrm{Nu}}_h$ can, thus, be interpreted as the ratio of the "resistance" to energy diffusion across a *fluid* layer of thickness L to the actual resistance to energy diffusion across the prevailing thermal boundary layer in the fluid "wetting" the surface of the object ($\bar{\delta}_{\mathrm{fluid}}/k_{\mathrm{fluid}}$ or $1/\bar{h}_{\mathrm{fluid}}$).

When *local* heat fluxes are available, they can be reported as *local* heat-transfer coefficients; e.g., $[\dot{q}''_w/(T_w - T_\infty)]_{\mathrm{local}}$ would be the *local, dimensional* heat-transfer coefficient, and

$$\left\{\frac{\dot{q}''_w}{[k(T_w - T_\infty)/L]}\right\}_{\mathrm{local}} \equiv \mathrm{Nu}_h \qquad (5.3\text{-}4)$$

a corresponding *local dimensionless* heat-transfer coefficient.[5]

For a solid, nonablating surface exposed to a (locally) nonreacting gas, the local (normal component of this) energy flux in the solid is the same as that in the adjacent gas, and hence can be written:

$$\dot{q}''_w = -\left(k\frac{\partial T}{\partial n}\right)_{n=0^+} \qquad (5.3\text{-}5)$$

[5] Sometimes other lengths are used to define/report *local* dimensionless heat-transfer coefficients, e.g., the length *along* the surface up to the point in question.

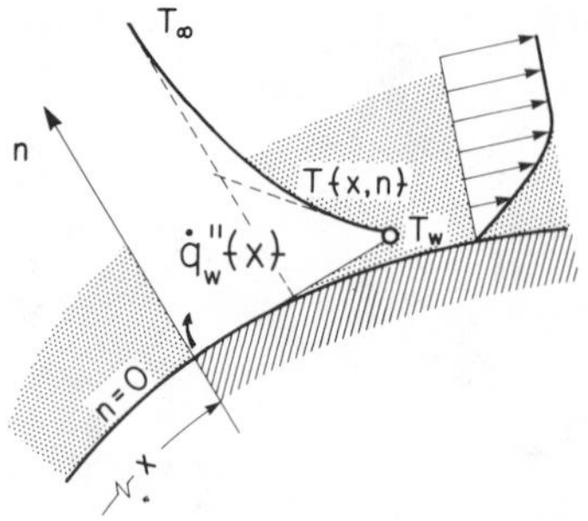

Figure 5.3-1 Thermal boundary layer near a curved solid surface.

(even though this may not be the way $\dot{q}''_w$ is experimentally determined). Here n is a coordinate measured normal to the surface and $n = 0^+$ denotes evaluation on the $+$, or fluid, side of the surface $n = 0$ (Figure 5.3-1).

As a specific example, briefly consider the steady, axially symmetric flow of a fluid of temperature T_∞ past an isolated isothermal hot *sphere* of diameter d_w (Figure 5.3-2). As a result of energy diffusion (conduction) *and* convection, some temperature field, $T\{r, \theta\}$, will be established in the fluid-filled space $r > (d_w/2)$ about the sphere. In the absence of radiation, at the sphere surface there will be local radial heat fluxes given by:[6]

$$\dot{q}''_w = -k\left(\frac{\partial T}{\partial r}\right)_{r=d_w/2} \tag{5.3-6}$$

and these will depend upon position (hence θ) along the sphere surface ($r = d_w/2$). The total heat-flow rate from the solid to the fluid, loosely called the "convective"

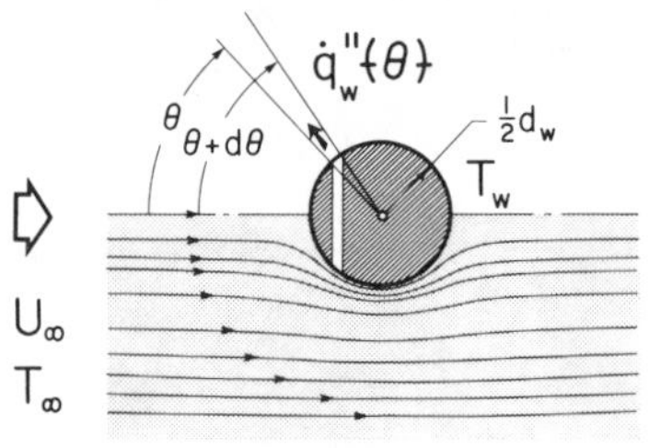

Figure 5.3-2 Axisymmetric steady flow past an isolated isothermal solid sphere of diameter d_w.

[6] Note that there is no actual radial *convection* across the surface $r = d_w/2$ since $v_r = 0$ on this (solid) surface.

heat-flow rate,[6] will be the area-weighted sum of these fluxes, that is,

$$\dot{q}_w = \iint \dot{q}''_w \, dA_w = \int_0^\pi \left(-k\frac{\partial T}{\partial r}\right)_{r=d_w/2} \cdot (\pi/2)(d_w^2 \sin\theta)\, d\theta. \tag{5.3-7}$$

It is usually simple to measure the *total* rate of heat transfer $\dot{q}_w$ needed to maintain an entire sphere (say) at some constant temperature T_w; hence one would report:

$$\overline{\mathrm{Nu}}_h \equiv \frac{[\dot{q}_w/(\pi d_w{}^2)]}{[k(T_w - T_\infty)/d_w]}. \tag{5.3-8}$$

If one could measure or calculate *local* heat fluxes $\dot{q}''_w\{\theta\}$, then one could, of course, report even the *local* heat-transfer coefficients:

$$\mathrm{Nu}_h\{\theta\} \equiv \frac{\dot{q}''_w\{\theta\}}{[k(T_w - T_\infty)/d_w]}. \tag{5.3-9}$$

But since:

$$\dot{q}_w = \int_0^\pi \dot{q}''_w\{\theta\}\frac{\pi}{2} d_w^2 \sin\theta \, d\theta \tag{5.3-10}$$

these two dimensionless heat-transfer coefficients must be interrelated by:

$$\overline{\mathrm{Nu}}_h = \frac{1}{2}\int_0^\pi \mathrm{Nu}_h\{\theta\}\sin\theta\, d\theta; \tag{5.3-11a}$$

that is, *local* heat-transfer data are sufficient to calculate the *total* (surface-averaged) heat-transfer coefficients $\overline{\mathrm{Nu}}_h$ under identical conditions (Re, Pr), but not *vice versa.*

Measurements of the angular distribution of the dimensionless heat-transfer coefficient Nu_h are most easily made for the important special case of a nonrotating *cylinder* in cross-flow, the behavior of which is similar to the sphere. Figures 5.3-3 and 5.3-4, which exhibit results at $\mathrm{Pr} \approx 0.7$ (air) over the Reynolds'-number range $4 \leqslant \mathrm{Re} \leqslant 2.19 \times 10^5$, reveal the increasing relative importance of the "rear" ($\pi/2 \leqslant \theta \leqslant \pi$) of the cylinder as Re increases. Surface (perimeter)-averaged heat-transfer coefficients, $\overline{\mathrm{Nu}}_h\{\mathrm{Re}, \mathrm{Pr}\}$, can be obtained by appropriate θ-integrations of these data (see Exercise 5.14) or by direct measurements (e.g., using a conductive cylinder as a transient calorimeter). Such results will be discussed below (Sections 5.5.3 and 5.6.4) and displayed in Figure 8.1-1.

We will see that, in certain simple cases, temperature fields and their corresponding surface heat-transfer rates and coefficients can be *predicted* by mathematical or numerical solution of the pertinent balance equations. More commonly, however, recourse is made to full-scale or model energy-transfer *experiments* (which often circumvent the temperature field entirely by measuring directly local or total surface energy-flow rates). Owing to the difficulty of making temperature

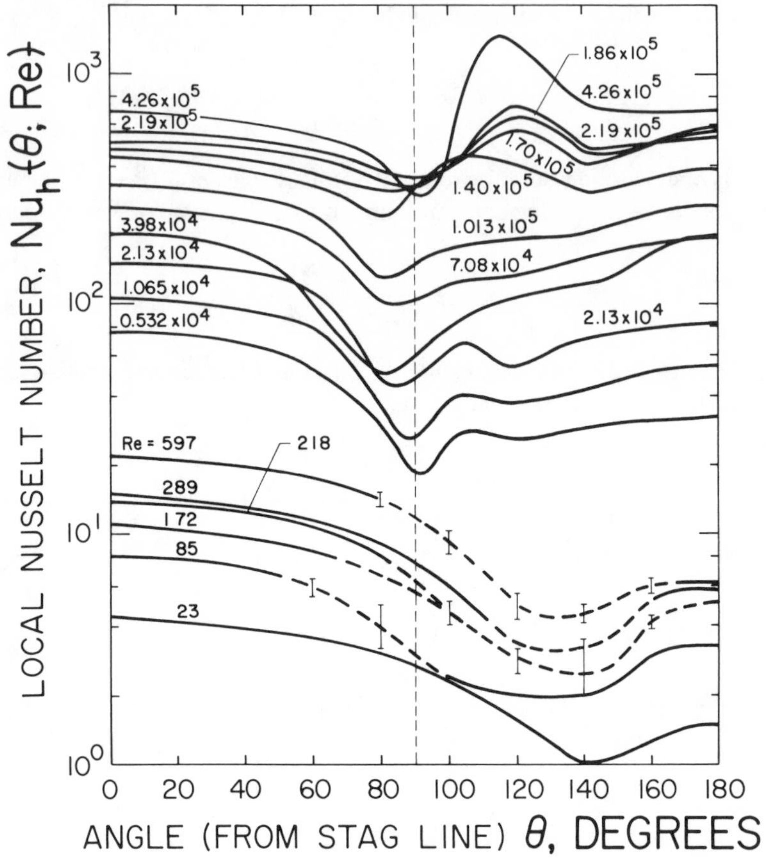

Figure 5.3-3 Experimentally determined *local* Nusselt numbers for cross-flow of a Newtonian fluid ($Pr = 0.7$) about a circular cylinder at various Reynolds' numbers (adapted from Giedt, W.H. *Trans., ASME* **71**, 378 (1949) and Van Meel (1962)).

measurements very close to a solid (or liquid) surface, rarely is the surface-energy flux experimentally obtained from Eq. (5.3-6).

Perhaps the simplest technique for obtaining $\overline{Nu}_h$-data in external-flow forced convective systems is the "transient calorimeter" technique, in which the fluid-to-solid heat-transfer rate $(-\dot{q}_w)_{fs}$ is measured by following the rate of temperature change of a nearly isothermal calorimeter fashioned in the desired shape, suddenly immersed in a fluid of the desired Prandtl number, at the desired orientation and Reynolds' number (see Chapter 7). In this case, the macroscopic energy balance (Eq. (2.3-3)) applied to a control volume enveloping the entire calorimeter, simplifies to:

$$\left(\rho c V \frac{dT}{dt}\right)_s = (-\dot{q}_w)_{fs} = A_w \cdot \overline{Nu}_h \cdot \left\{\frac{k_f(T_\infty - T_s)}{L}\right\}, \qquad (5.3\text{-}11b)$$

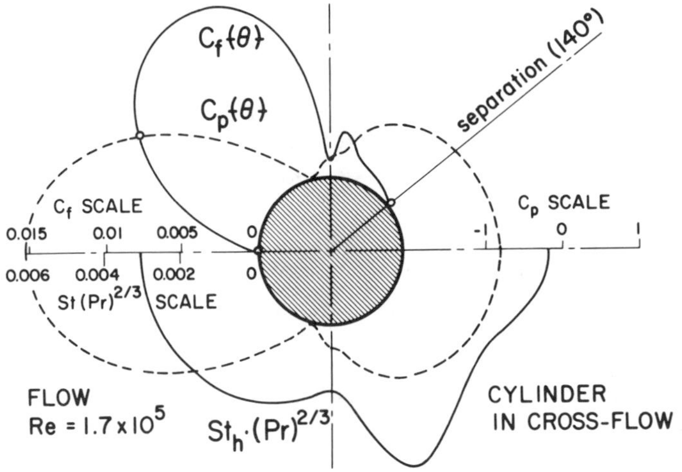

Figure 5.3-4 Polar plot of experimentally determined local $\mathrm{St}_h\mathrm{Pr}^{2/3}$, C_f and C_p-distributions for cross-flow of a Newtonian fluid about a circular cylinder at $\mathrm{Re} = 1.7 \times 10^5$ (adapted from Fage and Falkner (1931) and Giedt (1959)).

where V_s and A_w are, respectively, the volume and "wetted" area of the solid calorimeter, L its reference length (used to define $\overline{\mathrm{Nu}}_h$ and Re), $(\rho c)_s$ its heat capacity per unit volume, and $T_s\{t\}$ the recorded calorimeter temperature. Since this equation would require differentiation of $T_s\{t\}$-data to infer $\overline{\mathrm{Nu}}_h$-values, it is preferable to exploit its integrated form. If $(\rho c)_s$, k_f, and $\overline{\mathrm{Nu}}_h$ are approximately T-independent, then we readily find:

$$\frac{T_\infty - T_s\{t\}}{T_\infty - T_s\{0\}} = \exp\left\{-\frac{t}{t_h}\right\}, \tag{5.3-11c}$$

where

$$t_h \equiv \frac{(V_s/A_w)}{L} \cdot \frac{(\rho c)_s}{(\rho c)_f} \cdot \frac{1}{\overline{\mathrm{Nu}}_h} \cdot \frac{L^2}{\alpha_f} \tag{5.3-11d}$$

is the effective *time-constant* for the exponential decay of the fraction of the initial temperature difference $T_\infty - T_s\{0\}$ still "unaccomplished" at time t. Here $(V_s/A_w)/L$ is a shape-dependent pure number ($=1/6$ for a sphere), and α_f is the *fluid* thermal diffusivity, $k_f/(\rho c)_f$. Taking the logarithm of both sides of Eq. (5.3-11c) we see that t_h^{-1} can be determined as the slope of a plot of

$$\ln\{[T_\infty - T_s\{0\}]/[T_\infty - T_s\{t\}]\} \quad vs. \quad t,$$

and $\overline{\mathrm{Nu}}_h$ can then be determined from Eq. (5.3-11d). [Another common application of Eq. (5.3-9d) is the estimation of the *response-time* of a temperature-measuring device (e.g., thermocouple bead or resistance wire "thermometer") in a nonsteady flow (see Exercise 2.1)]. Representing the entire calorimeter by a single temperature at each time implies that temperature nonuniformities *within* the calorimeter material (usually copper or aluminum) must be negligible compared to temperature nonuniformities within the fluid "film" (boundary layer) wetting the calorimeter. Based on the concepts discussed in Section 5.4.3, this would be valid only if $\mathrm{Bi}_h \ll 1$, where:

$$\mathrm{Bi}_h = \frac{(V_s/A_w)/k_s}{(1/\bar{h})} = \frac{(V_s/A_w)}{L} \cdot \frac{k_f}{k_s} \cdot \overline{\mathrm{Nu}}_h .$$

This group, called the *Biot number* for energy *transfer*, is the ratio of the "internal" (calorimeter) resistance to energy diffusion to the "external" (fluid film) resistance to energy diffusion. Incidentally, transient calorimeter inferred values of Nu_h can be considered "quasi-steady" (QS) if t_h is large compared to a fluid-film characteristic "adjustment" time $(L/\overline{\mathrm{Nu}}_h)^2/\alpha_f$. This means that transient calorimeter measurements must be made under conditions such that

$$\frac{L}{(V_s/A_w)} \cdot \frac{(\rho c)_f}{(\rho c)_s} \cdot \frac{1}{\overline{\mathrm{Nu}}_h} \ll 1.$$

This QS condition is easiest to satisfy by testing in flowing *gases*, since gas ρc-values at STP are smaller than those of most solids by over three decades.

5.3.2 Conservation Equation Governing the Temperature Field, Typical Boundary Conditions, and Outline of Solution Methods

To *predict* temperature fields, one must solve a single partial differential equation (PDE) subject to conditions specified initially (IC) and along the boundaries of the region of interest (BC). The PDE governing the T-field is readily obtained from the local energy conservation equation (Eq. (2.3-4)), when combined with constitutive equations appropriate to the medium in question. The resulting equation can, however, be quite complicated because of:

a. unsteadiness (transients, including "turbulence");
b. flow effects (convection and viscous dissipation);
c. variable properties of the medium (including those associated with variable chemical composition);
d. energy release associated with homogeneous *chemical reactions*;[7] and
e. possible coupling with the coexisting "photon gas" (see Section 5.9.4).

[7] Given by the sum of all $-h_i\dot{r}_i'''$ terms (Eq. (5.3-12)).

For example, neglecting only the latter (e), in a chemically reacting mixture of perfect gases it is straightforward (but tedious) to show that $T\{\mathbf{x},t\}$ satisfies the PDE:

$$\rho c_p\left[\frac{\partial T}{\partial t}+\mathbf{v}\cdot\mathbf{grad}\,T\right]=-\sum_i h_i\dot{r}_i'''+\dot{q}'''-\mathrm{div}\left\{\dot{\mathbf{q}}''-\sum_i \mathbf{j}_i''h_i\right\}-\mathbf{grad}\,T\cdot\sum_i c_{p,i}\mathbf{j}_i''$$
$$+\left(\frac{\partial p}{\partial t}+\mathbf{v}\cdot\mathbf{grad}\,p\right)+(\mathbf{T}:\mathbf{grad}\,\mathbf{v})+\sum_i \mathbf{j}_i''\cdot\mathbf{g}_i, \qquad (5.3\text{-}12)$$

where the notation $\mathbf{T}:\mathbf{grad}\,\mathbf{v}$ symbolizes a particular scalar (calculated from sums of the products of components of the local tensors $\mathbf{T}$ and $\mathbf{grad}\,\mathbf{v}$), called the local *viscous dissipation rate*. This PDE for $T\{\mathbf{x},t\}$ is derived by combining Eqs. (2.3-4) and (2.2-4) of Chapter 2, together with the thermodynamic properties of ideal (gas) mixtures. The quantity c_p is the specific heat of the prevailing mixture; that is, $c_p = \sum_i \omega_i c_{p,i}$.

The simplest possible PDE governing the T-field is the degenerate form of Eq. (5.3-10), when none of the above-mentioned complications arise. In that case $T\{\mathbf{x}\}$ satisfies the (Laplace) equation:

$$\mathrm{div}\{\mathbf{grad}\,T\} = 0 \qquad (5.3\text{-}13)$$

(also written $\nabla^2 T = 0$), which can be expressed as an explicit linear PDE for the steady-state temperature field $T\{\mathbf{x}\}$ in any convenient coordinate system. Physically this equation simply states that the net outflow (per unit volume) of energy (transferred by the mechanism of Fourier diffusion) must vanish at each point in such a medium. We will return to this equation in Section 5.4.

Boundary conditions (BCs) on $T\{\mathbf{x},t\}$ are frequently of three types:

1. T-specified everywhere along each boundary surface;
2. Heat flux specified along each surface; or
3. Some combination of T and local heat flux specified along each surface,

with Type 3 being the most general, and Type 1 the simplest to deal with. In the latter case, resulting heat-transfer coefficients are called *isothermal surface* heat-transfer coefficients. It is often assumed that such *coefficients* can be applied, with reasonable accuracy, even to immobile surfaces that are not quite isothermal. Most surfaces, including those experiencing *spatially constant heat flux*, are indeed generally nonisothermal.

Solution methods vary widely depending upon the degree of complexity of the problem. Most versatile are approximate *numerical* methods (finite difference, finite element) in which one *algebraically* solves for the temperatures only at certain node points within the domain.[8] However, some problems [simple domain, simple flow field (including no flow), constant properties, few participating phenomena] can be formally solved exactly by mathematical methods which often reduce the problem of solving one PDE to the problem of solving one or more *ordinary* differential equations (ODEs). These now-classical methods include:

a. separation of variables (Fourier) (cf. Section 5.4.6.2 and Appendix 5.1);
b. combination of variables ("self-similarity" methods; Sections 5.4.6.1, 5.5.2);
c. transform (Laplace, Mellin, etc.) methods.

Of course, steady-state problems in one space dimension (see Sections 5.4.3 and 5.4.4) lead immediately to an ODE for the temperature field. Many heat-transfer problems yield to more than one exact or approximate method, and these cases are often used to test ("calibrate") approximate methods intended for still more complex problems.

5.4 TEMPERATURE DISTRIBUTIONS AND SURFACE HEAT TRANSFER FOR QUIESCENT MEDIA OF UNIFORM COMPOSITION

5.4.1 Relevance

It might be thought that the conditions of "quiescence" (no fluid motion) and uniform composition are so special as to be of little interest in engineering applications, but this is not the case. Consider the following examples:

a. firebrick inner wall of a steady-flow (continuous) furnace;
b. low-volatility droplet (e.g., fuel) after having been accelerated to approximately the surrounding gas velocity;
c. water-cooled cylinder wall of a reciprocating piston (IC) engine;
d. gas turbine blade-root (attachment) combination (Figure 5.4-1);

etc. The first of these is an example of steady, one-dimensional (planar) heat flow. The second is an example of quasi-steady (QS) spherically symmetric heat flow (Section 5.4.4); and the third is an example of unsteady (periodic) heat flow in a solid (see also Figure 5.4-2) of complex geometry and with a complicated boundary condition on the inner wall (heat flux-temperature relation).

---

[8] E.g., on a uniform cartesian mesh, Laplace's equation can be shown to be satisfied approximately if each nodal temperature is the *arithmetic mean of its surrounding* (*"nearest neighbor"*) *nodal temperatures*. This condition can be satisfied by a systematic iterative procedure called "relaxation." Interestingly enough, this "finite-difference analog" condition (Appendix 8.2) is also identical to that describing a "drunkard's walk" on a cartesian mesh (i.e., the probability of wandering to a particular mesh point in terms of the probability of being at each of the adjacent mesh points). Thus, energy (and mass) diffusion problems are also amenable to numerical solution by so-called "Monte Carlo" techniques (see, e.g., Howell, J.R. (1968)).

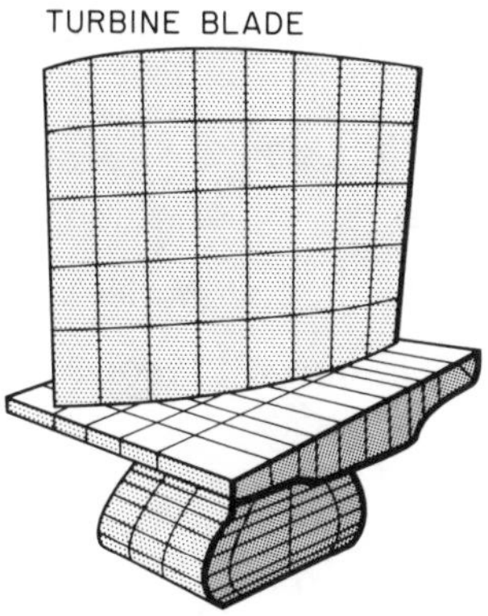

Figure 5.4-1 Three-dimensional heat-conduction model for gas turbine blade.

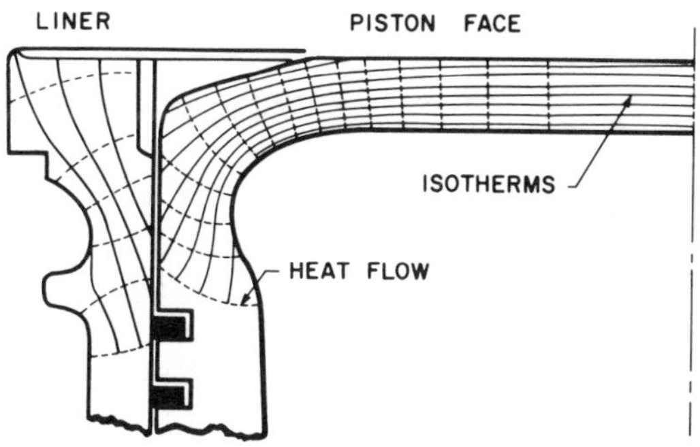

Figure 5.4-2 Heat diffusion in the wall of a water-cooled IC engine (adapted from Steiger and Aue (1964)).

5.4.2 Criteria for Quiescence

Stationary *solids* can usually be considered quiescent; however, with viscous fluids (liquids or gases), even in the absence of "forced" convection, one can encounter "*natural*" *convection* in a body force field due to the heat-transfer process itself. Whatever the cause, the slightest amount of convection can be important; hence it is first useful to state quantitative "*criteria for quiescence*."

This can be done based on the following simple arguments for laminar fluid flows; i.e., *convective* energy flow can be neglected if and only if:

$$\left\{\frac{\text{characteristic convective energy flux}}{\text{characteristic diffusive energy flux}}\right\}^{1/2} \ll 1. \tag{5.4-1}$$

Thus, we can write condition (5.4-1) in the explicit form:

$$\left\{\frac{\rho v_{\text{ref}} c_p (T_w - T_\infty)}{k(T_w - T_\infty)/L_{\text{ref}}}\right\}^{1/2} \ll 1. \tag{5.4-2}$$

Now consider the following two important special cases:

a. Forced Convection with "Imposed" Velocity U. Taking $v_{\text{ref}} \equiv U$ (e.g., the relative velocity between the droplet and surrounding gas in example (b) above (Section 5.4.1)) we find that *forced convection can be neglected if*:

$$\left(\frac{UL}{\alpha}\right)^{1/2} \ll 1, \tag{5.4-3}$$

where the group UL/α is sometimes called the *Peclet number*, written Pe. In terms of the more familiar groups:

$$\text{Re} \equiv \frac{UL}{\nu} \quad \text{and} \quad \text{Pr} \equiv \frac{\nu}{\alpha}, \tag{5.4-4a,b}$$

criterion (5.4-3) for the neglect of *forced* convection can be written:

$$(\text{Re} \cdot \text{Pr})^{1/2} \ll 1. \tag{5.4-5}$$

b. "Natural" Convection. In a natural convection problem the heat-transfer process itself causes a density difference $(\Delta\rho)_h$ and, therefore, the pressure difference *causing* the flow[9] is $(\Delta\rho)_h gL$. This can be set equal to $(\frac{1}{2})\rho_\infty v_{\text{ref}}^2$ to give the "inviscid" velocity estimate:

$$(v_{\text{ref}})_{\substack{\text{nat}\\\text{conv}}} \cong \left(2\frac{(\Delta\rho)_h}{\rho_\infty} gL\right)^{1/2}. \tag{5.4-6}$$

If we now write:

$$(\Delta\rho)_h/\rho_\infty = \beta_T \Delta T \approx \beta_{T\infty}(T_w - T_\infty) \tag{5.4-7}$$

(where

$$\beta_T \equiv \frac{1}{v}\left(\frac{\partial v}{\partial T}\right)_p = -\frac{1}{\rho}\left(\frac{\partial \rho}{\partial T}\right)_p \tag{5.4-8}$$

is the thermal expansion coefficient of the fluid), then our quantitative *sufficient*

---

[9] In many engineering applications the effective value of g is set by centrifugal acceleration rather than the Earth's gravity (see, e.g., Section 2.6.2).

criterion for the neglect of natural convection assumes the form:

$$(\mathrm{Ra}_h \cdot \mathrm{Pr})^{1/4} \ll 1, \tag{5.4-9}$$

where we have dropped the factor $2^{1/4} = 1.19$ (near unity) and introduced[10]

$$\mathrm{Ra}_h \equiv \frac{g\beta_T(\Delta T)L^3}{\nu^2} \cdot \frac{\nu}{\alpha} (\equiv \mathrm{Gr}_h \cdot \mathrm{Pr}), \tag{5.4-10}$$

the so-called *Rayleigh number for heat transfer*. The *Prandtl number*, Pr, is, again, the Newtonian fluid *property ratio*: ν/α.

Thus, for heat-transfer problems in the absence of simultaneous mass transfer (phase change), the important criteria to be satisfied in order to *neglect forced- or free-convection with respect to heat diffusion* are, respectively, inequalities (5.4-5) and (5.4-9). A "convective" augmentation of heat-transfer results (when these smallness criteria are *not* satisfied) will be discussed in Section 5.5.

5.4.3 Steady-State Heat Conduction

For steady-state energy diffusion across a constant-property planar slab of thickness L the condition $\mathrm{div}\{\dot{\mathbf{q}}''\} = 0$ combined with Fourier's heat-flux law (Eq. (5.2-2)) immediately provides the following simple ODE for $T\{x\}$ (a "degenerate" form of Laplace's equation):

$$\frac{d}{dx}\left(\frac{dT}{dx}\right) = 0; \tag{5.4-11}$$

dT/dx is therefore seen to be constant—i.e., the temperature profile will be a straight-line segment connecting the two unequal boundary temperatures T_1 and T_2 (Figure 5.4-3a). Thus:

$$\frac{T\{x\} - T_1}{T_2 - T_1} = \frac{x}{L}, \tag{5.4-12}$$

and the heat flux at any station, including the two boundary surfaces $x = 0$ and $x = L$, is simply given by:

$$\dot{q}''_w = -k\left(\frac{T_2 - T_1}{L}\right). \tag{5.4-13}$$

---

[10] The group Gr_h, called the *Grashof* group, can be regarded as the ratio of two *forces* (buoyancy force/viscous force). Gr_h appears in the dimensionless form of the momentum equation for a viscous fluid in natural convection. Note that $\mathrm{Gr}_h = (v_{\mathrm{ref}}L/\nu)^2$, where $v_{\mathrm{ref}} = (g\beta\,\Delta TL)^{1/2}$, so that $\mathrm{Gr}_h{}^{1/4} = (v_{\mathrm{ref}}L/\nu)^{1/2}$; indeed, $\mathrm{Gr}_h{}^{1/4}$ is often found to play the same role (in free-convection correlations) as $\mathrm{Re}^{1/2}$ plays in forced convection correlations.

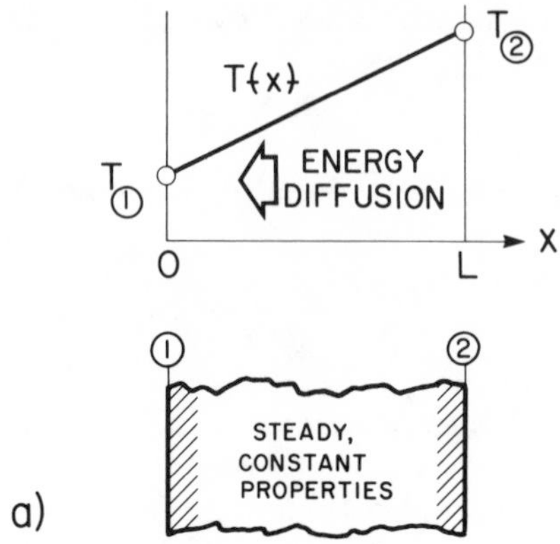

Figure 5.4-3a Heat diffusion through a planar slab with constant thermal conductivity.

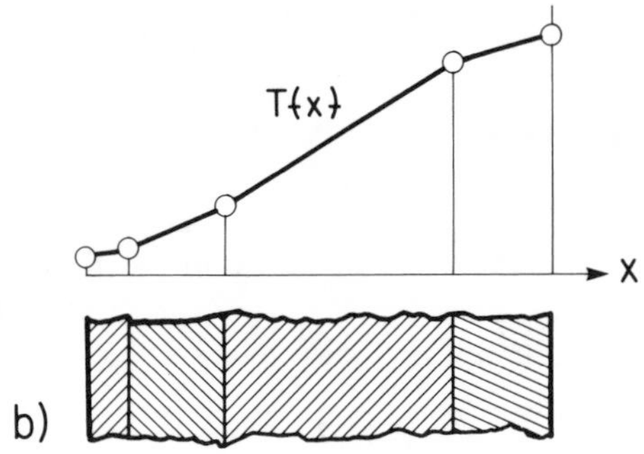

Figure 5.4-3b Heat diffusion through a composite planar slab (piecewise constant thermal conductivity).

The corresponding *dimensionless heat-transfer coefficient* Nu_h is readily found to be *unity*, indicating that in this case the actual wall heat flux $\dot{q}''_w$ is identical to the reference value $-k(T_2 - T_1)/L$.

The temperature distribution in a *composite* wall is likewise comprised of straight-line segments. Since each layer "in series" transmits the *same* heat flux one finds that:

$$(\dot{q}''_w)_{\substack{\text{composite}\\ \text{wall}}} = \frac{(\Delta T)_{\text{overall}}}{\sum_\ell (L/k)_\ell}, \tag{5.4-14a}$$

where $(L/k)_\ell$ is said to be the "thermal resistance" of the ℓth layer (cf. Figure 5.4-3b). This is the *thermal analog* of the series resistance *electrical* problem, with ΔT playing the role of voltage drop and $\dot{q}''_w$ playing the role of current. The reciprocal of the overall resistance is, accordingly, called the *overall conductance*, sometimes written U.

When the thermal conductivity is *not* constant, but appreciably temperature-dependent, Eq. (5.4-11) generalizes to:

$$\frac{d}{dx}\left(k\{T\}\frac{dT}{dx}\right) = 0. \tag{5.4-15}$$

In that case Eq. (5.4-13) still applies, except that the previously constant k-value must be replaced by the mean value of k over the relevant temperature interval; that is,

$$\bar{k} \equiv \frac{1}{T_2 - T_1} \cdot \int_{T_1}^{T_2} k\{T\}\, dT. \tag{5.4-16}$$

One-dimensional energy diffusion can, of course, also occur in systems with cylindrical or spherical symmetry. Indeed, source-free, steady-state *radial* heat flow across *composite cylinders* is frequently encountered in the case of insulated pipe heat loss. For that geometry the energy conservation condition $\text{div}\,\dot{\mathbf{q}}'' = 0$ is equivalent to $2\pi r\, \dot{q}_r'' = \dot{q}_r' = \text{const}$, where $\dot{q}_r'$ is the total radial heat flow rate *per unit length of cylinder* (see Table 5.4-1). For "nested" cylinders the same value of $\dot{q}_r'$ will apply to each layer and, by analogy to Eq. (5.4-14a), in the absence of *interfacial* energy resistances, we find:

$$\dot{q}_r' = \frac{(\Delta T)_{\text{composite solid}}}{\sum_\ell \left\{\dfrac{\ln(r_{\ell+1}/r_\ell)}{2\pi k_{\ell,\ell+1}}\right\}}, \tag{5.4-14b}$$

where each thermal resistance (cf. denominator) now involves the logarithm of the

Table 5.4-1 Steady-State, Source-Free Energy Diffusion in One Dimension

Geometry	*Quantity that is Constant*	*Typical "Resistance"*	*Resistance of Inner or Outer "Film"*[a]
Planar (x)	$\dot{q}_x''$	$\dfrac{x_2 - x_1}{k_{12}}$	$\dfrac{1}{h_f}$
Cylindrical (r)	$\dot{q}_r'(=2\pi r \dot{q}_r'')$	$\dfrac{\ln(r_2/r_1)}{2\pi k_{12}}$	$\dfrac{1}{2\pi r_w h_f}$
Spherical (r)	$\dot{q}_r(=4\pi r^2 \dot{q}_r'')$	$\dfrac{\left(\dfrac{1}{r_1} - \dfrac{1}{r_2}\right)}{4\pi k_{12}}$	$\dfrac{1}{4\pi r_w^2 h_f}$

[a] Where $h_f \equiv (\dot{q}_w/A_w)/(\Delta T)_{\text{film}} = (k_{\text{film}}/\delta_{h,\text{eff}})$.

radius ratio.[11] If the "driving force" (ΔT) overall includes the temperature drop across *fluid* films (at the inner and/or outer radius of the composite cylinder), then the equivalent thermal resistance of such fluid films, $(2\pi r_w h_f)^{-1}$ (see Table 5.4-1), where h_f is the *dimensional* heat-transfer coefficient) must be added to the denominator of Eq. (5.4-14b). The corresponding situation for the nested-*sphere* geometry is summarized in the third row of Table 5.4-1; however, for spherically symmetric systems we focus attention instead on a more commonly encountered special case, *viz*., steady-state energy diffusion *outside* of a single isothermal sphere (Section 5.4.4).

5.4.4 Steady-State Heat Conduction Outside of an Isothermal Sphere

The *procedure* indicated above for a slab can be readily carried over to somewhat more complicated geometries; i.e., one solves the relevant PDE or ODE with its boundary conditions (BC) for the *temperature field* and then evaluates the heat *flux* at the surface of interest *via* local application of Fourier's law. The relevant local heat-transfer *coefficient* then follows from its definition.

An important and instructive special case is that of heat conduction outside an isothermal sphere (temp T_w, diam $d_w = 2a_w$) in a quiescent medium of constant thermal conductivity k and distant temperature T_∞. In this case there is spherical symmetry and the energy-balance condition $\operatorname{div}(\dot{\mathbf{q}}'') = 0$ reduces to the second order, linear, homogeneous ODE:

$$\frac{1}{r^2}\cdot\frac{d}{dr}\left(r^2 k\frac{dT}{dr}\right) = 0, \tag{5.4-17}$$

equivalent to the physical condition that the total radial heat flow rate be constant; that is,

$$\dot{q}_r \equiv 4\pi r^2 \dot{q}'' = 4\pi r^2\left(-k\frac{dT}{dr}\right) = \text{const} \tag{5.4-18}$$

(cf. Column 2 in Table 5.4-1). This is itself a separable linear ODE for $T(r)$, with the solution:

$$T(r) = (T_w - T_\infty)\left[\frac{r}{d_w/2}\right]^{-1} + T_\infty. \tag{5.4-19}$$

---

[11] For steady radial heat flow across a single cylindrical annulus, the radial temperature profile is *not* linear. Yet $\dot{q}_r'$ can be expressed as the product of $k(\Delta T)/(\Delta r)$ and a *mean* area per unit length, where (see Eq. (5.4-14b)) the appropriate mean area per unit length is the "logarithmic mean radius," $(r_2 - r_1)/\ln(r_2/r_1)$. This type of mean occurs often in transport theory (see, e.g., Section 5.5.4).

The corresponding heat flux at $r = d_w/2$, readily found from $-k(dT/dr)_{r=d_w/2}$, is

$$\dot{q}''_w = 2\left[k \cdot \frac{(T_w - T_\infty)}{d_w}\right]. \tag{5.4-20}$$

If the reference length is chosen to be the sphere *diameter*, d_w, this flux (5.4-20) corresponds to the *dimensionless heat-transfer coefficient*:

$$\overline{\text{Nu}_h} = 2. \tag{5.4-21}$$

Since conditions are uniform over the sphere surface, we can also write:

$$\overline{\text{Nu}_h} = 2 \tag{5.4-22}$$

for the *surface-averaged heat-transfer coefficient.*

We conclude that, *for an isolated, isothermal, solid sphere in an infinite fluid,* $\overline{\text{Nu}_h} \to 2$ *in the limit of negligible forced or natural convection*; i.e., when $(\text{Re} \cdot \text{Pr})^{1/2} \to 0$ and $(\text{Ra}_h \cdot \text{Pr})^{1/4} \to 0$.

It is instructive to consider here two important generalizations of these well-known results. The first deals with a thermal conductivity, k, which is temperature-dependent. In this case, it is easy to show that Eq. (5.4-20) is simply replaced by:

$$\dot{q}''_w = \frac{2}{d_w}[\phi_T\{T_w\} - \phi_T\{T_\infty\}], \tag{5.4-23}$$

where the function $\phi_T\{T\}$, sometimes called the heat flux "potential" (see Chapter 3, Exercise 3.1(f)), is defined by:

$$\phi_T\{T\} \equiv \int_{T_{\text{ref}}}^{T} k\{T\}\, dT. \tag{5.4-24}$$

Note that in Eq. (5.4-23) the dimensionless heat-transfer *coefficient* ($\text{Nu}_h = 2$) remains unaltered, but the "driving force for energy transfer" has changed (from $k(T_w - T_\infty)$ to $\phi_T\{T_w\} - \phi_T\{T_\infty\}$), which, as shown explicitly below, is *not* simply proportional to $\Delta T \equiv T_w - T_\infty$. For example, if, as in the case of most nonreacting gases (Chapter 3, Section 3.3.4),

$$k\{T\} \approx k_\infty \cdot \left(\frac{T}{T_\infty}\right)^{\varepsilon} \tag{5.4-25}$$

(where ε is a parameter usually between 0.5 and 1.0), then:

$$\dot{q}''_w = 2 \cdot \frac{k_\infty T_\infty}{d_w}\left\{\frac{1}{1+\varepsilon}\left[\left(\frac{T_w}{T_\infty}\right)^{1+\varepsilon} - 1\right]\right\}. \tag{5.4-26}$$

It can be verified that Eq. (5.4-26) can be "recovered" from (5.4-20) by merely inserting the temperature-averaged thermal conductivity:

$$k_{\text{avg}} = \frac{1}{T_{\text{w}} - T_\infty} \cdot \int_{T_\infty}^{T_{\text{w}}} k\{T\}\, dT = \frac{[\phi_T\{T_{\text{w}}\} - \phi_T\{T_\infty\}]}{T_{\text{w}} - T_\infty}. \tag{5.4-27}$$

The use of ϕ_T-differences as the "driving force" for energy diffusion can also be extended to chemically reacting gas mixtures. Then $\phi_T\{T; p\}$ is more complex than a simple power law (such as $\sim T^{1+\varepsilon}$ above) but readily calculable (see, also, Section 5.8 and Exercise 5.19).

The second important generalization of this classical sphere heat-transfer problem deals with the effect on Nu_h when *radial fluid convection* is introduced (e.g., by steadily forcing fluid mass through a (porous) "solid" sphere). This adds the *convective term* $\rho c_{\text{p}} \mathbf{v} \cdot \mathbf{grad}\, T = \rho c_{\text{p}} v_r (dT/dr)$ to the energy-balance equation (5.3-12). Regarding the radial variation of v_r, we note that conservation of *mass* yields:

$$\dot{m} = 4\pi r^2 \dot{m}'' = 4\pi r^2 (\rho v_r) = \text{const} = 4\pi a_{\text{w}}^2 \dot{m}_{\text{w}}'' = 4\pi a_{\text{w}}^2 \rho_{\text{w}} v_{r,\text{w}} \tag{5.4-28}$$

so that, if $\rho = \text{const}$,

$$v_r\{r\} = v_{r,\text{w}} \cdot \left(\frac{r}{a_{\text{w}}}\right)^{-2}. \tag{5.4-29}$$

This simple convective-energy diffusion problem is also readily solved for the steady-state fluid temperature distribution $T\{r\}$ and the corresponding diffusion (Fourier) heat flux at the wall (where $r = a_{\text{w}}$). The result can be put in the interesting form:

$$\dot{q}_{\text{w}}'' = (\dot{q}_{\text{w}}'')_{\text{no "blowing"}} \cdot F\{\text{blowing}\} \tag{5.4-30}$$

or

$$\text{Nu}_h = \text{Nu}_{h,0} \cdot F\{\text{blowing}\}, \tag{5.4-31}$$

where $(\dot{q}_{\text{w}}'')_{\text{no blowing}}$ is again given by Eq. (5.4-18), and the "correction factor" $F\{\text{blowing}\}$ is given by:

$$F\{\text{blowing}\} = \frac{\text{Pe}_{\text{w}}}{\exp\{\text{Pe}_{\text{w}}\} - 1}, \tag{5.4-32}$$

where

$$\text{Pe}_{\text{w}} \equiv \frac{v_{\text{w}} \delta_{h,0}}{\alpha} = \frac{v_{\text{w}} d_{\text{w}}}{\alpha \text{Nu}_{h,0}} \tag{5.4-33}$$

and, in this special case, $\mathrm{Nu}_{h,0} = 2$ (Eq. 5.4-21). Similar correction factors, which can be considerably less than unity for "strong wall blowing," have been worked out for more complicated flow geometries (indeed, blowing (called "transpiration") is an effective method for reducing convective heat-transfer coefficients to objects (such as turbine blades) exposed to "hostile" thermal environments). It is interesting that Eqs. (5.4-32) and (5.4-33) provide a reasonable first approximation for many such cases. These results also formally apply when v_w is *negative*—i.e., when there is wall "suction" rather than "blowing." In those cases, of course, energy-transfer coefficients are *increased.* Thus, in effect, wall "blowing" *increases* the effective thermal boundary layer (BL) thickness, whereas "suction" decreases it.

In the light of these comments about the effect of wall "blowing" on energy-transfer coefficients, recall that, in Section 4.5.2, we commented that wall blowing is a technique that has been used to reduce *momentum*-transfer (skin-friction) coefficients. It will also be seen (in Chapter 6) that blowing and suction have corresponding effects on solute *mass-transfer coefficients* in such geometries/flows.

5.4.5 Steady-State, Quasi-One-Dimensional Heat Conduction

Energy-diffusion situations of engineering importance are, of course, usually multidimensional; however, in many cases the energy diffusion is predominantly in one direction, and the *quasi-one-dimensional approach* (exploited in Section 4.3.1 for convective *momentum* transfer in ducts of slowly varying area, $A(x)$) proves to be quite useful. This is particularly true for the analysis of *energy diffusion* within slender "fins" often added to the gas-side of primary heat-transfer surfaces (e.g., fluid-containing tubes, see, e.g., Figure 5.4-4) so as to "pack" more heat-transfer area into every unit volume of heat exchanger.[12]

To design such a system, it is necessary to size each fin (length, width, shape) such that the added area is indeed "effective"—i.e., most of the fin should not operate at a surface temperature very different from the fin root (attachment or bond-area) temperature. The quantitative calculation of *fin efficiency* (*effectiveness*) *factors*, defined by:

$$\eta_{\text{fin}} \equiv \frac{\begin{array}{c}\text{Actual heat dissipation}\\ \text{rate by fin surface}\end{array}}{\begin{array}{c}\text{Ideal heat dissipation}\\ \text{rate if entire fin at } T_{\text{root}}\end{array}}, \tag{5.4-34}$$

has been fruitfully approached using quasi-one-dimensional methods, as briefly outlined below.

[12] "Compact" heat exchangers usually have more than about 2.5×10^2 m^2 (surface area) per cubic meter of exchanger volume (Kays and London (1964)). In extreme cases, they contain more than $4 \times 10^3\,\mathrm{m}^2/\mathrm{m}^3$, most of which is "fin" area (see, also, Section 5.5.4 for the estimation of required heat-exchanger surface area).

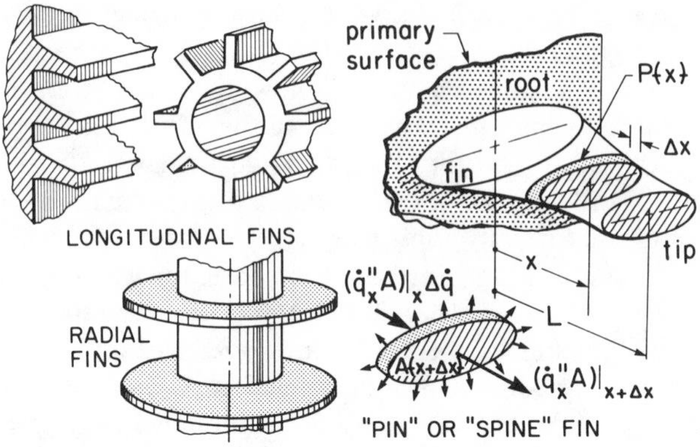

Figure 5.4-4 Extended surfaces ("fins") for augmented energy transfer; types and nomenclature.

A typical simple example is the so-called pin ("peg" or "spine") fin of slowly varying area $A\{x\}$, "wetted" perimeter $P\{x\}$, and length L, losing heat by convection to a surrounding fluid of uniform temperature T_∞ over its entire outer surface (see Figure 5.4-4c).

Suppose we focus attention on the cross-sectional area-averaged *fin material temperature*, $T\{x\}$, neglecting temperature nonuniformities transverse to the predominant (x-)direction of solid-state energy diffusion. If the fin/fluid heat-exchange rate for the slice of fin material between x and $x + \Delta x$ is formally written:

$$\Delta\dot{q} \cong \bar{h}\{x,\ldots\} \cdot [T\{x\} - T_\infty] \cdot P\{x\}\Delta x, \tag{5.4-35}$$

where $\bar{h}\{x,\ldots\}$ is the *dimensional* perimeter-mean heat-transfer coefficient at location x (Section 5.3.1), then a steady-flow energy balance on the semidifferential control volume $A(x) \cdot \Delta x$ takes the form:

$$(\dot{q}_x''A)|_x - (\dot{q}_x''A)|_{x+\Delta x} = \bar{h}\{x,\ldots\} \cdot [T\{x\} - T_\infty]P\{x\}\,\Delta x. \tag{5.4-36}$$

Dividing both sides of this equation by Δx, passing to the limit $\Delta x \to 0$, and introducing the Fourier energy-diffusion law (Section 3.3.1):

$$\dot{q}_x'' = -k\left(\frac{dT}{dx}\right) \tag{5.4-37}$$

leads to the conclusion that in the steady state, $T\{x\}$ must satisfy the second-order ODE:

$$\frac{d}{dx}\left(kA\frac{dT}{dx}\right) = \bar{h}\{x,\ldots\} \cdot [T\{x\} - T_\infty] \cdot P\{x\}, \tag{5.4-38}$$

subject to a prescribed root temperature, $T\{0\}$, and some appropriate condition on $T\{x\}$ imposed at the fin tip ($x = L$). This comprises a well-defined two-point boundary-value problem (BVP), from which $T\{x\}$ and, hence, η_{fin}, can be obtained. For example, if the *tip* area heat loss is negligible,[13] then $(dT/dx)_{x=L} = 0$ and

$$\eta_{\text{fin}} = \frac{\displaystyle\int_0^L \bar{h}\{x,\ldots\} \cdot [T\{x\} - T_\infty] P\{x\}\, dx}{\displaystyle\int_0^L \bar{h}\{x,\ldots\} \cdot [T\{0\} - T_\infty] P\{x\}\, dx}, \tag{5.4-39}$$

where the numerator could, alternatively, have been written:

$$-\left(kA\left(\frac{dT}{dx}\right)\right)_{x=0^+},$$

since the total heat lost by the fin to the fluid (per unit time) must enter through the fin *root*.

A simple but important special case is that for which k, $\bar{h}$, A, and P are each constant (x-independent), which leads to the instructive closed-form results:[14]

$$\frac{T\{x\} - T_\infty}{T\{0\} - T_\infty} = \frac{\cosh\{\Lambda \cdot (1 - (x/L))\}}{\cosh\{\Lambda\}} \tag{5.4-40}$$

and, *via* Eq. (5.4-39):

$$\eta_{\text{fin}} = \frac{\tanh \Lambda}{\Lambda} = \frac{1}{\Lambda}\left[\frac{\exp\{\Lambda\} - \exp\{-\Lambda\}}{\exp\{\Lambda\} + \exp\{-\Lambda\}}\right], \tag{5.4-41}$$

where Λ is the governing dimensionless parameter:

$$\Lambda \equiv \left(\frac{\bar{h}P}{kA}\right)^{1/2} \cdot L = \left(\frac{4\bar{h}}{kd_{\text{eff}}}\right)^{1/2} L \tag{5.4-42}$$

(which arises naturally in writing the nondimensional form of Eq. (5.4-38)), and $d_{\text{eff}} \equiv 4A/P$ is the "effective diameter" of the fin. Thus, for such a fin to operate at above 90 percent efficiency, say, it is necessary that conditions be arranged such

---

[13] Alternatively, it may be more realistic to include heat loss from the tip of area $A\{L\}$, in accord with a heat-transfer coefficient appropriate to this location.

[14] Here $\cosh\{\ \} \equiv \frac{1}{2}[\exp\{\ \} + \exp\{-\ \}]$, etc. By regarding a thermocouple or thermometer well as a "pin fin," *measuring* $T\{L\}$, $T\{0\}$, and estimating $\bar{h}$, k, Eq. (5.4-40) can be used to calculate the prevailing *fluid* temperature, T_∞, thereby correcting for the probe longitudinal conduction error (cf. Section 1.6.5 and Exercise 5.4).

that $\Lambda \leqslant 0.58$. Many other cases of engineering interest[15] (cf. Figure 5.4-4) have been analyzed in this way,[16] or numerically (see, e.g., the survey of Kern and Kraus (1972)). It is interesting to note that, if conditions are such that a quasi-one-dimensional energy-diffusion analysis is itself questionable, chances are that the advantages of extending the primary heat-transfer surface by adding the fin in the first place are also marginal.

5.4.6 Transient Heat Diffusion: Concept of a Thermal Boundary Layer

The simplest *transient* heat-diffusion problems are those in which a region initially at some uniform temperature T_0 is suddenly altered by changing the boundary temperature or heat flux.

5.4.6.1 *The Method of Combination-of-Variables ("Self-Similarity")*

Two important special cases are:

a. semi-infinite wall with sudden change in boundary temperature to a new constant value (from T_0 to $T_w > T_0$, say); and
b. semi-infinite wall with *periodic heat flux* imposed at boundary.

In both transient cases, because there is only one *spatial* dimension, we are led to a simple PDE for the temperature field $T(x, t)$. In the absence of convection, volume heat sources and variable properties, the relevant PDE is:

$$\frac{\partial T}{\partial t} = \alpha \frac{\partial^2 T}{\partial x^2}, \qquad (5.4\text{-}43)$$

where α is the "thermal diffusivity" of the medium ($\alpha \equiv k/(\rho c)$). $T(x, t)$ must be found in the "open" domain $x \geqslant 0$, $t > 0$ subject to the initial condition $T(x, 0) \equiv T_0$ and a boundary condition (at $x = 0$) specified below.

a. Sudden Change in Boundary Temperature. If the boundary temperature is suddenly raised (or lowered) to a constant value $T_w (t > 0)$, (i) How long does it take interior points to "realize" this? (ii) What are the instantaneous temperature profiles within the material? and (iii) What are the corresponding surface heat

[15] Fin theory has also been used to analyze the thermal performance of "fins" on animals (both living and long extinct). For a quantitative treatment of the thermal role of the plates on the dinosaur *stegosaurus*, see Farlow, J., *et al.* (1976).

[16] The previously stated results apply equally well to a long *longitudinal* fin (Figure 5.4-4) of constant cross section. In that case, by a limiting process, we find that A/P is half of the fin thickness.

fluxes? This problem has been solved exactly by noting[17] that the resulting temperature profiles are always "self-similar." That is, $[T_w - T\{x, t\}]/[T_w - T_0]$ depends upon x and t only through their *combination*:

$$\eta \equiv \frac{x}{(4\alpha t)^{1/2}}. \tag{5.4-44}$$

Indeed, one finds that Eq. (5.4-43) and its side conditions are satisfied by:

$$\frac{T_w - T\{x, t\}}{T_w - T_0} = \frac{2}{\sqrt{\pi}} \cdot \int_0^{\eta} \exp\{-\xi^2\}\, d\xi \equiv \operatorname{erf}\{\eta\}, \tag{5.4-45}$$

obtained by transforming Eq. (5.4-43) to a separable ODE and integrating twice, subject to the above-mentioned boundary/initial data.

An interesting and important general feature of this solution is that the thermal effects of an altered boundary temperature are confined to a thermal "boundary layer" of nominal ("penetration") thickness:

$$\delta_h \cong 4(\alpha t)^{1/2}, \tag{5.4-46}$$

which "propagates" into the medium. Interior points deeper than this experience negligible temperature change, until overtaken by this "thermal wave."

When $t \to 0$, $\delta_h \to 0$ and the corresponding heat flux at the surface $x = 0$ becomes infinitely large. Indeed, the wall *heat flux* always behaves like $k(T_w - T_0)/\delta_h$, falling off with elapsed time as $t^{-1/2}$. The corresponding accumulated heat flow up to time t grows with elapsed time, being directly proportional to $t^{1/2}$.

If, instead of T_w, a constant wall *heat flux*, $\dot{q}''_w$, is suddenly "imposed" and maintained at $x = 0^+$, it is remarkable that the local Fourier flux $\dot{q}''$ satisfies the same PDE/BVP as T (above), so that $(\dot{q}''_w - \dot{q}'')/\dot{q}''_w = \operatorname{erf}\{\eta\} \equiv 1 - \operatorname{erfc}\{\eta\}$. Upon integration, this provides an expression for the resulting temperature field in terms of $\int_\eta^\infty \operatorname{erfc}\{\xi\}\, d\xi$. (In particular, $T_w - T_0$ is found to rise like $2(\dot{q}''_w/k) \cdot (\alpha t/\pi)^{1/2}$.) If

[17] A simple method of recognizing that such problems admit solutions by the method of combination-of-variables is the following: If one re-expresses this boundary-value problem (BVP) in *dimensionless* terms (see Ch. 7), provisionally introducing an arbitrary *reference time*, t_{ref}, and *reference length*, x_{ref}, one finds that the PDE governing $[T_w - T\{x, t\}]/[T_w - T_0] \equiv T^*$ can be rendered parameter-free by selecting $t_{ref} = y_{ref}^2/\alpha$. One would then formally conclude that:

$$T^* = \operatorname{fct}\left\{\frac{x}{x_{ref}}, \frac{\alpha t}{x_{ref}^2}\right\}.$$

However, there is no physical reference length in this problem, so that T^* must depend on a *combination* of x/x_{ref} and $\alpha t/x_{ref}^2$, which is independent of x_{ref}. Obviously, $(x/x_{ref})/[4\alpha t/x_{ref}^2]^{1/2} \equiv \eta$ (Eq. (5.4-44)) is such a combination. For another (related) approach as applied to a problem involving diffusion with *convection*, see, e.g., Rosner (1963).

neither T''_w nor $\dot{q}''_w$ is "imposed," but rather a BC of the more general form:

$$h_{\text{fluid}}[T_\infty - T_w(t)] = -k\left(\frac{\partial T}{\partial x}\right)$$

(where h_{fluid} is the (dimension*al*) outer fluid heat-transfer coefficient, assumed constant), a "self-similar" solution (*via* the method of combination-of-variables) is no longer possible. However, an exact analytic solution to this linear problem can still be found using (Laplace) transform methods.

b. Periodic Heat Flux at x = 0. This case gives insight into the behavior of the cylinder walls in a reciprocating (IC) engine, and similar periodic situations. Here the *thermal penetration depth* is found to be frequency-dependent, being on the order of:

$$\delta_h \approx \left(\frac{2\alpha}{\omega}\right)^{1/2}, \qquad (5.4\text{-}47)$$

where ω is the "circular" frequency $2\pi f$ of the imposed heat flux. Note that if we insert the properties of aluminum ($\alpha \approx 0.92\ \text{cm}^2/\text{s}$) and a frequency corresponding to, say, 3000 rpm, then δ_h is only about 1 mm.

Even finite-thickness objects behave as though they are semi-infinite in extent under conditions such that the thermal BL thickness δ_h remains small compared to the actual transverse dimension. When this inequality breaks down, the temperature field becomes "nonself-similar" and must be obtained by methods other than the method of "combination-of-variables."

5.4.6.2 The Fourier Method of Separation of Variables: Transient Energy Diffusion in a Solid of Finite Thickness

As mentioned earlier, even for finite-thickness objects, the above-mentioned method (sometimes called the method of "*combination*" of variables) and solutions apply when the penetration depth of the thermal boundary layer is sufficiently small. The situation for longer times (or lower frequencies, cf. Section 5.4.6.1, part (b)) is perhaps best illustrated by considering transient-energy diffusion in a *finite* slab, $-b < x < b$, of initial temperature T_0, whose outer surfaces $x = \pm b$, are suddenly brought to T_w at time $t = 0^+$. Of course, the PDE Eq. (5.4-43) governing $T(x, t)$, which is a statement of local energy conservation, still applies, but now the boundary conditions are:

$$T(x, 0) = T_0, \qquad (5.4\text{-}48a)$$

$$T(\pm b, 0^+) = T_w\ (\text{const}), \qquad (5.4\text{-}48b)$$

$$\left(\frac{\partial T}{\partial x}\right)_{x=0} = 0, \qquad (5.4\text{-}48c)$$

where the latter condition (imposed at the centerplane $x = 0$ and sometimes erroneously called a "symmetry" condition) is a consequence of the absence of heat sources (or sinks) along $x = 0$. Boundary-value problems (BVPs) of this sort for *linear* PDEs can be solved by a clever method, called the method of *separation of variables*, invented by J. Fourier [1822]. The now-classical method is outlined below for the particular case of energy diffusion in a finite slab, and described in more general terms in Appendix 5.1. Since *mass* diffusion problems are often governed by identical PDEs/BVPs (when written in terms of appropriate nondimensional dependent and independent variables—Section 6.3.3) the method of separation of variables is also widely used for solving *linear mass-transfer* problems on simple domains (see, e.g., Crank (1975)). Effects of nonlinearities and/or complicated domains are usually treated numerically using the methods of weighted residuals (Appendix 8.4), finite-differences (Appendix 8.2), or finite-elements (Appendix 8.3).

Before proceeding, it will be convenient in this problem (and subsequent problems) to introduce nondimensional variables (see, also, Chapter 7). In the present case, this is done by comparing (a) all temperature differences (from T_w) to the "reference" temperature rise[18] $T_w - T_0$, (b) all actual times to the reference time b^2/α, and (c) all actual positions to the reference length b; that is, we introduce:

$$T^*(x^*, t^*) \equiv \frac{T_w - T(x,t)}{T_w - T_0} \qquad (0 \leqslant T^* \leqslant 1), \tag{5.4-49}$$

$$x^* \equiv x/b \qquad (-1 \leqslant x^* \leqslant 1), \tag{5.4-50}$$

$$t^* \equiv \alpha t/b^2 \qquad (0 \leqslant t \leqslant \infty). \tag{5.4-51}$$

In these terms the present BVP becomes that of finding the function $T^*(x^*, t^*)$, which satisfies the parameter-free PDE:

$$\frac{\partial T^*}{\partial t^*} = \frac{\partial^2 T^*}{\partial x^{*2}} \qquad (\text{on } -1 \leqslant x^* \leqslant 1, \quad 0 \leqslant t^* < \infty), \tag{5.4-52}$$

subject to the initial condition (IC): $T^*(x^*, 0) = 1$ and the boundary conditions: $T^*(\pm 1, t^*) = 0$, $(\partial T^*/\partial x^*)_{y^*=0} = 0$. These requirements are conveniently summarized in Figure 5.4-5, which should be kept in mind in considering the steps that follow.

Fourier's idea was to propose and find a solution of the "separable" form:

$$T^*(x^*, t) = \sum_{n=-\infty}^{n=\infty} X_n(x^*) \cdot \theta_n(t^*), \tag{5.4-53}$$

[18] Subject to the restriction of constant thermophysical properties, all results in what follows apply irrespective of the sign of $T_w - T_0$; therefore they apply to "quenching" as well as to sudden heating.

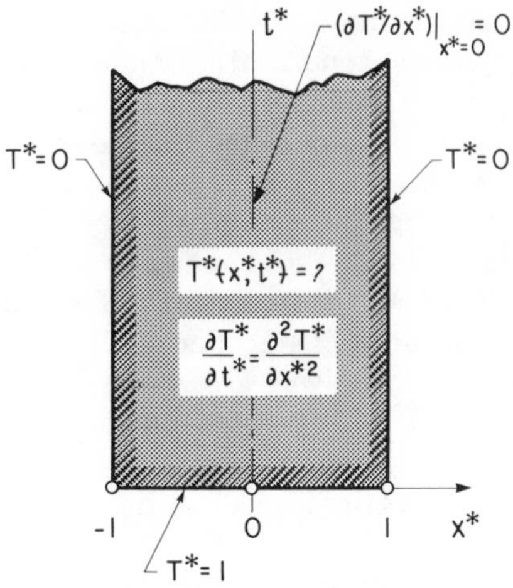

Figure 5.4-5 Domain of (nondimensional) independent variables; slab heat diffusion problem.

which would satisfy all of the above-mentioned conditions. At first glance this approach might not seem promising, since the problem of finding one function has been replaced by finding an infinite number of functions $X_n(x^*)$ and $\theta_n(t^*)$, $-\infty < n < \infty$. However, there is now considerable flexibility to be exploited, and at least each of these latter functions depends on only *one* variable, shown below to satisfy simple *ordinary* differential equations (ODEs). Indeed, if Eq. (5.4-53) is inserted into the PDE (5.4-52), we find that:

$$\sum_{n=-\infty}^{\infty} X_n(x^*)\frac{d\theta_n}{dt^*} = \sum_{n=-\infty}^{\infty} \theta_n \frac{d^2X_n}{dx^{*2}}. \tag{5.4-54}$$

This equation will certainly be satisfied if corresponding terms on the LHS and RHS are equal; i.e., *for each integer n* we require:

$$X_n(x^*)\frac{d\theta_n}{dt^*} = \theta_n \cdot \frac{d^2X_n}{dx^{*2}} \tag{5.4-55}$$

or, collecting like terms:

$$\frac{1}{\theta_n}\cdot\frac{d\theta_n}{dt^*} = \frac{1}{X_n}\cdot\frac{d^2X_n}{dx^{*2}}. \tag{5.4-56}$$

By hypothesis, the LHS of Eq. (5.4-56) is (at most) a function of t^* alone, whereas the RHS is (at most) a function of x^* alone. Therefore, the only way Eq. (5.4-56) can be satisfied is if each side is equal to the same constant, shown below to be negative. Therefore, we replace Eq. (5.4-56) by the sets of ODEs:

$$\frac{1}{\theta_n} \cdot \frac{d\theta_n}{dt^*} = -c_n^2 \qquad (n = 0, \pm 1, \pm 2, \ldots) \tag{5.4-57}$$

and

$$\frac{1}{X_n} \cdot \frac{d^2 X_n}{dx^{*2}} = -c_n^2 \qquad (n = 0, \pm 1, \pm 2, \ldots). \tag{5.4-58}$$

According to the first-order Eq. (5.4-57), for this BVP the time-dependent functions θ_n are then all decaying exponentials:

$$\theta_n\{t^*\} = \theta_n\{0\} \cdot \exp\{-c_n^2 t^*\} \tag{5.4-59}$$

[disallowing growing exponentials rules out positive values of the RHS of Eqs. (5.4-57) and (5.4-58)] whereas, according to the linear second-order Eq. (5.4-58), the space-dependent functions are trigonometric; that is,

$$X_n\{x^*\} = B_n \sin\{c_n x^*\} + C_n \cos\{c_n x^*\}. \tag{5.4-60}$$

It is now necessary to select the constants B_n, C_n, $\theta_n\{0\}$, and c_n to satisfy the remaining conditions of this BVP. (Thus far, only the PDE has been satisfied.) The centerplane condition, which implies $dX_n/dx^* = 0$ at $x^* = 0$, will be satisfied if we take $B_n = 0$. Moreover, the boundary condition $T^*\{\pm 1, t^*\} = 0$ will be satisfied if we select for the c_n-values the "*special*" (*eigen-*)*values*:

$$c_n = \left(n + \frac{1}{2}\right)\pi \qquad (n = 0, \pm 1, \pm 2, \ldots) \tag{5.4-61}$$

(since $\cos[(n + \frac{1}{2})\pi] = 0$ when n is any integer). Combining these results we are left with the problem of finding a single set of coefficients D_n such that the initial condition (IC) is satisfied, with the complete solution now in the form of an infinite series:

$$T^*(x^*, t^*) = \sum_{n=0}^{n=\infty} D_n \cdot \cos\left[\left(n + \frac{1}{2}\right)\pi x^*\right] \cdot \exp\left[-\left(n + \frac{1}{2}\right)^2 \pi^2 t^*\right] \tag{5.4-62}$$

over all positive integers.[19] The selection of the coefficients ("amplitudes") D_n such

[19] Explicitly, $D_n = \theta_n\{0\} \cdot C_n + \theta_{-(n+1)}\{0\} \cdot C_{-(n+1)}$, where we have made use of the fact that for every positive integer n there is an equal corresponding term [at $-(n+1)$].

that the IC is satisfied, that is,

$$T^*(x^*, 0) = 1 = \sum_{n=0}^{\infty} D_n \cdot \cos\left[\left(n + \frac{1}{2}\right)\pi x^*\right], \tag{5.4-63}$$

is a special case of the problem of expanding (or representing) an arbitrary function in terms of a series of (here) trigonometric functions—i.e., the selection of Fourier coefficients in a Fourier series expansion. By exploiting the following ("orthogonality") property of the cosine function:

$$\int_{-1}^{1} \cos\left(\left(m + \frac{1}{2}\right)\pi x^*\right) \cos\left(\left(n + \frac{1}{2}\right)\pi x^*\right) dx^* = \begin{cases} 0 & \text{if } m \neq n, \\ 1 & \text{if } m = n, \end{cases} \tag{5.4-64}$$

Fourier showed that in the present case the D_n should be chosen as follows:

$$D_n = \frac{2(-1)^n}{\left(n + \frac{1}{2}\right)\pi}, \tag{5.4-65}$$

completing the specification of T^* (Eq. (5.4-62)).

It is interesting to note that, despite the alternating signs (cf. Eq. (5.4-65)), the convergence of this series is usually excellent, except for times such that $\delta_h < b$, in which case $T^* \approx \operatorname{erf}(\eta)$ (where η is now given by $(b - x)/(4\alpha t)^{1/2}$—see Section 5.4.6.1).

The Fourier method of separation of variables can also be applied to situations for which:

1. The initial conditions are not uniform; e.g., $T^*(x^*, 0)$ is specified.
2. The boundary conditions do not directly specify the value of T, but rather some relationship between T and its derivative normal to the surface (e.g., specified heat flux); see below.
3. The boundaries of the domain are coordinate surfaces, but in curvilinear coordinate systems (e.g., cylindrical,[20] spherical, parabolic, etc.). (See, e.g., Carslaw and Jaeger (1959), Crank (1975), and Appendix 5.1.)

However, the method is clearly limited to *linear* PDEs, since only then is the *sum* of solutions also a solution to the original PDE.

A commonly encountered nonexplicit BC example (cf. generalization (2) above) is one in which the heat flux from a surrounding fluid is approximated *via* a specified (dimensional) heat-transfer coefficient, h, in which case the condition

[20] The space-dependent functions $R_n(r^*)$, which appear for the simplest diffusion problems in cylindrical coordinates, are, in fact, called cylinder (Bessel-) functions.

$T\{\pm b, t\} = T_w$ (specified) is replaced by the linear interrelation:

$$k\left(\frac{\partial T}{\partial x}\right)_{x=\pm b} = h_{\text{fluid}} \cdot [T_\infty - T\{\pm b, t\}]. \tag{5.4-66}$$

In such cases the resulting temperature field $T^*\{x^*, t^*\}$ within the solid will also depend on the dimensionless parameter (Biot number):

$$\text{Bi}_h \equiv \frac{(b/k)}{(1/h_{\text{fluid}})} \tag{5.4-67}$$

i.e., the ratio of the thermal resistance of the semi-slab to the thermal resistance of the external (fluid) film. Indeed, Eq. (5.4-62) would be recovered in the limit $\text{Bi}_h \to \infty$ (see Section 5.3.2 for the simpler case $\text{Bi}_h \to 0$).

A.B. Newman (1931) has proved that transient (and steady-state) diffusion in many "hybrid" geometries can be described using solutions applicable to their simpler "parts." For example, the temperature field in a *finite* cylindrical solid $(-L/2 \leqslant z \leqslant L/2; 0 \leqslant r \leqslant R)$ can be written in the product form:

$$\frac{T_\infty - T\{r, z, t\}}{T_\infty - T_0} = \left[T^*\left\{\frac{z}{L/2}, \frac{\alpha_z t}{(L/2)^2}, (\text{Bi}_h)_z\right\}\right]_{\text{slab}} \cdot \left[T^*\left\{\frac{r}{R}, \frac{\alpha_r t}{R^2}, (\text{Bi}_h)_r\right\}\right]_{\text{cylinder}}, \tag{5.4-68}$$

where Eqs. (5.4-62 and -65) describe the "slab" part for the limiting case $(\text{Bi}_h)_z \to \infty$. It is noteworthy that this result holds even if the solid material is anisotropic (that is, $k_z \neq k_r$; cf. Section 3.3.1) and the fluid-side heat-transfer coefficient at the cylinder *ends* (appearing in $(\text{Bi}_h)_z$ is different from the heat-transfer coefficient assumed for the cylinder surface (appearing in $(\text{Bi}_h)_r$. Clearly, Newman's theorem amplifies considerably the value of existing solutions to diffusion problems in simple geometries.

5.5 TEMPERATURE DISTRIBUTIONS AND SURFACE HEAT-TRANSFER COEFFICIENTS IN STEADY LAMINAR FLOWS

5.5.1 Introduction

When either $(\text{Re} \cdot \text{Pr})^{1/2}$ or $(\text{Ra}_h \text{Pr})^{1/4}$ is not negligibly small, energy convection and diffusion are *both* important. However, if Re or Ra_h are below so-called "transition" (threshold) values the resulting flows remain stable to inevitable small disturbances. If the boundary conditions (e.g., wall temperatures) are time-independent, such flows remain steady everywhere and "laminar" (nonturbulent). Three instructive surface geometries are briefly discussed below: (a) the flat plate, (b) the isolated sphere, and (c) the straight circular duct. The first two flow situations are said to be "external," the latter "internal." We conclude this section with a brief discussion of simultaneous energy convection and diffusion in the absence of physical surfaces, our example being the "point" heat source in a uniform stream.

5.5.2 Thermal Boundary Layer Adjacent to a Flat Plate

a. Forced Convection (Constant Properties, Newtonian Fluid). If the Reynolds' number (based on overall plate length) Re_L satisfies the condition $\mathrm{Re}_L{}^{1/2} \gg 1$ and yet $\mathrm{Re}_L \lesssim 10^6$, then the flow field in the vicinity of an unheated plate is well described by the Blasius function $f(\eta)$ discussed in Chapter 4. If the fluid properties are nearly constant, then this flow field also applies in the presence of heat transfer. For the case of an isothermal hot plate, there will be a thin thermal boundary layer in the immediate vicinity of the plate. Within this boundary layer $T(x, y)$ satisfies the approximate PDE:

$$\rho c\left[u\frac{\partial T}{\partial x} + v\frac{\partial T}{\partial y}\right] = k\frac{\partial^2 T}{\partial y^2}, \tag{5.5-1}$$

where we have neglected *streamwise* (compared to transverse) heat diffusion. Otherwise this equation guarantees a local balance (per unit volume) between the net outflow of energy by convection and the net inflow of energy by Fourier diffusion. The two coefficient functions appearing in Eq. (5.5-1), *viz.*,

$$u(x, y) = Uf'(\eta) \tag{5.5-2}$$

and

$$v(x, y) = U \cdot \mathrm{Re}_x{}^{-1/2} \cdot [\eta f' - f(\eta)] \tag{5.5-3}$$

are known (Blasius) functions of the similarity variable

$$\eta(x, y) \equiv \frac{1}{2} \cdot \frac{y}{x}(\mathrm{Re}_x)^{1/2}. \tag{5.5-4}$$

Pohlhausen [1921] showed that the temperature field in such a thermal boundary layer will also be a function of $\eta(x, y)$ alone, provided $T_\infty \equiv T(x, \infty)$ and $T_w \equiv T(x, 0)$ are constants. Indeed, in the absence of viscous dissipation:[21]

$$\frac{T - T_w}{T_\infty - T_w} = \frac{\int_0^\eta \exp[-\mathrm{Pr}\int_0^\eta f(\eta)\,d\eta]\,d\eta}{\int_0^\infty \exp[-\mathrm{Pr}\int_0^\eta f(\eta)\,d\eta]\,d\eta}. \tag{5.5-5}$$

[21] When $\mathrm{Pr} = 1$, $(T - T_w)/(T_\infty - T_w) = u/U$, quantitatively expressing an "analogy" between energy transport and x-momentum transport.

The wall heat fluxes, readily computed from $-k(\partial T/\partial y)_{y=0}$, are found to be described by the local dimensionless heat-transfer coefficient (based on running length x):

$$\mathrm{Nu}_h\{x\} \cong 0.332\,\mathrm{Re}_x^{1/2}\,\mathrm{Pr}^{1/3} \qquad (\mathrm{Pr} \geqslant 0.7). \tag{5.5-6}$$

In *forced* convection surface-transfer problems, there is another reference heat flux that can be used to define a dimensionless heat-transfer coefficient, *viz.*:

$$\dot{q}''_{\mathrm{ref}} \equiv \rho U c_{\mathrm{p}} \cdot (T_{\mathrm{w}} - T_{\infty}). \tag{5.5-7}$$

Comparing actual heat fluxes to *this* reference heat flux defines the dimensionless *Stanton number* for heat transfer, written St_h. In this case, Eq. (5.5-6) is equivalent to:

$$\mathrm{St}_h\{x\} = 0.332\,\mathrm{Re}_x^{-1/2}\,\mathrm{Pr}^{-2/3}. \tag{5.5-8}$$

It is no accident that, when $\mathrm{Pr} = 1$, $\mathrm{St}_h\{x\}$ becomes identical to the dimensionless *skin-friction coefficient* $c_{\mathrm{f}}/2$ (Chapter 4). This "coincidence" follows from the fact that when $\mathrm{Pr} = 1$ the x-momentum and temperature fields become identical; i.e., there is a *strict "analogy" between momentum and heat-transfer for forced-convection flows in the absence of appreciable streamwise pressure gradients.* Equation (5.5-8) extends this analogy to fluids for which the momentum diffusivity (kinematic viscosity) and thermal diffusivity are *not* equal (recall that, for air, $\alpha \approx \nu/(0.7)$, implying that the thermal boundary layer will be somewhat thicker than the x-momentum density (or "vorticity") boundary layer).

b. Natural Convection. For laminar, steady *"free" convection* of a Newtonian fluid past a vertical isothermal surface, it is also possible to obtain $\mathrm{Nu}_h\{x\}$ theoretically,[22] except in this case the velocity and temperature fields are coupled, necessitating a simultaneous solution for both. This can readily be seen in the simplest limiting case, in which all thermophysical properties of the fluid are nearly constant. Then the pressure gradient "driving" such a flow will, to a first approximation, satisfy the static balance condition: $0 = -\mathbf{grad}\,p + \rho_{\infty}\mathbf{g}$, so that the "buoyancy force" $-\mathbf{grad}\,p + \rho\mathbf{g}$ appearing in the local momentum-balance equation can be written: $-(\rho_{\infty} - \rho)\mathbf{g} \approx \beta_T \cdot (T - T_{\infty})\mathbf{g}$, where β_T is the fluid's thermal-expansion coefficient (Eq. (5.4-8)). Thus, the momentum density field cannot be obtained without simultaneous consideration of the (thermal) energy density field, and *vice versa*. For this case the local dimensionless wall heat flux is given by:

$$\mathrm{Nu}_h\{x\} \approx 0.517(\mathrm{Ra}_h)^{1/4} \qquad (\mathrm{Pr} = 0.73, \quad 10^4 \leqslant \mathrm{Ra}_h \leqslant 10^9) \tag{5.5-9a}$$

a result predicted by K. Pohlhausen (see, e.g., Schlichting (1979)), using the method of

---

[22] In view of the absence of an imposed (reference) velocity, *Stanton* numbers are not used to describe "pure" natural convection surface heat (or mass) transfer rates.

self-similar solutions (after making the "boundary layer" (BL) approximations to simplify the *coupled* PDEs satisfied by $\mathbf{v}(\mathbf{x})$ and $T(\mathbf{x})$). Here, both Nu_h and Ra_h are based on the "running" length x measured up the plate (to the point in question) so that Eq. (5.5-9a) corresponds to a thermal BL (slope) thickness $\delta_h(x)$ which grows like $x^{1/4}$. It follows that the area-averaged heat-transfer coefficient $\overline{\mathrm{Nu}}_h$ on a vertical plate of total height L is given by:

$$\overline{\mathrm{Nu}}_h \approx 0.689(\mathrm{Ra}_{h,L})^{1/4} \quad (\mathrm{Pr} \approx 0.73), \tag{5.5-9b}$$

where the coefficient is (4/3)(0.517) (see, e.g., solution to Exercise 5.10). Except for heat transfer through Newtonian fluids with low Prandtl numbers (e.g., liquid metals), these pre-multipliers of $(\mathrm{Ra}_h)^{1/4}$ are expected to be Pr-insensitive since Eq. (5.5-9a) predicts the correct Pr-dependence for the laminar heat-transfer coefficient Nu_h in the asymptotic limit $\mathrm{Pr} \to \infty$ (Morgan and Warner (1956)).

5.5.3 Convective Heat Transfer from/to an Isolated Sphere

a. Forced Convection (Constant Properties, Newtonian Fluid). While Froessling showed theoretically that, over a restricted Re-range, $\overline{\mathrm{Nu}}_h - 2$ should be proportional to $(\mathrm{Re})^{1/2}(\mathrm{Pr})^{1/3}$, the flow field about even an isolated sphere is sufficiently complex to encourage greater reliance on experimental data. Actually, the simple relation:

$$\overline{\mathrm{Nu}}_h \approx 2\underbrace{[1 + 0.276(\mathrm{Re})^{1/2}(\mathrm{Pr})^{1/3}]}_{\substack{\text{Forced convection}\\ \text{augmentation}\\ \text{("wind"-) factor}}} \tag{5.5-10}$$

provides an adequate fit to these data for the parameter range: $\mathrm{Re} < 10^4$ and $\mathrm{Pr} \geqslant 0.7$, say. Here d_w is the reference length in the definition of $\overline{\mathrm{Nu}}_h$ and Re.

Though it was developed to describe energy transfer to/from an isolated *solid* sphere, this result is frequently applied to nearly isolated *liquid* droplets in a spray, provided the effects of internal circulation (surface shear-induced) and surface evaporation (Stefan flow; see Section 6.4.3) are negligible.

Analogous Nu_h correlations are available for an isolated circular *cylinder* in cross-flow ((see Figure 8.1-1 for data covering: $10^{-1} \leqslant \mathrm{Re} \leqslant 3 \times 10^5$). It is noteworthy that *laminar BL theory*, which predicts a $\mathrm{Re}^{1/2}$-dependence, is valid only over an intermediate portion (perhaps $10^2 < \mathrm{Re} < 10^4$) of this Re-range).

b. Natural Convection. For $\mathrm{Ra}_h < 10^9$ a correlation similar in structure to Eq. (5.5-10) describes available area-averaged heat-transfer data for isolated spheres in the *absence of forced convection*, i.e.,

$$\overline{\mathrm{Nu}}_h \approx 2\underbrace{[1 + 0.3(\mathrm{Gr}_h)^{1/4}(\mathrm{Pr})^{1/3}]}_{\substack{\text{Natural convection}\\ \text{augmentation factor}}}. \quad (5.5\text{-}11)$$

Note that both of the above correlations are for the *average* heat flux from/to the sphere rather than the *distribution* of local heat flux around it (cf. Section 5.3.1). The latter is highly nonuniform, behaving differently in the wake region (rear) than in the region upstream of the point of vanishing shear stress ("separation"). $\overline{\mathrm{Nu}}_h$ data for a horizontal *cylinder* will be summarized in Figure 7.2-1.

Note that both of the above convective heat-transfer correlations (Eqs. (5.5-11) and (5.5-10) for an isolated sphere), and their counterparts for an isolated circular *cylinder* (see Figures 7.2-1 and 8.1-1, respectively) describe the *average* heat flux $\dot{q}_w/A_w$ from/to the solid rather than the *distribution* of local heat flux $\dot{q}''_w$ over the "wetted" surface (cf. Section 5.3.1). As for the cylinder (Figure 5.3-3), the *local* Nu_h-values are highly position-dependent, behaving differently in the region of separated flow ("leeward") than in the ("windward") region upstream of the location of vanishing wall shear stress ("separation point").

Of course, "buoyancy" forces ($-\mathbf{grad}\, p + \rho \mathbf{g}_{\mathrm{eff}}$) can play a role even in cases which, at first sight, appear to be pure "forced" convection systems. Comparison of Eqs. (5.5-11) and (5.5-10) suggests that the buoyancy force influence will not be negligible unless the parameter $\mathrm{Gr}_h^{1/4}/\mathrm{Re}^{1/2}$ is small ($\ll 1$), a conclusion that has been borne out in experiments directed at elucidating "mixed" natural- and forced-convection behavior. Buoyancy-influenced forced convective transfer is commonly encountered in, e.g., large-scale combustion systems (e.g., fossil fuel-fired power stations (see Exercise 7.2)) and in small flow reactors used to chemically vapor deposit (CVD) semiconductor films for the electronics industry (see Exercise 6.8).

5.5.4 Convective Heat Transfer to/from the Fluid Flowing in a Straight Circular Duct ($\mathrm{Re} < 2 \times 10^3$)

Viscous fluids are often heated (or cooled) by passing them steadily *through a section* of heated straight duct of circular cross section. Of interest is the duct length required, say, to raise the mean temperature of the fluid, initially at T_0, to some desired level[23] less than T_w. The average axial fluid velocity will be written U, corresponding to the total mass flow rate $\rho_0(\pi d_w^2/4)U$.

[23] One finds that the *average* fluid temperature cannot attain T_w in any finite length of duct (cf. Eq. (5.5-21a)).

In the simplest case the fluid velocity profile $v_z(r)$ is parabolic, corresponding to *constant property* flow of a Newtonian fluid (Poiseuille) far downstream of the "hydrodynamic" inlet. The temperature field $T(r, z)$ downstream of the "thermal inlet" (at $z = 0$) will then approximately satisfy the linear, second-order PDE:

$$2U \cdot \left[1 - \left(\frac{2r}{d_w}\right)^2\right] \cdot \frac{\partial T}{\partial z} = \frac{\alpha}{r} \cdot \frac{\partial}{\partial r}\left(r \frac{\partial T}{\partial r}\right), \tag{5.5-12}$$

where, again, axial conduction has been neglected compared with radial conduction in the local energy balance in cylindrical coordinates. If the boundary conditions are:

$$T(r, 0) \equiv T_0 = \text{const},$$
$$T(d_w/2, z) \equiv T_w = \text{const}, \tag{5.5-13a,b,c}$$
$$\left(\frac{\partial T}{\partial r}\right)_{r=0} = 0,$$

then $T(r, z)$ can be found by the Fourier method of "separation-of-variables" (cf. Section 5.4.6.2; formally leading to an infinite sum of products of Bessel-like functions of r and decaying exponentials of z). The axial distribution of *wall heat fluxes* can then be found from $-k(\partial T/\partial r)_w$, by term-by-term differentiation with respect to radius, and evaluation at $r = d_w/2$. Local dimensionless heat-transfer coefficients, Nu_h, in such situations are based on comparing actual heat fluxes to the reference *local* heat flux:

$$\dot{q}''_{\text{ref}} \equiv k\left[\frac{T_w - T_b(z)}{d_w}\right], \tag{5.5-14}$$

where $T_b(z)$ is the so-called "mixing-cup average temperature" at axial station z (that is, the fluid temperature that would result from severing the duct at z and adiabatically mixing the effluent).[24] Figure 5.5-1 shows the resulting Nu_h as a function of the relevant dimensionless axial variable (Graetz [1885]):

$$\xi_h \equiv \frac{1}{\text{Re Pr}} \cdot \frac{z}{d_w}. \tag{5.5-15}$$

Figure 5.5-1 reveals that:

$$\mathrm{Nu}_h \approx \text{const}\, \xi_h^{-1/3} \qquad \text{for } \xi_h \to 0, \tag{5.5-16}$$

[24] Also called the "bulk" temperature of the fluid at station z (hence the subscript b), $T_b(z)$ can be calculated from the equality:

$$\dot{m} c_p(T_b)[T_b(z) - T_{\text{ref}}] \equiv \int_A \dot{m}''_z c_p(T) \cdot [T - T_{\text{ref}}]\, dA.$$

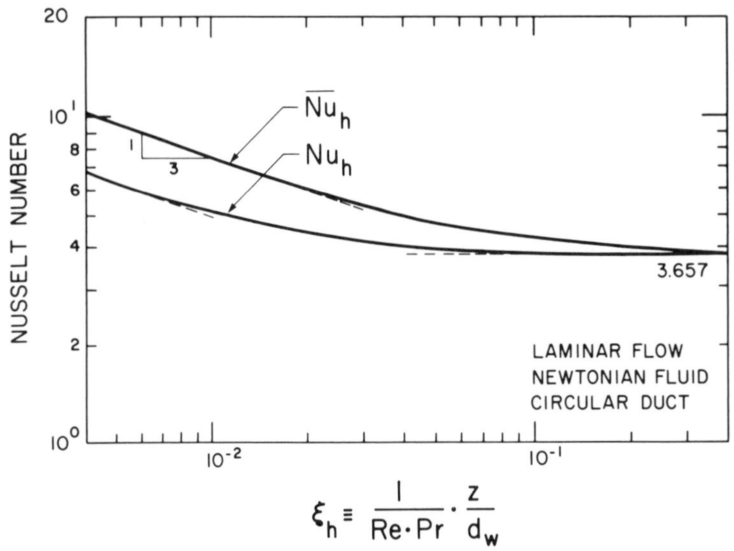

Figure 5.5-1 Nusselt number distribution in a straight circular duct with fully developed viscous (Newtonian) fluid flow.

and

$$Nu_h \to 3.657 \qquad \text{for } \xi_h \to \infty. \tag{5.5-17}$$

Thus, while Nu_h approaches infinity at the thermal inlet, far downstream of the thermal inlet, Nu_h in laminar viscous duct flow approaches a calculable *constant value*, said to be the "thermally fully developed" or asymptotic Nusselt number. Of importance in calculating $T_b(z)$ in applications is $\overline{Nu}_h(\xi_h)$ defined by:

$$\overline{Nu}_h(\xi_h) \equiv \frac{1}{\xi_h} \cdot \int_0^{\xi_h} Nu_h(\xi'_h)\, d\xi'_h. \tag{5.5-18}$$

This axial *average heat-transfer coefficient* can be written in the product form:

$$\overline{Nu}_h(\xi_h) = \overline{Nu}_h(\infty) \cdot \left[F\begin{pmatrix}\text{thermal}\\ \text{entrance}\end{pmatrix} \right] \tag{5.5-19}$$

where $\overline{Nu}_h(\infty) = 3.657$ and the Graetz–Nusselt numerical values can be curve-fit by (Churchill (1978)):

$$F(\text{thermal entrance}) \approx [1 + (7.60\xi_h)^{-8/3}]^{1/8}. \tag{5.5-20}$$

An overall energy balance on the total fluid contained in the Eulerian CV between 0 and z then leads to the useful interrelation:

$$\frac{T_w - T_b(z)}{T_w - T_b(0)} = \exp\{-4\xi_h \overline{\mathrm{Nu}}_h(\infty) \cdot F(\text{thermal entrance})\}, \qquad (5.5\text{-}21a)$$

from which the length required to achieve any prescribed T_b can be obtained numerically.

The following alternative equations are fully equivalent to Eq. (5.5-21a). Suppose one asks what choice of $(\Delta T)_{\text{mean}}$ will ensure that:

$$\frac{\dot{q}/A_w}{\left\{\dfrac{k(\Delta T)_{\text{mean}}}{d_w}\right\}} = \overline{\mathrm{Nu}}_h(\xi_{\text{max}}), \qquad (5.5\text{-}21b)$$

where

$$\dot{q} = \dot{m}c_p[T_b(L) - T_b(0)]$$

and

$$A_w = \pi d_w L.$$

Equation (5.5-21a) reveals that the appropriate choice is:

$$\Delta T_{\text{mean}} \equiv \frac{(T_w - T_b(0)) - (T_w - T_b(L))}{\ln\left[\dfrac{T_w - T_b(0)}{T_w - T_b(L)}\right]} \equiv \text{LMTD}. \qquad (5.5\text{-}21c)$$

That is, the "logarithmic-mean" temperature difference (LMTD) has this important property.[25] Indeed, Eq. (5.5-21b, c) is often used to calculate values of $\overline{\mathrm{Nu}}_h$ from experimental values of bulk temperature change and wall temperature. Provided $T_w(z)$ is nearly uniform and the fluid properties are nearly constant, $\overline{\mathrm{Nu}}_h$-values computed *via* LMTD will be no different from those defined by Eqs. (5.5-14) and (5.5-18).

Other important internal-flow heat-transfer problems exhibit the same general features. Thus:

[25] LMTD also plays an important role in the overall design (sizing) of co-current or counter-current two-stream heat exchangers (see, e.g., Kays and London (1958)). In such cases, provided the overall conductance U is nearly constant within the unit, the total required transfer area is found to be related to the total amount of heat exchange by $A = \dot{q}/[U \cdot (\text{LMTD})]$, where LMTD is the logarithmic-mean temperature difference between the *two streams* calculated in terms of ΔT at the *inlet* and *outlet* of the heat exchanger.

Table 5.5-1 Heat Transfer and Friction for Fully Developed Laminar Newtonian Flow through Straight Ducts of Specified Cross-Section[a] (after Shah and London (1978))

Geometry	b/a	Nu_T	Nu_{H1}	Nu_{H2}	$C_f Re$	$\frac{j_{H1}^{\dagger}}{C_f}$	$\frac{Nu_{H1}}{Nu_T}$
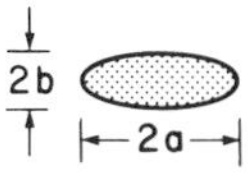	1	3.657	4.364	4.364	16.000	0.307	1.19
	1/2	3.74	4.88		18.24	0.301	1.30
	1/8	3.72	5.09		19.15	0.299	1.37
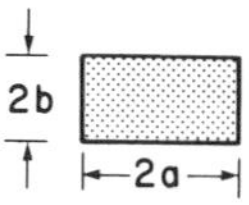	1	2.976	3.608	3.091	14.227	0.286	1.21
	1/2	3.391	4.123	3.017	15.548	0.299	1.22
	1/4	4.439	5.331	2.930	18.233	0.329	1.20
	1/8	5.597	6.490	2.904	20.585	0.355	1.16
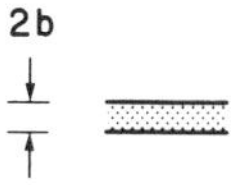	0	7.541	8.235	8.235	24.000	0.386	1.09
	0	4.861	5.385	5.385	24.000	0.253	1.11
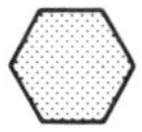		3.34*	4.002	3.862	15.054	0.299	1.20
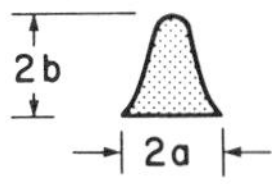	$\sqrt{3}/2$	2.39*	3.014	1.474	12.630	0.269	1.26
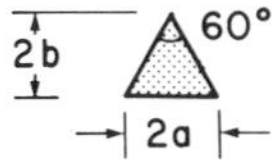	$\sqrt{3}/2$	2.47	3.111	1.892	13.333	0.263	1.26

* Interpolated values.
† This heading is the same as $Nu_{H1}\, Pr^{-1/3}/C_f Re$ with $Pr = 0.7$; Nu, C_f, Re based on $d_{eff} \equiv 4A/\text{Perimeter}$.
[a] The subscripts T, H1, and H2 on the relevant Nusselt numbers mean, respectively:
T: pertaining to constant wall temperature;
H1: pertaining to axially constant heat flux;
H2: pertaining to peripheral and axial constancy of heat flux.

a. Ducts of noncircular cross section have different values of $\overline{Nu}_h\{\infty\}$ (see Table 5.5-1), after Shah and London (1978)).
b. If the axial (not perimeter) *wall heat flux* (rather than the *wall temperature*) is constant, then, for a circular duct, $\overline{Nu}_h\{\infty\}$ is found to be 4.364 (see Table 5.5-1).

c. If the velocity "profile" is uniform ("plug flow") rather than parabolic, then $\overline{\mathrm{Nu}_h(\infty)}$ for a circular duct is 5.78 (not shown in Table 5.5-1).
d. If the fluid flow in the heat-transfer section is not "fully developed" (but rather z-dependent), then the fluid *viscosity* will explicitly enter the problem[26] and calculations must be repeated for fluids of various Prandtl numbers (curve B, Figure 5.5-1).

5.5.5 Heat Exchange between a Fluid and a Porous Medium or "Packing"

Adopting the viewpoint of Section 4.7, we consider a packed duct to be the equivalent of many tortuous "interstitial" ducts defined by the spaces between the bed "particles." By analogy with the case of steady-flow heat exchange in a single straight duct of noncircular cross section, we then propose an appropriate dimensionless energy-transport coefficient, $\mathrm{Nu}_{h,\mathrm{bed}}$, and leave to *experiment* the determination of its dependence on the arguments $\mathrm{Re}_{\mathrm{bed}}$ and Pr. Here, as before:

$$\mathrm{Re}_{\mathrm{bed}} \equiv \frac{G_0 d_{\mathrm{p,eff}}}{\mu(1-\varepsilon)}, \tag{5.5-22}$$

and $G_0(\equiv \dot{m}/A_0)$ is the so-called "superficial" fluid mass velocity.

Consider an element dz of column height, which includes a total volume $A_0\,dz$ and a packing (accessible) surface area of $a'''A_0\,dz$. If an amount of energy $d\dot{q}$ is transferred from the packing to the fluid within this increment of column height, then the average heat flux is $d\dot{q}/(a'''A_0\,dz)$, and an appropriate *dimensionless heat-transfer coefficient* (Nusselt number) would be:

$$\mathrm{Nu}_{h,\mathrm{bed}} = \mathrm{const}\cdot\frac{d\dot{q}/(a'''A_0\,dz)}{[k(T_w - T_b)]/d_{i,\mathrm{eff}}}. \tag{5.5-23a}$$

A common choice, consistent with this form, is:

$$\mathrm{Nu}_{h,\,\mathrm{bed}} \equiv \frac{d\dot{q}/(a'''A_0\,dz)}{[k(T_w - T_b)]/[(\varepsilon/1-\varepsilon)d_{\mathrm{p,eff}}]} = \frac{(1/A_0)/[d\dot{q}/d(z/d_{\mathrm{p,eff}})]}{[6(1-\varepsilon)^2]/\varepsilon\cdot[k(T_w - T_b)]/d_{\mathrm{p,eff}}} \tag{5.5-23b}$$

and we anticipate that $\mathrm{Nu}_{h,\mathrm{bed}}$, so defined, will depend upon $\mathrm{Re}_{\mathrm{bed}}$ and Pr.

Now consider, for simplicity, the heating of a fluid passing steadily through an isothermal bed (temperature T_w). For the steady flow of a nonreacting fluid mixture,

[26] Cf. the previous (fully developed flow) example in which only the ν-independent *product* Re · Pr, the so-called Peclet number, Ud_w/α, appears.

the heat exchanged, $d\dot{q}$, will cause a bulk temperature rise dT_b given by:

$$d\dot{q} = \dot{m}c_p dT_b, \tag{5.5-24}$$

if we neglect the axial *diffusion* of energy from adjacent (upstream and downstream) volume elements. Then Eq. (5.5-23b) is seen to be equivalent to a separable, first-order ODE governing the bulk fluid temperature distribution, $T_b(z)$. An instructive yet common limiting case is that for which all properties (bed and fluid coefficients) are constant (with respect to z). The bulk temperature $T_b(z)$ then approaches the packing temperature T_w in accord with the exponential decay law:[27]

$$\frac{T_w - T_b(z)}{T_w - T_b(0)} = \exp\left\{-6\cdot(1-\varepsilon)\left[\frac{\mathrm{Nu}_{h,\mathrm{bed}}(\mathrm{Re}_{\mathrm{bed}}, \mathrm{Pr})}{\varepsilon\,\mathrm{Re}_{\mathrm{bed}}\cdot\mathrm{Pr}}\right]\cdot\frac{z}{d_{\mathrm{p,eff}}}\right\}. \tag{5.5-25}$$

This law has been used to infer Nu_h-values from systematic T_b-measurements over a range of $\mathrm{Re}_{\mathrm{bed}}$, ε and Pr-values, for packings of various shapes (spheres, short cylinders, rings, saddles). A representative set of results is plotted in Figure 5.5-2 (after Whittaker (1972)), which reveals that the function

$$\mathrm{Nu}_h\mathrm{Pr}^{-0.4} \simeq 0.4\mathrm{Re}_{\mathrm{bed}}^{1/2} + 0.2\mathrm{Re}_{\mathrm{bed}}^{2/3} \tag{5.5-26}$$

provides a reasonable curve-fit[28] (in the range $3 \leqslant \mathrm{Re}_{\mathrm{bed}} \leqslant 10^4$, $0.6 \leqslant \mathrm{Pr}$, $0.48 \leqslant \varepsilon \leqslant 0.74$). Having determined this correlation, Eq. (5.5-25) now becomes a tool for making rational engineering estimates of:

the column height (depth)[29] required to cause a prescribed fluid temperature rise (design problem); and

the outlet fluid temperature for a prescribed column height (depth).

It is interesting to note that Eqs. (5.5-25 and -26) apply not only to ordinary "fixed" beds but also to:

dense, staggered tube bundles ($\varepsilon_{\mathrm{eff}} < 0.65$; $\geqslant 20$ rows deep), and/or

"fluidized" beds (with $\varepsilon(\mathrm{Re}_{\mathrm{bed}})$ before the onset of appreciable bubbling.

Moreover, analogous expressions (to be discussed in Section 6.4.4) hold for species *mass* exchange between a fluid and a fixed fluidized bed, or tube bundle. Of

---

[27] By convention, the quantity in the square brackets is called the bed *Stanton number*, $\mathrm{St}_{h,\mathrm{bed}}$.

[28] By introducing empirical "shape factors" for particular types of packings, the scatter of this type of correlation can be reduced further.

[29] This height L must, for self-consistency, satisfy the inequality $L/d_{\mathrm{p,eff}} \gg 1$, due to our assumptions of (a) quasi-homogeneous flow model (cf. Section 2.6.3) and (b) neglect of axial (streamwise) diffusion (cf. axial convection).

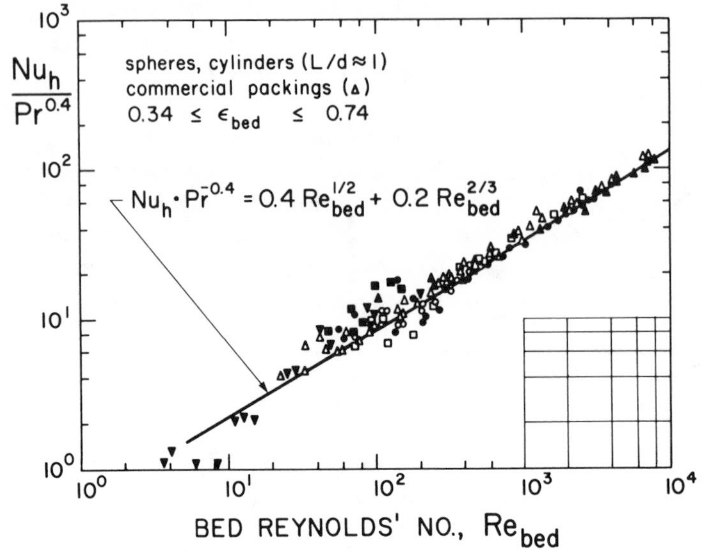

Figure 5.5-2 Experimentally determined dependence of packed-bed Nusselt number, $Nu_{h,\,bed}$, on the bed Reynolds' number, Re_{bed}, and fluid Prandtl number, Pr (adapted from Whittaker (1972)).

course, in many cases of engineering interest, momentum, energy, and mass exchange processes occur *simultaneously*. An important example is the *fixed-bed catalytic reactor*, in which an exoergic chemical reaction takes place on (and within) catalyst pellets randomly loaded within, say, a cylindrical vessel of constant cross section. While the preliminary design of such equipment is based on the "plug flow" approach given in this section, in practice refinements are needed to take care of inevitable "nonidealities," such as radially nonuniform (local) G_0-values through the packing, nonuniform packing temperature, variable fluid properties, etc. (See, e.g., Froment and Bischoff (1979).)

5.5.6 Diffusion from a Steady Point Source in a Uniform Moving Stream

Predicting the "thermal wake" of a point heat source of strength $\dot{q}$ in a constant-property, uniform, nonturbulent stream is another fundamental problem in the theory of convective diffusion. Indeed, this solution is of such a simple form that it has been used to experimentally determine thermal (and mass) diffusivities in combustion-heated gases. Moreover, the downstream effects of various *distributed* heat sources can be constructed by superposing these *point*-source solutions (exploiting the linearity property of the basic PDE).

In this case (Figure 5.5-3) energy is *convected* downstream ($+z$-direction) and simultaneously *diffuses* both radially and axially. Accordingly, the axially symmetrical temperature field (independent of θ) will satisfy the local energy-balance equation:

$$U\frac{\partial T}{\partial z} = \alpha\left[\frac{1}{r}\cdot\frac{\partial}{\partial r}\left(r\frac{\partial T}{\partial r}\right) + \frac{\partial^2 T}{\partial z^2}\right] \qquad \text{(if } \alpha = \text{const).} \qquad (5.5\text{-}27)$$

Despite the fact that we are using the cylindrical coordinates r, z, it is also convenient to introduce the *spherical radius*

$$s \equiv (r^2 + z^2)^{1/2} \qquad (5.5\text{-}28)$$

measured from the point source. One then seeks a solution to the PDE (5.5-27) satisfying the following "boundary" conditions:

$$\lim_{s\to\infty} T = T_\infty \qquad \text{(specified constant),} \qquad (5.5\text{-}29)$$

$$\lim_{s\to 0}\left[-(4\pi s^2)k\left(\frac{\partial T}{\partial s}\right)\right] = \dot{q} \qquad \text{(specified source strength),} \qquad (5.5\text{-}30)$$

$$\lim_{r\to 0}\left(\frac{\partial T}{\partial r}\right) = 0 \qquad \text{for } z > 0 \qquad (5.5\text{-}31)$$

(the last of these is sometimes erroneously called a "symmetry" condition). It is easy to verify that

$$T\{r, z\} - T_\infty = \frac{\dot{q}}{4\pi k s}\cdot\exp\left[\frac{-U(s - z)}{2\alpha}\right] \qquad (5.5\text{-}32)$$

satisfies these conditions and the PDE (5.5-22) exactly. While exhibiting nonphys-

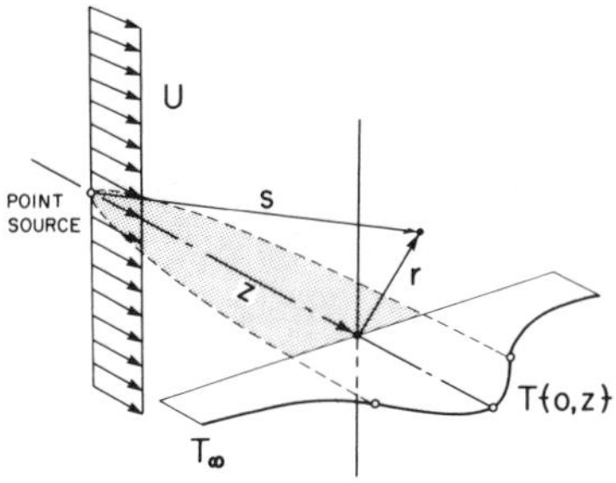

Figure 5.5-3 Energy convection and diffusion from a continuous point source in a uniform stream.

ical behavior ($T \to \infty$) in the immediate vicinity of the point source, this relation even describes the behavior of *finite*-size sources at distances downstream that are very large compared to the source size. At any fixed $z \gg r$, one finds that the radial temperature *profiles are Gaussian* in shape, with

$$\frac{T(r,z) - T_\infty}{T(0,z) - T_\infty} \approx \exp\left[-\frac{Uz}{4\alpha}\left(\frac{r}{z}\right)^2\right]. \tag{5.5-33}$$

Along the axis ($r = 0$), Eq. (5.5-32) takes a particularly simple form, *viz.*:

$$T(0,z) - T_\infty = \frac{\dot{q}}{4\pi k z}. \tag{5.5-34}$$

This result can be used to determine k-values based on *axial* temperature decay measurements.

In this case note that we do not calculate/report any "heat-transfer coefficient" (no physical boundary surfaces are present in the flow). In all of the cases discussed above, we have assumed that there are no volumetric heat sources acting within the flow field. The analogous convective-diffusion problem involving a point *mass* source in a fluid stream plays an important role in air pollution modeling (see Sections 5.6.5, 6.3.3, and 6.5.1).

5.6 TIME-AVERAGED TEMPERATURE DISTRIBUTIONS AND SURFACE HEAT-TRANSFER COEFFICIENTS IN "STEADY" TURBULENT FLOWS

5.6.1 Criteria for Transition to Turbulence and the Effects of Turbulence

The upper limits on Re and Ra_h for each of the Nu_h (or St_h) relations given in Section 5.5 pertain to the onset of *turbulence* within the boundary layer, i.e., localized nonsteadiness associated with *enhanced* time-averaged transport rates of momentum (Section 2.6.3), energy, and mass (Section 6.5). Above these so-called "transition" Reynolds' or Rayleigh numbers[30] the dependence of the heat-transfer coefficient on Re and Ra_h differs from the corresponding laminar case. Indeed, "blind" extrapolations of the laminar heat-transfer coefficient laws well into the turbulent region would lead to serious underestimates of surface transport rates (see Figure 5.6-1).

[30] In simple viscous-flow configurations, transition values of Re or Ra_h can be *predicted* theoretically, based on the predicted *stability* of the corresponding laminar flow with respect to small disturbances. In most cases, however, this information is obtained from experiments.

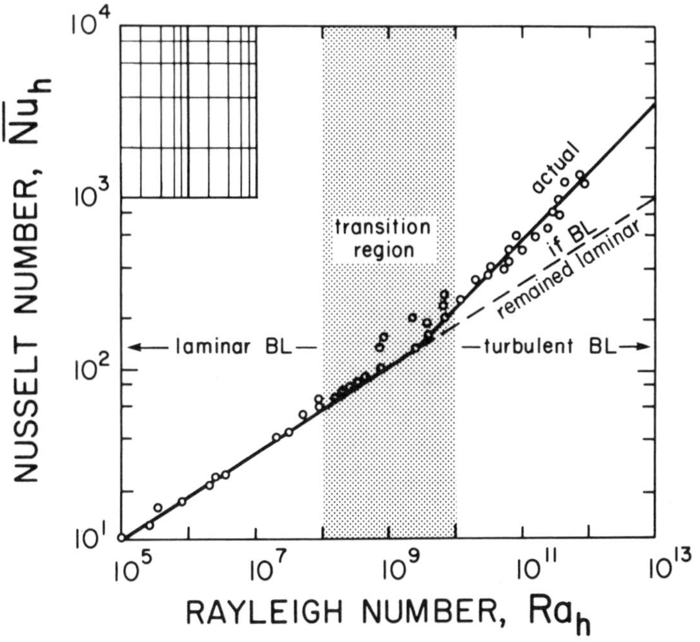

Figure 5.6-1 Area-averaged natural convective heat-transfer data for vertical flat surfaces in an otherwise quiescent Newtonian fluid. Note transition to turbulence (within the thermal BL) at Ra_h-values (based on plate length) above *ca.* 10^9 (adapted from Eckert and Jackson (1950)).

Enhanced surface transport rates of *momentum* associated with turbulence are usually undesired, since they manifest themselves as increased *drag* (or pressure drop in duct flows). However, the turbulent enhancement in *heat* (and mass) transport is usually welcome, leading to smaller required heat-exchange surface area (and hence equipment of smaller capital cost). We will see that, unfortunately, steps taken to further increase heat-transport (e.g., surface roughening) usually are associated with somewhat larger (but adverse) effects on drag and/or pressure drop.

5.6.2 Forced Convective Turbulent Heat Transfer from/to Straight, Smooth Ducts

The qualitative behavior discussed in Section 5.5.4 for *laminar* flow in straight ducts of circular cross section carries over to the turbulent ($Re > 2.1 \times 10^3$) case, but with the following significant modifications:

a. The fully developed Nusselt numbers $Nu_h\{\infty\}$, $\overline{Nu_h}\{\infty\}$ are not constants but, rather, dependent upon both Re and Pr. Experimentally, one finds (Colburn,

Dittus, Boeltler):

$$\mathrm{Nu}_h\{\infty\} = \overline{\mathrm{Nu}}_h\{\infty\} \cong 0.023\mathrm{Re}^{0.8}\mathrm{Pr}^{1/3} \tag{5.6-1}$$

or equivalently:

$$\mathrm{St}_h\{\infty\} = \overline{\mathrm{St}}_h\{\infty\} \cong 0.023\mathrm{Re}^{-0.2}\mathrm{Pr}^{-2/3} \tag{5.6-2}$$

provided $\mathrm{Re} > 10^4$ and $\mathrm{Pr} \geqslant 0.7$. (Interestingly enough, these results apply surprisingly well to ducts of *noncircular* cross section if they are treated as *circular* ducts with an *equivalent* (*effective*) *diameter* of four times the (cross-sectional area), divided by the wetted perimeter of the cross section).)

b. The thermal entrance effect (Eq. (5.5-19)) is modified to:

$$F(\text{thermal entrance}) \approx 1 + \left(\frac{z}{d_w}\right)^{-2/3} \qquad (\mathrm{Re} > 10^4,\ \mathrm{Pr} > 0.6). \tag{5.6-3}$$

Actually, more detailed analyses of turbulent pipe flow have revealed that the above-mentioned Re and Pr-dependencies (Eqs. (5.6-1, 5.6-2, and 5.6-3)) are oversimplifications. These analyses have also clarified the close relationship that exists between $\mathrm{St}_h(\infty)$ and the pipe *friction factor* $\frac{1}{2}C_f\{\mathrm{Re}\}$ (extended Reynolds' analogy). This is mentioned here not only because of the technological importance of turbulent pipe flow, but also because many of the methods, and even results, can be carried over to other turbulent BL flows near solid surfaces (Section 5.6.3).

5.6.3 Turbulent Thermal Boundary Layers on a Smooth Flat Plate

For $\mathrm{Re}_L > 10^5$ the viscous flow within a flat plate *forced convection* boundary layer (BL) itself becomes turbulent somewhere along the plate length. In that case heat transfer occurs across a *laminar* BL for $x < x(\text{transition})$ and across a turbulent BL for $x(\text{transition}) \leqslant x \leqslant L$. The situation is simpler if the turbulence in the BL is "triggered" near the leading edge so that the turbulent portion dominates the heat-transfer properties of the entire plate. In that case, in the range $10^5 \leqslant \mathrm{Re}_L \leqslant 10^7$ the dimensionless *skin friction* (*coefficient*) *distribution* is well approximated by:

$$\frac{\tau_w\{x\}}{\rho_\infty U^2} \equiv \tfrac{1}{2}c_f\{x\} = 0.0288(\mathrm{Re}_x)^{-1/5}. \tag{5.6-4}$$

If it is assumed that St_h bears the same relation to $\frac{1}{2}c_f$ as for fully developed turbulent flow in a straight pipe, then we are immediately led to the corresponding local

Stanton-number distribution:

$$\mathrm{St}_h\{x\} \cong 0.0288(\mathrm{Re}_x)^{-1/5}\mathrm{Pr}^{-2/3} \tag{5.6-5}$$

or, in terms of the dimensionless Nusselt number ($\dot{q}''_{\mathrm{w}}/\dot{q}''_{\mathrm{w,ref}}$ with $\dot{q}''_{\mathrm{w,ref}} \equiv k(T_{\mathrm{w}} - T_\infty)/x$):

$$\overline{\mathrm{Nu}}_h\{x\} \cong 0.0288(\mathrm{Re}_x)^{4/5}\mathrm{Pr}^{1/3}. \tag{5.6-6}$$

Averaging over the entire plate length, but on only one side, gives

$$\overline{\mathrm{Nu}}_h \cong 0.036(\mathrm{Re}_L)^{4/5}\mathrm{Pr}^{1/3} \tag{5.6-7}$$

where $0.036 = (5/4)(0.0288)$.

These semi-empirical results correspond to a *thermal boundary-layer slope thickness* δ_h which grows like $x^{1/5}$ (turbulent BL) rather than $x^{1/2}$ (laminar BL). Thus, *time-averaged heat fluxes along an isothermal plate in turbulent forced convection BL flow fall off with distance from the leading edge approximately as* $x^{-1/5}$.

For rough estimates Eqs. (5.6-5) and (5.6-6) are sometimes used to estimate turbulent forced-convection heat-transfer coefficients on *nonisothermal, curved* surfaces (e.g., turbine blades, along nozzle surfaces, etc.). In such cases, x is interpreted as the distance *along the surface* (measured, say, from the forward stagnation line). It has been found that forced-convection *turbulent* BL heat-transfer coefficients are less sensitive to pressure gradients, and the nature of the thermal boundary condition, than their *laminar* counterparts.

Turbulence within the boundary layer can also set in for *natural* convection from a vertical isothermal surface. Thus, for Newtonian fluids with Prandtl numbers near or equal to that of air, it has been found experimentally that, above Ra_h-values of about 10^9 (see Figure 5.6-1):

$$\mathrm{Nu}_h\{x\} \cong 0.10(\mathrm{Ra}_h)^{1/3} \qquad (\mathrm{Pr} \approx 0.7), \tag{5.6-8}$$

where, again, Nu_h and Ra_h are based on the length measured from the leading edge of the plate. Thus, while in *laminar* BL natural-convection flow the local heat fluxes fall off as $x^{-1/4}$, for *turbulent BL natural-convection flow, the local heat fluxes are nearly constant along the plate surface.* It follows that Eq. (5.6-8) also implies

$$\overline{\mathrm{Nu}}_h \cong 0.10(\mathrm{Ra}_h)^{1/3} \qquad (\text{for } \mathrm{Pr} \approx 0.7, \quad \mathrm{Ra}_{h,L} > 10^9), \tag{5.6-9}$$

where the reference length in both $\overline{\mathrm{Nu}}_h$ and Ra_h is now the total plate height L (which, in this particular (turbulent BL) case, drops out of the actual calculation of the heat flux (loss or gain)).

It is interesting and appropriate to note here that an effect very much like turbulent natural convection is produced by the localized surface ("nucleate") *boiling* of a liquid in contact with a sufficiently hot surface. Most often the liquid and its vapor are chemically inert and the phase change is brought about by external heating of the surface. (However, this situation can also arise as the result of the exoergic chemical reaction of the fluid with the solid—as in the case of surface-catalyzed hydrogen peroxide or hydrazine decomposition.) In such cases many small vapor bubbles are nucleated in the locally superheated region at the fluid/solid interface. Upon breakaway and buoyant rise, these bubbles agitate the surrounding fluid and dramatically increase the effective heat-transfer coefficients. Not surprisingly, the onset of this kind of "turbulence" (associated with localized phase change) is *not* dictated by a transition Ra_h-value (cf. Section 5.6.1) but, rather, by a local critical "superheat" ($T_w - T_{sat}(p)$) dependent on the particular fluid/solid combination (see, e.g., Lienhard (1981)). This phenomenon and the associated additional latent-heat transport are exploited in many applications calling for high heat-transfer coefficient at a nearly constant temperature (slightly above $T_{sat}(p)$). However in any configuration there is a sharp upper limit to the rate of vapor escape from the hot surface, and, therefore, the corresponding heat flux. Since *higher* heat fluxes can ordinarily be accommodated only at *much* higher surface temperature, one usually chooses to operate safely below this so-called "boiling-crisis" point.

5.6.4 Time-Averaged Total Heat Transfer Rate from/to an Isolated Sphere at High Re

For Reynolds' numbers much above 3×10^5, turbulence within the thermal BL should also amplify the total heat-transfer coefficient, $\overline{Nu}_h$, for an isothermal sphere. Experimental data up to $Re = 10^5$ appear to be well correlated by (Richardson (1968), Whitaker (1977)):

$$\overline{Nu}_h = 2 + [0.4(Re)^{1/2} + 0.06(Re)^{2/3}](Pr)^{0.4} \qquad (5.6\text{-}10)$$

(for $Pr \geqslant 0.7$), which differs from the Froessling relation (5.5-10) mainly *via* the presence of an explicit $(Re)^{2/3}$ term to account for the increasing importance of the *wake region* (centered about the *rear* stagnation point). Reliable experiments and associated correlations valid in the Re-range above 3×10^5 do not appear to be available.[31] Unfortunately, this situation cannot be remedied by recourse to transport *theory* because of the great complexity of the separated flow about even a smooth sphere in this Reynolds'-number range. As mentioned earlier (see Section 5.5.3), corresponding results for an isolated circular *cylinder* in cross-flow up to $Re = 3 \times 10^5$ are shown in Figure 8.1-1.

[31] However, it is well known that the heat-transfer coefficient, Nu_h, does not exhibit the sharp drop exhibited by the momentum (drag) coefficient C_D at $Re \approx (2\text{–}4) \times 10^5$ (cf. Figure 4.4-2).

It should be remarked that the turbulence effects discussed in this Section (5.6) for external flow geometries (plate, sphere, etc.) set in even if the mainstream is completely nonturbulent; i.e., the turbulence is confined to the shear layer adjacent to the solid surfaces. Even at lower Reynolds' numbers, however, all of the previously mentioned transfer coefficients (Section 5.5) can be augmented by *mainstream turbulence* (intensity and, perhaps, scale). For this purpose, empirical correction factors have been developed which introduce the new parameters:

$I_{t,\infty} \equiv$ mainstream turbulence "*intensity*" (ratio of root-mean-square (rms) velocity fluctuation to time-averaged approach velocity),

and

$L_{t,\infty}/L \equiv$ ratio of mainstream turbulence (macro-) *scale* to body dimension,

the latter parameter being unnecessary if $L_{t,\infty}/L \ll 1$. Mainstream turbulence intensity can affect integrated momentum, energy, and mass transport coefficients in two distinct ways, by:

a. triggering earlier transition to turbulence of the laminar boundary layer on the object (causing a larger fraction of the object to experience a turbulent boundary layer), and
b. modifying the time-averaged transport across the "laminar" boundary layer itself.

Laminar boundary layers in the presence of a large pressure gradient are found to be especially sensitive to mainstream turbulence if it is above some "threshold" intensity. For example, at the forward stagnation line of a cylinder in cross-flow (at *ca.* $\text{Re} = 10^5$) Nu_h appears to be almost unaffected up to $I_{t,\infty} = 0.007 = 0.7\%$, but then increases by *ca.* 10% at an $I_{t,\infty}$-value of only 1.5% (see, e.g., Schlichting (1979)).

5.6.5 Continuous Point Heat Source in a Uniform Turbulent Stream

If we consider a steady point source of energy in a spatially *uniform turbulent stream* [with constant effective thermal diffusivity α_t ($\equiv k_t/(\rho c_p)$)], then the "downwind" time-averaged temperature, $T(r, z)$, will be identical to that given in Section 5.5.5 with the replacements $\alpha \to \alpha_t$ and $k \to k_t$. This implies that transverse temperature profiles far downwind of such a source become *Gaussian* in shape. If we denote by $r_{1/2}$ the radial location where

$$\frac{T(r_{1/2}, z) - T_\infty}{T(0, z) - T_\infty} = \frac{1}{2}, \qquad (5.6\text{-}11)$$

then we readily find:

$$r_{1/2} = 1.6651\left(\frac{\alpha_t z}{U}\right)^{1/2}. \tag{5.6-12}$$

Thus, far downstream of the source the time-averaged isotherms become axially symmetric, "nested" paraboloids, with the "thermal wake" spreading like $z^{1/2}$. As in the laminar case, temperature measurements downstream of a continuous point energy source can be used to determine the relevant "diffusivity"—in this case, α_t in isotropic turbulent flows.

In meteorological (e.g., pollutant dispersion) modeling there is considerable interest in situations for which:

a. α_t is different in the two transverse directions x and y (i.e., the turbulence is *anisotropic*), and
b. the mean streamwise velocity is not uniform but exhibits a spatial *gradient* in the vertical (say y)-direction.

These features, as well as *buoyancy*, cause significant departures from axial symmetry in the resulting time-averaged temperature fields; i.e., spread rates are different in the x- and y-directions, and "top–bottom" $(+y, -y)$ symmetry is destroyed.

5.7 ANALOGIES BETWEEN ENERGY AND MOMENTUM TRANSPORT

5.7.1 Fully Turbulent Jet Flow into a Co-Flowing Surrounding Stream

Since the "eddy diffusion" mechanism transports both linear momentum and energy, we might expect the relevant diffusivities ν_t and α_t to be nearly equal; that is,

$$\mathrm{Pr}_t \equiv \nu_t/\alpha_t \approx 1, \tag{5.7-1}$$

where Pr_t, the so-called *turbulent* Prandtl number, is not a *fluid* property, but a local *flow* property. This near-equality has the following important consequences in "fully" turbulent flows ($\nu_t \gg \nu, \alpha_t \gg \alpha, \ldots$), such as those occurring in free jets and wakes (i.e., "separated" flows) at sufficiently high Reynolds' numbers.

For definiteness, consider a heated round *jet* (velocity U_j, diameter d_j) issuing into a co-flowing "secondary" stream (velocity U_s) under conditions such that:

$$\frac{(U_j - U_s)d_j}{\nu} \gg 3 \times 10^4. \tag{5.7-2}$$

In the absence of appreciable viscous dissipation and chemical heat sources, the time-averaged temperature field $T(r, z)$ will satisfy the approximate[32] linear PDE:

$$\rho c_p\left[v_z\frac{\partial T}{\partial z}+v_r\frac{\partial T}{\partial r}\right]=\frac{1}{r}\frac{\partial}{\partial r}\left(r\rho c_p\alpha_t\frac{\partial T}{\partial r}\right), \tag{5.7-3}$$

subject to:

$$T(r,0)=\begin{cases}T_j & \text{for } r<d_j/2,\\ T_s & \text{for } r>d_j/2,\end{cases} \tag{5.7-4a,b}$$

and

$$T(\infty,z)=T_s, \tag{5.7-5a}$$

$$\left(\frac{\partial T}{\partial r}\right)_{r=0}=0. \tag{5.7-5b}$$

(Note that, for convenience, we have dropped the overbar to denote time-averaged quantities.)

If the secondary stream is not confined (see below), then there will exist a negligible streamwise pressure gradient. In that case the time-averaged *axial velocity* field (specific linear momentum in the z-direction) $v_z(r, z)$ will satisfy the approximate *nonlinear* PDE:

$$\rho\left[v_z\frac{\partial v_z}{\partial z}+v_r\frac{\partial v_z}{\partial r}\right]=\frac{1}{r}\frac{\partial}{\partial r}\left(r\rho\nu_t\frac{\partial v_z}{\partial r}\right), \tag{5.7-6}$$

subject to:

$$v_z(r,0)=\begin{cases}U_j & \text{for } r<d_j/2\\ U_s & \text{for } r>d_j/2\end{cases} \tag{5.7-7a,b}$$

$$U_z(\infty,z)=U_s, \tag{5.7-8a}$$

$$\left(\frac{\partial v_z}{\partial r}\right)_{r=0}=0. \tag{5.7-8b}$$

Comparing with Eqs. (5.7-3 and -6) and their corresponding boundary conditions (BCs), it is now obvious that if $\mathrm{Pr}_t = 1$ (5.7-1) and c_p is nearly constant, then *$v_z(r, z)$ and $T(r, z)$ satisfy PDEs and BCs that are identical in mathematical form.*

[32] Again, we neglect axial diffusion compared to radial diffusion. This is loosely called a "boundary" layer (BL) approximation even in this problem—which is free of the influence of solid boundaries!

Suppose now that measurements were made of the streamwise velocity field and cast in the dimensionless form:

$$u^* \equiv \frac{v_z - U_s}{U_j - U_s} = \text{fct}\left(\frac{2r}{d_j}, \frac{z}{d_j}\right). \tag{5.7-9}$$

One can conclude that the time-averaged *temperature* field in this jet, when cast in the form of contours of constant,

$$T^* \equiv \frac{T - T_s}{T_j - T_s} = \text{fct}\left(\frac{2r}{d_j}, \frac{z}{d_j}\right), \tag{5.7-10}$$

should involve the *same function* as in (5.7-9); that is, *contours of constant u^* and T^* should be identical.* Both quantities will start out with the same "top-hat" profiles (1 for $r < d_j/2$ and 0 for $r \geqslant d_j/2$ at $z/d_j = 0$), both vanish at large transverse distances, and both will have a vanishing radial derivative along the axis.

Note that this "analogy" between energy and momentum transport in fully turbulent jet (or wake) flow does not require that ν_t and α_t be spatially constant,[33] only that $\nu_t \approx \alpha_t$ everywhere. To the extent that the latter is true, streamwise velocity fields can be used to predict temperature fields in such flows, and *vice versa.* We will see (Section 6.5) that a similar conclusion can be reached for "tracer" *mass* transport in fully turbulent jet (or wake) flows, since D_t should be comparable to both ν_t and α_t.

For flows that are *not* fully turbulent (lower Re jets and wakes, as well as flows near solid boundaries, where turbulence is locally damped), Eqs. (5.7-3) and (5.7-6) will involve the effective diffusivities:

$$\nu_{\text{eff}} \equiv \nu + \nu_t, \tag{5.7-11}$$

$$\alpha_{\text{eff}} \equiv \alpha + \alpha_t. \tag{5.7-12}$$

Thus, even if $\nu_t = \alpha_t$ (i.e., the *turbulent* Prandtl number were unity), in general we would *not* have $\nu = \alpha$; that is, most fluids, including gases, have Prandtl numbers which differ from unity. Accordingly, this strict analogy between streamwise momentum and energy transport breaks down at lower Reynolds' numbers, unless it happens that the *fluid* has $\text{Pr} \approx 1$ (e.g., pure steam).

Careful measurements in free-turbulent jets (cf. Figure 5.7-1) reveal that, in practice, u^* and T^* are not quite equal, even in "fully turbulent" flows, in the sense that $r_{1/2}$ for the T^*-field grows somewhat faster than $r_{1/2}$ for the u^*-field. This implies that Pr_t is somewhat *less* than unity (α_t is somewhat greater than ν_t). Engineering predictions for such flows are therefore often based on the estimate: $\text{Pr}_t \approx 0.5$ *everywhere in the flow field,* but in compressible turbulent flows it is unlikely that Pr_t is actually constant.

[33] Experimentally it is found that, for such jets, ν_t is nearly constant *across* the jet at the value: $\nu_t \approx 0.026[\bar{v}_z(0, z) - U_s] \cdot r_{1/2}(z)$. (See Section 4.6.4.)

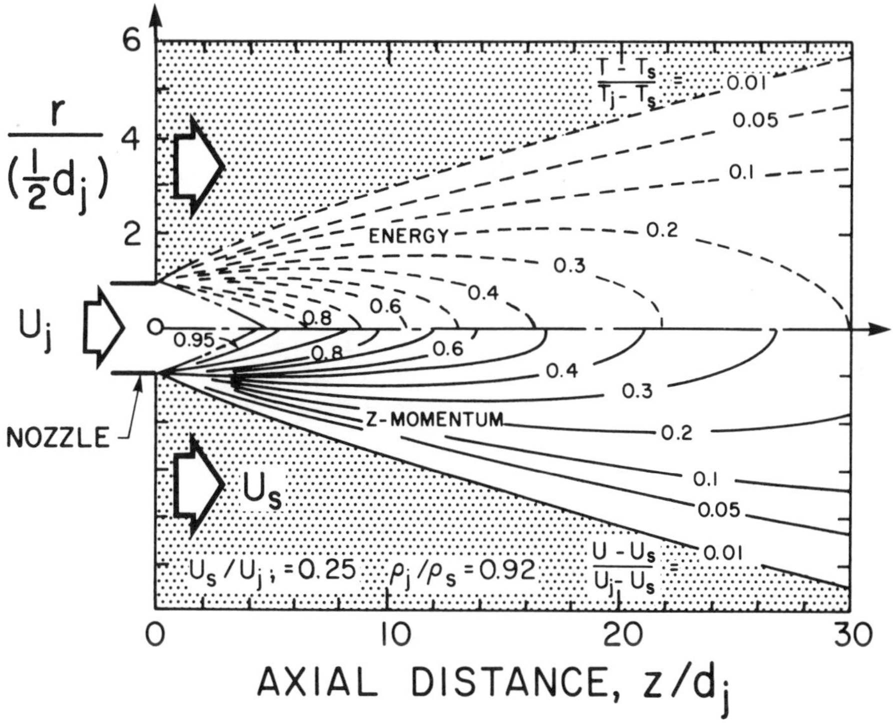

Figure 5.7-1 Constant velocity and temperature contours for a turbulent round jet in a co-flowing stream of velocity, U_s, and temperature T_s (adapted from Forstall and Shapiro (1950)).

5.7.2 Analogies between Momentum and Energy Transfer for Solid-Wall Boundary-Layer Flows: Extended Reynolds' Analogy for Surface-Transfer Coefficients in Fluids with Pr ≠ 1

For turbulent flows near solid boundaries, especially in the absence of appreciable streamwise pressure gradients, the physical and mathematical similarity between streamwise *momentum* and *energy* transport implies an interesting connection between the corresponding dimensionless *surface-transfer coefficients* $\frac{1}{2}c_f$ and St_h. Even though turbulence is damped in the immediate vicinity of the wall ($y = 0$) and most Newtonian fluids have a Prandtl number, ν/α, different from unity, this "analogy" can be made quantitative by exploiting the results of careful $\bar{v}_x(y)$ measurements, usually made in "fully developed" turbulent pipe flow.[34] However,

[34] For the present purpose we are reverting to the use of the Eulerian coordinate x to define *streamwise* locations along the surface transferring momentum and/or energy (rather than z, used in Sections 5.5 and 5.6).

by casting these results into a "universal" form (which makes no explicit reference to the pipe geometry), conclusions can be drawn that are approximately valid for more general turbulent boundary layers—e.g., in the surfaces of nozzles, turbine blades, wings, etc. This procedure and its implications are outlined below.

By combining pressure drop data (to obtain the wall shear stress $\bar{\tau}_w$) with impact-pressure derived time-averaged *velocity profile* measurements in turbulent pipe flow, one can determine and plot:

$$u^+ \equiv \frac{\bar{v}_x}{(\bar{\tau}_w/\rho)^{1/2}} = \text{fct}\left(\frac{(\bar{\tau}_w/\rho)^{1/2} y}{\nu}\right) \equiv u^+(y^+). \quad (5.7\text{-}13)$$

The resulting "universal velocity profile" is valid in the pipe Re-range $10^4 < \text{Re} < 10^6$ (see Figure 5.7-2). Now, the time-averaged x-momentum equation is approximately:

$$\frac{d}{dy}\left\{\rho[\nu + \nu_t(y)]\frac{d\bar{v}_x}{dy}\right\} = 0, \quad (5.7\text{-}14)$$

which merely states that the time-averaged stress $\bar{\tau}_{yx}$ is approximately independent of distance from the wall. For a constant density fluid it follows that:

$$\left[1 + \frac{\nu_t}{\nu}\right] \cdot \frac{du^+}{dy^+} = 1 \quad (5.7\text{-}15)$$

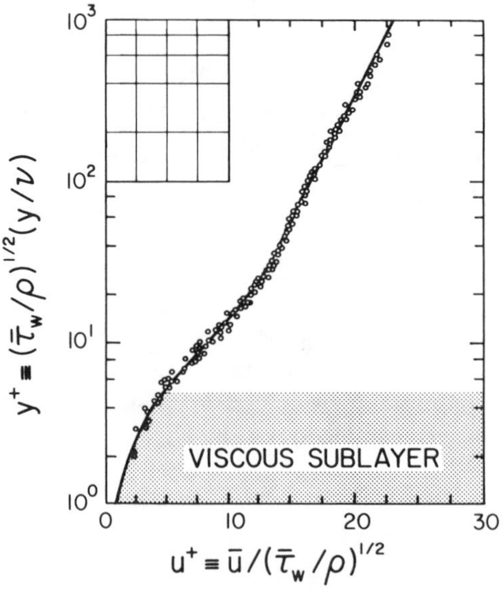

Figure 5.7-2 Universal velocity profile near the wall for fully developed turbulent pipe flow of a Newtonian fluid.

or

$$\frac{\nu_t}{\nu} = \left(\frac{dy^+}{du^+} - 1\right). \qquad (5.7\text{-}16)$$

This means that a *knowledge of the* $u^+\{y^+\}$ *or* $y^+\{u^+\}$ *relation is equivalent to a knowledge of the distribution* of ν_t/ν everywhere within *the turbulent BL.*

If we now assume that $\mathrm{Pr}_t = 1(\alpha_t = \nu_t)$, there is sufficient information to find the time-averaged temperature profile $\bar{T}\{y\}$ when the wall temperature differs from the bulk fluid temperature. The corresponding energy-conservation equation for $\bar{T}(y)$ is $\overline{\dot{q}''_y} = \text{const}$, or:

$$\frac{d}{dy}\left\{\rho c_p[\alpha + \alpha_t]\frac{d\bar{T}}{dy}\right\} = 0, \qquad (5.7\text{-}17)$$

which now can be solved to give $\bar{T}\{y\}$ and, hence, the corresponding wall heat flux $\overline{\dot{q}''_w}$. This problem was first attacked by Prandtl and Taylor, and later by Von Karman, Van Driest, and Deissler (1955) (in the light of improved $u^+\{y^+\}$ data close to the wall).

Prandtl's model was simply that the BL could be roughly divided into two adjacent layers, across which $\bar{u}$, $\bar{T}$, and $\bar{\tau}_{yx}$, $\overline{\dot{q}''_y}$ are continuous; that is,

a. a *laminar* "sublayer" in which $u^+ = y^+$ up to $y^+ \cong 5$ (here $\nu_t = 0$, as indicated by Eq. (5.7-16)), and
b. a fully turbulent "outer" layer in which:[35]

$$u^+ = c_1 \ln y^+ + c_2 \qquad (y^+ \geqslant 5)$$

(where the constants c_1 and c_2 were determined by a best-fit to Figure 5.7-2).

On the basis of this simple turbulent BL flow model, it is straightforward to derive the so-called "extended" Reynolds' analogy:

$$\mathrm{St}_h \simeq \frac{\frac{1}{2}C_f}{1 + 5(\frac{1}{2}C_f)^{1/2}(\mathrm{Pr} - 1)}, \qquad (5.7\text{-}18)$$

which interrelates *surface* heat- and momentum-*transfer coefficients* even for fluids with $\mathrm{Pr} \neq 1$. For heat transfer through turbulent *gas* boundary layers, later refinements (e.g., inclusion of an intermediate or "buffer" layer) are numerically of minor importance. Equation (5.7-18) reveals that, if $\mathrm{Pr} < 1$ (as for air, in which $\alpha \cong$

[35] This result implies that momentum transfer in turbulent shear layers behaves as if the eddy mean-free-path (mixing length) is simply proportional to the distance from the wall.

$\nu/(0.7)$), then St_h will *exceed* $(1/2)C_f$ by a small amount which is itself C_f-dependent (hence Re-dependent).

The ratio $\mathrm{St}_h/[(1/2)C_f]$ is sometimes called the "Reynolds' analogy factor." Reference to our previous discussion of the flat-plate *laminar* boundary layer (Eq. (5.5-6)) reveals that this factor is approximately $\mathrm{Pr}^{-2/3}$ (or about 1.27 for air). Actually, this is often a sufficient approximation, even for *turbulent* boundary layers (cf. Eq. (5.6-5)) provided Pr is not too different from unity.

On a curved body for which the velocity distribution $\bar{u}_e\{x\}$ at the outer edge of the thin BL is such that the parameter:

$$\Lambda \equiv \frac{(\nu/\bar{u}_e)[(1/\bar{u}_e)\cdot(d\bar{u}_e/dx)]}{(C_f/2)^2} \tag{5.7-19}$$

is not small compared to unity, the above-mentioned relation between St_h and $C_f/2$ will break down. This is due to the importance of dp/dx in the *streamwise momentum* equation, a term which has no counterpart in the corresponding *energy* equation. For this reason the *frictional* contribution to the overall drag coefficient $C_D\{\mathrm{Re}\}$ of "bluff bodies" is different from $\mathrm{St}_h\mathrm{Pr}^{2/3}$ at high Reynolds' numbers.

For turbulent BL flow, surface roughness (mean height $> 5\nu/u_*$) increases St_h (and St_m (Ch. 6)) but, unfortunately, increases the corresponding momentum-transfer coefficient C_f even more. Very approximately, it has been found (see, e.g., Norris (1970)) that $\mathrm{St}_h/\mathrm{St}_{h,\,\mathrm{smooth}} \approx (C_f/C_{f,\,\mathrm{smooth}})^{1/2}$.

In retrospect, it should be noted that the above semi-empirical treatment of energy exchange between a turbulent viscous fluid and a solid surface made use of the *time-averaged* momentum and energy equations (see Section 2.6.2) and a simple postulate regarding the effective eddy *diffusivities* α_t and ν_t. More recent detailed measurements have revealed the presence of spatially well-ordered, relatively large-scale, vortex-like motions in the viscous sublayer, which, in effect, repetitively exchange fluid between the high energy "core" and wall regions. These observations have spawned (or revived) what might be called "statistical microflow" treatments which, in contrast, start from an idealized structural model of the flows near the surface and attack the *transient* (or quasi-steady) equations governing such "microflows." The resulting *time-average* transport rate is then calculated from the postulated "population dynamics" of these local flows at the fluid/solid interface (see, e.g., Section 6.7.5, and the comprehensive critical review of Sideman and Pinczewski (1975)). In principle, such statistical microflow models can provide greater physical insight than the now-classical "eddy diffusivity approach" and, accordingly, they may ultimately prove useful in treating more general turbulence problems involving interfacial[36] transport of momentum and energy (or mass)—(e.g., inclusion of simultaneous chemical reaction, flexible (fluid–fluid) interfaces, interfacial resistance and capacitance effects, dynamical nonequilibrium between suspended particles and their "host" eddies (see Section 6.6.3), etc.).

[36] Analogous statistical microflow ("structural") models are also being proposed to deal with chemically reacting turbulent *jet* flows at "low" Reynolds' numbers (see, e.g., Spalding (1979).

5.7.3 Remarks on Recent Developments in Turbulence Modeling

The extent to which turbulent BL predictions are based on *experimental* data (e.g., pipe-flow velocity-profile data, round-jet velocity-profile data, etc.) is already clear from the preceding discussion. Our examples dealt with the simplest of the following three methods of "modeling" turbulent transport in new situations; *viz.*:

a. Algebraic Modeling of Double Correlations, with Pr_t = const (Prandtl [1928], Von Karman [1930]) In this approach the double correlations (e.g., Reynolds' stress components; cf. Section 2.6.3) appearing in the time-averaged balance equations are taken to be simple functions of either the time-mean-flow variables themselves and/or position **x**. As perhaps the best-known example, consider the case of turbulent shear flow of an incompressible fluid near a smooth wall. By analogy with gas-kinetic theory, Prandtl proposed that it should be possible to write the effective momentum diffusivity, ν_t, defined by:

$$-\overline{\rho v'_x v'_y} \equiv (\bar{\tau}_{yx})_{\text{Reynolds}} \equiv \rho \nu_t \left(\frac{\partial v_x}{\partial y} + \frac{\partial \bar{v}_y}{\partial x} \right), \tag{5.7-20}$$

as the product of a "mean-free-path" or "mixing-length," ℓ_{mix} (for a turbulent eddy) and an x-direction fluctuation velocity. If the latter is taken as $\ell_{\text{mix}}|\partial \bar{v}_x/\partial y|$, and ℓ_{mix} is taken to be linearly proportional to the distance y from the wall, then available $\bar{v}_x(y)$ data can indeed be reproduced if $\ell_{\text{mix}} \cong 0.36y$ (see Section 5.7.2). This turbulence "model" can be extended to embrace *energy* transfer most simply by assuming that the turbulent energy diffusivity is locally proportional to ν_t; that is,

$$\alpha_t = \frac{\nu_t}{\text{Pr}_t}, \tag{5.7-21}$$

where Pr_t, the turbulent Prandtl number, is an empirically chosen constant near unity.

Another example of an algebraic turbulence model, appropriate to free, "fully" turbulent, axisymmetric jets (and wakes)—(and also associated with Prandtl)—has been discussed in Section 4.6. In such cases ν_t, appearing in the time-averaged z-momentum equation, can be taken to be proportional to the product of the local velocity difference across the jet and jet "half-width," that is,

$$\nu_t(z) = 0.026\,|\bar{v}_z(0, z) - \bar{v}_z(\text{“}\infty\text{”}, z)| \cdot r_{1/2}(z). \tag{5.7-22}$$

Again, extension to the turbulent "diffusion" of *energy* is simplest by assuming that α_t and ν_t are locally proportional (cf. Eq. (5.7-21)); however, Pr_t need not be taken to be the same constant as for turbulent flow near a wall.

Note that, in these (and other) "algebraic" models of turbulence, no additional PDEs need be solved to "predict" the time-averaged flow field.

b. Modeling of the Triple Correlations in Terms of the Double Correlations and/or Local Mean-Flow Variables (Donaldson (1972), Mellor (1973)). This procedure, called "higher-order closure," implies that all of the double correlations (e.g., Reynolds' fluxes, etc.) are treated as new dependent-variables governed by their own PDEs and BCs. The triple correlations appearing in their PDEs are, in turn, modeled *via* these local dependent variables introducing, hopefully, constants which are more universal to provide "closure." While potentially more general, this procedure is computationally impractical for most three-dimensional turbulent flows, especially those involving several reacting species or phases.

c. Two-PDE Models of Turbulence (Rotta (1972), Launder and Spalding (1972), Wolfshtein et al. (1975), Rodi (1982)). This class of methods, intermediate in complexity and generality between approaches (a) and (b) above, involves the solution of additional convective–diffusion PDEs for *two* scalar quantities assumed to adequately characterize the local turbulence. These might be the local "fields" of:

i. time-averaged *kinetic energy*/mass associated with turbulence, *viz.* $\overline{\mathbf{v}' \cdot \mathbf{v}'}/2 \equiv \kappa_t$, and
ii. time-averaged turbulent *energy dissipation rate* per unit mass,[37] written ε.

These local field quantities (κ_t, ε) are, in turn, used to model double correlations appearing in all of the time-averaged balance equations. For turbulent "diffusion" of momentum far from the influence of solid-wall "damping," this is equivalent to introducing an isotropic local turbulent *diffusivity* $\nu_t = \text{const} \cdot (\kappa_t^2/\varepsilon)$. For initially unmixed turbulent flows with rapid homogeneous chemical reactions, the corresponding characteristic local mixing- (eddy breakup-) *time* κ_t/ε plays an important role in determining the time-averaged chemical reaction rates (see below), and the effective eddy mass-diffusivity of "heavy" particles (e.g., dust particles or droplets) suspended in the turbulent flow (see Section 6.6.2). While more flexible than approach (a), and now numerically feasible for rather complex flows, even this approach requires some seven empirical "constants," which are not, in principle, "universal." Moreover, the required new boundary and initial data (e.g., on $\kappa_t(\mathbf{x}, t)$ and $\varepsilon(\mathbf{x}, t)$) are often unavailable, and must be inferred from less direct information.

Turbulence in chemically reacting fluids poses some particularly challenging modeling problems, some of which are briefly described in Chapter 6, dealing with species mass-transfer by convection and diffusion. The source of the difficulty lies in the fact that a knowledge of the local time-averaged species densities $\bar{\rho}_i (i =$

---

[37] Given by $\nu\,\overline{(2\,\mathbf{Def}\,\mathbf{v}' : \mathbf{grad}\,\mathbf{v}')}$ in terms of the velocity fluctuation fields $\mathbf{v}'(\mathbf{x}, t)$. To mathematically model *chemically reacting* turbulent flows (see below) it is also necessary to follow, in addition to κ_t and ε, the distribution of a local "unmixedness" variable g, related to the turbulence intensity of a "passive scaler," Ω (e.g., a normalized inert tracer mass fraction satisfying the convective diffusion PDE without source terms); i.e., $g(\mathbf{x}, t) \equiv \overline{\Omega'\Omega'}$ (see Section 6.5.5.6 for a typical application). In fact, such 3-PDE (κ-ε-g) turbulence models are now viewed as containing the minimum level of information needed to adequately "predict" turbulent, chemically reacting flows.

$1, 2, \ldots, N$) and time-averaged temperature $\bar{T}$ is not sufficient to define the *time-averaged reaction rate*, even when the "instantaneous" reaction rate laws $\dot{r}_i'''\{\rho_1, \rho_2, \ldots, T\}$ are completely known. While this problem arises even in nearly isothermal systems (e.g., turbulent liquids; see below) a significant difference between $\dot{r}_i'''$ and $\dot{r}_i'''\{\bar{\rho}_1, \bar{\rho}_2, \ldots, \bar{T}\}$ in nonisothermal turbulent systems (e.g., turbulent premixed flames) can result from the strong nonlinear dependence of chemical rate "constants" on instantaneous temperature. Generally of comparable importance, however, are the effects of molecular-level reactant "unmixedness" or "segregation," which can occur even in isothermal situations, especially when the reactants are initially separated. For example, clearly a local bimolecular elementary reaction between species A and B, locally present in relative concentrations $\omega_A = \bar{\omega}_A + \omega_A'$ and $\omega_B = \bar{\omega}_B + \omega_B'$, need not occur at a local time-averaged reaction rate proportional to the product of the *time-averaged* mass fractions: $\bar{\omega}_A \cdot \bar{\omega}_B$. This is easiest to see in the case where both $\bar{\omega}_A$ and $\bar{\omega}_B$ are locally nonzero but A and B never actually coexist *at the same time*. Such local "unmixedness" effects must be included in predictions of pollutant emission (e.g., NO(g)) from turbulent combustors.

5.8 CONVECTIVE ENERGY TRANSPORT IN CHEMICALLY REACTING SYSTEMS

In most chemically reacting (multicomponent) systems, the laws of *species* mass transfer (Chapter 6) are also needed to predict *energy* transfer rates. This is easiest to see from Eq. (2.5-8), for the *energy-diffusion flux vector*, that is,

$$\dot{\mathbf{q}}'' = -k\,\mathbf{grad}\,T + \sum_{i=1}^{N} \mathbf{j}_i'' h_i, \tag{5.8-1}$$

the normal component of which governs the local energy flux $\dot{q}_w''$ at a solid boundary. There are two simple limiting cases, *viz.*:

a. Local Thermochemical Equilibrium (LTCE)BL. In this case the local compositions that govern the $\mathbf{j}_i''$ are themselves linked to the temperature field, allowing Eq. (5.8-1) to be written in the deceptively simple form (Brokaw (1957, 1960))

$$\dot{\mathbf{q}}'' = -k_{\text{LTCE}}\,\mathbf{grad}\,T, \tag{5.8-2}$$

where the *effective* thermal conductivity, k_{LTCE}, can greatly exceed the value of k expected from a locally "inert" mixture of the same composition. Such energy-transfer problems reduce to their pseudo-pure fluid counterparts, except that the effective transport properties are now highly variable. Indeed, useful estimates of convective heat-transfer rates can be made by combining the use of nonreactive transfer *coefficients*, Nu_h, with *heat-flux potential difference* "driving forces" ($\Delta\phi_T$), where, in the presence of LTCE chemical reaction,

$$\phi_T = \int_{T_{\text{ref}}}^{T} k_{\text{LTCE}}\{T; p\}\, dT$$

(Rosner, D.E., *Amer. Rocket Soc. J.* **31**, 816–818 (1961)). For a simple illustrative calculation of $k_{\text{LTCE}}\{T; p\}$, applicable to equilibrium dissociating–recombining gaseous hydrogen ($H_2 \rightleftarrows 2H$), see Exercises 3.14 and 5.19.

b. Chemically "Frozen" Boundary Layer (CFBL) (Rosner (1959)). In this case chemical reactions do not take place *within* the BL and we can formally write:

$$-\left(k\frac{\partial T}{\partial n}\right)_{\text{w}} = \rho_e u_e \text{St}_h \cdot (h_{T,\text{w}} - h_{T,e}) \tag{5.8-3}$$

(where h_T is the "sensible" or "thermal" enthalpy of the local mixture), and, in the absence of appreciable thermal (Soret) diffusion:

$$j''_{i,\text{w}} = \rho_e u_e \text{St}_{m,i} \cdot (\omega_{i,\text{w}} - \omega_{i,e}), \tag{5.8-4}$$

where the $\text{St}_{m,i}$ are *mass-transfer Stanton numbers* (cf. Section 6.3). Combining these relations to form $\dot{q}''_{\text{w}}$ gives:

$$\dot{q}''_{\text{w}} = \rho_e u_e \text{St}_h \cdot \left\{\Delta h_T + \sum_{i=1}^{N} r_{\text{chem},i}(\Delta\omega_i)h_{i,\text{w}}\right\}, \tag{5.8-5}$$

where we have introduced the notation:

$$r_{\text{chem},i} \equiv \frac{\text{St}_{m,i}}{\text{St}_h} \approx (\text{Le}_i)^{2/3}, \tag{5.8-6}$$

called the *recovery factors for mainstream chemical energy* (Rosner (1961, 1975)) and the BL "difference operator:"

$$\Delta \equiv (\quad)_{\text{w}} - (\quad)_e. \tag{5.8-7}$$

We conclude that energy-transfer predictions for the CFBL case with heterogeneous chemical reaction (e.g., surface catalysis) or phase change can be made by:

a. retaining the usual ("nonreactive") heat-transfer *coefficient* $\text{St}_h\{\text{Re}_x, \text{Pr}\}$;
b. replacing the usual "driving force" $c_p \Delta T$ (Eq. (5.5-7)) by the *generalized driving force*:

$$\Delta h_T + \sum_{i=1}^{N} r_{\text{chem},i} h_{i,\text{w}}\, \Delta\omega_i. \tag{5.8-8}$$

This technique provides immediately useful results when the kinetics of the heterogeneous reactions (or phase change) are fast enough to ensure LTCE *at station* w. More generally, independent species-mass balances, involving heterogeneous chemical kinetics, are needed to complete the prediction of energy transfer in systems with chemical reactions (cf. Section 6.5.3).

It is interesting to note that, irrespective of *homogeneous* chemical kinetics, if all of the Fick diffusion coefficients D_i are approximately equal to the mixture thermal diffusivity α (i.e., all the $\mathrm{Le}_i \approx 1$), then Eq. (5.8-1) simplifies to:

$$\dot{\mathbf{q}}'' \approx -(k/c_p)\,\mathbf{grad}\,h, \tag{5.8-9}$$

where h is the *specific enthalpy of the local mixture* (including chemical contributions). In such cases, irrespective of both homogeneous and heterogeneous kinetics,

$$\dot{q}''_w \approx \rho_e u_e \mathrm{St}_h\{\mathrm{Re}_x, \mathrm{Pr}\} \cdot (h_w - h_e). \tag{5.8-10}$$

This result,[38] which again involves the "nonreactive" heat-transfer coefficient, is especially useful for estimating maximum possible energy fluxes in the presence of homogeneous and/or heterogeneous chemical reactions.

The strategy of preserving St_h and including the effects of additional phenomena in a generalized "driving force" also allows estimates of the effects of BL *viscous dissipation* on wall energy fluxes. This is accomplished by merely adding

$$r_{KE}\{\mathrm{Re}, \mathrm{Pr}\} \cdot \frac{u_e^2}{2}$$

to the driving force shown in Eq. (5.8-5), or, less accurately, by interpreting h_e in Eq. (5.8-10) as the local *stagnation* enthalpy. The function $r_{KE}(\mathrm{Pr}; \mathrm{Re})$ appearing above, called the "recovery factor" (for mainstream kinetic energy), is somewhat smaller than unity for viscous fluids with $\mathrm{Pr} < 1$. r_{KE}-values are near $(\mathrm{Pr})^{1/2}$ for laminar BL flows and near $(\mathrm{Pr})^{1/3}$ for turbulent BL flows.

The approach of predicting wall energy fluxes, $-\dot{q}''_w$, using *enthalpy* (or "recovery" enthalpy (Section 6.5.4)), driving forces and energy-transfer coefficients obtained under simpler conditions (e.g., no homogeneous chemical reaction) can also be extended to more complex but technologically important cases in which there is net mass transfer $\dot{m}''_w$ at the wall due to either: (a) transpiration cooling using a coolant dissimilar to the fluid in the external stream, and which may, indeed, react chemically with it, or (b) self-regulating chemical "ablation" of the surface (as in the gasification rate of graphite). However, in either case it is essential to realize that $-\dot{q}''_w$ is then the nonradiative energy-diffusion flux in the local *mixture*[39] which

---

[38] Actually, even in variable-property *nonreactive* energy transport, Stanton numbers based on:

$$\dot{q}''_{\mathrm{ref}} \equiv \rho_e u_e \cdot (h_w - h_e)$$

(instead of $\rho_e u_e c_{p,e}(T_w - T_e)$) are widely used.

[39] For example, in the case of transpiration cooling, $-\dot{q}''_w$ is *not* the same as the product of the transpiration coolant flux and the *coolant's* enthalpy rise.

prevails on the fluid (e.g., gas) side of the fluid/solid interface (see Eq. (5.8-1))—(cf. Rosner (1975)). This has been clearly demonstrated for the experimentally studied case of H_2- or He-transpiration-cooled porous plates exposed to the turbulent boundary-layer flow of H-atom-rich products of oxyacetylene combustion (see, e.g., Rosner (1975)). On the basis of these demonstrations and theoretical considerations, many complex situations involving transpiration cooling or ablation with chemical reaction and variable thermophysical properties can be estimated using the nonreactive, constant-property transfer coefficients (experimental and/or theoretical) discussed (earlier) in Chapter 5 and to be treated further in Chapter 6.

5.9 REMARKS ON THE IMPLICATIONS AND TREATMENT OF RADIATION-ENERGY TRANSFER

Anyone who has stood near a large fire, or potter's kiln, can testify to the radiative-energy transfer received therefrom. This kind of transfer plays an important part in:

a. furnace energy transfer (kilns, boilers, etc.), and
b. the spread of combustion into unburned fuel-rich areas.

For example, in combustion applications the primary radiation sources are:

1. hot, opaque, solid surfaces (usually bounding the combustion space);
2. particulate matter suspended in the combustion products (soot, fly-ash);
3. polyatomic gaseous combustion-product molecules with permanent dipoles (e.g., CO_2, H_2O, etc.);
4. radiation from molecular fragments (e.g., C_2, CH, OH, etc.) "pumped" into excited states as a result of localized homogeneous exoergic chemical reactions.

A detailed treatment of the laws of radiation emission and absorption from each of these "sources" is beyond the scope of this text. Here we will highlight only a few key points concerning sources (1)–(3), which usually dominate the emission with respect to *energy transport* (as opposed to visibility).

5.9.1 Radiation Emission from, and Exchange between Opaque Solid Surfaces

The maximum possible rate of radiation *emission* from each unit area of an opaque surface at temperature T_w (expressed in kelvins) is:

$$\dot{e}_b'' = \sigma_B T_w^4 = 56.72\left(\frac{T_w}{1000}\right)^4 \ \frac{\text{kW}}{\text{m}^2} \qquad (5.9\text{-}1)$$

Table 5.9-1 Black-Body Radiant Emission[a] from Surfaces at Various Temperatures

Temp. (K)	*Relevance*	$\dot{e}_b''$ (kW/m^2)	λ_{max} (μm)[b]
300	"Room" temperature	0.459	9.66
1000	Many engrg. surfaces	56.72	2.9
3000	Tungsten lamp filament	4594	0.966
6000	Sun's outer temperature	73510	0.483

[a] Cf. solar "constant" (1.35 kW/m^2).
[b] *Via* Wien relation $\lambda_{max} \cdot T_w = 2897.6$ μm-K.

(Stefan–Boltzmann "black-body" radiation law). This radiation is distributed over all directions and wavelengths (Planck distribution function), with the maximum amount occurring at the wavelength

$$\lambda_{max} = \frac{2897.6}{T_w} \quad \mu\text{m} \tag{5.9-2}$$

(Wein "displacement" law). Table 5.9-1 collects some representative values of $\dot{e}_b''$ and λ_{max}. It is interesting to note that (i) a surface of temperatures only 393 K can put out energy at the same rate as solar radiation incident upon our planetary atmosphere (the solar "constant," 1.35 kW/m^2); and (ii) λ_{max} values for $1000 < T_w < 2000$ K are in the near infrared (invisible) region of the spectrum, from 2.9 μm to 1.45 μm. Real surfaces will radiate only some fraction, $\varepsilon_w\{T_w\}$, of $\dot{e}_b''$ (Eq (5.9-1)), where $\varepsilon_w\{T_w\}$ is the so-called *total hemispheric emittance* of the surface (see Figure 5.9-1 and Table 5.9-2).

Table 5.9-2 Approximate Temperature Dependence[a] of Total Radiant-Energy Flux from Heated Solid Surfaces (cf. Rosner (1964))

Surface	*Temperature range* (K)	*n*
Opaque quartz (vitreosil)	530-1090	3.5
Alumina ceramic	530-1090	2.5
Thoria ceramic	530-1090	2.3
Magnesia ceramic	530-1090	2.5
Iron	700-1300	5.55
Nichrome	325-1310	4.1
Polished tungsten	420-530	5.4
Polished magnesium	420-530	5.1
Molybdenum	420-530	5.2
98% pure polished aluminum	420-530	5.4
Nickel	463-1280	4.65
Silver	610-980	4.1
Platinum	640-1150	5.0

[a] $n \equiv 4 + d(\ln \varepsilon_w)/d(\ln T_w)$

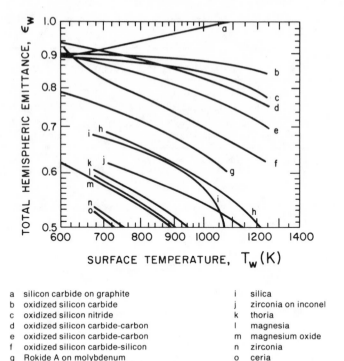

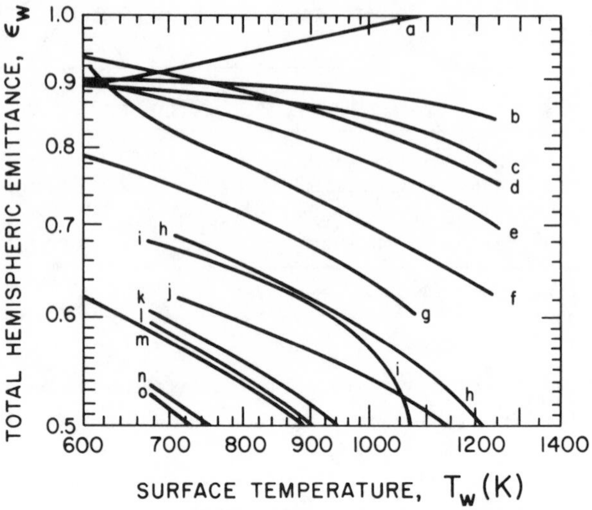

a	silicon carbide on graphite	i	silica
b	oxidized silicon carbide	j	zirconia on inconel
c	oxidized silicon nitride	k	thoria
d	oxidized silicon carbide-carbon	l	magnesia
e	oxidized silicon carbide-carbon	m	magnesium oxide
f	oxidized silicon carbide-silicon	n	zirconia
g	Rokide A on molybdenum	o	ceria
h	alumina on inconel		

Figure 5.9-1 Dependence of total hemispheric emittance on surface temperature for several refractory materials (log-log scale) (adapted from Rosner (1964)).

Two surfaces of areas A_i and A_j separated by an IR-transparent gas will *exchange* radiation at a net rate:

$$\dot{q}_{\mathrm{rad},ij} = A_i \cdot F_{ij}\{\varepsilon_i, \varepsilon_j, \text{geometry}\}\sigma_B(T_i^4 - T_j^4) \tag{5.9-3}$$

where F_{ij}, the so-called "grey-body view factor," accounts for the facts that (a) area j "sees" only a portion of the radiation emitted by area i (and *vice versa*);[40] (b) neither area is capable of emitting at the maximum ("black-body") rate;[41] and (c) area j *reflects* some of the incident energy back to area i, and *vice versa.*

An important application of Eq. (5.9-3) is the case of an isothermal emitter of area A_w in a partial enclosure of temperature $T_{\mathrm{enclosure}}$ filled with an IR-transparent moving gas. Then:

$$(\dot{q}_{\mathrm{w,rad}}/A_{\mathrm{w}}) = F\{\varepsilon_{\mathrm{w}}, \varepsilon_{\mathrm{encl.}}, \text{geometry}\}\sigma_{\mathrm{B}}(T_{\mathrm{w}}^4 - T_{\mathrm{encl.}}^4). \tag{5.9-4}$$

Since the local gas at temperature T_∞ is IR-transparent, the surface will also be

---

[40] Since $\dot{q}_{ij} = \dot{q}_{ji}$, one notes the "reciprocity" relation: $A_iF_{ij} = A_jF_{ji}$. View factors for some important black body emitter-receiver combinations in axisymmetric reactors are shown in Figures 5.9-2 and 5.9-3.

[41] That is, each surface has an *emittance*, ε, less than unity.

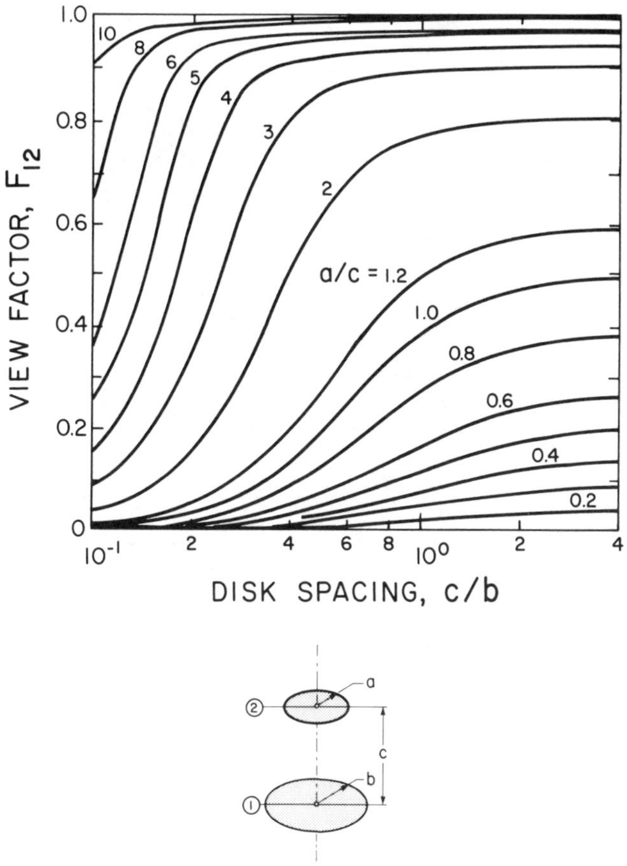

Figure 5.9-2 Predicted view-factors between two parallel coaxial disks (adapted from Sparrow and Cess (1978)).

losing energy by "convection" at the average flux:

$$(\dot{q}_{w,\,conv}/A_w) = \overline{Nu}_h\{Re, Pr\} \cdot k_\infty \left(\frac{T_w - T_\infty}{L}\right). \tag{5.9-5}$$

The *total* net average heat flux from the surface will be the algebraic *sum* of these. Put another way, *radiation* contributes, in effect, the following additive term to the ordinary (convective) heat-transfer *coefficient*:

$$\overline{Nu}_{h,\,rad} \equiv \frac{F\{\varepsilon_w, \varepsilon_{encl.}, \text{geometry}\}\sigma_B(T_w^4 - T_{encl.}^4)}{k_\infty(T_w - T_\infty)/L}. \tag{5.9-6}$$

The relative importance of radiation then follows immediately from a comparison between $\overline{Nu}_{h,\,rad}$ (Eq. (5.9-6)) and $\overline{Nu}_h$ based on convection alone (Sections 5.5 and 5.6). In general:

$$\overline{Nu}_{h,\,eff} = \overline{Nu}_{h,\,conv.} + \overline{Nu}_{h,\,rad} \tag{5.9-7}$$

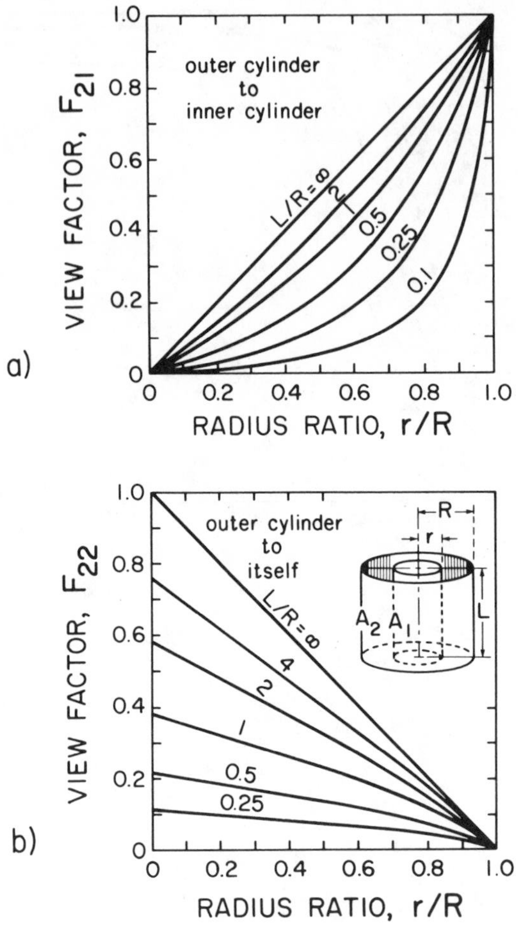

Figure 5.9-3 Predicted view-factors for the concentric cylinder geometry: (a) outer cylinder to inner cylinder; (b) outer cylinder to itself.

and $\overline{\mathrm{Nu}}_{h,\mathrm{rad}}$ will be seen to be relatively important *in high-temperature systems* and/or systems for which $\mathrm{Nu}_{h,\mathrm{conv.}}$ is low (e.g., "natural" convection).

5.9.2 Radiant Emission and Transmission by Dispersed Particulate Matter

The laws governing radiant emission from dense clouds of small particles are complicated by the facts that the particles are usually:

a. themselves small compared to $\lambda_{\max}$ (Eq. (5.9-2));

b. not opaque;
c. generally at temperatures different from the local "host" gas.

When the cloud is so dense that the photon mean-free-path,[42] ℓ_{photon}, is *small* compared to macroscopic lengths of interest, then radiation can be approximated as a "diffusion" process (the Roesseland or "optically thick" limit). For a pseudo-homogeneous medium, this leads to an additive ("photon") contribution to the apparent thermal conductivity:

$$(k_{\text{rad}})_{\text{eff}} = \frac{16}{3} n_{\text{eff}}^2 \ell_{\text{photon}} \sigma_{\text{B}} T^3, \tag{5.9-8}$$

where n_{eff} is the effective *refractive index* of the medium. Except for complication (a) above, the physical situation here is similar to the augmentation in the effective thermal conductivity of a high-temperature packed bed (see Ex. 5.22).

Furnace flames are usually deliberately run under "sooting" conditions to maximize radiant emission to the desired heat "sinks" (products, boiler tubes, etc.); for maximum furnace efficiency, however, this soot must be burned up prior to reaching the flue (exit) duct(s).

5.9.3 Radiant Emission and Transmission by IR-Active Vapors

An isothermal, hemispherical gas-filled "dome" of radius L_{rad} contributes the incident flux ("irradiation")

$$\dot{q}''^{(-)}_{\text{rad, g-w}} = \varepsilon_{\text{g}}\{X_1, X_2, \ldots, T_{\text{gas}}\} \cdot \sigma_{\text{B}} T_{\text{g}}^4 \tag{5.9-9}$$

to a unit area of surface centered at its base, where

$$X_i \equiv p_i L_{\text{rad}} \text{ ("optical depth" of radiating species } i\text{)}, \tag{5.9-10}$$

and the "total emissivity" of the gas mixture, $\varepsilon_{\text{g}}\{X_1, X_2, \ldots, T_{\text{g}}\}$, may be determined from direct overall energy-transfer experiments (see Figure 5.9-4 for the single species H_2O(g)) or from spectral information for each contributing gas, summed over all appropriate wavelength intervals. More generally, the gas actually "viewed" by any element of surface is neither hemispherical nor isothermal, and Eq. (5.9-9) must be replaced by:

$$\dot{q}''^{(-)}_{\text{rad, g-w}} = \frac{1}{\pi} \cdot \int_{\Omega} \int_{X} \sigma_{\text{B}} \cdot T_{\text{g}}^4 \frac{\partial \varepsilon_{\text{g}}}{\partial X_i} \cdot \cos\theta \cdot d\Omega \, dX_i \tag{5.9-11}$$

[42] E.g., the length over which the intensity of incident radiation would drop by the factor e^{-1} due to absorption.

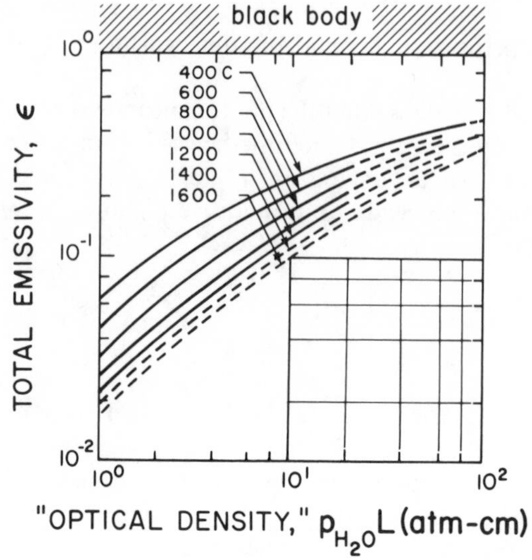

Figure 5.9-4 Total effective emissivity of water vapor as a function of "optical depth" $p_{H_2O} \cdot L$ (adapted from Eckert (1937)).

for the special case of one dominant emitting species *i*. Here $T_g(\theta, \phi, X_i)$ is the temperature in the gas at a position defined by the angles θ (measured from the normal), ϕ, and the running "optical depth" $\int_0 dX_i$. In practice, Eq. (5.9-11), which also involves integration over all solid angles Ω, is usually approximated by:

$$\dot{q}''^{(-)}_{\text{rad, g-w}} \approx \varepsilon_g((p_i L_{\text{rad}})_{\text{eff}}, T_{\text{g, eff}}) \cdot \sigma_B T^4_{\text{g, eff}}, \tag{5.9-12}$$

where $(p_i L_{\text{rad}})_{\text{eff}}$ is some effective optical depth and L_{eff} the equivalent dome radius for the particular gas configuration "seen" by the surface area in question.

Furnace enclosure problems in which the radiation couples only weakly with the contained gas are usually solved approximately by noting that each unit area of the bounding surface sees not only the *incoming radiation* $\dot{q}''^{(-)}_{\text{rad, g-w}}$ emitted by the gas, but also a certain fraction of the radiation emitted by distant *surface* elements (cf. Eq. (5.9-3)), scaled down by the *transmittance*, τ, of the intervening gas (for this emitted radiation). By subdividing the total surface into N_w quasi-isothermal segments, and writing/solving N_w such coupled energy balances, it is possible to estimate all resulting steady-state surface temperatures (say, for adiabatic surfaces), irradiances, and corresponding radiosities (total outgoing radiation fluxes).[43]

It is interesting to note that:

a. L_{eff} for each unit area of a very long cylinder containing an isothermal, radiating gas is close to the cylinder *diameter*. Values of L_{eff} for other common

[43] Given by the sum of $\varepsilon_w \dot{e}''_{b,w}$ and the *reflected* portion, $\rho_w \dot{q}''^{(-)}_{\text{rad}}$, of the IR-radiation $\dot{q}''^{(-)}_{\text{rad, w}}$, for each unit area of surface (here ρ_w is the wall *reflectivity*).

gas geometries are given in the literature (e.g., Hottel and Sarofim (1967), Edwards, Denny, and Mills (1979), or Sparrow and Cess (1967)).

b. For the pure vapors $H_2O(g)$ and $CO_2(g)$ the effective isothermal total emissivity, ε_g, does not extrapolate to unity at infinite optical depth, but rather to values $\varepsilon_g(\infty)$ given, respectively, by 0.9 and 0.23.

c. Because of interactions between the IR bands of CO_2 and H_2O, simply adding ε_g-values computed for each pure vapor will overestimate ε_g(mixture). For example, at "infinite" optical depth $CO_2 + H_2O$ mixtures have an effective emittance of "only" about 0.98.

d. High-pressure rocket motors and large nonsooting ("nonluminous") fires often have optical depths in excess of 10^3 atm-cm (10 atm-m), at which it is fortunate that ε_g becomes insensitive to inevitable gas-temperature uncertainties.

5.9.4 Remarks on the Quantitative Treatment of Radiation in High-Temperature Chemical Reactors

Three distinct approaches (outlined below) have been used to predict coupled radiation–convection–conduction energy transport. Each of these approaches can be regarded as a numerical scheme for approximately satisfying a *photon population balance equation*, which is coupled to the usual conservation equations governing the coexisting "material" phase. The photon population balance equation, in turn, governs the evolution of photon "densities" in an augmented "space" which necessarily includes frequency and spatial *direction* (cf. Section 1.2).

a. "Net Interchange via Action-at-a-Distance" Method. Here one extends equations like (5.9-3) to include the net radiant interchange between distant Eulerian control *volumes* of participating gas, with each such volume interacting with *all* other volumes. Clearly the extent of each such interaction will depend upon absorption and scattering of radiation *along* the relevant *intervening paths*.

b. Six-Flux (Differential) Model of Net Radiation Transfer. Here the radiation field is presumed to be represented by six fluxes at each point in space, *viz.*:

$$\left.\begin{array}{l} I_x^{(+)}, I_x^{(-)} \\ I_y^{(+)}, I_y^{(-)} \\ I_z^{(+)}, I_z^{(-)} \end{array}\right\} \quad \text{in a cartesian coordinate system.}$$

In each of these directions, each flux is assumed to change according to *local* emission (coefficient ε) and absorption + scattering (coefficients α, σ, respectively); for example,

$$\frac{\partial I_x^{(+)}}{\partial x} = \varepsilon \sigma_B T^4 - (\alpha + \sigma) I_x^{(+)} + \frac{\sigma}{5}\left(\sum \text{(five other fluxes)}\right), \qquad (5.9\text{-}13)$$

and five similar first-order PDEs for the remaining fluxes. These six PDEs (or three equivalent second-order PDEs for the arithmetic-mean fluxes $\frac{1}{2}(I_x^{(+)} + I_x^{(-)})$, etc.) are then solved, subject to appropriate BCs at the combustor walls.

Note that in this representation, the *radiation* energy flux vector has the x-component $\dot{q}''_{\text{rad},x} = I_x^{(+)} - I_x^{(-)}$, with similar equations for the y- and z-components.

c. Monte-Carlo Calculations of Photon "Bundle" Histories. Here one computes the progress of large numbers of "photon bundles," each containing the same amount of energy, and infers wall-energy fluxes by counting photon-bundle arrivals in the areas of interest. Photon-bundle histories are generated in accordance with the known statistical laws of photon interaction (absorption, scattering, etc.) with gases and/or surfaces, including their dependence on local number density, temperature, frequency, and angle. Most often this method is used to estimate the net local radiative source (or sink) terms in the energy-conservation PDE governing the material phase, the latter being treated by conventional finite-difference methods (Appendix 8.2) after each iteration. Computations are terminated when convergence is achieved, *and* calculated energy fluxes have an adequate probability of lying within a few percent of the mean radiative energy fluxes computed (cf., e.g., Howell, J.R. (1968), and Siegel and Howell (1981)).

Method (a) yields *integrodifferential* equations (IPDEs), which are numerically cumbersome for high-resolution multi-zone combustor calculations. In contrast, method (b) leads to a system of PDEs analogous to those already considered in Chapter 2 (except for the absence of transient and "convective" terms). For this reason some form of the *differential* six-flux method will probably be the method of choice for high-resolution engineering predictions of coupled radiation transfer and gas dynamics in chemical reactors (e.g., combustors)—(see F. Lockwood (1980)).

Since the photon "phase" energy density (at most $4\sigma_B T^4/c$, where c is the prevailing speed of light) is usually accumulating at a negligible rate, then the energy equation governing the photon phase immediately reveals that the net source term $\dot{q}'''$ (given by the difference between the local rates of energy absorption and emission) appearing in the energy equations governing the coexisting *material phase* (cf. Eqs. (2.3-3, 2.3-4, 3.2-6, and 5.3-10)) can be simply identified with

$$-\text{div}\,\dot{\mathbf{q}}''_{\text{rad}}.$$

For this reason a material phase is said to be in "local (radiative) equilibrium" (LRE) with a coexisting photon phase if

$$\text{div}\,\dot{\mathbf{q}}''_{\text{rad}} = 0.$$

SUMMARY

- Convection and diffusion usually occur together and their relative contributions to the local energy flux will vary within a flow field. Even when the surface heat flows are by pure diffusion, the overall process may still be called "*convective* heat transfer" because convection elsewhere has influenced the magnitude of the local diffusion fluxes of energy.

- Radiation can usually not be treated as a *diffusive* phenomenon—in fact, the photon mean-free-path is usually *large* enough to allow radiative transfer to be uncoupled from simultaneous energy transport by Fourier diffusion and convection.
- Dimensionless heat-transfer "coefficients" ($\mathrm{Nu}_h, \mathrm{St}_h$) are merely ratios of *actual* heat fluxes to suitable *reference* heat fluxes; thus, for Nu_h:

$$\dot{q}''_{\mathrm{ref}} \equiv k\left(\frac{\Delta T}{L}\right),$$

whereas in forced convection systems we can also use St_h, for which

$$\dot{q}''_{\mathrm{ref}} \equiv \rho U c_{\mathrm{p}} \Delta T.$$

- In simple cases, Nu_h and St_h can be *predicted* (analytically or numerically) from first principles. More often, they are measured by techniques that circumvent the problem of measuring fluid temperatures very close to the surface.
- There are *thermal* BLs just as there are viscous BLs. In steady-state problems they are thin only if:

$$(\mathrm{Re} \cdot \mathrm{Pr})^{1/2} \gg 1 \qquad \text{(forced convection)}$$

or

$$(\mathrm{Ra}_h \cdot \mathrm{Pr})^{1/4} \gg 1 \qquad \text{(free- (natural-) convection).}$$

This simplifies the analysis of such regions *via* Prandtl BL approximations. These *groups* (when sufficiently small) also dictate when convection can be neglected. Recall that energy can be *convected* in a moving *solid*. What distinguishes a *fluid* is its continuous rate of deformation (angle + volume).

- In the absence of large, streamwise pressure gradients, the energy-transfer coefficient St_h can be related to the corresponding momentum-transfer coefficient $c_{\mathrm{f}}/2$ (skin friction)—(when $\mathrm{Pr} = 1$ the two coefficients are equal, even in laminar flow); however, this so-called "Reynolds' analogy" breaks down for sufficiently large streamwise pressure gradients.
- The information needed to solve problems in *turbulent* heat transfer usually comes, in part, from *experiments* on simple turbulent flows—e.g., turbulent pipe flow combined with the assumption that the "eddies" transport energy with about the same effective diffusivity as for linear momentum.
- The effects of both viscous dissipation ("recovery of mainstream KE") and chemical reaction (recovery of mainstream ChE) can be approximately taken into account by merely using a total enthalpy "driving force," while retaining the same (nonreactive) transfer coefficient.

Appendix 5.1

Outline of Fourier Method of "Separation of Variables" and Eigenfunction Expansions

Nonself-similar diffusion problems (cf. Sections 5.4.6 and 5.5.2) governed by a *linear* PDE + BC on a domain bounded by simple coordinate surfaces [e.g., members of the class of the commonly used orthogonal coordinate systems (see, e.g., Morse and Feshback (1953))] can be solved by a method (devised by J. Fourier) called "separation of variables" outlined below. In effect, the method reduces the problem of solving the linear *PDE* to the simpler problem of solving two or more linear ODEs (one for each independent variable).

Diffusion problems governed by *nonlinear* PDEs (owing to the "source" term and/or variable properties), nonlinear BCs, or on domains of more complex shape,[44] are usually solved by approximate ("finite-difference"-type) numerical methods outlined in Appendix 8.2. In the latter class of methods, *algebraic* equations are generated and solved for the temperatures (or mass fractions) only at a finite number of discrete node points.

Steps:

1. Write the *linear* diffusion problem (PDE + BC) in the most natural orthogonal coordinate system (such that the linear BC are imposed along coordinate surfaces ξ_1, ξ_2, ξ_3).
2. Suitably nondimensionalize (and normalize, where possible) independent and dependent variables; i.e., choose T^*, $t^* \equiv \alpha t/b^2$, etc.
3. Seek a solution $T^*(t^*, \xi_1^*, \xi_2^*, \xi_3^*)$ in the form of an *infinite series* of "products," i.e.:

$$T^* = \sum_n \theta_n(t^*) \cdot X_{1,n}(\xi_1{}^*) X_{2,n}(\xi_2{}^*) X_{3,n}(\xi_3{}^*).$$

Determine the OD-equation satisfied by each of these special functions by back-substitution into the PDE. These equations will contain a common (eigenvalue) parameter or "separation constant." The corresponding spatial functions $X_{1,n}, X_{2,n}, X_{3,n}$ are called "eigenfunctions." (They may be sine, cos, Bessel

---

[44] For other examples and details, see: Carslaw and Jaeger (1959); Crank (1956); Morse and Feshback (1953).

functions, etc. Frequently the "time-eigenfunctions" $\theta\{t^*\}$ are decaying exponentials of the form $\theta_n\{t^*\} = \text{const} \cdot \exp(-c_n^2 t^*)$; for steady-state diffusion, $\theta = 1$.)

4. Select the eigenvalues and eigenfunctions such that all *boundary* conditions are satisfied (at $\xi_1^*, \xi_2^*, \xi_3^* = 0$, specified constants).
5. Select all remaining undetermined constant coefficients in the general solution such that *initial* conditions are satisfied by the resulting "eigenfunction expansion" of the separable initial condition $T^*\{0, \xi_1^*, \xi_2^*, \xi_3^*\}$. The coefficients ("generalized Fourier coefficients") are found using the orthogonality properties of the $X_{1,n}, X_{2,n}, X_{3,n}$-functions on the domains of interest.

The resulting eigenfunction series expansion is then an *exact solution* to the BVP initially posed. Any desired accuracy can be obtained by including a sufficiently large number of terms.

TRUE/FALSE QUESTIONS

5.1 T F In engineering applications the photon mean-free-path is usually large enough to allow radiative transfer to be uncoupled from simultaneous energy transport by Fourier diffusion and convection.

5.2 T F A solidification "wave" violates the Second Law of Thermodynamics, since the result of the transformation (a(poly)crystalline solid) has a lower entropy than the melt from which it was formed.

5.3 T F Heat *diffusion* plays no role in the determination of *convective* heat-transfer coefficients.

5.4 T F The use of temperature difference as a "driving force" for convective heat transfer ("Newton's law of cooling/heating") has its origin in Fourier's linear law of heat diffusion.

5.5 T F Variable fluid properties (e.g., the viscosity–temperature dependence) are more likely to influence convective heat-transfer coefficients for Newtonian *liquids* than for (Newtonian) *gas* mixtures.

5.6 T F Solids are incapable of "convecting" energy.

5.7 T F "Fins" are more often placed on the liquid side of gas/liquid heat exchangers because that is the side of maximum heat "conductance."

5.8 T F The emitted radiation from all real surfaces increases with the fourth power of the absolute surface temperature.

5.9 T F Increasing the thickness of the air gap between two panes of glass will inevitably increase the effective thermal resistance of a composite "window."

5.10 T F Entropy can be *convected*, but it cannot *diffuse*.

5.11 T F Localized boiling can dramatically augment heat-transfer coefficients as a result of both latent heat transport and the "stirring" action of the vapor bubbles produced.

5.12 T F If energy is to be conserved across an interface, the temperature on either side of the interface must be the same—i.e., the temperature must be continuous across the interface.

5.13 T F For a propagating phase boundary, the normal component of the energy diffusion flux vector must be the same on both sides of the boundary.

5.14 T F In a system undergoing chemical reaction and/or phase change, there are often important contributions to the energy diffusion flux vector other than the *Fourier* ($-k\,\mathbf{grad}\,T$) contribution.

5.15 T F An understanding of the heat-transfer properties of straight ducts can be used in the formulation of "economical," general correlations for packed-bed heat transfer, but actual heat-transfer coefficients in such "tortuous-duct" systems are best determined by direct experimentation.

EXERCISES

5.1 a. Estimate the thermal resistance of a single sheet of glass of thickness 0.3175 cm (1/8 inch).

b. Estimate the thermal resistance of a series "stack" of two such sheets of gas separated by a quiescent air gap of 0.3175 cm.

c. If the air "films" on the hot and cold sides of the composite double-pane window are, respectively, of equivalent thickness 2.8 mm and 0.69 mm, what is the *overall* thermal resistance to heat flow?

d. If the interior (ambient) room temperature is 25.5°C (70°F) and the exterior (ambient) temperature is −0.78°C (30°F), calculate the heat loss in kW/m^2.

e. Estimate the critical gap thickness (cm) at the onset of natural convection, and the heat-loss rate at this condition of incipient natural convection.

5.2 A rigid, low-density (*ca.* 8.5 lb/ft^3) cellular (bubble) glass insulation material has been developed (Pittsburgh CORNING "Foamglas"®), which is impermeable to vapors, liquids, and resistant to degradation by heat (up to *ca.* 900°F) or cold (down to −450°F). The effective thermal conductivity is reported to be 0.27 (BTU in/ft^2 hr °F) at −100°F, 0.32 at 0°F, 0.36 at 100°F, and 0.42 at 200°F. Consider the application of this insulation to prevent water-vapor condensation on a 12″ (O.D.), −20°F, chemical-process system pipe of emittance 0.4 in a 10 mph environment of 90°F, 80 percent relative humidity air. The equilibrium water-vapor pressure (inches Hg(ℓ)) is 1.2587×10^{-2} at −20°F and 1.4219 at 90°F.

Estimate the required insulation thickness and compare your conclusion to the nominal value 2.5 in., suggested by the manufacturer (Brochure F1-160, 6/81). Is the manufacturer's recommendation "on the safe side?" Would the next smaller thickness available (2.0 in.) suffice?

5.3 Closely examine the "radiator" (finned channel or finned tube *convective* heat exchanger) on a water-cooled automobile, and answer the following questions about it:

a. What kind of "fins" are used to extend the basic (primary) heat-transfer surface? How are they attached? Why is the attachment important?

b. Estimate the *order-of-magnitude* of the ratio of fin area to primary (unfinned) area. Do the fins (approximately) satisfy our assumptions (cf. Section 5.4.5)?

c. If cost, weight, fabricability, and corrosion considerations were temporarily set aside, and fin materials of various thermal conductivities were compared, would copper fins necessarily be much better than steel fins?

d. In determining the wall *thickness* of the corrugated fin material, what important competing factors must be considered (say, for a given channel or tube spacing)?

e. What are the effects of "fouling" on: (i) Fin and primary air-side surfaces (due to bugs, dirt, etc.)? (ii) Water-side surfaces (e.g., rust, antifreeze deposits, etc.)?

f. Why is a fan utilized "behind" the heat exchanger?

g. Why are the fins placed on the "air-side" rather than the "water-side" of the heat exchanger? Relate this decision to the relative magnitude of the heat-transfer "resistances" in series (air-side, wall material, water-side).

5.4 To measure the temperature of heated air flowing through a pipe, a thermometer pocket (or "well") of the form shown in Figure 5.4E (with $L = 7.62$ cm, $d_w = 0.95$ cm, wall thickness 0.8 mm) is screwed into the wall of the pipe and projects into an air stream (of velocity 15.2 m/s). The measured temperature of the wall of the duct is 93.4°C and the thermocouple reads 182°C, but, owing to heat conduction along the walls of the thermometer well, the thermocouple is expected to indicate a temperature below the true air temperature.

a. By treating the thermometer well as a "fin," estimate the true temperature of the air. The heat transferred from the air stream to a cylinder placed at right angles to the direction of flow can be calculated from the correlation shown in Figure 8.1-1 and for the air in the duct the properties may be estimated from Table 8.1-1. For the material of the thermometer well use $k = 1.21 \times 10^{-1}$ kW/m-K (after J.R. Simonson (1975)).

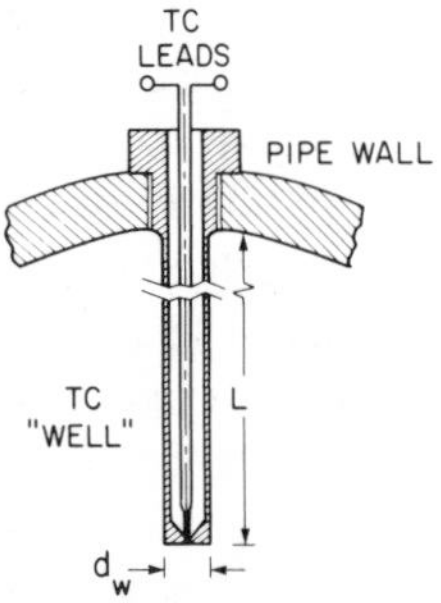

Figure 5.4E

$k = 1.21 \times 10^{-1}$ kW/m-K (after J.R. Simonson (1975)).

b. Are the assumptions underlying your "fin" results applicable to this case? Evaluate quantitatively where possible.

c. How would you estimate the time-constant (thermal response) of this assembly?

5.5 A gas turbine blade, cooled at the root, is 10.2 cm long and has a nearly constant cross-sectional area 1.93 cm^2, and a perimeter of 7.6 cm. Gas at 815°C streams across it, and one end (the root) is cooled to 483°C. The mean heat-transfer coefficient for the gas-flow conditions can be assumed constant over the surface at 0.284 kW/(m^2K), and the thermal conductivity of the material of the blade is 26×10^{-3} kW/(mK).

a. Find the heat passing to the cooled end of the blade per unit time (neglect the heat gained from the uncooled end of the blade) (after J.R. Simonson (1975)).

b. If the entire blade could be maintained at the root temperature (483°C), what would the heat-transfer rate have been?

c. Estimate the actual blade-tip temperature under the conditions of part (a). If this value is too high for the material being considered, what options are open to the designer?

5.6 a. Estimate the "thermal time constant," t_h, of a 0.5 mm diam. Pt/Pt–Rh thermocouple bead immersed in an airflow with the nominal properties: $T_\infty = 1400$ K, $p = 1$ atm, $U = 5$ m/s. If the airflow were not steady, up to about what frequency of fluctuations could this thermocouple be used to "track" gas-temperature fluctuations? What options are available to reduce the thermal (inertia) time constant of such a thermocouple bead?

b. The idea that temperature nonuniformities can be neglected *within* an object with thermal inertia requires that the internal resistance to energy diffusion be negligible compared to the corresponding external (fluid "film") resistance; i.e., the heat transfer *Biot number*, Bi_h, must be small compared to unity. Calculate Bi_h for the thermocouple bead situation of Part (a) to confirm the validity of the "thermal time constant" analysis above.

5.7 Lead "shot" having an average diameter of 5.1 mm is at an initial temperature of 204°C. To quench the shot, it is added to a quenching oil bath held at 32.2°C and falls to the bottom with a residence time of 15 s. Under conditions such that the average convection heat-transfer coefficient, $\bar{h}$, is 200 W/$m^2\cdot$K and neglecting radial temperature nonuniformities within each particle, estimate the temperature of the shot after the fall. (For lead, $\rho \cong 11370$ kg/m^3 and $c_p \cong 0.138$ kJ/kg·K, after C. Geankoplis, Ex. 5.2-2.)

5.8 In a test facility a long, ferrous-metal alloy cylinder ($d_w = 2$ in., $k =$ 20 BTU/hr ft °F, $\rho = 488$ lb/ft^3, $c = 0.11$ BTU/lb_m°F) at $T_0 = 70$°F is to be suddenly thrust into a gas jet with the following properties:

$$\left.\begin{aligned} p &= 1\text{ atm} \\ T_\infty &= 1700\text{F} \\ U &= 400\text{ ft/s} \end{aligned}\right\}\quad \begin{aligned}&\text{Treat as "hot air"}\\ &\text{(neglect effect of } H_2O,\ CO_2).\end{aligned}$$

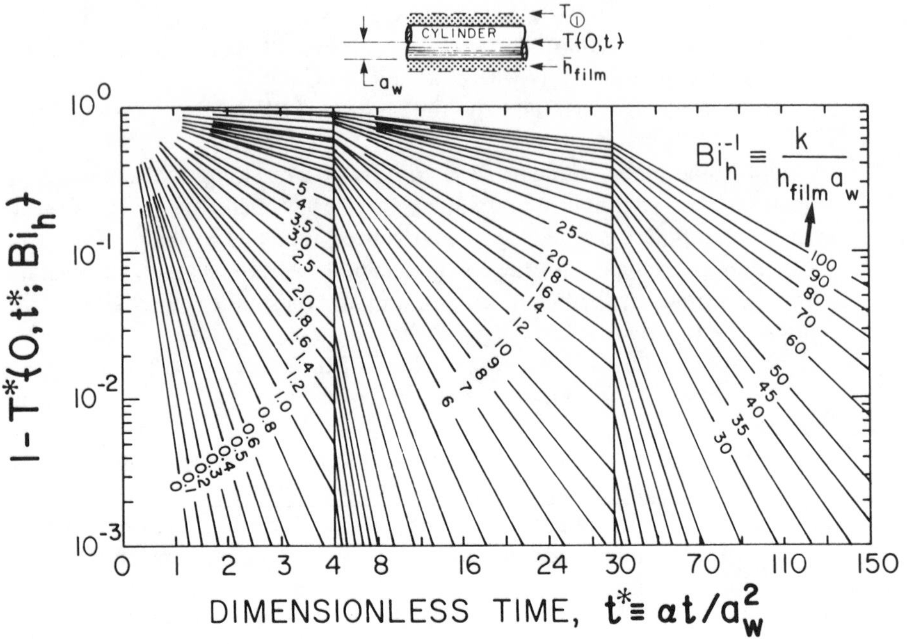

Figure 5.8E

Neglecting radiative heat exchange:

a. Estimate the time it will take the centerline temperature to increase to 1668°F.[45]

b. Compare this to the time that would have been required if $k \to \infty$ (i.e., if $Bi_h \to 0$) and discuss the reason(s) for the difference. Calculate Bi_h if the cylinder were made of *copper*.

c. If the cylinder diameter were doubled, by what factor would the required time be increased? Discuss and compare to the corresponding expectations in the limits $Bi_h \to 0$, $Bi_h \to \infty$.

d. Itemize and consider the validity of the assumptions underlying your estimates.

e. Could this ferrous-metal alloy rod, if suitably instrumented, be easily used to infer experimental gas-side heat-transfer coefficients? What would you recommend to determine Nu_h (or h_{film}) more easily?

5.9 A moving, locally planar solidification "front" at $x = \xi\{t\}$ (see Figure 5.9E) separates a quiescent melt, $x > \xi\{t\}$, from a crystallized region, $x < \xi\{t\}$. Suppose the energy flux vector within each adjacent phase can be written in

---

[45] Use the attached plot (Figure 5.8E) of $1\text{-}T^*\{0, t^*; Bi_h\}$ *vs.* t^*, summarizing results obtained by the Fourier method of "separation of variables" applied to an infinite *cylinder* (after Heisler [1947]).

the simplest (Fourier) form $\dot{\mathbf{q}}'' = -k\,\mathbf{grad}\,T$, where k is the (scalar) thermal conductivity. If k_m and k_s are, respectively, the thermal conductivities of the melt and solid, and L is the *latent heat of fusion* of the solid (per unit mass), then in the heat-conduction literature, one-dimensional transient solidification problems are quantitatively treated using the following conditions imposed at the moving "front":

$$\rho_m L \cdot \frac{d\xi}{dt} = k_s \left(\frac{\partial T_s}{\partial x}\right)_{x=\xi(t)} - k_m \left(\frac{\partial T_m}{\partial x}\right)_{x=\xi(t)} \tag{1}$$

$$T_m = T_s \quad at\ x = \xi(t) \tag{2}$$

a. Show that each of these relations is compatible (not inconsistent) with the general "jump condition" derived from the energy conservation equation for a moving discontinuity (in this case, a moving *phase boundary* with $\dot{m}''_{\pm} \neq 0$) (see Eq. (2.6-10)).

b. Itemize and, if possible, defend using typical numbers (see Part (d) below) the assumptions that must be made to obtain Eqs. (1, 2) for a one-dimensional solidification wave.

c. Under what conditions *would* the conductive *heat flux* be continuous across a *moving* phase boundary? Consider, e.g., a simple change in crystal structure.

d. If a solidification front in quiescent molten iron is propagating at the rate of 5 cm/s, what must be the instantaneous heat flux *away* from the front? ($V_s(T_{mp}) = 7.66\ \text{cm}^3 \cdot (\text{g-mole})^{-1}$, $V_m(T_{mp}) = 7.96\ \text{cm}^3 \cdot (\text{g-mole})^{-1}$, $ML = 3.30$ kcal/mole, $T_{mp} = 1809$ K.)

e. Propagating solidification fronts are observed, yet less entropy *convectively* "flows" out of them (crystalline solid) than "flows" in (*via* the melt). Does such a "front," therefore, *violate Clausius*' inequality? Explain.

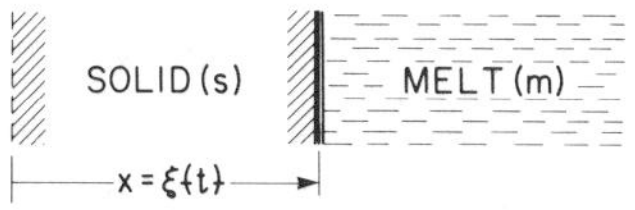

Figure 5.9E

5.10 A manufacturer/supplier of fibrous 90% Al_2O_3–10% SiO_2 insulation board (0.5 inches thick, 70% open porosity) does not provide direct information about its thermal conductivity, but does report hot- and cold-face temperatures when it is placed in a vertical position in 80°F still air, heated from one side and "clad" with a thermocouple-carrying thin stainless steel plate (of total hemispheric emittance 0.90) on the "cold" side.

a. Given the following table of hot- and cold-face temperatures for an 18″ high specimen, estimate its thermal conductivity (when the pores are filled with air at 1 atm). (Express your result in $(BTU/ft^2\text{-}s)/(°F/in)$ and $(W/m \cdot K)$, and itemize your basic assumptions.)

b. Estimate the "R"-value of this insulation at a nominal temperature of 1000°F in air at 1 atm.

c. If this insulation were used under vacuum conditions, would its thermal resistance increase, decrease, or remain the same? (Discuss.)

T ("*hot*" *face*)	*T* ("*cold*" *face*)
3000°F (service limit)	790°F
2400°F	670°F
1600°F	480°F

5.11 In a typical honeycomb-type automobile exhaust catalytic converter, hot IC engine combustion products are passed through a parallel array of small channels and then are discharged through an ordinary tailpipe. In a preliminary engineering design it is desired to estimate the gas temperature *exiting* the channels for various assumed channel lengths, under the following conditions:

inlet gas temperature	700 K
inlet gas pressure	1 atm
inlet gas composition (mole fractions)[46]	$y(N_2) = 0.93$, $y(CO) = 0.02$, $y(O_2) = 0.05$
inlet gas velocity	10^3 cm/s
channel cross-section dimensions	1.5 mm × 1.5 mm (each channel)
assumed channel *wall* temp.	500 K

For simplicity, assume that no *gas phase* chemical reaction (e.g., CO oxidation) occurs, and make/defend whatever remaining assumptions you deem reasonable under the circumstances.

Answer the following specific questions:

a. What is the approximate gas viscosity (Table 8.1-1), density, and corresponding Reynolds' number for this "internal" flow based on the mean gas velocity and the effective (equivalent) channel diameter $d_{eff} \equiv 4 \cdot$ (cross-sectional area)/(wetted perimeter). Is the flow laminar or turbulent under these conditions?

b. What is the *approximate* thermal conductivity (expressed in cal cm^{-1} s^{-1} K^{-1}) and Prandtl number of the gas mixture under these conditions? (Cf. Table 8.1-1.)

[46] We explicitly neglect the inevitable presence of $H_2O(g)$ and $CO_2(g)$, lumping them into an "effective" N_2 composition.

c. Assuming fully developed laminar flow within each channel, and using the results discussed in Section 5.5.4, estimate the mixing-cup average (bulk) gas temperature that would exit each such channel if the channel *length* were 0.2 m, 0.4 m, 0.6 m, 0.8 m, or ∞. Plot these first four results on (log) $[T_b(0) - T_b(z)]/[T_b(0) - T_w]$ *vs.* (linear) L/d_{eff} coordinates.
d. What would be the perimeter-mean *heat flux* to the channel walls at the $L = 0.4$ m and $L = 0.8$ m downstream positions? (Express results in cal cm^{-2} s^{-1} and Watt/m^2. (Note: 1 cal = 4.184 Joule, 1 Watt $\equiv$ 1 Joule/s.)
e. Has the flow become "thermally fully developed" by the $L = 0.8$ m station? What does this imply about the downstream distribution of the heat-transfer *coefficient* and heat *flux*?
f. Would the procedures and assumptions you have used to make these estimates have been different if the fluid were a Newtonian *liquid* rather than *gaseous* combustion products? Why or why not?

5.12 Consider the estimation of *heat-transfer coefficients* needed for the preliminary design of a pressurized *water*-cooled nuclear reactor core. In the proposed heat exchanger configuration, 60°F water (pressurized to prevent localized boiling at the wall temperature of 300°F) will enter each 1 in. (I.D.) straight circular tube used to cool the reactor. If the flow rate in each tube is 30 gal/min, then, at an axial location where the bulk fluid temperature is estimated to be 100°F, calculate the expected:

a. Reynolds' number (based on tube diameter and mean axial velocity);
b. Friction factor (dimensionless wall shear stress) for a smooth wall;
c. Heat transfer Stanton number (dimensionless heat-transfer coefficient) at the prevailing Reynolds' and Prandtl numbers. (Compare to the result you would have predicted if you had erroneously used the *laminar* skin-friction coefficient.)
d. Maximum roughness height at which the tube walls would still be characterized by the estimate of $C_f(\text{Re}, \ldots)$ and $\text{St}_h(\text{Re}, \ldots)$. For *rougher* tube surfaces would St_h remain proportional to C_f?

Justify your choice of heat-transfer correlation, and the manner in which you have evaluated fluid properties (see Table 5.12E). Could Reynolds' analogy between turbulent forced-convection *heat* and *momentum* transfer for $\text{Pr} \simeq 1$ fluids be accurately applied to water under these conditions? Are there any quantitatively useful "analogies" for turbulent *natural* convection?

Table 5.12E Selected Properties of $H_2O(l)$

T(°F)	$\rho(\text{lb}_m/\text{ft}^3)$	$10^5\nu(\text{ft}^2/\text{s})$	$\text{Pr}(\equiv \nu/\alpha)$
32	62.4	1.93	13.7
60	62.3	1.22	8.07
100	62.1	0.736	4.51
200	60.1	0.342	1.91
300	57.3	0.227	1.22

5.13 (*Pebble Bed Regenerative Gas Heater Revisited*) In our preliminary "black box" design of the fixed (pebble) bed regenerative gas heater (cf. Exercise 2.10), we knowingly assumed: negligible pressure drop ($-\Delta p$) for the gases passing through the bed (see supplementary Question 4); and, more significantly:[47] the exiting gas (temperature) achieves the instantaneous bed temperature (A2). We are now in a position to critically and quantitatively investigate these preliminary design assumptions for a proposed pebble-bed depth (height) of, say, 1.5 ft.:

a. Estimate $-\Delta p$ for $\dot{m}_a = 1000\ \text{lb}_\text{m}/\text{hr}$ of 70°F air passed through the pebble bed when it is near 1000°F (i.e., near the start of the gas supply-time interval).

b. Based on your knowledge of gas-property variations, if the above calculation were repeated when the bed is near 500°F, would $-\Delta p$ increase or decrease? Why?

c. Is this (unrestrained) bed near the condition of "incipient fluidization"?

d. When the pebble bed is at 1000°F, estimate the actual *air* temperature exiting from the bed at the start of the "blow-down" (hot gas supply) cycle.

e. Why is the result of part (d) different from 1000°F, and how will this type of disparity influence the actual bed design requirements?

f. Suppose the inlet air contained, in suspension, 0.1 μm diam. $SiO_2(s)$ dust in dilute amounts (for which we can use the Stokes–Einstein equation to estimate the effective particle diffusivity in air). Estimate the maximum possible effectiveness with which this bed would act as a "depth" filter, reducing the outlet dust concentration below the inlet value—i.e., calculate the expected filter "efficiency" or its complement, the percentage "penetration." (See Section 6.5.1.)

g. Can you quantitatively assess (A1)—*viz*., is it reasonable to expect the bed temperature to be *spatially uniform* at each instant? Can this assumption be validated in terms of an effective Biot number? What energy-transfer mechanisms act to keep the bed temperature spatially uniform at each instant? What causes the tendency to bed temperature *non*-uniformity?

h. List and critically discuss your most important underlying assumptions. Are they valid under the present pebble-bed heater conditions? How would the phenomena considered in Parts (d) and (g) influence your heater design (e.g., amount of bed material required)?

5.14 a. For convective heat transfer to/from a circular *cylinder*, derive the relationship between the dimensionless coefficients $\overline{\text{Nu}}_h\{\text{Re}, \text{Pr}\}$ and $\text{Nu}_h\{\theta; \text{Re}, \text{Pr}\}$. That is, derive the counterpart of Eq. (5.3-11a) for the *cylinder* geometry.

---

[47] For reasons discussed previously, we anticipate that correcting for pressure drop will have less of an effect on the heater design parameters than correcting for *incomplete bed/gas heat exchange*.

b. In deriving this relation, was it necessary to make any assumptions about (i) the "mechanism" of convection (forced, "free," or "mixed" (force + free))? (ii) The nature of the fluid (Newtonian or non-Newtonian)? (iii) The absence of nearby cylinders? or (iv) The surface temperature distribution $T_w(\theta)$ around the cylinder?

c. At Re-values of 0.708×10^5 and 2.19×10^5, calculate the *surface-averaged* dimensionless heat-transfer coefficient $\overline{\mathrm{Nu}_h}$ (Eq. (5.3-2)) by an appropriate numerical integration of the graphical $\mathrm{Nu}_h(\theta; \mathrm{Re}, \mathrm{Pr} = 0.7)$ data plotted in Figure 5.3-3 and compare your results to those displayed in Figure 8.1-1 (Chapter 8) or "predicted" by a correlation (cf. (5.6-10)). If there are significant disparities, to what do you attribute them?

d. Using laminar boundary-layer *theory* (Sections 4.5.2 and 5.5.2), it is possible to predict the following result for the forced flow of a Newtonian fluid *at the forward stagnation "point"* (*line*):

$$\mathrm{Nu}_h(0; \mathrm{Re}, \mathrm{Pr}) \approx 0.570\left(\frac{d_w}{U}\frac{du_e}{dx}\right)_{x=0}^{1/2} \cdot (\mathrm{Re})^{1/2}(\mathrm{Pr})^{0.4} \qquad (\text{for } \mathrm{Pr} \geqslant 0.7),$$

where the velocity gradient $(du_e/dx)_{x=0}$ is given by $4 \cdot (U/d_w)$ for steady, incompressible, *inviscid* flow past a circular cylinder (Section 4.3.3). Compare this $\mathrm{Nu}_h(0; \mathrm{Re}, \mathrm{Pr})$ prediction to the measured values at the above-mentioned Re-values.

5.15 Solutions to the convective-diffusion boundary-layer equations are often based on the local use of a Taylor-series expansion for the tangential velocity near a solid body; i.e.,

$$u(x, y) = \left(\frac{\partial u}{\partial y}\right)_{y=0} y + \frac{1}{2!}\left(\frac{\partial^2 u}{\partial y^2}\right)_{y=0} y^2 + \cdots$$

a. Does this expansion of $u(x, y)$ satisfy the "no-slip" condition $u(x, 0) = 0$?

b. Show that, at high Reynolds' numbers, the coefficient $(\partial u/\partial y)_{y=0}$ can be replaced by $\tau_w(x)/\mu$, where $\tau_w(x)$ is the local shear stress $\tau_{yx}(x, 0)$.

c. Show that, at high Reynolds' numbers, the coefficient $(\partial^2 u/\partial y^2)_{y=0}$ can be replaced by $(dp/dx)/\mu$, where dp/dx is the local streamwise static *pressure* gradient.

d. For large Pr or Sc, the (heat/mass) diffusion boundary layer is thin and, sufficiently far upstream of the separation point (where $\tau_w = 0$), only the first term above is sometimes retained to derive the spatial distribution of transfer coefficient. Near separation, however, the second term may dominate, invalidating the above-mentioned approximation. Assuming that experimental values are available of the following coefficients for a cylinder of diameter d_w in cross-flow:

$$\left.\begin{aligned} C_f &\equiv \frac{\tau_w}{\frac{1}{2}\cdot \rho U^2}, \\ C_p &= \frac{(p - p_\infty)}{\frac{1}{2}\cdot \rho U^2}, \\ \mathrm{Nu}_h &\equiv \frac{h d_w}{k} \end{aligned}\right\} \quad \text{as a function of polar angle } \theta \text{ measured from the forward stagnation line.}$$

Show that the ratio of the second to the first term in the above Taylor series for $u(x, y)$, when evaluated at $y = \delta_h$, can be expressed

$$\frac{(dC_p/d\theta)}{C_f \cdot \mathrm{Nu}_h(\theta)} \qquad (\theta \text{ in radians})$$

and evaluate this ratio as a function of θ at $\mathrm{Re} = 1.7 \times 10^5$ using the data graphed in Figure 5.3-4.

e. On this basis, up to what angle would you expect transfer coefficients based on $u(x, y) = (\tau_w(x)/\mu)y$ to be valid? (Conversely, this defines the domain over which measurements of convective heat or mass-transfer coefficients might be used to infer the distribution of $C_f(x)$.)

f. Based on retention of only the leading term, the following transfer-coefficient results can be derived for a cylinder:

$$\left.\begin{matrix} \mathrm{St}_h \mathrm{Pr}^{2/3} \\ \text{or} \\ \mathrm{St}_m \mathrm{Sc}^{2/3} \end{matrix}\right\} = 0.53837\ \mathrm{Re}^{-1/3} \cdot \frac{(C_f)^{1/2}}{\left[\int_0^\theta (C_f)^{1/2}\, d\theta\right]^{1/3}}.$$

How well does this agree with the experimental *heat*-transfer data for air ($\mathrm{Pr} = 0.7$) mentioned above? Discuss.

g. Consider a flow region in which $C_f(\theta; \mathrm{Re})$ values are proportional to $\mathrm{Re}^{-1/2}$. What then is the predicted Re-dependence of the local heat/mass-transfer coefficient Nu?

h. Suppose only the pressure coefficient $C_p(\theta; \mathrm{Re})$ were available at the Reynolds' number of particular interest to you (say $\mathrm{Re} = 1 \times 10^4$). Using laminar boundary-layer theory, outline how you could estimate the distribution of skin friction $C_f(\theta; \mathrm{Re})$ appearing in the equations (cf. Part (f)) for the heat/mass-transfer Stanton numbers.

i. Why are the results of Part (f) so different from the simple "flat-plate" form of the extended Reynolds' analogy (Chilton–Colburn)? That is:

$$\frac{C_f}{2} \cong \begin{cases} \mathrm{St}_h \mathrm{Pr}^{2/3} \equiv j_h \\ \mathrm{St}_m \mathrm{Sc}^{2/3} \equiv j_m. \end{cases}$$

5.16 Figure 5.3-3 reports experimentally observed angular distributions of the local Nusselt number, $\mathrm{Nu}_h(\theta; \mathrm{Re}, 0.7)$ for an isothermal circular cylinder in cross-flow (air, $\mathrm{Ma}^2 \ll 1$, $\mathrm{Kn} \ll 1$) at Reynolds' numbers (Ud/ν-values) between 2.3×10^1 and 2.19×10^5. Of interest in a number of applications is the *fraction of the total transport that occurs on the forward-facing portion of the cylinder* ($-\pi/2 \leqslant \theta \leqslant \pi/2$). Using these data, calculate and plot the "forward fraction" at a function of Reynolds' number, Re. Estimate the Re-value at which this fraction drops to $\frac{1}{2}$. Why does an increasing fraction of the transfer occur "at the rear" as the Reynolds' number is increased? Are these observations consistent with the structure of Eq. (5.6-10), which is a particular curve-fit to *total* transfer-coefficient data for a *sphere* (cf. Figure 8.1-1)?

5.17 Select any problem in *turbulent*-flow heat- (and/or mass-)transfer in the engineering literature for which a quantitative "prediction" has been made, and answer as many of the following questions as possible:

a. How have the *turbulent* transport terms (i.e., Reynolds' fluxes and associated diffusivities) been mathematically "modeled"?

b. What experimental data have been used in the turbulent transfer law (e.g., what and how many "universal" constants have been used)? In what ways are the predicted configurations similar to the configuration for which the empirical constants have been derived?

c. What assumption has been made concerning the turbulent diffusivity *ratios* (i.e., *turbulent Prandtl* numbers)? With what justification?

d. Are there laminar regions embedded somewhere within the upstream or near-wall regions of flow field? If so, at what point has "transition to turbulence" been assumed to occur? What criterion has been used to define transition to turbulence?

e. Estimate a representative *scale* of turbulence in your problem. How does it compare to macroscopic geometric dimensions appearing in the problem? How does it compare with the prevailing intermolecular spacing or mean-free-path?

f. Can the formulation accommodate (or utilize) known upstream and/or boundary conditions on the free-stream turbulence level, or turbulence intensity?

g. Are there realizable conditions when the turbulent-transport model used would predict *no* turbulent mixing?

h. Do the eddy-transport coefficients unrealistically vanish (or "blow up") anywhere in the flow field? Is it necessary or possible to avoid this difficulty?

i. If a second-order irreversible homogeneous chemical reaction were occurring in the flow field, how might you estimate the local correlation $\overline{n'_{\mathrm{A}} n'_{\mathrm{B}}}$ appearing in the expression for the time-averaged reaction rate, $-\bar{\dot{r}}'''_{\mathrm{A}}$? Can the "intensity" of temperature and/or concentration changes be "predicted" or related to the "intensity" of velocity disturbances?

5.18 Assume that the true rate "constant" of a homogeneous chemical reaction has the Arrhenius temperature dependence: $k = A \exp\{-E/(RT)\}$ (where A

and E/R are constants).

a. If, in a stationary *turbulent* flow, the instantaneous temperature can be written $T = \bar{T} + T'$, then derive an expression for the time-averaged rate constant, $\bar{k}$, in terms of the time-averaged temperature $\bar{T}$ and the associated turbulence "intensity": $(\overline{T'^2})^{1/2}/\bar{T}$.

b. If $E = 40$ kcal/mole, estimate the amount by which $\bar{k}$ will exceed $k\{\bar{T}\}$ at a location where the turbulent temperature fluctuations of the fluid mixture amount to 5 percent and $\bar{T} = 1000$ K.

c. What is meant by local "unmixedness," and how does it affect the time-averaged chemical reaction rate?

5.19 Using the data and results of Exercise 3.14 (effective thermal conductivity of equilibrium-dissociating hydrogen gas) and Section 5.8 (use of $\Delta\phi_T$ as "driving force" to estimate *convective* energy transfer in LTCE-mixtures), estimate the rate of natural convection energy loss in Watts per cm length of 60 μm-diameter horizontal tungsten filament maintained at 3000 K in 300 K, 1 atm hydrogen. As a first approximation, assume that $\Delta H \approx$ const $= \Delta H\{3000\text{ K}\} = 109.84$ kcal/g-mole. Itemize and defend all other assumptions you introduce. Note that natural convection heat-transfer coefficients, $\overline{\mathrm{Nu}}_h$, for a horizontal *cylinder* in a nonreacting Newtonian fluid are summarized in Figure 7.2-1. (See, also, Exercise 6.13 for an alternative approach to this same problem.)

5.20 Measurements indicate that the sun radiates approximately as a black body with the maximum emission at $\lambda = 0.5$ μm (5000 Å). Moreover, the mean distance $\bar{r}_{12}$ between the Sun and Earth, together with their respective mean radii R_s, R_E are given in Figure 5.20E. Using these data, estimate:

a. The amount of energy S, of all wavelengths, incident in unit time on a unit area of a surface placed at right angles to the Sun's rays at the Sun–Earth mean distance, $\bar{r}_{12}$. (This energy flux, called the "solar constant," plays a central role in the fields of meteorology and solar energy utilization.)

b. The fraction, F_{SE}, of total solar radiation intercepted by the Earth (the Sun–Earth "view factor").

c. The Earth's expected mean radiation-equilibrium[48] temperature, T_E, assuming that (i) the average reflectivity ("albedo") of the Earth

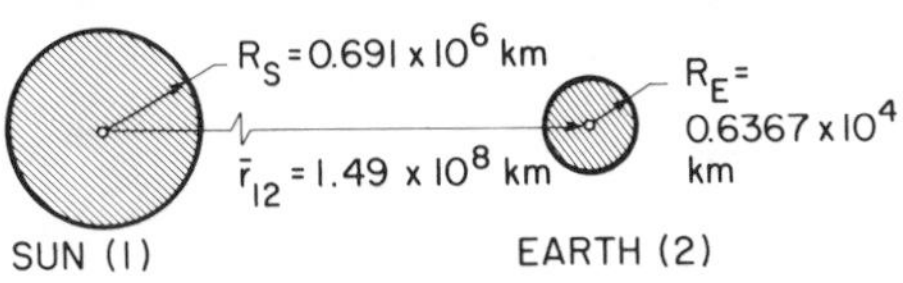

Figure 5.20E

[48] That is, the temperature at which it loses energy by radiation as rapidly as it gains energy (in this case also by radiation).

(+ atmosphere) is 0.36, and (ii) the influence of the Earth's moon is negligible.

d. In considering the design of a solar collector system for homes, what limits on required collector area may immediately be set by the value of the solar constant referred to above? (Suppose the average individual power requirement in the USA is *ca.* 5 kW.)

5.21 Microwave heating is finding increasing uses in the chemical-process industry and in food preparation (aircraft, home, restaurant, etc.). However, radiation exposure above *ca.* 10^{-2} W/cm^2 in the vicinity of such equipment is considered hazardous to human health. If we define the microwave "region" as electromagnetic radiation falling in the wavelength interval 10^{-1}-10 mm, compare the above-mentioned radiation flux threshold to that portion of the solar flux[49] falling in the same wavelength region. Use the fact that, according to the Planck black-body distribution law (cf. Eq. (5.9-1)):

$$\frac{d\dot{e}''_b}{d\lambda} \equiv \dot{e}''_{b\lambda} = 2\pi c^2 h \lambda^{-5}\left\{\exp\left(\frac{ch}{\lambda k_B T}\right) - 1\right\}^{-1},$$

where $h \equiv$ Planck's constant and λ is the wavelength.

5.22 Due to scattering and absorption processes, one can define a photon (radiation) mean-free-path, ℓ_{rad} analogous to the molecular mean-free-path in a low-density gas. Indeed, when $\ell_{\text{rad}} \ll L$, radiation can be fruitfully treated as photon "diffusion" and, as a corollary, radiation contributes to the effective thermal conductivity k_{eff} of engineering materials.

a. For a high-temperature packed bed of particles with diameter d_p and emittance ε_{rad} derive an approximate expression for this radiative contribution, k_{rad} in terms of T_p, d_p, and any other relevant quantities.

b. Inserting typical numbers for porous alumina, when would you expect radiation to contribute appreciably to the energy flux in the bed?

c. Typically, how large is ℓ_{rad} and can we assume $\ell_{\text{rad}} \ll d_p$ in the above example? (Suppose d_p = *ca.* 5 mm.)

d. Does radiation contribute to the effective thermal conductivity of porous alumina itself?

5.23 To predict the performance of a particular high-temperature axisymmetric chemical reactor, a design engineer requires not only the view factors given in Figure 5.23E, but also the view factor between an *annular* ring and a coaxial circular area at the reactor base (see Figure 5.23E)..

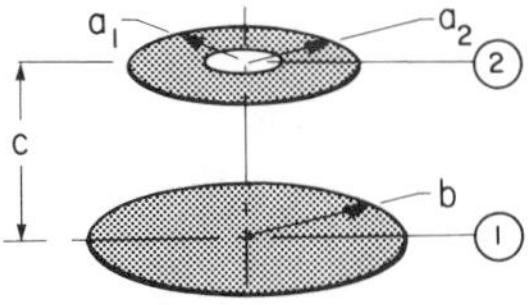

Figure 5.23E

[49] An accessible account of solar-energy applications is given in Daniels (1975).

a. Using "view-factor algebra" and "reciprocity," can you relate the necessary view factor $F_{21}\{c/b;\ a_2/c,\ a_2/a_1\}$ to the view factors $F_{12}\{c/b,\ a/c\}$ plotted in Figure 5.9-2?[50]

b. Numerically evaluate F_{21} for the particular case:

$$a_2 = 0.11\ \text{m} \qquad b = 0.14\ \text{m}$$
$$a_1 = 0.080\ \text{m} \qquad c = 0.50\ \text{m}$$

c. Calculate the net rate of interchange of radiant energy (in kW) if $T_2 = 1800$ K, $T_1 = 350$ K, and both surfaces are considered "black" ($\varepsilon_1 = 1, \varepsilon_2 = 1$).

REFERENCES

Brokaw, R.S., *J. Chem. Phys.* **32**, pp. 1005–1006 (1960); see, also, *ibid.* **26**, pp. 1636–1643 (1957).

Deissler, R.G., *NACA Report 1210* (1955).

Donaldson, C. DuP., *AIAA J.* **10**, pp. 4–12 (1972).

Eckert, E.R.G., VDI-Forschung, H. **387**, 1–20 (1937).

Eckert, E.R.G., and T.W. Jackson, *NACA RFM 50D25* (July, 1950).

Eckert, E.R.G., and E. Soehngen, *ASME Trans.* **74**, p. 346 (1952).

Fage, A., and V.M. Falkner, Aeronautical Research Council (England) Reports and Memoranda. No. 1369 (1931).

Farlow, J., C.V. Thompson, and D.E. Rosner, *Science* (*AAAS*) **192**, pp. 1123–1125 (1976).

Forstall, W., Jr., and A.L. Shapiro, *J. Appl. Mech.* **72**, pp. 339–408 (1950).

Froment, G.F., and K.B. Bischoff, *Chemical Reactor Analysis and Design*. New York: J. Wiley (1979).

Giedt, W.H., *ASME Trans.* **71**, pp. 378–381 (1949).

Heisler, M.P., *ASME Trans.* **69**, p. 227 (1947).

Kays, W.M., and A.L. London, *Compact Heat Exchangers* (Second Ed.). New York: McGraw-Hill (1964).

Launder, B.E., in *Turbulence* (Second Ed.), P. Bradshaw, ed. Berlin: Springer-Verlag, p. 231 (1978).

Lockwood, F., and N.B. Shah, in *Eighteenth Symposium* (*Internat.*) *on Combustion*, The Combustion Institute, Pittsburgh, PA, pp. 1405–1414 (1981).

Mellor, G.L., and H.J. Herring, *AIAA J.* **11**, pp. 590–599 (1973).

Morgan, G.W., and W.H. Warner, *J. Aeron. Sci.* **23**, pp. 937–948 (1956).

Newman, A.B., *Trans. AIChE* **27**, p. 310 (1931).

Norris, R.H., in "Augmentation of Convective Heat and Mass Transfer," *Proc. ASME Symposium* (A. Bergles and R. Webb, Eds.); pp. 16–26 (Dec. 2, 1970).

---

[50] For this "opposed disk" configuration ($a_2 = a_1$) $F_{12} = \frac{1}{2}[Z - (Z^2 - 4X^2Y^2)^{1/2}]$, where

$$X \equiv a/c, \qquad Y \equiv c/b, \qquad Z \equiv 1 + (1 + X^2)Y^2.$$

Verify this by predicting F_{12} for $a = 0.11$ m, $b = 0.14$ m, $c = 0.5$ m, and comparing the results to that read off the graph.

Prandtl, L., *Zeit. Physik* **29**, p. 487 (1928); see also *Zeit. Physik* **2,** p. 1072 (1910).
Richardson, P.D., *WADD TN-59-1* (1968).
Rodi, W., *The Prediction of Free Turbulent Boundary Layers by Use of a Two-Equation Model of Turbulence*, Ph.D. Dissertation, Univ. of London (1972).
Rodi, W., *AIAA J.* **20**, pp. 872–879 (1982).
Rosner, D.E., *Amer. Rocket Soc. J.* **29**, pp. 215–216 (1959).
Rosner, D.E., *Amer. Rocket Soc. J.* **31,** pp. 816–818 (1961).
Rosner, D.E., *Int. J. Heat Mass Transfer* **6**, pp. 793–804 (1963).
Rosner, D.E., in *Supersonic Flow, Chemical Processes, and Radiative Transfer*. London: Pergamon Press, pp. 439–483 (1964).
Rosner, D.E., *Comb. Sci. and Technol.* **10**, pp. 97–108 (1975).
Rotta, J.C., *Turbulente Strömungen*. Stuttgart: Teubner (1972).
Spalding, D.B., in *17th (Int.) Symposium on Combustion*, The Combustion Institute, Pittsburgh, PA, pp. 431–440 (1979).
Steiger, A., and G.K. Aue, *Proc. Inst. Mech. Engrs. 177* Part 3C, 68 (1964–1965).
Thompson, W.P., Part 7 (Heat Flux Gauges), in *Methods of Experimental Physics*, Vol. 18B, *Fluid Dynamics* (R. Emrich, ed.). New York: Academic Press (1981); pp. 663–685.
Van Meel, D.A., *Int. J. Heat Mass Transfer* **5**, pp. 715–722 (1962).
Whitaker, S., *AIChE J.* **18**, pp. 361–371 (1972).
Wolfshtein, M., D. Naot, and A. Lin, Ch. 1 in *Topics in Transport Phenomena* (C. Gutfinger, ed.). New York: J. Wiley, pp. 3–45 (1975).

BIBLIOGRAPHY

Elementary

Daniels, F., *Direct Use of The Sun's Energy*. New York: Ballantine Books (1975) (Paperback 23794).
Edwards, D.K., V.E. Denny, and A.F. Mills, *Transfer Processes*. New York: McGraw-Hill (Hemisphere); Second Edition (1979).
Giedt, W.H., *Principles of Engineering Heat Transfer*. New York: D. Van Nostrand Co. (1957).
Levenspiel, O. *Engineering Flow and Heat Exchange*. New York: Plenum Press (1984).
Welty, J.R., C.E. Wicks, and R.E. Wilson, *Fundamentals of Momentum, Heat, and Mass Transfer* (Third Ed.). New York: J. Wiley (1984).

Intermediate

Bennett, C.O., and J.E. Meyers, *Momentum, Heat, and Mass Transfer*. New York: McGraw-Hill (1972). Third edition (1982).
Bird, R.B., W. Stewart, and E.N. Lightfoot, *Transport Phenomena*. New York: J. Wiley (1960).
Eckert, E.R.G., and R.M. Drake, Jr., *Analysis of Heat and Mass Transfer*. New York: McGraw-Hill (1972).
Gray, W.G., J.K. Kilham, and R. Muller, *Heat Transfer from Flames*. Salem N.H.: Elek Science Publishers (1976).

Hottel, H.C., and A.F. Sarofim, *Radiative Transfer*. New York: McGraw-Hill (1967).
Kays, W.M., *Convective Heat and Mass Transfer*. New York: McGraw-Hill (1966).
Kern, D.Q., and A.D. Kraus, *Extended Surface Heat Transfer*. New York: McGraw-Hill (1972); see also Kraus, A.D., *Analysis and Evaluation of Extended Surface Thermal Systems*, N.W.: Hemisphere Press (1982).
Launder, R.E., and D.B. Spalding, *Mathematical Models of Turbulence*. New York: Academic Press (1972).
Leontiev, A., ed., *Théorie des Echanges de Chaleur et de Masse*. Moscow: Editions MIR (1985). (Trans. of 1979 Russian edition.)
Lienhard, J., *A Heat Transfer Textbook*. Englewood Cliffs, NJ: Prentice-Hall (1981).
Shah, R.K., and A.L. London, *Laminar Flow Forced-Convection in Ducts*. New York: Academic Press (1978).
Simonson, J.R., *Engineering Heat Transfer*. Washington, D.C.: Hemisphere Press (1975).
Sparrow, E.M., and R.D. Cess, *Radiation Heat Transfer*. New York: Brooks-Cole (1967).
White, F.M., *Heat Transfer*. Reading, MA: Addison-Wesley (1984).
Whitaker, S., *Fundamental Principles of Heat Transfer*. New York: Pergamon Press (1977).

Advanced

Carslaw, H.S., and J.D. Jaeger, *Conduction of Heat in Solids*. London: Oxford Univ. Press (1959) (Second Edition).
Cebeci, T., and P. Bradshaw, *Physical and Computational Aspects of Convective Heat Transfer*. New York: Springer-Verlag (1985).
Crank, J., *The Mathematics of Diffusion* (Second Ed.). Oxford: Oxford Univ. Press (1975).
Howell, J.R., "Application of Monte Carlo to Heat Transfer Problems," in Vol. 5, *Advances in Heat Transfer*. New York: Academic Press (1968), pp. 1–54.
Howell, J.R., *A Catalog of Radiation Configuration Factors*. New York: McGraw-Hill (1982).
Howell, J.R., and R. Siegel, "Thermal Radiation Heat Transfer," Vol. II, NASA, Office of Technology Utilization, Washington, D.C. (1969).
Leslie, D.C., *Developments in the Theory of Turbulence*. Oxford: Clarendon Press—Oxford Science Publications (1983). (Paperback edition of 1973 original.)
Libby, P.A., and F.A. Williams, eds., *Turbulent Reacting Flows*, Vol. 44, Topics in Applied Physics. Berlin: Springer (1980).
Morse, P.M., and H. Feshback, *Methods of Theoretical Physics*. New York: McGraw-Hill (1953).
Schlichting, H., *Boundary-Layer Theory*. (Seventh Ed.). New York: McGraw-Hill (1979).
Sideman, S., and W.V. Pinczewski, "Turbulent Heat and Mass Transfer at Interfaces: Transport Models and Mechanisms," Ch. 2 in *Topics in Transport Phenomena* (C. Gutfinger, ed.). Washington, D.C.: Hemisphere (1975), pp. 47–212.
Siegel, R., and J.R. Howell, *Thermal Radiation Heat Transfer* (Second Ed.). New York: McGraw-Hill (1981).
Spalding, D.B., in *Comb. Sci. Technology* (Special Issue on *Turbulent Reactive Flows* (1976)).

Additional Sources of Information

Students are encouraged to examine the following journals to become familiar with current research and trends in the general area of heat transport:

International Journal of Heat and Mass Transfer (Pergamon Press).

ASME Transactions, Series C, *Journal of Heat Transfer.*

Heat Transfer—Japanese Research.

Heat Transfer—Russian Research.

International Journal of Heat and Fluid Flow (Butterworths).

The following series of books also have useful review articles on important topics in heat transfer:

Advances in Heat Transfer (16 volumes through 1984), Academic Press (New York).

Proc. International Heat-Transfer Conferences (eight conferences through 1986), Hemisphere Publishing Corp. (Washington, D.C.).

6

Mass Transport Mechanisms, Rates, and Coefficients

6.1 RELEVANCE

6.1.1 Transport-Controlled Situations

Especially in applications where the reagents are initially separated, *mass-transfer* rates play a decisive role in determining chemical reactor behavior. For example, in many combustors (cf. Section 1.1.2), the fuel and oxidizer must "find" each other before local exoergic oxidation reactions can ensue. In fact, chemical reaction rates and associated reactor *volumes* can often be adequately predicted without *any* knowledge of the intrinsic chemical kinetics of the reagent combination—with combustion examples being spray devices (fuel droplets) and/or fuel vapor jets at atmospheric or elevated pressures (Section 1.1.2). When this is true the observed reaction rate is said to be *transport-controlled*, with transport of momentum (Chapter 4), energy (Chapter 5), *and* mass (this chapter) usually playing important roles (as shown in Section 6.5.5.5).

6.1.2 Kinetically Limited Situations

In *premixed* fuel/oxidizer systems, or even diffusion-flame situations (near extinction, or when interest is focused on pollutant emission), *chemical kinetics* plays an essential role. Yet, by virtue of the "law of mass action" governing all elementary reaction steps, local reaction rates are determined by the prevailing local reactant concentrations, reflecting the simultaneous influence of *mass* transport. Moreover, when the chemical kinetic laws are themselves obtained from independent laboratory chemical reactor experiments, the laws of mass transport are essential to interpret the data. This is especially true for "nonideal reactors" characterized by the presence of *gradients* in concentration and temperature. For example, composition profile data from a low-pressure flat flame (Figure 1.2-5) cannot be used to extract reliable chemical kinetic data without accurate corrections for the prevailing species mass fluxes (diffusion and convection).

6.1.3 "Ideal" Steady-Flow Chemical Reactors (PFR, WSR) (cf. Figure 6.1-1)

Whether intrinsic kinetics and/or transport limitations govern the local effective volumetric reaction rate $-\dot{r}_A'''$ for some key reagent A, say, the *overall reactor volume V* required to convert reagents to valuable products (or energy, etc.) is dependent both upon the apparent chemical kinetics $-\dot{r}_A'''$ and the reagent-product *contacting pattern* within the reactor. For *steady-flow* reactors there are two important limiting ("ideal") cases that are very easily understood and treated, *viz.*, the so-called "plug-flow" reactor (PFR) and the "well-stirred" reactor (WSR).

6.1.3.1 Plug-Flow Reactor (PFR)

In a PFR the reacting fluid mixture moves through the vessel (e.g., long tube) in one predominant (ξ) direction, with negligible recirculation or backmixing (e.g., negligible streamwise (ξ-direction) diffusion). In such a case the governing conservation equations can be written in their quasi-one-dimensional form (Section 4.3.1) and, in particular, we find $\dot{m} = \rho v_z A = \text{const}$, and the species A mass balance becomes the ODE:

$$\frac{\dot{m}}{A} \cdot \frac{d\omega_A}{d\xi} \equiv -j''_{A,w} \cdot \frac{P(\xi)}{A(\xi)} + \dot{r}_A'''. \qquad (4.3\text{-}5)$$

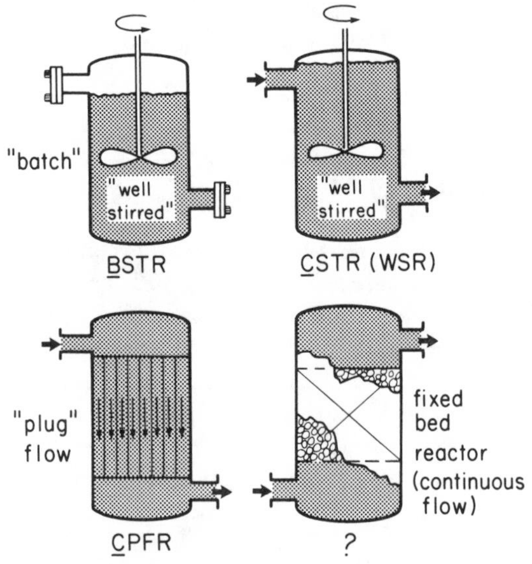

Figure 6.1-1 Basic chemical reactor types.

Here ω_A is the cross-section area-averaged reactant mass fraction[1] at streamwise station ξ, $j''_{A,w}$ is the wall diffusion flux of species A associated with possible heterogeneous (sidewall) chemical reactions, and $P\{\xi\}$ is the local "wetted" perimeter of the flow area $A\{\xi\}$. If we lump the heterogeneous term into an *effective* (*pseudo-*) *homogeneous* term:

$$-\dot{r}'''_{A,\mathrm{eff}} \equiv -\dot{r}'''_{A} + j''_{A,w} \cdot \frac{P\{\xi\}}{A\{\xi\}} \tag{6.1-1}$$

and note that $A\,d\xi = dV$ (increment in reactor volume), then Eq. (6.1-1) becomes simply:

$$\dot{m}\left(-\frac{d\omega_A}{dV}\right) = -\dot{r}'''_{A,\mathrm{eff}}. \tag{6.1-2}$$

Finally, if $-\dot{r}'''_{A,\mathrm{eff}}$ can be uniquely related to the local reagent composition ω_A (e.g., first-order reactant consumption in an isothermal reactor, etc.),[2] then Eq. (6.1-2) immediately provides the following useful result for the vessel volume required to reduce the reagent composition from $\omega_{A①}$ (feed) to $\omega_{A②}$ (reactor exit):

$$V_{\mathrm{PFR}} = \dot{m} \cdot \int_{\omega_{A②}}^{\omega_{A①}} \frac{d\omega_A}{[-\dot{r}'''_{A,\mathrm{eff}}\{\omega_A\}]}; \tag{6.1-3}$$

i.e., the required reactor volume per unit mass flow rate of feed is equal to the area under the $[-\dot{r}'''_{A,\mathrm{eff}}\{\omega_A\}]^{-1}$ *vs.* ω_A curve between the composition limits $\omega_{A②} \to \omega_{A①}$ (see Figure 6.1-2). Clearly, to reduce the reactor volume required for a specified reactant conversion it is necessary to reduce the mass "throughput" rate and/or accelerate the effective reaction kinetics.

6.1.3.2 Well-Stirred Reactor (WSR)

In contrast, for a WSR in steady flow there *is* backmixing, and it is so intense as to eradicate all internal (intra-vessel) composition nonuniformities. In that (limiting) case the chemical reaction takes place at a single composition that will be negligibly different from the *exit* stream composition, $\omega_{A②}$. The mass balance equations then simplify to: $\dot{m}_{①} = \dot{m}_{②} = \mathrm{const} \equiv \dot{m}$ and:

$$\dot{m}(\omega_{A①} - \omega_{A②}) = -\dot{r}'''_{A,\mathrm{eff}}\{\omega_{A②}\} \cdot V_{\mathrm{WSR}} \tag{6.1-4}$$

---

[1] *Transverse* nonuniformities in composition are frequently neglected in PFR models.
[2] Such a relation can also be established for adiabatic chemical reactors (no heat loss) since the resulting fluid temperature rise, $T - T_{①}$, is then still uniquely related to the change in reagent composition, $\omega_{A①} - \omega_A$.

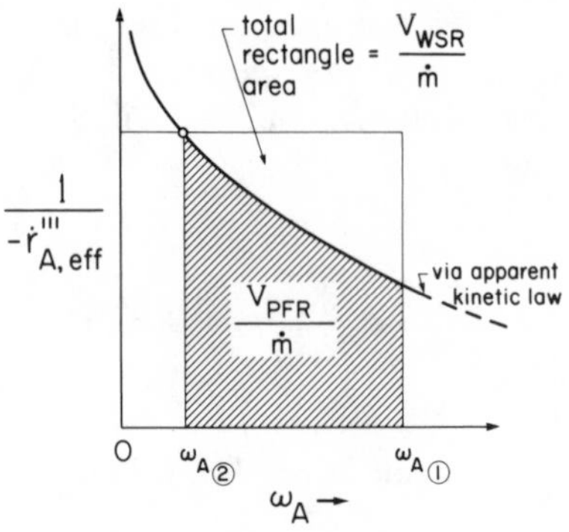

Figure 6.1-2 Reciprocal reaction rate *vs.* reagent composition plot to determine ideal reactor volume required (per unit mass flow of feed).

Therefore, in contrast to Eq. (6.1-3), we find:

$$V_{\text{WSR}} = \dot{m} \cdot \frac{\omega_{\text{A}①} - \omega_{\text{A}②}}{[-\dot{r}'''_{\text{A, eff}}\{\omega_{\text{A}②}\}]}, \tag{6.1-5}$$

which also has a simple geometric interpretation (see rectangular *area* on Figure 6.1-2). For the case illustrated ($-\dot{r}'''_{\text{A, eff}}$ increases monotonically with ω_{A}), note that $V(\text{WSR}) > V(\text{PFR})$.

It should be remarked that physical appearance alone does not betray whether a real chemical reactor more closely approximates a PFR or a WSR. Thus:

a. At sufficiently low gas pressures, a short straight tube with *axial* flow will behave like a WSR (due to molecular backmixing);
b. At higher pressures, effective mixing can also be accomplished without any mechanical agitators *via* the turbulent stirring action produced by reactant jet injection (as in portions of an aircraft gas turbine combustor (cf. Figure 1.1-10)); and
c. A *radial*-flow, thick annular bed can perform like a PFR.

Instead, inert tracer stimulus–response experiments (Section 6.7) are frequently used to diagnose the contacting pattern in a vessel.

Of course, real chemical reactors may depart appreciably from either of these two extreme cases and the *mass* transport laws discussed below, taken together with the momentum and energy transport laws of Chapters 4 and 5, will be seen to be necessary to design or improve them, or scale them up to a much larger size (e.g., production rather than laboratory). Even the effective, pseudohomogeneous reaction rate laws, $-\dot{r}'''_{\text{A, eff}}$, usually involve transport factors, especially for *multiphase* chemical reactors, as will be illustrated in Section 6.4.4 for transport to/reaction within porous pellets (containing dispersed catalyst microparticles) packed within the vessel. Often a "real" chemical reactor can be successfully modeled (represented mathematically) as a network of ideal reactors (WSRs, PFRs) suitably interconnected (with recycle streams, etc.), as will be discussed in Section 6.7. Scale-model testing and/or scale-up aspects of chemical reactors will be more fully discussed in Chapter 7.

6.2 MECHANISMS OF MASS TRANSPORT AND THEIR ASSOCIATED TRANSPORT PROPERTIES

The three *mechanisms* of mass transport, i.e.,

1. convection,
2. diffusion, and
3. free-molecular flight,

are analogous to the mechanisms of energy transport (Section 5.2). Since engineering applications are usually well into the continuum regime[3] ($\mathrm{Kn} \ll 1$), we will focus our attention here on convection and diffusion, which usually "collaborate."

6.2.1 Convection

It is conceptually useful to distinguish here between two types of convection, *viz.*, convection due to motion of the host ("carrier") fluid, and convection associated with drift of the solute through the host fluid as a result of net forces applied directly to the solute.

6.2.1.1 Host Fluid Convection

Consider a fluid mixture (moving through Eulerian space) with the local mass flux vector:

$$\dot{\mathbf{m}}'' = \rho \mathbf{v} \tag{6.2-1}$$

(also equal to the local linear-momentum density of the fluid mixture). If this mixture contains some chemical species i with local *mass fraction* ω_i, then the corresponding local *convective mass flux vector* for that species is

$$(\dot{\mathbf{m}}_i'')_{\mathrm{conv}} = \omega_i \dot{\mathbf{m}}'' = \rho_i \mathbf{v}. \tag{6.2-2}$$

---

[3] An exception would be the structure of a strong detonation wave (whose thickness is not large compared to the prevailing molecular mean-free-path).

6.2.1.2 Solute Convection ("Phoresis")[4] Resulting from Forces Acting Directly on Solute

Consider the common situation in which each local "solute" population *drifts* with a quasi-steady velocity $\mathbf{c}_i$ (relative to the prevailing local mixture velocity $\mathbf{v}$) in response to a local *applied force* (e.g., gravitational, electrostatic, etc.). Whatever the cause, such phenomena contribute the term:

$$(\dot{\mathbf{m}}_i'')_{\text{drift}} = \rho_i \mathbf{c}_i \tag{6.2-3}$$

to the total mass flux vector for that chemical species.

6.2.2 Concentration Diffusion (Fick, Brownian, or "Eddy")

In the presence of spatial gradients in local composition, small-scale "random-walk"-type motions (due to molecular impacts or "eddy" motion) will also be associated with a net "drift" of that species, contributing:

$$(\dot{\mathbf{m}}_i'')_{\text{diffusion}} \equiv \mathbf{j}_i'' \cong -D_{i,\text{eff}}\rho\,\mathbf{grad}\,\omega_i \tag{6.2-4}$$

to its flux through space. On a molecular level, this is called the *Fick diffusion flux*, after the physiologist A. Fick [1855]. For particles "suspended" in a host fluid, this is called the *Brownian diffusion flux*, after the botanist R. Brown [1827]. In a local turbulent flow this time-averaged mass flux is called the *eddy diffusion flux*, and is really the result of time-averaging the species *convective flux* (in accordance with O. Reynolds' procedure (Section 2.6.3)).

In summary, for the cases of greatest interest to us, *the total mass flux vector* for each species can therefore be written:

$$\dot{\mathbf{m}}_i'' = \omega_i \dot{\mathbf{m}}'' + \rho_i \mathbf{c}_i - D_{i,\text{eff}}\rho\,\mathbf{grad}\,\omega_i. \tag{6.2-5}$$

6.2.3 Free-Molecular "Flight"

If there were no collisions with other molecules, each molecule would travel according to its *molecular speed* across any intervening empty space. The *net* flux due to such motion would be the algebraic sum of the fluxes associated with the different directions and speeds of molecules of that species. This mechanism of *mass transport* is quite analogous to that of energy transport by photons through a nonabsorbing/nonscattering medium.

---

[4] The word element "phore" implies "bearer" or "carrier."

6.2.4 Solute Diffusivities—Real and Effective

The quantity $D_{i,\mathrm{eff}}$, defined by Eq. (6.2-4), is called the effective "mass diffusivity" of species i in the prevailing (homogeneous or pseudohomogeneous) medium. There are important situations in which it is *not* a scalar (single number), but rather a tensor defined by nine (or six independent) local numbers, because diffusion may be easier in some *directions* than others. This is true for solute diffusion in:

a. anisotropic solids (e.g., single crystals), and
b. anisotropic fluids (e.g., turbulent shear flow).

For such cases diffusion is not "down the concentration gradient" but, rather, skewed with respect to $-\mathbf{grad}\,\omega_i$. Fortunately, however, $D_{i,\mathrm{eff}}$ can often be treated as a single scalar coefficient, applying equally to mass diffusion in any direction (e.g., in a fine-grained polycrystalline metal or quiescent fluid (gas or liquid)).

Experimental values of $D_{i,\mathrm{eff}}$, as well as theories that provide useful insight about their dependence on local environmental and/or fluid dynamic parameters, are discussed in Section 3.4.4.

6.2.5 Drift Velocities due to Solute-Applied Forces

In Chapter 2 we considered each unit mass of species i to be acted upon by the force $\mathbf{g}_i$. Without specifying the *origin* of this force ($m_i\mathbf{g}_i$ on each molecule or particle mass m_i), it will clearly cause a quasi-steady drift speed $\mathbf{c}_i$ given by

$$\mathbf{c}_i = \frac{m_i\mathbf{g}_i}{f_i}, \tag{6.2-6}$$

where f_i is the prevailing *friction coefficient* (inverse of the "mobility") relating the drag force to the local slip (drift) velocity. For example, the Stokes' "no-slip" friction coefficient is simply $3\pi\mu\sigma_{i,\mathrm{eff}}$ for solutes that are large compared to the local solvent mean-free-path.

6.2.5.1 Sedimentation

When $\mathbf{g}_{i,\mathrm{eff}}$ is the *gravitational body force* (with an Archimedes correction for the weight of the displaced fluid), then $\mathbf{c}$ is called the settling or "sedimentation" speed of the particle, written $\mathbf{c}_{i,s}$.

6.2.5.2 Electrophoresis

When $\mathbf{g}_i$ is due to the presence of an *electrostatic field* and either a net charge on species (particle) i, or an induced dipole on the neutral species (particle) i, then the resulting migration is called "electrophoresis" and the drift velocity written $\mathbf{c}_{i,e}$. This is especially important in determining the motion of ions and/or charged particles in field-containing regions (e.g., electrostatic precipitator for fly-ash removal).

6.2.5.3 Thermophoresis

Heavy molecules and particles in a temperature gradient experience a force (usually in the direction from hot-to-cold, i.e., proportional to $-\mathbf{grad}(\ln T)$), which produces a net transport $\rho_i \mathbf{c}_{i,T}$ called *thermophoresis.* Thermophoretically driven mass transport is important in some gaseous systems (e.g., flame "structure" and surface combustion of hydrogen/air) and most aerosol systems (soot and ash deposition on cooled heat-exchanger surfaces (Section 8.2), sampling targets, and/or turbine blades)—(Rosner (1980); Rosner and Fernandez de la Mora (1982)).

Other causes of "phoresis" exist, but, for example, in combustion applications they are usually negligible compared to those explicitly discussed above. "Diffusiophoresis," "photophoresis," etc., discussed in treatises on aerosols, liquid electrolytes, or colloids in liquids, fall into this category.

6.2.6 Particle (or Heavy Molecule) "Slip," Pressure Diffusion, and Inertial Separation

Heavy molecule-containing flows and even dusty gas flows can be treated as "single-phase" flows only if the heavy particles follow the host fluid sufficiently closely. A criterion for this can be stated in terms of the so-called characteristic "*stopping time*," t_p, of an individual particle, compared to the characteristic acceleration/deceleration time, $t_{\text{flow}}(L/U)$, of the prevailing host fluid flow. Here t_p is the time required for an injected particle's velocity to drop by the factor e^{-1} in the prevailing viscous fluid. For a particle of intrinsic mass density, ρ_p, and diameter d_p, Stokes' drag law gives (since $t_p = m_p/f(\text{Stokes})$),

$$t_p = \frac{m_p}{f(\text{Stokes})} = \frac{\rho_p d_p^{\,2}}{18\mu}. \tag{6.2-7}$$

In any case, the *Stokes' number*, Stk, is defined by:

$$\text{Stk} \equiv \frac{t_p}{t_{\text{flow}}}; \tag{6.2-8}$$

i.e., a sort of inverse Damköhler number governing "dynamical-"(not chemical-) nonequilibrium. For, if $\text{Stk} \ll 1$, then the particles follow the host flow closely and small departures can be treated as the result of concentration diffusion, pressure diffusion, etc. For $\text{Stk} > 10^{-1}$, this "diffusion" approximation breaks down and separate momentum equations must be written for each coexisting "phase."[5]

[5] Indeed, the so-called Stefan–Maxwell laws governing *multicomponent concentration diffusion* in mixtures can be derived from *species* momentum-balance considerations (see C. Truesdell (1962)) in the limit of close momentum coupling among the drifting species.

6.3 OBTAINING CONCENTRATION FIELDS AND CORRESPONDING SURFACE-TRANSFER RATES/COEFFICIENTS

6.3.1 Relation between Concentration Fields, Wall Mass-Transfer Rates, and Mass-Transfer Coefficients

Let $\omega_i(\mathbf{x}, t)$ denote the mass fraction field for chemical species i (or particle size class i). In some cases this can be "mapped" using either:

a. "immersion" methods (e.g., very small "suction" probes), or
b. "remote" or "nonintrusive" methods (e.g., spectroscopic, or light scattering).

More often, however, measurements are made of only the corresponding *local* fluxes $\dot{m}''_{i,\mathrm{w}}$ at important boundary surfaces (e.g., the local rate of naphthalene sublimation into a gaseous stream), or perhaps only the *average flux* for the entire surface, $\dot{m}_i/A_\mathrm{w} \equiv \overline{\dot{m}''_i}$.

As in the case of heat flow, maximum use of mass-transfer data can be made by comparing the *observed* diffusional contribution to the mass fluxes, $\overline{j''_{i,\mathrm{w}}}$, to either one of the two *reference* values:

$$\dot{m}''_{i,\,\mathrm{ref}} \equiv \begin{cases} D_i\rho(\omega_{i,\mathrm{w}} - \omega_{i,\infty})/L, \quad \text{or} & (6.3\text{-}1\mathrm{a}) \\ \rho U(\omega_{i,\mathrm{w}} - \omega_{i,\infty}). & (6.3\text{-}1\mathrm{b}) \end{cases}$$

Use of the first of these defines the dimensionless *Nusselt number* for mass transport,[6] i.e.:

$$\overline{\mathrm{Nu}}_{m,i} \equiv \frac{(j_{i,\mathrm{w}}/A_\mathrm{w})}{D_i\rho(\omega_{i,\mathrm{w}} - \omega_{i,\infty})/L}. \qquad (6.3\text{-}2)$$

This dimensionless diffusional-flux parameter is widely used for quiescent systems, as well as for both forced *and* natural convection systems. The second reference mass flux (used only for *forced* convection systems) leads to the dimensionless *Stanton number for mass transport*, $\overline{\mathrm{St}}_{m,i}$, that is,

$$\overline{\mathrm{St}}_{m,i} \equiv \frac{(j_{i,\mathrm{w}}/A_\mathrm{w})}{\rho U(\omega_{i,\mathrm{w}} - \omega_{i,\infty})}. \qquad (6.3\text{-}3)$$

Here $j_{i,\mathrm{w}}$ will be an area-weighted average of the normal component of the diffusional portion of the flux, *viz.*:

$$j''_{i,\mathrm{w}} = -[D_{i,\,\mathrm{eff}}\rho\,\mathbf{grad}\,\omega_i \cdot \mathbf{n}]_\mathrm{w} \qquad (6.3\text{-}4)$$

---

[6] Often called the *Sherwood number* by American chemical engineers (after T.K. Sherwood (1903–1976)). The group $\mathrm{Pe}_m \equiv \mathrm{Re} \cdot \mathrm{Sc}$ is sometimes called the Bodenstein number.

for concentration diffusion. However, $j''_{i,w}$ is rarely *experimentally* determined in this way (e.g., it is simpler to *weigh* a naphthalene specimen before and after a short time increment!). In contrast, if it were possible to *predict* the $\omega_i(\mathbf{x}, t)$ field from the species i conservation PDE subject to appropriate IC/BCs, then Eq. (6.3-4) could be used to calculate *local* and *perimeter* (area)-averaged mass-transfer coefficients.

When the host fluid properties are highly variable, it is necessary to carefully specify where ρ and D_i appearing in $\dot{m}''_{\text{ref}}$ (Eq. (6.3-1)) are evaluated (e.g., at upstream infinity?, at the wall?, etc.). Either choice can be made, with the former being more common.

One can also discuss mass transfer in terms of *dimensional* coefficients—i.e., *the mass flux obtained per unit "driving force."* There are many such coefficients in use, some based on $\Delta\omega_i$, some based on Δn_i, and, for gases, some based on Δp_i. Accordingly, great care is needed in using such *dimensional* coefficients in making engineering predictions in any particular unit system (English, MKS, etc). For this reason, here we use only their *dimensionless* counterparts (see, also, Section 6.4.2).

When the species i is contained in the mainstream ("feed"), an additional *dimensionless mass-transfer coefficient* is in widespread use for forced convection systems, *viz.*, the so-called "*capture fraction,*" η_{cap}. This is the ratio of the actual collection rate $-\dot{m}_{i,\text{m}}$, to the rate at which that mainstream species flows through the *projected* ("frontal") area of the target, i.e.:[7]

$$\eta_{\text{cap}} \equiv \frac{-\dot{m}_{i,\text{w}}}{\rho U \omega_{i,\infty} A_{\text{w,proj}}}. \tag{6.3-5}$$

Note that when $\omega_{i,\text{w}} \ll \omega_{i,\infty}$ so that $-\dot{m}_{i,\text{w}} = -j_{i,\text{w}}$, this coefficient is closely related to the area-averaged mass-transfer Stanton number. Indeed:

$$\overline{\text{St}}_m = \left(\frac{A_{\text{w,proj}}}{A_{\text{w}}}\right) \cdot \eta_{\text{cap}}, \tag{6.3-6}$$

which provides a simple physical interpretation for $\overline{\text{St}}_m$. Note that when $\omega_{i,\text{w}} \ll \omega_{i,\infty}$, $\overline{\text{St}}_m$ will generally be smaller than $(1/2)\eta_{\text{cap}}$, since $A_{\text{w,proj}} \leqslant (1/2)A_{\text{w}}$.

6.3.2 Conservation Equations Governing the Concentration Fields, Typical Boundary Conditions, and Outline of Solution Methods

The concentration fields $\omega_i(\mathbf{x}, t)$ $(i = 1, 2, \ldots, N)$ are generally *coupled* by the facts that:[8]

---

[7] It should be remarked that capture efficiencies are *not* limited to 100 percent! This is easy to see by considering capture on a single thin plate aligned with the approach flow. Such a plate would present virtually no projected area but, if sufficiently long, could capture mass at an appreciable rate.

[8] In nondilute systems with three or more components, the *diffusion flux laws* may also couple these PDEs.

a. homogeneous reaction rates involve many local species—that is, $\dot{r}_i'''$ depends on chemical species compositions *other* than ω_i alone;
b. all of the local mass *fractions* must sum to unity; hence only $N - 1$ such equations are actually independent of the equation of *total mass* conservation (6.3-8, below).

Irrespective of the Eulerian coordinate system used, the local species i mass-conservation (balance) condition can be written:

$$\frac{\partial \rho_i}{\partial t} + \text{div}(\dot{\mathbf{m}}_i'') = \dot{r}_i''' \qquad (i = 1, 2, \ldots, N), \tag{6.3-7}$$

where $\dot{\mathbf{m}}_i''$ is given by Eq. (6.2-5) and $\dot{r}_i'''$ is the local mass rate of production of chemical species i *via* homogeneous chemical reactions (cf. Eq. (2.1-7)). Since $\rho_i = \rho\omega_i$ and, by virtue of total mass conservation:

$$\frac{\partial \rho}{\partial t} + \text{div}(\dot{\mathbf{m}}'') = 0, \tag{6.3-8}$$

Eq. (6.3-7) can be rewritten in the revealing form:

$$\rho\left(\frac{\partial \omega_i}{\partial t} + \mathbf{v} \cdot \mathbf{grad}\,\omega_i\right) = -\text{div}(\mathbf{j}_{i,\text{diff}}'' + \rho\omega_i\mathbf{c}_i) + \dot{r}_i''', \tag{6.3-9}$$

where the left-hand side is proportional to the (Lagrangian) rate of change of ω_i *following a fluid parcel.* On the right-hand side (RHS), the flux $\rho\omega_i\mathbf{c}_i$ is sometimes considered the forced "diffusion" flux, that is, $\mathbf{j}_{i(\text{forced})}''$. We conclude that, *if these "diffusion" fluxes are zero (or spatially constant) and/or* $\dot{r}_i''' = 0$, ω_i *will stay constant following each fluid parcel (e.g., in steady flow along a streamline).*

These PDEs for the $\omega_i(\mathbf{x}, t)$ are not only coupled to each other, but also coupled to the PDEs governing the linear momentum density field, $\rho\mathbf{v}(\mathbf{x}, t)$—(Chapters 2 and 4)—and the temperature field $T(\mathbf{x}, t)$—(Chapter 5)—which strongly influences functions of local thermodynamic state, and, especially, the magnitude of $\dot{r}_i'''$ (*via* the Arrhenius chemical rate constants). Consequently, all fields must usually be obtained *simultaneously*—i.e., they must be self-consistent.

The simplest possible PDE governing a composition field, say, $\omega_A(\mathbf{x}, t)$, is the (Laplace) equation:

$$\text{div}(\mathbf{grad}\,\omega_A) = 0, \tag{6.3-10}$$

(sometimes abbreviated: $\nabla^2\omega_A = 0$), which holds when there are no:

a. transients ("accumulation"),
b. flow (convection) effects,

c. variations in fluid properties ($D_A \rho$, etc.),
d. homogeneous chemical reactions involving species A,
e. "forced diffusion" effects (electrophoresis,...).

Physically, this second-order linear PDE merely states that the net outflow (per unit volume) of species A mass *via* Fick (Brownian) diffusion must vanish at each point in the medium (cf. Eq. (5.3-13) for the *temperature* field under similar conditions). In cartesian coordinates (x, y, z), Eq. (6.3-10) takes the familiar form:

$$\frac{\partial^2 \omega_A}{\partial x^2} + \frac{\partial^2 \omega_A}{\partial y^2} + \frac{\partial^2 \omega_A}{\partial z^2} = 0. \qquad (6.3\text{-}11)$$

Boundary conditions (BCs) on the $\omega_i(\mathbf{x}, t)$ are usually one of two types; i.e., one specifies either:

1. ω_i (e.g., LTCE value) along each boundary surface, or
2. some interrelation between the flux $\dot{m}''_{i,w}$ and $\omega_{i,w}$ (e.g., *via heterogeneous* chemical kinetics).

A simple example of type 2 will be considered in Section 6.5.3.

Solution methods are either *numerical* or *analytical*, "exact" or approximate, as already outlined for the analogous case of the energy-transfer PDE. Indeed, in certain simple cases (see Section 6.3.3) previously discussed *solutions* can be carried over to their *mass-transfer* counterparts.

6.3.3 Analogies between Mass and Energy Transfer and the Processes that "Break" These Analogies

We have discussed (Chapter 5) many *energy-transfer* problems with and without convection, which are now quantitatively well understood based on available experiments and/or computations. For the most part these are for HAC situations[9] in which:

1. The fluid composition is spatially uniform (no homogeneous chemical reaction, no mass transfer, etc.);
2. The boundary conditions are simple, e.g., $T_w = \text{const}$;
3. The fluid properties (ρ, μ, c_p, k, etc.) are nearly constant;
4. All volumetric heat sources, including viscous dissipation (cf. Section 3.2.4) and chemical reaction (cf. Eq. (5.3-12)) are negligible;

[9] HAC stands for *heat*-transfer analogy condition.

and, in most of these cases we know the *coefficients* $\overline{\text{St}_h}$ or $\overline{\text{Nu}_h}$ (as a function of the key dimensionless parameters Re, Pr, Ra_h, etc.) as well as, perhaps, the corresponding temperature field:

$$\frac{T_w - T}{T_w - T_\infty} \equiv T^*\{\mathbf{x}^*, t^*; \text{Re}, \text{Pr}, \ldots\}. \tag{6.3-12}$$

Now it is interesting and important to notice that, subject to the "analogy conditions" stated as (1)–(4) below, the rescaled chemical species *concentration* variable:

$$\frac{\omega_{A,w} - \omega}{\omega_{A,w} - \omega_{A,\infty}} \equiv \omega^*\{x^*, t^*; \text{Re}, \text{Sc}, \ldots\}, \tag{6.3-13}$$

and the corresponding coefficients $\overline{\text{St}_m}$, or $\overline{\text{Nu}_m}$ will be identical functions of their respective arguments. Reference to Eq. (6.3-9) and its BCs reveals that the important MAC-conditions are:

1. Species A concentration is dilute ($\omega_A \ll 1$);
2. $\omega_{A,w}$ specified constant along surface;
3. Negligible forced diffusion (phoresis);
4. No homogeneous chemical reaction.

When (1) is satisfied, the net mass transfer at (subliming, or dissolving) surfaces of substance A is small enough so that the condition $(\mathbf{v} \cdot \mathbf{n})_w = 0$ is approximated even in a problem with net mass transfer. Condition 2 ensures that the boundary conditions on T^* and ω^* will be the same (in dimensionless form). This remarkably useful *analogy between mass and heat transfer* holds even where:

a. the fluid flow is itself "caused" by the transfer process (i.e., *natural convection* in a body force field), and
b. the analogy to linear *momentum* transfer breaks down due to the presence of streamwise pressure gradients.

This will be discussed further in Sections 6.5.1 and 7.2.

Subject to these conditions we need not reproduce for mass transfer the discussions of Chapter 5 for heat transfer. In brief, the Schmidt number $\text{Sc} \equiv \nu/D_A$ plays the same role in mass transfer as does Pr for heat transfer.[10] Moreover, the mass-transfer analog of the *Rayleigh number*, Ra_h, is:

$$\text{Ra}_m \equiv \frac{g\beta_\omega(\omega_{A,w} - \omega_{A,\infty})L^3}{\nu^2} \cdot \frac{\nu}{D_A} \equiv \text{Gr}_m \cdot \text{Sc}. \tag{6.3-14}$$

[10] Indeed, there is considerable justification for calling the diffusivity ratio ν/D_A the *mass transfer Prandtl number*, Pr_m (Russian practice).

Here β_ω defines the dependence of local fluid density on the composition variable ω_A; that is,

$$\beta_\omega \equiv \frac{1}{v}\left(\frac{\partial v}{\partial \omega_A}\right)_{p,T} = -\frac{1}{\rho}\left(\frac{\partial \rho}{\partial \omega_A}\right)_{p,T}, \tag{6.3-15}$$

and we assume $|\beta_\omega(\omega_{A,w} - \omega_{A,\infty})| \ll \rho_\infty$.

Actually, there are cases other than those described above for which we can also establish valuable relationships between $T^*\{\mathbf{x}^*, t\}$ and some composition variable. This includes situations in which:

a. Species A is *not* dilute, so that $(\mathbf{v}\cdot\mathbf{n})_w$ may be nonzero (equivalent to "blowing" or "suction" at the wall), corresponding to a nonzero *convective* contribution to *both* heat and mass transfer *at the wall* (Section 6.4.3).
b. Two chemical species A and B (e.g., fuel and oxidizer) may *react* in some fixed (stoichiometric) ratio, releasing a fixed amount of heat for each unit mass of species A (say) concerned. Then, if $D_{A,\text{eff}} \approx D_{B,\text{eff}} \approx \alpha_{\text{eff}} \equiv [k/(\rho c_p)]_{\text{eff}}$, a relation between temperature and chemical compositions can be established *irrespective of the chemical kinetic law* (Schvab and Zeldovich [1938]; cf. Section 6.5.5).

For the present, however, we exploit analogy conditions (1)–(4) above, and the simple mass-heat-transfer analogies that result. When AC conditions (1)–(4) are *not* met, for either *h*eat or *m*ass transfer, the above-mentioned analogies break down, and should not be used (even for estimates) without further justification or the corrections discussed below.

6.3.4 Mass-Transfer Coefficient Correction Factors to Account for Analogy-Breaking Phenomena

Two commonly occurring mass transport phenomena, which break the above-mentioned analogy, are *forced diffusion* (e.g., sedimentation, electrophoresis, or thermophoresis) and *homogeneous chemical reaction* (i.e., frequently, these have no counterpart in the energy equation for $T\{\mathbf{x}, t\}$).

Consider these phenomena in a BL-situation in which some substance A, present in the mainstream at $\omega_{A,\infty}$, is being transported *to the wall*, where $\omega_{A,w} \ll \omega_{A,\infty}$.

a. Phoresis toward the Wall. Phoresis caused by one or more of the above-mentioned phenomena will distort the concentration profiles and *increase* the wall diffusional flux[11] $-j''_{A,w}$. In such a case we can write:

$$\text{Nu}_m = F\{\text{suction}\}\cdot \text{Nu}_{m,0}, \tag{6.3-16}$$

[11] If $\omega_{A,w} \neq 0$, phoresis also adds a drift contribution $(\rho\omega_A)_w \mathbf{c}_A \cdot \mathbf{n}$ to the wall flux (cf. Eq. (6.2-3)).

where $\mathrm{Nu}_{m,0}$ is the corresponding mass-transfer coefficient without the phoretic enhancement ($\mathrm{Nu}_{m,0}$ and Nu_h *are* analogous) and $F\{\text{suction}\}$ is the "*suction augmentation factor*" associated with phoresis. We show below that $F\{\text{suction}\}$ is a simple function of a Peclet number based on the drift speed $-c$, the boundary-layer thickness $\delta_{m,0}$, and the diffusion coefficient D_{A}, *viz.*:

$$\mathrm{Pe}_{\text{suction}} \equiv \left[\frac{(-c)\delta_{m,0}}{D}\right]_{\mathrm{A}} = \left[\frac{(-c)L}{D\,\mathrm{Nu}_{m,0}}\right]_{\mathrm{A}}, \tag{6.3-17}$$

or, equivalently:

$$\mathrm{Pe}_{\text{suction}} = \frac{(-c)}{U} \cdot \frac{1}{\mathrm{St}_{m,0}}. \tag{6.3-18}$$

Indeed, in most cases:

$$F\{\text{suction}\} \cong \frac{\mathrm{Pe}_{\text{suction}}}{1 - \exp\{-\mathrm{Pe}_{\text{suction}}\}}. \tag{6.3-19}$$

This relation will be applied in Chapter 8 to the estimation of thermophoretically augmented small-particle mass-transfer rates to heat-exchanger surfaces immersed in hot, flowing combustion products.

b. Homogeneous Reaction within BL. If chemical species A is produced, say, by a homogeneous first-order reaction within the BL, then the ω_{A}-profile will be distorted,[12] again increasing the diffusional flux $-j''_{\mathrm{A,w}}$ at the wall. In this case we introduce a similar correction factor, $F\{\text{reaction}\}$, such that:

$$\mathrm{Nu}_m = F\{\text{reaction}\} \cdot \mathrm{Nu}_{m,0} \tag{6.3-20}$$

or

$$\mathrm{St}_m = F\{\text{reaction}\} \cdot \mathrm{St}_{m,0}. \tag{6.3-21}$$

It is easy to show that $F\{\text{reaction}\}$ depends primarily on the *Damköhler* (Hatta) *number*:

$$\mathrm{Dam} \equiv \frac{t_{\text{diff}}}{t_{\text{chem}}} = \frac{(\delta_{m,0}^2/D)}{(1/k''')} = \left(\frac{k'''\delta_{m,0}^2}{D}\right)_{\mathrm{A}}, \tag{6.3-22}$$

where k''' is the relevant *first-order rate constant* (with the dimensions of $(\text{time})^{-1}$).

[12] Note that this is not the case for the chemical-*element* mass fractions since, irrespective of the complexity of the chemical kinetics, element mass fractions satisfy source-free PDEs (Section 2.1.3).

This parameter can be rewritten using[13]

$$\delta_{m,0} = \frac{L}{\mathrm{Nu}_{m,0}} = \frac{D}{U\,\mathrm{St}_{m,0}}. \tag{6.3-23}$$

For an irreversible reaction, with $\omega_{\mathrm{A,w}} \ll \omega_{\mathrm{A},\infty}$:

$$F\{\text{reaction}\} \cong \begin{cases} (\mathrm{Dam})^{1/2}/\sin(\mathrm{Dam})^{1/2} & \text{(source)}, \\ (\mathrm{Dam})^{1/2}/\sinh(\mathrm{Dam})^{1/2} & \text{(sink)}, \end{cases} \tag{6.3-24a,b}$$

which has the property $F\{\mathrm{Dam}\} \to 1$ when $\mathrm{Dam} \to 0$.

In the chemical engineering literature, such *reaction "augmentation" factors* are sometimes called *Hatta* factors, after the investigator, S. Hatta [1932]. Physically, $F\{\text{reaction}\}$ can be regarded as the BL thickness ratio: $\delta_{m,0}/\delta_m$, where δ_m has, in effect, been altered by the presence of the homogeneous reaction (see, also, Sections 6.4.2 and 6.5.5.7).

We remark that if only *heterogeneous* chemical reactions occur, then we can often assume $\mathrm{Nu}_m = \mathrm{Nu}_{m,0}$. This does not mean that the wall mass flux is uninfluenced by a Damköhler parameter *relevant to the wall reaction*, only that the mass-transfer *coefficient* is approximately unaltered, and (in the absence of phoresis) obtainable *via* the above-mentioned mass-heat-transfer analogy (see Sections 6.5.3 and 6.5.4 for surface-catalyzed reactions).

6.4 CONCENTRATION DISTRIBUTIONS AND SURFACE MASS-TRANSFER COEFFICIENTS FOR QUIESCENT MEDIA

Transient or steady-state mass transfer through quiescent media occurs in many engineering applications. The condition of negligible convection, which now becomes:

$$(\mathrm{Re} \cdot \mathrm{Sc})^{1/2} \ll 1, \tag{6.4-1}$$

may be more difficult to achieve in *condensed fluid phases* because of the largeness of Sc (D_i much smaller than ν). A combustion example would be the evolution of composition *inside* a viscous, *multicomponent fuel droplet* after the liquid fuel droplet has been decelerated to nearly the local gas velocity. Another example is that of oxygen gas transport to the outer surface of a small solid carbon particle that is motionless with respect to the surrounding gas. In the latter case we can assume

[13] Incidentally, this equation also shows that $\mathrm{St}_{m,0}$ can be interpreted as $(U\delta_m/D)^{-1}$, i.e., the *inverse* of the Peclet number based on boundary-layer thickness.

$\overline{\mathrm{Nu}}_m \approx 2$ (cf. Eq. (5.4-22)) since conditions (6.4-1) and its natural convection analog (5.4-9) are usually satisfied. Our discussion in Section 6.4 will emphasize the parallels and differences with respect to Section 5.4, dealing with *heat* transfer in quiescent media.

6.4.1 Criteria for Quiescence

Based on Section 5.4, one would expect that two sufficient conditions to neglect convection are (6.4-1) and:

$$(\mathrm{Ra}_m)^{1/4} \cdot (\mathrm{Sc})^{1/4} \ll 1. \qquad (6.4\text{-}2)$$

However, when there is simultaneous heat transfer,[14] the natural *convective velocity* may still be due to $\beta_T \Delta T$; hence, it is also necessary that:

$$(\mathrm{Ra}_h)^{1/4} \cdot \left(\frac{\mathrm{Sc}}{\mathrm{Le}}\right)^{1/4} \ll 1. \qquad (6.4\text{-}3)$$

These conditions are, however, *not sufficient* because in nondilute systems the mass-transfer process itself can give rise to a convection *normal to the surface*, which has nothing to do with natural convection. This is easiest to see by considering a surface transferring chemical species A into a fluid with $\omega_{A,\infty}$. If the carrier fluid (B) does not penetrate the surface, then there must be a *convective outflow*—sometimes called the "Stefan flow" (or "blowing") sufficient to overcome the inevitable Fick influx of species B. This condition leads immediately to the estimate:

$$B_m \equiv \frac{v_w \delta_m}{D} = \frac{\omega_{A,w} - \omega_{A,\infty}}{1 - \omega_{A,w}}. \qquad (6.4\text{-}4)$$

Note here that v_w is the *fluid* velocity at the interface, not the interface velocity itself. Hence, an additional condition for the neglect of convection in mass transport systems is:

$$\frac{v_w \delta_m}{D} = \frac{\omega_{A,w} - \omega_{A,\infty}}{1 - \omega_{A,w}} \ll 1. \qquad (6.4\text{-}5)$$

In dilute systems, for which both $\omega_{A,w}$ and $\omega_{A,\infty}$ are small compared to unity, this condition is inevitably met. However, if $\omega_{A,w} \neq \omega_{A,\infty}$ and $\omega_{A,w} \to 1$, this Stefan-convection normal to the surface becomes extremely important (e.g., at surface

[14] For the treatment of natural convection caused by *multiple driving forces*, see Ostrach (1980).

temperatures approaching the boiling point of a liquid fuel surface). The effect of nonnegligible B_m on mass-(and energy-) transfer coefficients is discussed in Section 6.4.3.

6.4.2 Composite Planar Slabs

For a single quiescent layer with constant properties $\mathrm{Nu}_m = 1$, corresponding to a *linear* $\omega_A(x)$ profile (cf. Section 5.4.3). However, Eq. (5.4-14a) for a *composite* wall does not carry over to the mass-transfer case without the following essential modification: whereas T is continuous in going from layer to layer, the mass fraction ω_A *is not*, i.e., there are two unknown steady-state concentrations at each interface, not one (see Figure 6.4-1; what *is* continuous is the chemical potential or thermodynamic "activity" of species A). Thus, the linear-diffusion laws must be reformulated using a *continuous* concentration variable in order to be able to use the additive resistance concept, together with an overall driving force. This, of course, applies to nonplanar composite geometries as well (cf. Table 5.4-1).

Using the equivalent[15] α-phase mass fraction of solute A as the continuous composition variable, the solute A *mass flux* through a composite solid or quiescent liquid "membrane" can be expressed:

$$j''_{A,x} = \frac{(\Delta\omega_A)^{(\alpha)}_{\text{overall}}}{\sum_\ell \left(\frac{L}{D_A \rho}\right)_\ell \cdot \kappa_{\alpha,\ell}} \equiv K_A^{(\alpha)} \cdot (\Delta\omega_A^{(\alpha)})_{\text{overall}}. \tag{6.4-6}$$

Here the $\kappa_{\alpha,\ell}$ are the dimensionless *equilibrium solute* A "*partition*" *coefficients*, $(\omega_A^{(\alpha)}/\omega_A^{(\ell)})_{\text{LTCE}}$, between phase α and phase $\ell\,(=\alpha, \beta, \gamma, \delta, \ldots)$, and each $\kappa_{\alpha,\ell}$ is assumed to be concentration-independent.[16] As in the case of Eq. (5.4-14), external *fluid* films can be included as participating resistances by assigning to them effective stagnant film thicknesses, $\delta_{m,\text{eff}}$.

Dilute solute steady-state diffusional transfer between *two* contacting but immiscible *fluid* phases α, β occurs frequently in separation devices (e.g., "extraction" of a solute A from solvent β to α). The transport of solute A per unit area of $\alpha\beta$ interface is often modeled as though there were two *equivalent* "*stagnant*" *films* of thicknesses $\delta^{(\alpha)}_{m,\text{eff}}$ and $\delta^{(\beta)}_{m,\text{eff}}$, respectively, in *series*, with negligible interfacial resistance between them. In such a "two-film" theory (Lewis and Whitman [1924]), Eq. (6.4-6) applies, and the *overall* interphase mass-transfer coefficient (conduc-

[15] Thus $(\Delta\omega_A^{(\alpha)})$ overall can be regarded as the total species A mass fraction difference if the solute concentrations across the composite "membrane" were measured locally with an α-phase "equilibration" probe.

[16] Graphical or numerical methods can be used to treat problems characterized by non*linear* interfacial equilibrium relationships (Treybal (1968)). In such cases, "resistance additivity" (Eqs. 6.4-6 and -7) does *not* apply.

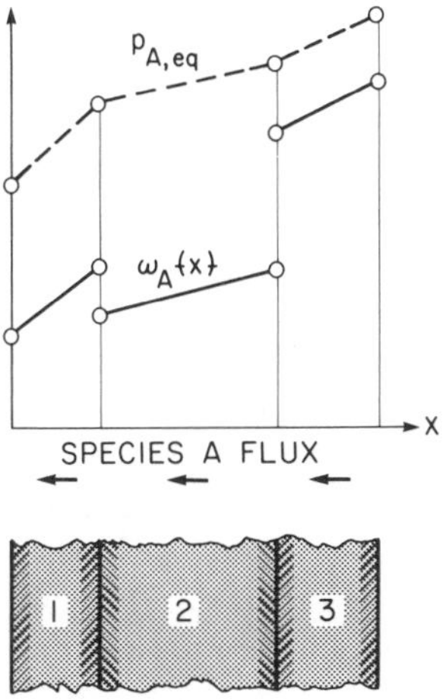

Figure 6.4-1 Mass transfer of substance A across a composite barrier: effect of piecewise discontinuous concentration (e.g., mass fraction $\omega_A\{x\}$).

tance), $K_A^{(\alpha)}$, simply satisfies the "additive-resistance" equation:

$$\underbrace{\frac{1}{K_A^{(\alpha)}}}_{\substack{\text{Overall}\\ \text{mass transfer}\\ \text{resistance}}} = \underbrace{\left[\frac{1}{(D_A\rho/\delta_{m,\,\text{eff}})^{(\alpha)}}\right]}_{\substack{\alpha\text{-phase}\\ \text{resistance}}} + \underbrace{\left[\frac{\kappa_{\alpha\beta}}{(D_A\rho/\delta_{m,\,\text{eff}})^{(\beta)}}\right]}_{\substack{\beta\text{-phase}\\ \text{resistance}}}. \qquad (6.4\text{-}7)$$

Note that, by making the symmetrical replacements $\alpha \to \beta$, $\beta \to \alpha$, Eq. (6.4-7) becomes an equally valid equation for $K_A^{(\beta)} \equiv j''_{A,x}/(\Delta\omega_A)^{(\beta)}_{\text{overall}}$.

In gas absorption (or the inverse operation, "stripping"), one of these two phases (say β) is the *vapor* phase and the relevant partition coefficient, $\kappa_{\alpha\beta}$, is inversely proportional to the so-called Henry constant, H, defined by

$$p_A = H \cdot (x_A)_{\text{solvent}} = H \cdot [(\omega_A M)_{\text{solvent}}/M_A],$$

where M is the solvent molecular weight and p_A, the partial pressure of species A in the vapor phase. Indeed, H can be regarded as a particular *dimensional* inverse partition (distribution) coefficient. In our notation:

$$\kappa_{\alpha\beta} = \left[\rho^{(\beta)}\frac{RT}{M^{(\alpha)}}\right] \cdot \frac{1}{H} \qquad (6.4\text{-}8)$$

if the β-phase (vapor mixture) obeys the perfect gas law.

In Eq. (6.4-7), each group $(D_A\rho/\delta_{m,\,\text{eff}})$ will be recognized as a *dimensional mass-transfer coefficient* (Section 6.3.1) for that phase, based on *solute mass fraction* as the choice of "driving force." When one of the contacting phases is a gas, chemical engineers most often use partial pressure (p_A-) difference as the "driving force" for a *molar* flux J''_A; that is, $K_A^{(g)}$ is defined as $J''_A/(\Delta p_{A,\text{eff}})_{\text{overall}}$.

Still other units have been used to describe the completely analogous rate phenomenon of gas dissolution and diffusion through rigid (or quiescent fluid) "membranes." While Henry's (linear) law may no longer describe the local gas/condensed phase equilibrium relation, the equilibrium solubility, S_i, of a vapor i in a condensed material is often expressed in terms of the volume (m^3 at STP (0°C, 1 atm)) of solute gas i per unit volume (m^3) of solvent per unit (atm) imposed partial pressure of gaseous species i. Thus, a quantity frequently used to characterize diffusional mass flux across a solvent layer is the *permeability* D_iS_i. When combined with thickness and partial pressure (difference) information, permeability data allow the calculation of diffusion fluxes provided local equilibrium can be assumed at each phase boundary. The student should distinguish between this permeability and the (unrelated) "permeability" used to characterize the flow of viscous fluids through porous media (Section 4.7).

Mass-transfer experiments performed on actual high-volumetric-efficiency contacting equipment (packed columns, spray towers, stirred bubbler tanks, etc.) can usually be interpreted to yield the *product* $K_A^{(\alpha)} a'''_{\alpha\beta}$, where $a'''_{\alpha\beta}$ is the effective (time-average) $\alpha\beta$ *interfacial area per unit volume of contactor*. This is a "pseudo-homogeneous" (volumetric) overall interphase mass-transfer coefficient. The interfacial (per unit area) coefficient $K_A^{(\alpha)}$ itself can be inferred only if an independent method for determining $a'''_{\alpha\beta}$ is invoked.

As already discussed (Section 6.3.4), the effective values of $\delta_{m,\text{eff}}^{(\alpha)}$ can often be dramatically reduced by the addition of suitable reagents to the solvent phase α, that is, simultaneous homogeneous chemical reaction increases liquid-phase mass-transfer coefficients, and thereby accelerates the rate of uptake of sparingly soluble (large H) gases (for which the liquid phase resistance is often "rate-limiting."[17]

If this additive "B" is present in sufficient excess that the homogeneous reaction in the solvent phase (α) can be considered pseudo-first-order (that is, $-\dot{r}'''_A$ is linearly proportional to the local concentration $\rho\omega_A$, with reaction rate constant k'''), then one can show that Eq. (6.4-7) still describes the overall mass-transfer coefficient $K_A^{(\alpha)}$, but now:

$$\delta_{m,\text{eff}}^{(\alpha)} = \frac{\delta_{m,\text{eff};0}^{(\alpha)}}{F\{\text{reaction}\}}, \tag{6.4-9}$$

where, in this case

$$F\{\text{reaction}\} = \frac{(\text{Dam})^{1/2}}{\tanh\{\text{Dam}\}^{1/2}} \tag{6.4-10}$$

[17] A commercially important example is the enhanced *absorption* of CO_2(g) into an aqueous solution of sodium carbonate (or potassium carbonate), according to the overall reaction: Na_2CO_3(ab) + CO_2(ab) + $H_2O(\ell)$ = $2NaHCO_3$(ab) (see, e.g., Sherwood, T.K., and R.L. Pigford (1952)). In some chemical processes the "solvent" is itself one reactant, as in the direct sulfonation of organic liquids by SO_3(g)—a step in the large-scale production of detergents.

and, as before:

$$\text{Dam} \equiv \frac{(\delta^2_{m,\,\text{eff};\,0}/D^{(\alpha)}_{\text{A}})}{(1/k''')}. \tag{6.3-22}$$

Cases in which depletion of the reactive additive (B) must be considered often require a numerical treatment due to the nonlinear coupling *via* the reaction rate $\rho^2 k''' \omega_{\text{A}} \omega_{\text{B}}$ appearing in each of the solvent-phase conservation equations for solute species A and B. A simple but important limiting case arises when the homogeneous chemical reaction is sufficiently rapid that the two reagents meet in their stoichiometric ratio at a thin reaction zone ("sheet"). In that case the distance between this reaction zone and the phase boundary plays the role of $\delta^{(\alpha)}_{m,\,\text{eff}}$ and we find:

$$F\{\text{reaction}\} \rightarrow 1 + \frac{1}{b} \cdot \left[\frac{D_{\text{B}}}{D_{\text{A}}} \cdot \frac{\omega_{\text{B},b}}{\omega_{\text{A},i}}\right]^{(\alpha)},$$

where $\omega_{\text{B},b}$ is the concentration of additive B in the bulk of the solvent, $\omega_{\text{A},i}$ is the concentration of transferred solute A at the solvent interface, and the reaction stoichiometry is such that b gm of species B are consumed with each gram of solute species A. The reader will recognize this situation as the isothermal analog of the "diffusion flame" discussed in Sections 1.1.3 and 6.5.5.7 (see Rosner (1971)).

A commercially important, but geometrically rather complex, example of multiple mass-transfer resistances in series occurs in three-phase "trickle-bed" reactors (e.g., used for the upgrading [desulfurization] of liquid fuel oils (cf. Problem 6.2) in which no liquid phase is present). In such devices the oil to be treated trickles down a solid packing (fixed bed) comprised of small, porous, catalyst-containing particles, while hydrogen-containing gas flows upward (as a continuous phase) through the bed. At any location within such a column the reactant hydrogen must pass from the *gas* phase, *through* the films of *liquid* fuel, and into the porous *solid* catalyst, where chemical reaction ultimately occurs with the sulfur-containing species adsorbed on the active catalyst. Conversely, the reaction product, H_2S, must then escape from the porous solid (where it was generated), through the liquid fuel, and into the gas phase (from which it is ultimately separated by selective condensation or absorption downstream of the reactor).

In treating the performance of interphase contacting equipment (e.g., steady-flow separators, multiphase reactors, etc.), it should be remembered that even very thin diffusion boundary layers in such *fluids* are not really "stagnant films" because of time-dependent interfacial distortion, local turbulence, flow development, etc. Hence $\delta^{(\alpha)}_{m,\,\text{eff}}$, $\delta^{(\beta)}_{m,\,\text{eff}}$, and their associated dimensional mass-transfer coefficients (Eq. (6.4-7)) are found to depend strongly on fluid-dynamical parameters (Reynolds' number, α/β phase flow-rate ratios, etc.) and also on a fractional power of the molecular diffusivity ratio $\nu/D_{\text{A}} \equiv \text{Sc}_{\text{A}}$ for the phase in question (see Section 6.5). In the remainder of this chapter we discuss the determination and use of *single-phase* mass-transfer coefficients.

6.4.3 Stefan Flow Effects on Mass- and Energy-Transport Coefficients

6.4.3.1 Quasi-Steady (QS)-State Diffusion Outside of an Isolated Sphere

We have already seen that in the completely quiescent case ($(\mathrm{Re}\cdot\mathrm{Sc})^{1/2} \ll 1$, $(\mathrm{Ra}_m)^{1/4} \ll 1, B_m \ll 1$) diffusional mass transfer to/from a sphere occurs at a rate corresponding to $\mathrm{Nu}_m = 2$. However, if $B_m \equiv v_w \delta_m / D$ is *not* negligible[18] the spherically symmetric problem described in the title does not lead to $\mathrm{Nu}_m = 2$ (cf. Eq. (5.4-22)), but rather:

$$\mathrm{Nu}_m = 2\frac{\ln\{1 + B_m\}}{B_m}, \tag{6.4-11}$$

where the reduction factor:

$$F_m\{\text{Stefan ``blowing''}\} \equiv \frac{\mathrm{Nu}_m}{\mathrm{Nu}_{m,0}} = \frac{\ln\{1 + B_m\}}{B_m} \tag{6.4-12}$$

results from the *radial outflow*[19] associated with the net mass-transfer flux $\dot{m}'' = \rho_w v_w$ across the phase boundary of the nontranslating sphere.

In practice, the quantity v_w is often an unknown being *sought* (see, e.g., Section 6.4.3.3); however, these QS results are equally valid if the velocity v_w were established by physically *blowing* ("transpiring") fluid through a *porous solid* sphere of the same diameter. For this reason the Stefan-flow parameter B_m is often called a "blowing" parameter. In *condensation* problems, v_w (hence B_m) is *negative*, so that the Stefan-flow effect is to *enhance* Nu_m due to the associated "suction."

If we define an alternative "blowing" parameter, $\mathrm{Pe}_{w,m}$ by

$$\mathrm{Pe}_{w,m} \equiv \frac{v_w \delta_{m,0}}{D} \tag{6.4-13}$$

(where $\delta_{m,0}$ is, again, the mass-transfer diffusion-layer thickness when $v_w = 0$), then Eq. (6.4-12) can alternatively be written:

$$F_m\{\text{Stefan blowing}\} \equiv \frac{\mathrm{Nu}_m}{\mathrm{Nu}_{m,0}} = \frac{\mathrm{Pe}_{w,m}}{\exp\{\mathrm{Pe}_{w,m}\} - 1}, \tag{6.4-14}$$

---

[18] In what follows, the subscript w denotes quantities evaluated on the exterior fluid (+) side of the phase boundary.

[19] An identical result is obtained for $F_m\{\text{Stefan blowing}\}$ for a *planar* (rather than spherically symmetric) BL.

which, apart from the sign of $\mathrm{Pe}_{w,m}$, will be recognized as equivalent to the *correction factor* to account for "phoretic suction" (Eq. 6.3-19).

6.4.3.2 Mass/Energy Transport Analogy in Presence of Stefan Flow

While the Stefan-flow effect on Nu_m is very similar to the *phoresis* effect discussed in Section 6.3.4, there is a significant distinction, *viz.*, while phoresis can strongly affect Nu_m for the transfer of a *dilute* species A, the corresponding *energy*-transfer coefficient Nu_h will be left unaltered. In contrast, when the Stefan-flow effect is important, it is *not* an analogy-breaking phenomenon, since the laws governing the associated change in Nu_h will be identical to those governing Nu_m. Thus, if we define the corresponding blowing parameters:

$$B_h \equiv \frac{v_w \delta_h}{\alpha}, \qquad \mathrm{Pe}_{w,h} \equiv \frac{v_w \delta_{h,0}}{\alpha} \tag{6.4-15}$$

where $\alpha(\equiv k/(\rho c_p))$ is the surrounding fluid *thermal* diffusivity, and rework the *energy*-transfer problem including the v_w-produced convection, we find:

$$F_n\{\text{Stefan blowing}\} \equiv \frac{\mathrm{Nu}_h}{\mathrm{Nu}_{h,0}} = \frac{\ln\{1 + B_h\}}{B_h} = \frac{\mathrm{Pe}_{w,h}}{\exp\{\mathrm{Pe}_{w,h}\} - 1}. \tag{6.4-16}$$

Thus, *the Stefan flow does not itself destroy the important analogy between energy and mass transport*; rather, this analogy now becomes:[20] Nu_h in *the same function of* B_h (or $\mathrm{Pe}_{w,h}$) *and Pr as* Nu_m *is of its corresponding arguments* B_m (or $\mathrm{Pe}_{w,m}$) *and Sc*. Thus, if $\mathrm{Pr} \neq \mathrm{Sc}$ the analogy does not require $\mathrm{Nu}_h = \mathrm{Nu}_m$; the latter equality is now seen to be valid only in completely quiescent (no forced, natural, or Stefan-flow convection) systems for which the analogy conditions of Section 6.3.3 are also satisfied.

6.4.3.3 QS Evaporation Rate of an Isolated Droplet

A typical and very important application of the above-mentioned concepts is the prediction of the QS evaporation rate of an isolated spherical droplet of chemical substance A in a hot gas—a model problem which involves simultaneous mass and energy transport. In this case, energy diffusion from the hotter gas supplies the latent heat required for vaporization, but at the problem outset we do not know either the droplet evaporation rate ($\dot{m}'' = \rho_w v_w$ or $\dot{m}_p = \pi d_p^2 \dot{m}''$) or the corresponding condition established at the vapor/liquid interface (e.g., $\omega_{A,w}$, T_w). Thus, we seek expressions for $\dot{m}''$ and $-\dot{m}_p$ for a given droplet size d_p in a gaseous environment of known temperature T_∞ and vapor mass fraction $\omega_{A,\infty}$, assuming all thermophysical

[20] In *forced* convection systems we can likewise state that $\mathrm{St}_m\{B_m, \mathrm{Re}, \mathrm{Sc}, \ldots\}$ is the same function as $\mathrm{St}_h\{B_h, \mathrm{Re}, \mathrm{Pr}, \ldots\}$.

properties of the gas mixture and liquid are known. For simplicity, we also assume that

1. vapor/liquid equilibrium (VLE) is established at the V/L interface;
2. the liquid[21] is pure ($\omega_A^{(\ell)} = 1$) and the surrounding gas is insoluble in it;
3. the droplet diameter d_p is very large compared to the gas mean-free-path;
4. forced and natural convection are negligible in the surrounding gas phase;
5. variable thermophysical property effects within the surrounding gas phase may be neglected;
6. species A diffusion in the vapor phase occurs according to Fick's (pseudobinary) law;
7. there is no chemical reaction of species A in the vapor phase;
8. the recession velocity of the droplet *surface* is negligible compared to the radial *vapor velocity*, v_w, established at the V/L phase boundary.

While numerous, none of these assumptions violates the basic conditions of Section 6.4.3.1; hence, all of the *transfer-coefficient* results stated therein are simultaneously valid. Therefore, we can short-circuit consideration of the profiles of $\omega_A\{r\}$ and $T\{r\}$ and go directly to the *interphase transport rates* and their relation to ω_A and T-values *at stations* ⓦ *and* ⓞⓞ.

The dimensionless "blowing" parameters $B_m \equiv v_w \delta_m / D_A$ and $B_h \equiv v_w \delta_h / \alpha$ might appear, at first sight, to be inconvenient since they involve not only the unknown $v_w (= \dot{m}''/\rho)$ but also the unknown diffusion layer thicknesses δ_m, δ_h (this is not the case for $Pe_{w,m}$ and $Pe_{w,h}$ since (owing to the stated assumption $\delta_{m,0}$ and $\delta_{h,0}$ are both equal to the droplet radius $d_p/2$). However, both B_m and B_h are readily related to thermophysical properties and state variables at stations ⓦ and ⓞⓞ (Spalding (1963)) *via* conservation conditions written for a "pillbox" straddling the L/V interface (Section 2.6.1). In this way (see below) we find:

$$B_m = \frac{\omega_{A,w} - \omega_{A,\infty}}{1 - \omega_{A,w}} \tag{6.4-17}$$

and

$$B_h = \frac{c_p(T_\infty - T_w)}{L_A}, \tag{6.4-18}$$

where, according to assumption (1), $\omega_{A,w} = \omega_{A,eq}\{T_w; p\}$. Because the structure and use of these *derived* results (see below), B_m and B_h are sometimes called dimensionless (mass, heat transfer) "driving force" parameters.

The species A mass balance (Eq. 2.1-9) is in this case simply that $\dot{m}''_{A,n}$ is continuous,[22] where, in each adjacent phase $\dot{m}''_A$ is given by Eq. (6.2-5). Since

[21] Unless otherwise specified all thermophysical properties pertain to the gas phase (g).
[22] For the sphere problem, the normal coordinate n is, of course, in the *radial* direction.

$\omega_A^{(\ell)} = 1$, in the absence of phoresis this condition simplifies to:

$$\dot{m}''_{A,n} = \dot{m}'' \cdot \omega_{A,w} + j''_{A,n,w}, \tag{6.4-19}$$

where use has also been made of the total mass balance condition $[\dot{m}''_n] = 0$. If we now insert $\dot{m}'' = \rho v_w$ and note that (according to assumption (6)),

$$j''_{A,n,w} = -\left[D_A\rho\left(\frac{\partial\omega_A}{\partial n}\right)\right]_w \equiv D_A\rho\frac{(\omega_{A,w} - \omega_{A,\infty})}{\delta_m}, \tag{6.4-20}$$

then we are immediately led to Eq. (6.4-4), of Section 6.4.1, which relates B_m directly to $\omega_{A,w}$ and $\omega_{A,\infty}$.

The *energy*-conservation condition at the V/L interface similarly provides the above-mentioned relation between B_h and $T_\infty - T_w$. Thus, Eq. (2.3-6) specializes to (neglecting contributions due to the rate at which work is done by the viscous stresses)

$$\dot{m}''L_A = -\dot{q}''_{n,w}(\text{Fourier}) \tag{6.4-21}$$

where L_A is the latent heat of evaporation:

$$L_A \equiv h_A\{T_w\} - h_A^{(\ell)}\{T_w; p\} \tag{6.4-22}$$

and, according to Eq. (3.3-3):

$$-\dot{q}''_n(\text{Fourier}) = \left[k\left(\frac{\partial T}{\partial n}\right)\right]_w \equiv k \cdot \frac{(T_\infty - T_w)}{\delta_h}. \tag{6.4-23}$$

Combining Eqs. (6.4-21) and (6.4-23) with the definition $B_h \equiv v_w\delta_h/\alpha$ immediately leads to Eq. (6.4-18).

In terms of these intermediate results for B_m and B_h, the remaining steps in the solution to our isolated-droplet evaporation problem may be stated as follows:

a. (6.4-17, -19) lead to the result:

$$\dot{m}'' = 2\frac{(D_A\rho)}{d_p} \cdot \ln[1 + B_m]. \tag{6.4-24}$$

b. The definition of Nu_h and Eqs. (6.4-18, -21, -23) lead to the equally valid result:

$$\dot{m}'' = \frac{\dot{q}''_n(\text{Fourier})}{L_A} = 2\frac{(k/c_p)}{d_p} \cdot \ln[1 + B_h]. \tag{6.4-25}$$

c. Compatibility between Eqs. (6.4-24, -25) requires that the "driving forces" be related by:

$$(1 + B_h) = (1 + B_m)^{\mathrm{Le}}, \tag{6.4-26}$$

where $\mathrm{Le} \equiv D_A/[k/(\rho c_p)]$ is the Lewis number (Section 3.4.4). In view of assumption (1) and Eqs. (6.4-17,-18), Eq. (6.4-26) then may be regarded as a transcendental equation for T_w, the solution of which then provides B_h as well as $\omega_{A,w} = \omega_{A,eq}(T_w; p)$ and, hence, B_m. (See Figure 6.4-2, constructed for the case $\mathrm{Le} = 1$, when $B_h = B_m$.)

In summary, the QS evaporation mass *flux* is then given by Eq. (6.4-25), where T_w is determined from Eq. (6.4-26). The corresponding single droplet (or "particle") evaporation rate $4\pi(d_p/2)^2\dot{m}''$ is thus:

$$\dot{m}_p = 2\pi d_p\left(\frac{k}{c_p}\right)\cdot\ln\left[1 + \frac{c_p(T_\infty - T_w)}{L_A}\right]. \tag{6.4-27}$$

Equating this to

$$-\frac{d}{dt}\left(\rho_\ell\frac{4\pi}{3}\left(\frac{d_p}{2}\right)^3\right) \qquad \text{(overall mass balance on the entire droplet)},$$

we readily find that in an environment with constant (time-independent) properties, while the temperature T_w will remain constant,[23] d_p^2 will decrease linearly with time, in accord with the law:

$$d_p^2 = d_{p,0}^2 - \left\{\alpha\left(\frac{\rho_g}{\rho_\ell}\right)\ln\left[1 + \frac{c_p(T_\infty - T_w)}{L_A}\right]\right\}t. \tag{6.4-28}$$

Formally[24] setting $d_p = 0$ gives the following important expression for the characteristic droplet lifetime:

$$t_{\text{life, vap}} = \frac{d_{p,0}^2}{\left\{8\alpha\left(\frac{\rho_g}{\rho_\ell}\right)\ln\left[1 + \frac{c_p(T_\infty - T_w)}{L_A}\right]\right\}}. \tag{6.4-29}$$

It is interesting to note the following generalizations of these important QS-continuum results:

1. When there is nonradial forced or natural convection (see Section 5.5.3),

[23] The temperature difference between a "dry" probe and an (evaporating) "wetted" probe is often used to experimentally establish the mainstream vapor content $\omega_{A,\infty}$. For this historic reason T_w is sometimes called the prevailing "wet-bulb" temperature.

[24] Of course, the continuum assumption (3) breaks down before this, but Eq. (6.4-28) still applies if the bulk of t_{vap} occurs in the continuum regime.

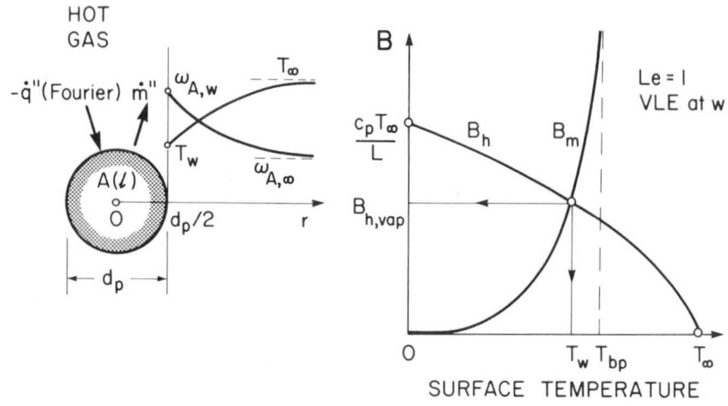

Figure 6.4-2 Simultaneous energy and mass transfer in the presence of an appreciable Stefan mass flow away from the phase boundary; QS evaporation of an isolated droplet.

evaporation rates can be estimated by replacing the factor 2 in Eq. (6.4-28) by $\overline{\mathrm{Nu}}_{m,B_m=0}$, and the factor 2 in Eqs. (6.4-24,-25) by $\mathrm{Nu}_{h,B_h=0}$. For this reason the exponent Le in Eq. (6.4-26) is replaced by the ratio

$$\mathrm{Le} \cdot \frac{\mathrm{Nu}_{m,B_m=0}}{\mathrm{Nu}_{h,B_h=0}}$$

(which, at high Reynolds' numbers, approaches $\mathrm{Le}^{2/3}$).

2. The eighth assumption (QS) breaks down when the quantity $(\rho_g/\rho_\ell)\ln\{1 + B_h\}$ is *not* small. In that event d_p^2 would no longer decrease linearly with time and Eq. (6.4-29) overestimates $t_{\mathrm{life,\,vap}}$, but $t_{\mathrm{life,\,vap}}$ remains proportional to $d_{p,0}^2$ (Rosner and Chang (1973)).
3. If species A is a fuel, F, and the ambient gas contains an oxidizer[25] O, then, upon "ignition," exothermic vapor-phase oxidation reactions occur in the diffusion boundary layer around the sphere, which can dramatically increase $\dot{m}_p$ and reduce the droplet lifetime (Section 1.1.3). When the homogeneous chemical kinetics are rapid enough for the reaction-zone (diffusion-flame) thickness to be small compared to the droplet radius and/or flame standoff distance, then a simple "two-film" diffusion theory is possible (Section 6.5.5.7), leading to simple estimates for the enhancement in $\dot{m}_p$ and reduction in t_{life}. When $\mathrm{Le} \cong 1$, the dominant effect will be seen to be the replacement of $B_{h,\,\mathrm{vap}}$ (Eq. (6.4-27)) in Eq. (6.4-28, -29) by:

$$B_{h,\,\mathrm{comb}} = \frac{c_p(T_\infty - T_w)}{L_A} + \frac{\omega_{O,\infty} f Q}{L_A} \qquad (6.4\text{-}30)$$

where f is the stoichiometric fuel/oxidant *mass* ratio (Table 6.5-1), Q is the

[25] This is the most commonly encountered situation; however, these results can, of course, equally well describe the "inverse" case of an oxidizer droplet in a vapor containing unburned fuel.

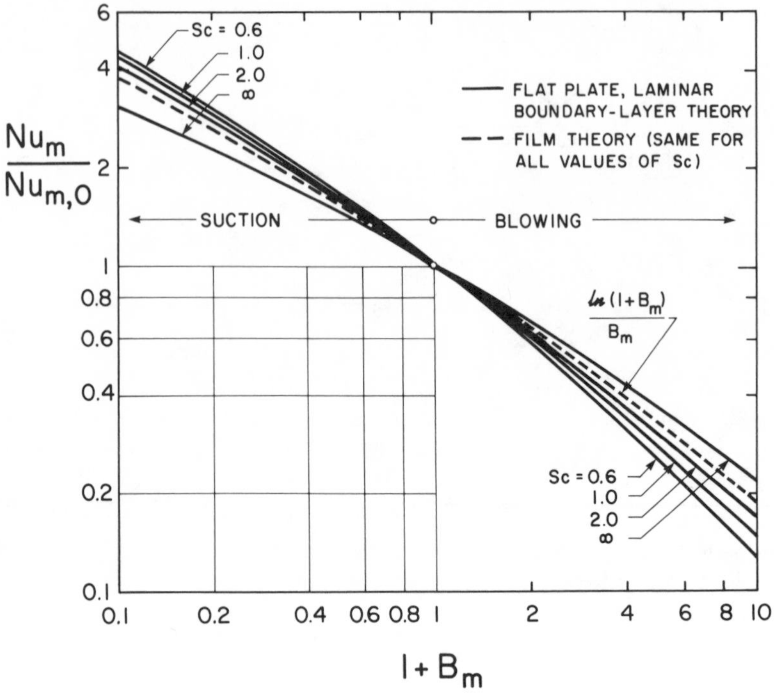

Figure 6.4-3 Transfer-coefficient reduction factor due to Stefan flow "blowing" ($B_m > 0$) and enhancement factor due to Stefan "suction" ($-1 \leqslant B_m < 0$) (after Bird, Stewart, and Lightfoot (1960)).

chemical heat release per unit *mass* of fuel consumed, and T_w is found to be close to the boiling point of the fuel at the prevailing pressure. This result carries over as well to the transient (non-QS) case (Rosner and Chang (1973)); however, Eq. (6.4-29) is no longer valid.

4. Most of these results carry over to solid/solid, liquid/liquid, and solid/liquid sphere phase-change problems (e.g., dissolution in nondilute systems).

6.4.3.4 Stefan-Flow Effects on Transfer Coefficients for Other Geometries and Flow Conditions

Equations (6.4-12,-16) for the Stefan flow ("blowing" or "suction") effect are frequently acceptable local approximations even in more general situations for which $\mathrm{Nu}_{B_m=0}$ varies over the transfer surface. Available calculations and measurements reveal that most *laminar* BLs are somewhat *more* sensitive to B than indicated by $B^{-1}\ln\{1 + B\}$, whereas turbulent BLs are somewhat *less* sensitive (see Figure 6.4-3, which includes numerically computed results for constant-property laminar flat-plate BLs).

Regardless of whether the simple $\ln\{1 + B_m\}$ factor[26] provides an adequate description of the dependence of species mass-transfer rate on concentration difference in any particular case, it should be cautioned that the nonlinearity in the mass-transfer law introduced by the Stefan flow destroys the validity of simple "resistance additivity" relationships (Sections 6.4.2 and 6.5.3) even when the interfacial kinetics (6.5.3) or phase equilibrium relationships (6.4.2) are linear.

6.4.4 Steady Mass Diffusion with Simultaneous Chemical Reaction: Catalyst Pellet "Effectiveness Factors"

As is well known, the time or volume required to carry out a chemical reaction can often be reduced by employing a catalyst. To avoid having to separate (recover) the catalyst from the reaction product, it is often impregnated throughout porous pellets. These pellets are packed into a "fixed bed" through which the reactant is passed. The volume requirement for such a "fixed-bed catalytic reactor" is then set by the ability of reactant(s) to diffuse into each porous pellet (and the ability of the product(s) to escape). If the pellet "cores" are inaccessible (because of the particular combination of pellet diameter, porosity, and catalytic activity (see below)), then (i) the expensive catalyst within this region is not utilized, and (ii) this portion of each pellet *volume* is not utilized, contributing to a larger-than-necessary volume for the overall chemical reactor.

A useful quantitative "continuum" model is that of a typical catalytic pellet for which we assume:

1. Steady-state diffusion and chemical reaction;
2. Spherical symmetry, including the (perimeter-mean) reactant A number density $n_{A,w}$ at $R = R_p$;
3. Radially uniform properties ($D_{A,\text{eff}}, k'''_{\text{eff}}, \rho \ldots$);
4. First-order irreversible pseudo-homogeneous chemical reaction *within* the pellet.

The steady-state $n_A\{r\}$-profile within such a pellet would then satisfy the local species A mass-balance equation (cf. Eq. (6.3-9)):

$$0 = -\operatorname{div}\{\mathbf{j}''_A\} + \dot{r}'''_{A,\text{eff}} \tag{6.4-31}$$

Since

$$\mathbf{j}''_A \cong -D_{A,\text{eff}}\,\rho\,\mathbf{grad}\,\omega_A \tag{6.4-32}$$

[26] Note that since $\ln\{1 + B\} = B[1 - \frac{1}{2}B + \frac{1}{3}B^2 - \ldots]$ the nonlinearity may be expected to set in whenever $\frac{1}{2}B$ is not negligible compared to unity.

and (assumption (4)),

$$-\dot{r}'''_{\text{A, eff}} = k'''_{\text{eff}}\,\rho\,\omega_{\text{A}}, \tag{6.4-33}$$

then $n_{\text{A}}(r)$ satisfies:[27]

$$\frac{1}{r^2}\cdot\frac{d}{dr}\left(r^2 D_{\text{A, eff}}\frac{dn_{\text{A}}}{dr}\right) = k'''_{\text{eff}}\,n_{\text{A}}, \tag{6.4-34}$$

where we have exploited assumption (2) and divided through by m_{A}/ρ.

The relevant boundary conditions on $n_{\text{A}}(r)$ will be taken to be:

$$n_{\text{A}}(R_{\text{p}}) = n_{\text{A,w}} \qquad \text{(considered known)} \tag{6.4-35}$$

and

$$(dn_{\text{A}}/dr)_{r=0} = 0. \tag{6.4-36}$$

The second of these[28] follows from applying the species A mass balance to a "microsphere" of radius ε. Upon examining the limit $\varepsilon \to 0$, one finds:

$$\lim_{\varepsilon\to 0} j''_{\text{A},r} = \lim_{\varepsilon\to 0}\left\{\frac{1}{4\pi\varepsilon^2}\cdot\int_0^{\varepsilon} \dot{r}'''_{\text{eff}}\,4\pi r^2\,dr\right\}, \tag{6.4-37}$$

which, for finite $\dot{r}'''_{\text{A}}(0)$, clearly leads to Eq. (6.4-36).

If $n_{\text{A}}(r)$ satisfying Eqs. (6.4-34, 6.4-35, 6.4-36) could be found, it would then be straightforward to calculate the following *catalyst utilization* (or "*effectiveness*") *factor*:

$$\eta_{\text{cat}} \equiv \frac{\text{actual reaction rate within pellet}}{\text{reaction rate within pellet if } n_A(r) = n_{\text{A,w}}} \tag{6.4-38}$$

or[29]

$$\eta_{\text{cat}} = \frac{\int_0^{R_{\text{p}}} k'''_{\text{eff}}\,n_{\text{A}}(r)4\pi r^2\,dr}{k'''_{\text{eff}}\,n_{\text{A,w}}\cdot\dfrac{4\pi R_{\text{p}}^3}{3}} = \frac{\left[D_{\text{A, eff}}\left(\dfrac{dn_{\text{A}}}{dr}\right)_{r=R_{\text{p}}}\right]\cdot(4\pi R_{\text{p}}^2)}{k'''_{\text{eff}}\,n_{\text{A,w}}\cdot\left(\dfrac{4\pi R_{\text{p}}^3}{3}\right)}. \tag{6.4-39}$$

---

[27] This ODE can also be obtained by writing Eq. (2.1-7a) for a *spherical shell*, dividing by the volume $4\pi r^2\,\Delta r$ and passing to the limit $4\pi r^2\,\Delta r \to 0$.

[28] It is not a "symmetry" condition, since radial profiles for which $(dn_{\text{A}}/dr)_{r=0} \neq 0$ would be equally symmetrical!

[29] The numerators of these expressions are equal since, in the steady state, the reactant consumed *within* the pellet must be entering the pellet *at* its surface.

Even prior to a complete mathematical solution, a "similitude analysis" (see Section 7.2) reveals that:

$$c \equiv \frac{n_A}{n_{A,w}} = \text{fct}\left\{\frac{r}{R_p}, \phi\right\} \equiv \text{fct}\{\zeta; \phi\}, \tag{6.4-40}$$

and, therefore:

$$\eta_{\text{cat}} = \frac{1}{\phi^2}\left(\frac{dc}{d\zeta}\right)_{\zeta=1} = \eta_{\text{cat}}\{\phi\}, \tag{6.4-41}$$

where the governing *dimensionless parameter* ϕ (E.W. Thiele modulus) is defined by:

$$\phi \equiv R_p \cdot \left(\frac{D_{A,\text{eff}}}{k'''_{\text{eff}}}\right)^{-1/2}. \tag{6.4-42}$$

This particular combination of pellet radius, catalytic activity, and effective diffusivity is thus seen to determine how close η_{cat} is to unity under the local conditions prevailing in the chemical reactor. Note that ϕ^2 can be regarded as the relevant Damköhler number for this problem; i.e., the ratio of the characteristic diffusion time, $(R_p^2/D_{A,\text{eff}})$, to the characteristic chemical reaction time, $(k'''_{\text{eff}})^{-1}$.

In this simple case a complete analytic solution is possible and very instructive. The normalized reactant-concentration variable $c\{\zeta;\phi\}$ is found to satisfy the second-order linear inhomogeneous ODE:

$$\frac{1}{\zeta^2}\frac{d}{d\zeta}\left(\zeta^2\frac{dc}{d\zeta}\right) = \phi^2 c \tag{6.4-43}$$

subject to the "split" BCs:

$$c\{1\} = 1, \tag{6.4-44}$$

$$(dc/d\zeta)_0 = 0. \tag{6.4-45}$$

The solution to this two-point BVP is readily found[30] to be:

$$c = \frac{\sinh\{\phi\zeta\}}{\zeta\sinh\{\phi\}} \tag{6.4-46a}$$

or, explicitly:

$$\frac{n_A\{r\}}{n_{A,w}} = \frac{\sinh\{\phi r/R_p\}}{(r/R_p)\sinh\{\phi\}}. \tag{6.4-46b}$$

[30] Remarkably, $\theta \equiv \zeta c$ satisfies the second-order linear ODE *with constant coefficients*: $d^2\theta/d\zeta^2 - \phi^2\theta = 0$, subject to $\theta\{1\} = 1$ and $\theta\{0\} = 0$.

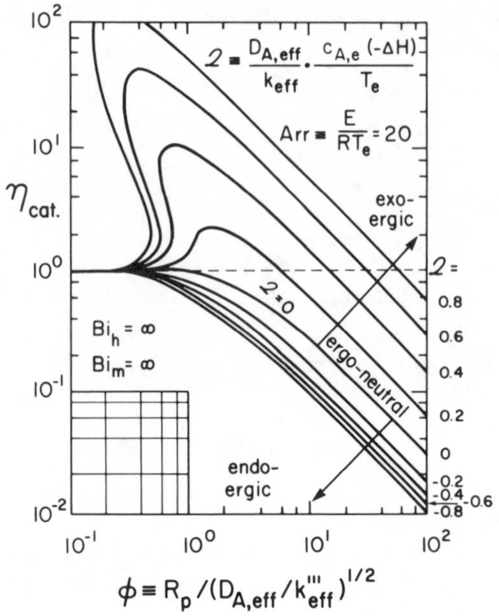

Figure 6.4-4 Catalyst effectiveness factor for first-order chemical reaction in a porous solid sphere (adapted from Weisz and Hicks (1962)).

Therefore, (cf. Eq. (6.4-41)) the catalyst-effectiveness factor is explicitly given by:

$$\eta_{\text{cat}} = \frac{3}{\phi} \cdot \left[\frac{1}{\tanh \phi} - \frac{1}{\phi} \right], \tag{6.4-47}$$

shown plotted ($\log \eta_{\text{cat}}$ *vs.* $\log \phi$ coordinates) as the $\mathscr{Q} = 0$ curve on Figure 6.4-4. Note that, whereas $\eta \to 1$ for $\phi \to 0$, $\eta_{\text{cat}} \sim 3/\phi$ for $\phi \gg 1$, the latter limit corresponding to reaction only in a thin "shell" near the outer perimeter of the pellet.[31]

An alternative presentation of η_{cat}, even more useful for many purposes, is obtained by exhibiting its dependence on the group $\Phi \equiv \phi^2 \eta$ (Wagner, Weisz) which is independent of the (often unknown) effective rate constant k'''_{eff}. Thus:

$$\Phi \equiv \frac{(-\dot{r}_{\text{A,obs}}/M_{\text{A}}) \Big/ \left(\frac{4\pi}{3} R_{\text{p}}^3 \right)}{(D_{\text{A,eff}} n_{\text{A,w}}/R_{\text{p}}^2)} = \frac{3}{4\pi} \cdot \frac{(-\dot{r}_{\text{A,obs}}/M_{\text{A}})}{(D_{\text{A,eff}} n_{\text{A,w}} R_{\text{p}})}. \tag{6.4-48}$$

This representation is used in Figure 6.4-5.

For chemical-reactor design purposes, η_{cat} results have been calculated for

[31] Indeed, ϕ can be regarded as the ratio of the pellet radius R_{p} to the characteristic *reactant penetration depth*: $(D_{\text{A,eff}}/k'''_{\text{eff}})^{1/2}$.

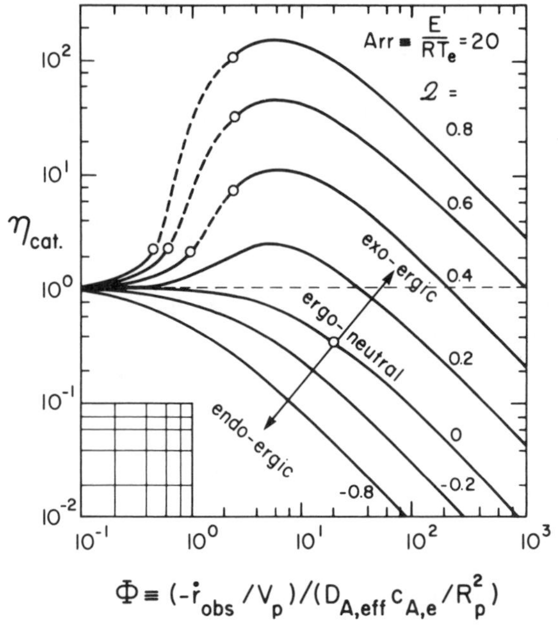

Figure 6.4-5 Catalyst effectiveness factor *vs.* experimentally observable (modified) Thiele modulus (adapted from Weisz and Hicks (1962)).

more complex pellet geometries,[32] and chemical kinetics, allowing also for local chemical heat release ("exoergicity"), a temperature-dependent chemical rate constant, and *external* (fluid-phase) resistances to mass and heat transport. Thus, the following additional parameters (discussed further below) influence the catalyst-effectiveness factor:

$$\mathrm{Arr} \equiv \frac{E}{RT_e}, \tag{6.4-49}$$

$$\mathcal{Q} \equiv \frac{D_{\mathrm{A,eff}}\, n_{\mathrm{A,e}}(-\Delta H)}{k_{\mathrm{eff}}\, T_e}, \tag{6.4-50}$$

$$n \equiv \text{reaction order}, \tag{6.4-51}$$

$$\mathrm{Bi}_m \equiv \left(\frac{D_{\mathrm{A,fluid}}}{D_{\mathrm{A,eff}}}\right) \cdot \frac{\mathrm{Nu}_m}{2} \tag{6.4-52}$$

$$\mathrm{Bi}_h \equiv \left(\frac{k_{\mathrm{fluid}}}{k_{\mathrm{eff}}}\right) \cdot \frac{\overline{\mathrm{Nu}}_h}{2}, \tag{6.4-53}$$

(Biot groups discussed below)

etc.

[32] Other geometries behave approximately like a sphere with an effective radius $3V_p/S_p$, where V_p is the pellet volume and S_p the pellet (exterior) area.

For example, when $\mathrm{Arr}=20$, $n=1$, $\mathrm{Bi}_m=\infty$, $\mathrm{Bi}_h=\infty$, the function $\eta_{\mathrm{cat}}(\Phi, \mathscr{Q})$ is displayed in Figure 6.4-5. Values in excess of unity are the result of local internal overheating (which increases k'''_{eff}) more than offsetting the reduction in local reactant concentration inside the pellet. However, values of η_{cat} are found to be near unity if[33] (Weisz):

$$\Phi \cdot \exp\left[\frac{\mathrm{Arr}\cdot\mathscr{Q}}{1+\mathrm{Arr}\cdot\mathscr{Q}}\right] < 1. \tag{6.4-54}$$

Criteria of this type are widely used by experimentalists seeking to infer the *intrinsic kinetics* of surface-catalyzed chemical reactions in the absence of appreciable "falsification" associated with the simultaneous presence of interpellet and intrapellet diffusion (mass, heat) limitations.

Regarding the parameters appearing in this now-classical reaction–diffusion problem, note that the above-mentioned Biot groups characterize the relative importance of *internal* and *external* transport resistances. When $\mathrm{Bi}_m \to \infty$ and $\mathrm{Bi}_h \to \infty$, $n_{\mathrm{A,w}} \to n_{\mathrm{A,e}}$ and $T_{\mathrm{w}} \to T_{\mathrm{e}}$, respectively, where the subscript e denotes conditions external to the fluid film (boundary) layer.

To deal quantitatively with simultaneous intrapellet mass and heat-transfer limitations, an interesting mathematical method has been used to reduce the nonisothermal catalyst-pellet problem to the solution of a single nonlinear ODE-BVP governing $c(\zeta)$ (or $T(\zeta)$). The method will be discussed in detail for the special case $D_{\mathrm{eff}} = \alpha_{\mathrm{eff}}$ in Section 6.5.5; however, in pure diffusion (no convection) steady-state cases, it can easily be generalized to allow for a $\alpha_{\mathrm{eff}} > D_{\mathrm{eff}}$, as is common for catalyst "supports."

The importance of the single-catalyst particle effectiveness factor, $\eta_{\mathrm{cat}}(\phi, \ldots)$, in the overall design of, say, a packed-bed chemical reactor is readily seen in the "plug-flow-reactor" (PFR-) limit already discussed in Section 6.1.3.1. In that case the required reactor volume was shown to be fixed by the mass throughput and the area under the function $[-\dot{r}'''_{\mathrm{A,eff}}(\omega_{\mathrm{A}}, \ldots)]^{-1}$ between the limits $\omega_{\mathrm{A}②}$ and $\omega_{\mathrm{A}①}$. We now find that, for a first-order surface-catalyzed reaction within the porous catalyst pellets comprising the fixed bed:

$$-\dot{r}'''_{\mathrm{A,eff}} = (1-\varepsilon_{\mathrm{bed}})\cdot\eta_{\mathrm{cat}}(\phi, \ldots)\cdot k'''_{\mathrm{eff}}\rho\omega_{\mathrm{A}}, \tag{6.4-55}$$

where the factor $(1-\varepsilon_{\mathrm{bed}})$ accounts for the absence of chemical reaction in the fluid between the catalyst pellets, and the factor $\eta_{\mathrm{cat}}(\phi, \ldots)$ accounts for the transport-induced intrapellet (and interpellet) reactant concentration and temperature nonuniformities. Thus, under conditions for which the local catalyst-effectiveness factors are considerably less than unity (cf. Figure 6.4-4), provision would have to be made for a correspondingly larger reactor volume for a given desired conversion, to compensate for the ineffectively used catalyst material in the "cores" of the catalyst pellets. Further details on the mathematical modeling of nonideal fixed-bed

[33] Typical values of the Arrhenius and heat-release parameters, Arr and $\mathscr{Q}$, are collected in Table 6.4-1.

Table 6.4-1 Representative Parameter Values for Some Heterogeneous Catalytic Reactions (after Hlavacek *et al.* (1969))

Chemical Reaction	$\mathscr{Q}$	Arr	$\mathrm{Bi}_m/\mathrm{Bi}_h$
Ethylene hydrogenation	0.0029	13.2	580
Ethylene hydrogenation	0.22–0.85	23–27	—
Benzene hydrogenation	0.03	15.7	150
Benzene hydrogenation	0.04–0.10	8–10	—
Methane oxidation	0.0056	21.5	310
Ethylene oxidation	0.068	—	45
Ethylene oxidation	0.13	13.4	—
Naphthalene oxidation	0.015	22.2	2.7
Sulfur dioxide oxidation	0.0031	—	105
Sulfur dioxide oxidation	0.01	14.8	—
Acrylonitrile synthesis	0.0019	9.9	4260
Vinyl chloride synthesis	0.254	6.5	—
Nitrous oxide dissociation	0.09	22	—
Oxidation of hydrogen	0.03–0.34	6.5–7.5	—
Oxidation of hydrogen	0.02–0.12	9	—

catalytic reactors, including instructive numerical examples, may be found in Chapter 11 of Froment and Bischoff (1979), and in H.H. Lee (1985).

In closing this quantitative discussion of reactant consumption and diffusion in a porous solid (catalyst "support"), it should be remarked that this formulation can also be adapted to describe the "penetration" of oxidizer vapor into a large "cloud" of individually burning fuel droplets (cf. Section 6.5.5.7), thereby predicting the extent to which outer fuel droplets influence the access of the inner droplets to the oxidizer (see, e.g., Labowsky and Rosner (1978)).

6.4.5 Transient Mass Diffusion: Concept of a Mass-Transfer (Concentration) Boundary Layer

The discussion of Section 5.4.6 carries over completely here, with the result that mass-transfer effects of the suddenly altered BC are felt in a growing boundary layer of thickness:

$$\delta_m \cong 4(Dt)^{1/2}. \tag{6.4-56}$$

Note that, if *both* thermal and mass-transfer BCs were suddenly altered, then the thermal layer would "outrun" the mass-transfer boundary layer with:

$$\frac{\delta_h}{\delta_m} = \left(\frac{\alpha}{D}\right)^{1/2} \equiv (\mathrm{Le})^{-1/2}, \tag{6.4-57}$$

where $D \ll \alpha$ ($\mathrm{Le} \ll 1$) for most solutes in condensed phases—especially metals. This thickness ratio holds equally well for the time-averaged penetration depth in the periodic BC case (cf. Eq. (5.4-49)).

6.5 CONVECTIVE MASS TRANSFER IN LAMINAR- AND TURBULENT-FLOW SYSTEMS

6.5.1 Analogies to Energy Transfer

Subject to the above-mentioned mass-heat transfer "analogy conditions" (1)–(4)—(see the paragraphs containing Eq. (6.3-12) and (6.3-13)—all results given in Chapter 5 for heat transfer in laminar or turbulent external or internal flows carry over to *mass* transfer by simply making the systematic replacements:

$$\begin{aligned} \alpha &\to D, \\ \mathrm{Pr} &\to \mathrm{Sc}, \\ \mathrm{Nu}_h &\to \mathrm{Nu}_m, \\ \mathrm{St}_h &\to \mathrm{St}_m, \\ \mathrm{Ra}_h &\to \mathrm{Ra}_m; \end{aligned} \tag{6.5-1a-e}$$

that is, to this level of approximation the two subject areas are completely equivalent, and experiments and/or computations in one provide results immediately useful in the other. For example, Eq. (5.5-33) also describes the concentration field downstream of a continuous point-source of species mass (e.g., a pollutant source) if the diffusivity α (or α_t) is replaced by D (or D_t) for a laminar (or turbulent) stream (see, e.g., Pasquill (1962)). Indeed, concentration measurements of this type have been used to infer D (or D_t-values) in high temperature gases (see, e.g., Westenberg (1966), Chapter 3). As examples of "internal" (duct-)flows consider the *mass transfer* of a dilute species A in a straight "*empty*" tube flow. The mass-transfer analog of Eq. (5.5-21a) becomes:

$$\frac{\omega_{\mathrm{A,w}} - \omega_{\mathrm{A},b}(z)}{\omega_{\mathrm{A,w}} - \omega_{\mathrm{A},b}(0)} = \exp\{-4\xi_m \cdot \overline{\mathrm{Nu}}_m(\infty) \cdot F(\text{entrance})\} \tag{6.5-2}$$

where

$$\xi_m \equiv \frac{1}{\mathrm{Re} \cdot \mathrm{Sc}} \cdot \frac{z}{d_{\mathrm{w}}} \, (\mathrm{Sc} \equiv \nu / D_{\mathrm{A-mix}}), \tag{6.5-3}$$

$$\overline{\mathrm{Nu}}_m(\infty) = \begin{cases} \text{Column 1 of Table 5.5-1} & \text{(laminar)}, \\ [0.023\,\mathrm{Re}^{0.8}\,\mathrm{Sc}^{1/3}] & \text{(turbulent)},^{34} \end{cases} \tag{6.5-4a,b}$$

[34] For turbulent transport, $\omega_{\mathrm{A},b}$ is to be interpreted as the *time-average* mass fraction of species A in the "bulk" stream.

and:

$$F\{\text{entrance}\} \cong \begin{cases} [1 + (7.60\,\xi_m)^{-8/3}]^{1/8} & \text{(laminar)}, \\ [1 + (z/d_w)^{-2/3}] & \text{(turbulent)}. \end{cases} \tag{6.5-5a,b}$$

Similarly, if the duct is "*packed*" (cf. Sections 4.7, 5.5.5), by analogy with Eq. (5.5-23b) we define:

$$\mathrm{Nu}_{m,\mathrm{bed}} \equiv \frac{dj_A/(a'''\,A_0\,dz)}{D_A\rho(\omega_{A,w} - \omega_{A,b})/[(\varepsilon/(1-\varepsilon))\cdot(d_{p,\mathrm{eff}})]}. \tag{6.5-6}$$

Since, in the absence of appreciable axial dispersion:

$$dj_A = \dot{m}\cdot d\omega_{A,b}, \tag{6.5-7}$$

we find:

$$\frac{\omega_{A,w} - \omega_{A,b}\{z\}}{\omega_{A,w} - \omega_{A,b}(0)} = \exp\left\{-6(1-\varepsilon)\cdot\left[\frac{\mathrm{Nu}_{m,\,\mathrm{bed}}\{\mathrm{Re}_{\mathrm{bed}}, \mathrm{Sc}\}}{\varepsilon\,\mathrm{Re}_{\mathrm{bed}}\cdot\mathrm{Sc}}\right]\cdot\frac{z}{d_{p,\,\mathrm{eff}}}\right\}, \tag{6.5-8a}$$

where (*cf*. Eq. (5.5-26) and Figure 5.5-2):

$$\mathrm{Nu}_{m,\mathrm{bed}}\mathrm{Sc}^{-0.4} \simeq 0.4\,\mathrm{Re}_{\mathrm{bed}}^{1/2} + 0.2\,\mathrm{Re}_{\mathrm{bed}}^{2/3} \tag{6.5-9}$$

if $3 \leqslant \mathrm{Re}_{\mathrm{bed}} \leqslant 10^4, 0.6 \leqslant \mathrm{Sc}, 0.48 \leqslant \varepsilon \leqslant 0.74$.

By convention, the quantity in square brackets (in Eq. (6.5-8a)) is called the *bed Stanton number for mass transfer*, written $\mathrm{St}_{m,\,\mathrm{bed}}$. Thus, Eq. (6.5-8a) can be rewritten in the transparent form:

$$\frac{\omega_{A,w} - \omega_{A,b}\{z\}}{\omega_{A,w} - \omega_{A,b}\{0\}} = \exp\{-\mathrm{St}_{m,\,\mathrm{bed}}a'''z\}, \tag{6.5-8b}$$

where, as before, a''' $(= 6(1-\varepsilon)/d_p)$ is the interfacial area per unit total volume of bed.

In all such transport problems involving the spatial approach to equilibrium, it is convenient to define a characteristic length, hereafter called the "*height of a transfer unit*" (HTU), by an equation of the form:

$$\frac{\omega_{A,w} - \omega_{A,b}\{z\}}{\omega_{A,w} - \omega_{A,b}\{0\}} \equiv \exp\left\{-\frac{z}{(\mathrm{HTU})}\right\}. \tag{6.5-10}$$

Here, HTU is the bed depth characterizing the exponential approach to mass-transfer equilibrium. Now we see that, in this case of single-phase fluid through a packed bed, this characteristic bed depth, HTU, is simply given by $(a'''\,\mathrm{St}_{m,\mathrm{bed}})^{-1}$.

Thus, dimensionless correlations for the quantity: HTU/d_p (often found in the ChE literature on packed beds) are the equivalent of correlations for $[d_p a''' \cdot \mathrm{St}_{m,\mathrm{bed}}]^{-1}$. More generally, when a second liquid phase (α) is wetting the packing although species A transport across the continuous-fluid phase (β, say vapor) is limiting the rate of solute A mass transport across the actual interfacial area $a'''_{\alpha\beta}$, a correlation for HTU/d_p is the equivalent of a correlation for $[d_p a'''_{\alpha\beta} \mathrm{St}_{m,\mathrm{bed}}]^{-1}$, where $a'''_{\alpha\beta}$ must be distinguished[35] from the "dry bed" value: $a''' = 6(1-\varepsilon)/d_p$.

These equations are widely used in the design of heterogeneous catalytic-flow reactors (Exercise 6.10) and physical separators—i.e., situations in which there is no chemical reaction *within* the fluid [cf. analogy condition (4), in paragraph containing Eq. (6.3-13)]. Moreover, provided a reasonable correlation for $\varepsilon\{\mathrm{Re}_{\mathrm{bed}}\}$ is introduced, Eq. (6.5-9) can also be used to predict the performance of *fluidized-bed* contactors, in the absence of appreciable recirculation or bypassing through "bubbles" (cf. Froment and Bischoff (1979)).

6.5.2 Analogies to Momentum Transport: High Schmidt-Number Effects

The mass-*momentum* analogy must be used with greater care because of the previously mentioned role of streamwise pressure gradient in "breaking" the relation between St and $c_f/2$. However, for laminar or turbulent flows with negligible pressure gradient (more precisely, negligible Λ, Eq. (5.7-18)), the Reynolds'–Chilton–Colburn relation:[36]

$$\mathrm{St}_m \approx \left(\frac{c_f}{2}\right) \cdot (\mathrm{Sc})^{-2/3} \qquad (6.5\text{-}11a)$$

provides a reasonable first approximation. For Sc-values near unity, as for "solute-gas" diffusion through most gaseous "solvents," Prandtl's form of the extended Reynolds' analogy (cf. Eq. (5.7-18) for $\mathrm{St}_h/(c_f/2)$) is acceptable. However, in many *mass*-transfer applications (e.g., aerosol particles, ions in aqueous solutions, etc.), the Fick diffusivity D is much smaller than the fluid momentum diffusivity ν (that is, $\mathrm{Sc} \gg 1$) and the corresponding relation for $\mathrm{St}_m/(c_f/2)$ would seriously underestimate St_m, especially for $\mathrm{Sc} > 10^2$. This is due to the increased importance, at high Sc-values, of accuracy deep within the viscous sublayer ($y^+ < 5$), where ν_t is not really zero. Indeed, recent measurements (e.g., Shaw and Hanratty (1977))

[35] For a recent review of correlations for $a'''_{\alpha\beta}/a'''$ (dry bed), see Laurent and Charpentier (1974).

[36] Chemical engineers call $\mathrm{St}_m(\mathrm{Sc})^{2/3}$ the "j_m-factor," expressing the extended Re-analogy in the simple approximate form: $j_m \cong j_h \cong (c_f/2)$. Equation (6.5-11b) indicates that, while the mass-heat-transfer portion of the "analogy" ($j_m \cong j_h$) remains adequate at high values of Sc or Pr, the momentum-transfer portion ($\cong c_f/2$) of the simple "analogy" breaks down when $\mathrm{Sc} \gg 1$ or $\mathrm{Pr} \gg 1$.

show that, for Sc ≫ 1:

$$\mathrm{St}_m \approx 0.08 \cdot \left(\frac{c_f}{2}\right)^{1/2} \cdot (\mathrm{Sc})^{-0.704}. \tag{6.5-11b}$$

While the Sc ≫ 1 dependence of St_m on $(c_f/2)^{1/2}$ (rather than $c_f/2$, as in the simple Reynolds–Chilton–Colburn relation above) is already contained in the extended von Karman relation (cf. Eq. (5.7-18)), the experimentally observed dependence of St_m on Sc when Sc ≫ 1 is seen to be significantly less sensitive than Sc^{-1}, with most earlier observers simply retaining $\mathrm{Sc}^{-2/3}$ (Friend and Metzner (1958), and Petukhov (1970)).

It should be remarked that *surface roughness*, when comparable to or greater in height than the ordinary thickness of the viscous sublayer ($\delta_{SL} \approx (c_f/2)^{1/2}(5\nu/U)$) *increases* both $c_f/2$ and the relevant Stanton numbers. Unfortunately, the effect of roughness in increasing St is not as great as its effect on the friction coefficient (hence pressure drop), as is also clear from Eq. (6.5-11b) at high Schmidt numbers.

6.5.3 Treatment of Mass-Transfer Problems with Chemical Nonequilibrium (Kinetic) Boundary Conditions (D.A. Frank-Kamenetskii (1969), D.E. Rosner (1964))

When a dilute species A reacts only at the fluid/solid interface, then, despite the change in the nature of the surface BC, the coefficient $\mathrm{St}_m\{\mathrm{Re}, \mathrm{Sc}\}$ still applies with acceptable accuracy.[37] In that case the mass flux of species A at the wall can be written:

$$-j''_{A,w} = \rho U\, \mathrm{St}_{m,A}\{\mathrm{Re}, \mathrm{Sc}_A\} \cdot (\omega_{A,\infty} - \omega_{A,w}), \tag{6.5-12}$$

and this flux will appear in the BC for species A at the fluid/surface interface. For example, if species A is being consumed at a local rate given by the (irreversible, first-order) chemical reaction:

$$-\dot{r}''_A \equiv \left\{\begin{matrix}\text{kinetic rate of consumption}\\ \text{of A at surface}\end{matrix}\right\} \cong k_w \rho \omega_{A,w}, \tag{6.5-13}$$

then the surface boundary condition or jump condition[38] (JC) will in this case take

[37] In some geometries/locations, the accuracy is actually perfect; e.g., laminar-flow stagnation point, rotating disk, etc. The error depends on the ratio St (constant flux BC)/St (constant driving force BC), which is not, however, unity in most situations (cf. Table 5.5-1).

[38] We assume here that there is no supply of species A from the condensed-phase side of the interface.

the simple form:

$$j''_{A,w} = \dot{r}''_{A}. \tag{6.5-14}$$

This JC provides an algebraic equation for the quasi-steady mass fraction, $\omega_{A,w}$, of species A established at the surface, which can therefore be inserted into Eq. (6.5-12) or (6.5-13) for the reaction rate. Doing this, one readily finds:

$$\frac{\omega_{A,w}}{\omega_{A,\infty}} = \frac{1}{1+C} \qquad \text{Local reactant "starvation"} \tag{6.5-15}$$

and

$$-j''_{A,w} = \left\{\frac{C}{1+C}\right\} \cdot [\rho U \, \mathrm{St}_{m,A} \omega_{A,\infty}], \tag{6.5-16}$$

where C (a surface Damköhler number (see Section 7.2.4), also called the "catalytic parameter") is *defined* by:

$$C \equiv \frac{k_w}{U \, \mathrm{St}_{m,A}} = \frac{k_w \delta_m}{D_A}. \tag{6.5-17}$$

Note that Eq. (6.5-16) expresses the reaction (or transfer) rate as a fraction of the maximum ("diffusion-controlled") rate. When $C \ll 1$, this fraction is small, and the rate approaches the "chemically controlled" value: $k_w \rho \omega_{A,\infty}$ (cf. Eq. (6.5-13)).

This "resistance-additivity" approach is usually adequate for engineering purposes when applied *locally* along a surface with x-dependent k_w, $\omega_{A,w}$, T_w, and St_m, etc. However, if these dependencies are not sufficiently slowly varying, more accurate methods (based on solving the PDE governing $\omega_A(\mathbf{x}, t)$, subject to its appropriate BCs) must be used.

If LTCE is achieved at station w as the result of rapid *heterogeneous* chemical reactions, then, in the absence of *homogeneous* chemical reactions, Eq. (6.5-12) applies with $\omega_{A,w} = \omega_{A,eq}(T_w, \ldots; p)$. More generally, an equation like (6.5-12) will apply to each chemical species i, with $\omega_{i,w} = \omega_{i,eq}(T_w, \ldots; p)$—a result often successfully used to estimate "chemical" vapor deposition (CVD) rates in multicomponent vapor systems with surface equilibrium (see, e.g., Rosner and Nagarajan (1985)). Even in the presence of rapid chemical reaction *in the gas phase*, a similar approach could be used to make useful quantitative estimates of CVD-rates by applying an equation like (6.5-12) to each *chemical element* (see Section 2.1.3). The principal uncertainty in this approach arises in estimating the effective Fick diffusion coefficient for each chemical element; e.g., $D_{(k)\text{-mix}}$ in the formal local diffusion flux expression:

$$\mathbf{j}''_{(k)} \equiv -D_{(k)\text{-mix}} \rho \, \mathbf{grad} \, \omega_{(k)} \qquad (k = 1, 2, \ldots, N_{\text{elem}})$$

(cf. Eq. (3.4-3)). However, in many vapor systems, adequate estimates of $D_{(k)\text{-mix}}$ *can* be made from a knowledge of $D_{i\text{-mix}}$-values (Section 3.4.4) for the principal vapor species that transport each chemical element. If this can be done, and if thermal (Soret) diffusion effects are secondary (see, e.g., Rosner (1980)), then CVD-rates in the presence of rather complex homogeneous chemical reactions can be estimated using the "nonreactive" convective mass-transfer coefficients: $\mathrm{St}_{m,i}\{\mathrm{Re}, \ldots, \mathrm{Sc}_i\}$ or $\mathrm{Nu}_{m,i}\{\mathrm{Re}, \ldots, \mathrm{Sc}_i\}$ discussed earlier in this chapter (or their heat-transfer analogs, Chapter 5).

6.5.4 Combined Energy and Mass Transport: Recovery of Mainstream Chemical and Kinetic Energy

Combining the discussions of Sections 5.8 and 6.5.3, we can answer the following important question: If a "thermometer" (thermocouple, etc.) is placed in a hot stream containing appreciable amounts of *kinetic energy* and *chemical energy*, what temperature will it read? If we can neglect radiation loss, the surface temperature will rise to a steady-state value at which the rate of "convective" heat *loss*:

$$(\dot{q}''_{\mathrm{w}})_{\mathrm{conv}} \cong \rho U \cdot \mathrm{St}_h\{\mathrm{Re}, \mathrm{Pr}\} \cdot [c_{\mathrm{p}} \cdot (T_{\mathrm{w}} - T_r)] \tag{6.5-18}$$

balances the rate of energy transport associated with chemical species A mass transport, that is,

$$(-\dot{q}''_{\mathrm{w}})_{\mathrm{diff}} \cong \rho U \cdot \mathrm{St}_m\{\mathrm{Re}, \mathrm{Sc}_{\mathrm{A}}\} \cdot \omega_{\mathrm{A},\infty} Q \tag{6.5-19}$$

(for a diffusion-controlled *heterogeneous* chemical reaction releasing energy Q per unit mass of A). Here T_r is the ordinary "*gas-dynamic recovery temperature*," that is,

$$T_r = T_\infty + r_{\mathrm{KE}}\{\mathrm{Pr}, \mathrm{Re}\} \cdot \frac{U^2}{2c_{\mathrm{p}}}. \tag{6.5-20}$$

Imposing the "adiabatic condition" $\dot{q}''_{\mathrm{w}} = 0$ by including *both* terms [(6.5-18) and (6.5-19)] leads immediately to the instructive result:

$$T_{\mathrm{w}} = T_\infty + r_{\mathrm{KE}}\{\mathrm{Pr}, \mathrm{Re}\}\frac{U^2}{2c_{\mathrm{p}}} + \frac{\mathrm{St}_m\{\mathrm{Re}, \mathrm{Sc}_{\mathrm{A}}\}}{\mathrm{St}_h\{\mathrm{Re}, \mathrm{Pr}\}} \cdot \frac{\omega_{\mathrm{A},\infty} Q}{c_{\mathrm{p}}}. \tag{6.5-21}$$

Note that, in forced convection systems the transport coefficient ratio $\mathrm{St}_m/\mathrm{St}_h$ plays the role of a *chemical-energy recovery factor* (Rosner (1975)), hereafter written r_{ChE}. For a *laminar* BL situation, with $r_{\mathrm{KE}} \approx \mathrm{Pr}^{1/2}$ and $r_{\mathrm{ChE}} \approx \mathrm{Le}^{2/3}$, Eq. (6.5-21) takes the simple explicit form:

$$T_w \approx T_\infty + (\mathrm{Pr})^{1/2}\frac{U^2}{2c_p} + (\mathrm{Le})^{2/3}\frac{\omega_{A,\infty}Q}{c_p}. \tag{6.5-22}$$

It is interesting to note that T_w can be higher or lower than the corresponding (thermodynamic) "total" temperature:

$$T_0 \equiv T_\infty + \frac{U^2}{2c_p} + \frac{\omega_{A,\infty}Q}{c_p}, \tag{6.5-23}$$

depending on the magnitudes of the *transport*-property ratios Pr and Le. In most *gas* mixtures, both r_{KE} and r_{ChE} are near unity,[39] so that such a probe records a temperature near T_0, not T_∞. The *chemical*-energy recovery term is important in measuring the temperature of gas streams that are out of *chemical* equilibrium, for if a catalytically active Pt thermocouple (say) is used, values of T_w far in excess of T_∞ or T_r can be recorded. Indeed, Eq. (6.5-21) defines a *generalized recovery temperature*, T_r', which includes the recovery of chemical as well as kinetic energy present in the mainstream. Thus, for nonadiabatic surfaces one can show that:

$$\dot{q}_w'' \approx \rho U \cdot \mathrm{St}_h\{\mathrm{Re}, \mathrm{Pr}\} \cdot c_p(T_w - T_r'); \tag{6.5-24}$$

i.e., the net rate of energy transport (in a forced-convection situation) will be linearly proportional to the "overheat" $(T_w - T_r')$.

6.5.5 "Analogies" in the Presence of a Homogeneous Exoergic Chemical Reaction: "Conserved Quantities" in Single-Phase Chemically Reacting Flows

The interrelationships discussed above between energy, mass, and occasionally momentum-transport coefficients, and their corresponding field densities, explicitly precluded the occurrence of homogeneous chemical reactions and the associated volumetric, "sensible" heat-generation rate (see Condition (4) for both Eqs. (6.3-12) and (6.3-13) of Section 6.3.3). Here we briefly explore the possibility of establishing useful approximate relationships between:

a. fluid temperature and chemical-species concentration fields (and then associated wall-transfer coefficients) *in the presence of homogeneous chemical reaction*, and

[39] Indeed, if Pr = 1 and Le = 1, then $T_w = T_0$. However, if the mainstream chemical energy is associated with a *mobile* species (e.g., H_2 in air), then Le > 1 and T_w can greatly *exceed* T_0, without any violation of the first Law of Thermodynamics!

b. flow-field variables, including chemical-species compositions with and without homogeneous chemical reaction.

For these purposes we restrict ourselves to single-phase, chemically reacting fluids, exploiting the conservation equations (PDEs) derived in Section 2.5. When dealing with turbulent flows, we will confine our attention to time-averaged quantities (Section 2.6.3).

Note that the interrelationships explored here will involve stable reactant and product *species* concentrations, and not merely chemical *element* concentrations,[40] for which examples of the types (a) and (b) above are far more obvious. Thus, since we have already noted that $\dot{r}'''_{(k)} = 0$ for each element 1, 2, 3, etc. (representing C, O, H, N, etc.) present in a chemically reacting mixture, each $\bar{\omega}_{(k)}\{\mathbf{x}, t\}$ will satisfy a time-averaged PDE,

$$\bar{\rho}\frac{D\bar{\omega}_{(k)}}{Dt} = \operatorname{div}[\bar{D}_{(k),\,\mathrm{eff}}\,\bar{\rho}\,\mathbf{grad}\,\bar{\omega}_{(k)}] \tag{6.5-25}$$

of the same form as the PDE governing that of an *inert tracer* molecule in the same chemically reacting ("hot") flow—that is,

$$\bar{\rho}\frac{D\bar{\omega}_I}{Dt} = \operatorname{div}[\bar{D}_{I,\,\mathrm{eff}}\,\bar{\rho}\,\mathbf{grad}\,\bar{\omega}_I]. \tag{6.5-26}$$

It follows that, however complex the chemical kinetics, if the BC/IC are similar, there will be an approximate interrelation between the $\bar{\omega}_{(k)}$ $(k = 1, 2, \ldots)$ and $\bar{\omega}_I$-fields in a "hot" flow; moreover, if chemical reaction does not drastically modify the $\bar{\mathbf{v}}\{\mathbf{x}, t\}$, $\bar{\rho}\{\mathbf{x}, t\}$ and $\bar{D}_{\mathrm{eff}}\{\mathbf{x}, t\}$ fields (appearing as coefficient functions in the PDEs), then one could perhaps predict each $\bar{\omega}_{(k)}$-field in terms of corresponding and far more readily accessible "cold" flow $\bar{\omega}_{\mathrm{A}}\{\mathbf{x}, t\}$ information.

In more restrictive, but important special cases (see below), a similar strategy can be adopted to describe both premixed and initially unmixed flows involving an overall-, stoichiometric-, irreversible homogeneous chemical reaction, and its thermal consequences.

6.5.5.1 "Conserved Variables" and Analogies for a Homogeneous Irreversible Fuel and Oxidant Stoichiometric Reaction at Arbitrary Rate

The essential feature exploited above was that, irrespective of the homogeneous kinetics, each $\bar{\omega}_{(k)}$-field satisfied a *source-free* PDE of the standard convective-diffusion form (6.5-25). Here we show that, if the following overall chemical

[40] Or groupings of atoms, called "radicals," which remain intact in chemical-reaction sequences.

reaction occurs locally:

$$1\,\text{gm O} + f\,\text{gm}\,F \rightarrow (1+f)\,\text{gram P} + fQ\,\text{cal}, \tag{6.5-27}$$

then certain linear combinations of the variables $\bar{\omega}_i (i = \text{O, F, P})$ and $\bar{T}$ also satisfy source-free PDEs of the convective-diffusion form. These combinations, analogous to element mass fractions, and called "conserved variables," will form the basis for the "hot-flow" and "cold-flow" interrelations we seek.

For simplicity, we will assume that the time-averaged ω_i and T fields (written hereafter without the overbars) themselves satisfy the coupled, inhomogeneous PDEs:

$$\rho\frac{D\omega_i}{Dt} = \text{div}[D_{i,\text{eff}}\,\rho\,\textbf{grad}\,\omega_i] + \dot{r}_i'''(i\ (i = \text{O, F, P}), \tag{6.5-28}$$

$$\rho c_p\frac{DT}{Dt} = \text{div}[\alpha_{\text{eff}}\,\rho\,c_p\,\textbf{grad}\,T] + \dot{q}_{\text{chem}}''' \tag{6.5-29}$$

where the coefficient functions $\mathbf{v}(\mathbf{x}, t)$ and $\rho(\mathbf{x}, t)$ must simultaneously satisfy the momentum and mass-conservation constraints considered explicitly in Section 6.5.5.3.

Inspection of Eq. (6.5-27) reveals that if reaction 6.5-27 is the only one occurring, then the source terms $\dot{r}_i'''$ and $\dot{q}_{\text{chem}}'''$ appearing above are necessarily interrelated by the sequence of equalities:

$$\frac{-\dot{r}_F'''}{f} = \frac{-\dot{r}_O'''}{1} = \frac{\dot{r}_P'''}{(1+f)} = \frac{\dot{q}_{\text{chem}}'''}{fQ}, \tag{6.5-30}$$

with the "outermost" equation corresponding to the obvious condition: $\dot{q}_{\text{chem}}''' = -\dot{r}_F'''Q$, where Q is the heat of reaction per unit mass of fuel vapor (see, e.g., Table 6.5-1). In particular, these source-interrelationships imply:

$$\dot{r}_O''' - \frac{1}{f}\dot{r}_F''' = 0, \tag{6.5-31}$$

$$\dot{q}_{\text{chem}}''' + fQ\dot{r}_O''' = 0, \tag{6.5-32}$$

and

$$\dot{r}_O''' + \frac{1}{1+f}\dot{r}_P''' = 0. \tag{6.5-33}$$

Now invoking the PDE (6.5-28) and Eq. (6.5-31), we see that, provided $D_{\text{O,eff}} = D_{\text{F,eff}} = D_{\text{eff}}$ locally, the combined variable $\omega_{\text{OF}} \equiv \omega_{\text{O}} - (1/f)\omega_{\text{F}}$ satisfies the *source-free* PDE:

$$\rho \frac{D\omega_{OF}}{Dt} = \operatorname{div}[D_{\text{eff}}\, \rho\, \mathbf{grad}\, \omega_{OF}]. \tag{6.5-34}$$

Hence, in this chemically reacting flow, ω_{OF} is a "conserved variable."

Invoking the source interrelation (6.5-32) and the PDEs (6.5-29) and (6.5-28), we observe that if $c_p \approx$ const and $\alpha_{\text{eff}} \approx D_{O,\text{eff}} = D_{\text{eff}}$, then the composite variable $c_pT + fQ\omega_O$ is a "conserved variable," also satisfying Eq. (6.5-34).

Finally, invoking Eqs. (6.5-33, and -28), we observe that if $D_{O,\text{eff}} \approx D_{P,\text{eff}} = D_{\text{eff}}$, then the linear combination $\omega_{OP} \equiv \omega_O + (1+f)^{-1}\omega_P$ is also a "conserved variable," satisfying Eq. (6.5-34).

For definiteness, consider the jet mixing problem discussed earlier (Section 5.7.1), but let the jet contain fuel vapor and the surrounding stream contain oxidizer. When chemical reaction occurs, we use each of the "conserved variables" to define normalized profiles:

$$\Omega \equiv \frac{(\ )_{\text{local}} - (\ )_s}{(\ )_j - (\ )_s}, \tag{6.5-35}$$

the station subscript j denotes the jet and s the surrounding fluid. Then, subject to the above-mentioned assumption, each of the variables[41]

$$\Omega_{OF} \equiv \frac{(\omega_{O,s} - \omega_O) + \dfrac{1}{f}\omega_F}{\omega_{O,s} + \dfrac{1}{f}\omega_{F,j}}, \tag{6.5-35a}$$

$$\Omega_{TO} \equiv \frac{c_p(T - T_s) - fQ(\omega_{O,s} - \omega_O)}{c_p(T_j - T_s) - fQ\omega_{O,s}}, \tag{6.5-35b}$$

$$\Omega_{OP} \equiv \frac{(\omega_{O,s} - \omega_O) - \dfrac{1}{1+f}\omega_P}{\omega_{O,s}} \tag{6.5-35c}$$

not only satisfies the same PDE:

$$\rho \frac{D\Omega}{Dt} = \operatorname{div}[D_{\text{eff}}\, \rho\, \mathbf{grad}\, \Omega] \tag{6.5-36}$$

but also *the same BC*. If the solution to this BVP is unique, then, despite the presence

[41] Assuming the jet contains no oxidizer, the surroundings no fuel vapor, and neither the jet nor surroundings contain the reaction product P(g).

Table 6.5-1 Physical and Combustion Properties of Selected Fuels in Air[a]

Fuel	*Mol. Wt.*	*Spec. Grav.*	T_{Boil} (°C)	*Heat of Vap.* (cal/g)	*Heat of Comb.* (kcal/g)	*Stoichiometry* % Vol.	*f*
Acetaldehyde	44.1	0.783	−56.7	136.1	—	0.0772	0.1280
Acetone	58.1	0.792	56.7	125.0	7.36	0.0497	0.1054
Acetylene	26.0	0.621	−83.9	—	11.52	0.0772	0.0755
Acrolein	56.1	0.841	52.8	—	—	0.0564	0.1163
Acrylonitrile	53.1	0.797	78.3	—	—	0.0528	0.1028
Ammonia	17.0	0.817	−33.3	328.3	—	0.2181	0.1645
Aniline	93.1	1.022	184.4	103.4	—	0.0263	0.0872
Benzene	78.1	0.885	80.0	103.2	9.56	0.0277	0.0755
Benzyl alcohol	108.1	1.050	205.0	—	—	0.0240	0.0923
n-Butane	58.1	0.584	−0.5	92.2	10.92	0.0312	0.0649
1-Butene	56.1	0.601	−6.1	93.3	10.82	0.0377	0.0678
Carbon disulfide	76.1	1.263	46.1	83.9	—	0.0652	0.1841
Carbon monoxide	28.0	—	−190.0	50.6	—	0.2950	0.4064
Cyclobutane	56.1	0.703	12.8	—	—	0.0377	0.0678
Cyclohexane	84.2	0.783	80.6	85.6	10.47	0.0227	0.0678
Cyclohexene	82.1	0.810	82.8	—	—	0.0240	0.0701
Cyclopentane	70.1	0.751	49.4	92.8	10.56	0.0271	0.0678
Cyclopropane	42.1	0.720	−34.4	—	—	0.0444	0.0678
trans-Decalin	138.2	0.874	187.2	—	—	0.0142	0.0692
n-Decane	142.3	0.734	174.0	86.0	10.56	0.0133	0.0666
Diethyl ether	74.1	0.714	34.4	83.9	—	0.0337	0.0896
Ethane	30.1	—	−88.9	116.7	11.34	0.0564	0.0624
Ethyl acetate	88.1	0.901	77.2	—	—	0.0402	0.1279
Ethano	46.1	0.789	78.5	200.0	6.40	0.0652	0.1115
Ethylene oxide	44.1	1.965	10.6	138.9	—	0.0772	0.1280
Furan	68.1	0.936	32.2	95.6	—	0.0444	0.1098
n-Heptane	100.2	0.688	98.5	87.1	10.62	0.0187	0.0661
n-Hexane	86.2	0.664	68.0	87.1	10.69	0.0216	0.0659
Hydrogen	2.0	—	−252.7	107.8	28.65	0.2950	0.0290
Iso-Propanol	60.1	0.785	82.2	158.9	—	0.0444	0.0969
Kerosene	154.0	0.825	250.0	69.5	10.30	—	—
Methane	16.0	—	−161.7	121.7	11.95	0.0947	0.0581
Methanol	32.0	0.793	64.5	263.0	4.74	0.1224	0.1548
n-Nonane	128.3	0.772	150.6	68.9	10.67	0.0147	0.0665
n-Octane	114.2	0.707	125.6	71.7	10.70	0.0165	0.0633
n-Pentane	72.1	0.631	36.0	87.1	10.82	0.0255	0.0654
1-Pentene	70.1	0.646	30.0	—	10.75	0.0271	0.0678
Propane	44.1	0.508	−42.2	101.7	11.07	0.0402	0.0640
Propene	42.1	0.522	−47.7	104.5	10.94	0.0444	0.0678
n-Propanol	60.1	0.804	97.2	163.9	—	0.0444	0.0969
Toluene	92.1	0.872	110.6	86.7	9.78	0.0227	0.0743
Gasoline (73 octane)	120.0	0.720	155.0	81.0	10.54	—	—
Gasoline (100 octane)	—	—	—	—	—	—	—
Jet fuel JP1	150.0	0.810	—	—	10.28	0.0130	0.0680
JP3	112.0	0.760	—	—	10.39	0.0170	0.0680
JP4	126.0	0.780	—	—	10.39	0.0150	0.0680
JP5	170.0	0.830	—	—	10.28	0.0110	0.0690

[a] Compiled using data presented in NACA 1300 (after Fristrom and Westenberg (1965)).

Flammability Limits (*% Stoich.*)		*Spont. Ign. Temp.* (°C)	*Fuel for Max. Flame Speed* (*% Stoichiom.*)	*Max. Flame Speed* (cm/s)	*Flame Temp. at Max. Fl. Speed* (K)	*Ign. Energy* (10^{-5} cal)		*Quenching Dist.* (mm)	
Lean	*Rich*					*Stoich.*	*Min.*	*Stoich.*	*Min.*
—	—	—	—	—	—	8.99	—	2.29	—
59	233	561.1	131	50.18	2,121	27.48	—	3.81	—
31	—	305.0	133	155.25	—	0.72	—	0.76	—
48	752	277.8	100	61.75	—	4.18	—	1.52	—
87	—	481.1	105	46.75	2,461	8.60	3.82	2.29	1.52
—	—	651.1	—	—	2,600	—	—	—	—
—	—	593.3	—	—	—	—	—	—	—
43	336	591.7	108	44.60	2,365	13.15	5.38	2.79	1.78
—	—	427.8	—	—	—	—	—	—	—
54	330	430.6	113	41.60	2,256	18.16	6.21	3.05	1.78
53	353	443.3	116	47.60	2,319	—	—	—	—
18	1,120	120.0	102	54.46	—	0.36	—	0.51	—
34	676	608.9	170	42.88	—	—	—	—	—
—	—	—	115	62.18	2,308	—	—	—	—
48	401	270.0	117	42.46	2,250	32.98	5.33	4.06	1.78
—	—	—	—	44.17	—	20.55	—	3.30	—
—	—	385.0	117	41.17	2,264	19.84	—	3.30	—
58	276	497.8	113	52.32	2,328	5.74	5.50	1.78	1.78
—	—	271.7	109	33.88	2,222	—	—	—	—
45	356	231.7	105	40.31	2,286	—	—	2.06	—
55	2,640	185.6	115	43.74	2,253	11.71	6.69	2.54	2.03
50	272	472.2	112	44.17	2,244	10.04	5.74	2.29	1.78
61	236	486.1	100	35.59	—	33.94	11.47	4.32	2.54
—	—	392.2	—	—	—	—	—	—	—
—	—	428.9	125	100.35	2,411	2.51	1.48	1.27	1.02
—	—	—	—	—	—	5.40	—	1.78	—
53	450	247.2	122	42.46	2,214	27.49	5.74	3.81	1.78
51	400	260.6	117	42.46	2,239	22.71	5.50	3.56	1.78
—	—	571.1	170	291.19	2,380	0.36	0.36	0.51	0.51
—	—	455.6	100	38.16	—	15.54	—	2.79	—
—	—	—	—	—	—	—	—	—	—
46	164	632.2	106	37.31	2,236	7.89	6.93	2.54	2.03
48	408	470.0	101	52.32	—	5.14	3.35	1.78	1.52
47	434	238.9	—	—	—	—	—	—	—
51	425	240.0	—	—	2,251	—	—	—	—
54	359	284.4	115	42.46	2,250	19.60	5.26	3.30	1.78
47	370	298.3	114	46.75	2,314	—	—	—	—
51	283	504.4	114	42.89	2,250	7.29	—	2.03	1.78
48	272	557.8	114	48.03	2,339	6.74	—	2.03	—
—	—	433.3	—	—	—	—	—	—	—
43	322	567.8	105	38.60	2,344	—	—	—	—
—	—	298.9	—	—	—	—	—	—	—
—	—	468.3	106	37.74	—	—	—	—	—
—	—	248.9	107	36.88	—	—	—	—	—
—	—	—	—	—	—	—	—	—	—
—	—	261.1	107	38.17	—	—	—	—	—
—	—	242.2	—	—	—	—	—	—	—

of homogeneous chemical reaction, these Ω-functions are everywhere equal to one another—moreover, if I denotes an *inert* tracer in the same ("hot") flow, and we define its rescaled counterpart:

$$\Omega_I(\mathbf{x}, t) \equiv \frac{\omega_I - \omega_{I,s}}{\omega_{I,j} - \omega_{I,s}}, \tag{6.5-37}$$

then, provided $D_{I,\mathrm{eff}} = D_{\mathrm{eff}}$ locally, each of the above-mentioned Ω-functions will also equal $\Omega_I(\mathbf{x}, t)$. (See, also, Section 6.5.5.2 below.)

These "hot-flow" interrelationships could be used to *reduce* either:

a. the number of independent composition and/or temperature *measurements* needed to characterize this chemically reacting flow,
b. the number of independent PDEs that have to be simultaneously solved in a *predictive* (e.g., finite difference; Appendix 8.2) scheme.

Thus, they significantly reduce the magnitude of the problem. However, by no means do they constitute its complete solution.

While illustrated here and in Sections 6.5.5.2–7 for initially unmixed ("diffusion flame") situations, these "conserved variables"[42] also simplify the prediction of *premixed* chemically reacting systems (e.g., flame structure and propagation speed (Sections 6.5.5.8 and 7.2.2.2)).

Consideration of the validity of the assumptions used to arrive at these "conserved variables" will be postponed to Section 6.5.5.6, after we present some important applications (prediction of diffusion flame length, Section 6.5.5.5).

6.5.5.2 The Diffusion Flame "Sheet" Limit: Implications for Burning Fuel Jets

While the validity of the previous results placed no restrictions on the reaction rate, it is often observed (especially in *laminar*-flow systems at or above atmospheric pressure) that the reaction-zone thickness[43] is very small compared to, say, the jet mouth diameter d_j; indeed, the flame zone often appears to be practically a mathematical surface or "sheet" occupying a negligible fraction of the jet "volume." In that limit, everywhere outside of this thin "sheet," the flow field is *nonreactive* because oxidizer and fuel vapor never coexist. Such a sheet can be considered a gas-dynamic discontinuity (Section 2.6.1) (a) where oxidizer and fuel vapor meet by diffusion in their stoichiometric ratio, and where (b) heat and product vapors are supplied to the fluid on either side of it.

If this diffusion-controlled flame-sheet limit is approximated in a system for which $D_{\mathrm{O,eff}} \approx D_{\mathrm{F,eff}} \approx \alpha_{\mathrm{eff}}$ (see Section 6.5.5.6), then the previous formulation

[42] Sometimes called Schvab–Zeldovich variables, after two Russian investigators who exploited them in early flame theories (Zeldovich [1944, 1950]).

[43] Defined, e.g., by the zone in which the reaction rate is greater than, say, 1% of its maximum value.

provides the following interesting results:

a. Since both ω_F and ω_O must vanish at such a "flame sheet" (subscript f) its location, $r_f(z_f)$ will be (implicitly) given by:

$$\Omega_I(r_f, z_f) = (\Omega_{OF})_f, \tag{6.5-38}$$

where $(\Omega_{OF})_f$ is readily found to be:

$$(\Omega_{OF})_f = [1 + \Phi]^{-1} \tag{6.5-39}$$

and

$$\Phi \equiv \frac{(\omega_{F,j}/\omega_{O,s})}{f}. \tag{6.5-40}$$

b. The flame sheet temperature T_f would be the solution of the equation:

$$(\Omega_{TO})_f = (\Omega_{OF})_f \tag{6.5-41}$$

that is, T_f would have the constant value:

$$T_f = T_s + [1 - (\Omega_{OF})_f] \cdot \frac{\omega_{O,s} f Q}{c_p} + (\Omega_{OF})_f (T_j - T_s). \tag{6.5-42}$$

Actually, Eq. (6.5-38) implicitly giving the flame shape is of greater theoretical than practical interest since $\Omega_I(r, z)$ is the "hot-flow" inert tracer profile. Only in cases for which $\Omega_I(r, z)$ is not very different from a well-known "cold-flow" (cf) profile (see Sections 6.5.5.3, -4, -5) can such information be used to "predict" flame shapes and lengths (Section 6.5.5.5).

6.5.5.3 Can a Component of the Fluid Momentum Density Be a Conserved Property?

Before considering possible relations between "hot" (reacting)-flows and their "corresponding" cold-flows, it is useful to consider the possibility that a component of the fluid linear *momentum density*[44] can also be a "conserved" variable. Recall that, in addition to the local mass-conservation condition (see Eq. (2.5-6b)), the specific momentum, $\mathbf{v}(\mathbf{x}, t)$, must satisfy the vector PDE:

$$\rho \frac{D\mathbf{v}}{Dt} = \text{div}[\nu_{\text{eff}}\, \rho((\mathbf{grad\, v}) + (\mathbf{grad\, v})^{\dagger})] + \mathbf{S} \tag{6.5-43}$$

[44] In two-dimensional viscous nonreacting fluid flows *with* pressure gradients and/or body forces, we have already shown that the fluid *vorticity* satisfies a convective diffusion equation of the form (6.5-36) with ν_{eff} replacing D_{eff}.

which contains the local "source:"

$$\mathbf{S} \equiv -\mathbf{grad}\, p + \mathrm{div}\left[\left(\kappa - \frac{2}{3}\mu\right)_{\mathrm{eff}} \cdot (\mathrm{div}\,\mathbf{v})\,\mathbf{I}\right] + \sum_{i=1}^{N} \rho_i \mathbf{g}_i. \tag{6.5-44}$$

Comparing Eq. (6.5-43) with Eq. (6.5-36), we observe that there can indeed be such cases if S is negligible and $\nu_{\mathrm{eff}} \approx D_{\mathrm{eff}}$. An important example is the component $v_z(r, z)$ in constant-pressure, round-jet mixing (see Sections 4.6.2 and 5.7.1) at high Reynolds' numbers, Re_{j}. Thus, the scalar variable:

$$\Omega_v \equiv \frac{v_z(r, z) - U_s}{U_j - U_s} \tag{6.5-45}$$

(written u^* in Section 5.7) indeed satisfies the source-free PDE

$$\rho \frac{D\Omega_v}{Dt} = \mathrm{div}[\nu_{\mathrm{eff}}\,\rho\,\mathbf{grad}\,\Omega_v] \tag{6.5-46}$$

in the presence (or absence) of homogeneous chemical reaction.

It can be concluded that in a chemically reacting ("hot"-) flow for which the momentum source function S is negligible (Eq. (6.5-44)), the streamwise momentum density is also a "conserved variable." Moreover, if the local diffusivities satisfy the condition $\nu_{\mathrm{eff}} \approx D_{\mathrm{eff}} \approx \alpha_{\mathrm{eff}}$, the rescaled momentum-density variable Ω_v (Eq. (6.5-45)) will be approximately equal to its counterparts Ω_I, Ω_{OF}, Ω_{TO} and Ω_{OP} for the boundary conditions considered here.

We reiterate that these extended "hot-flow" interrelationships simplify but by no means complete the task of characterizing or predicting laminar and/or turbulent chemically reacting flows. However, in those cases for which chemical reaction (and its associated heat release) do not dramatically alter the velocity and transport-property distributions, readily available "cold-flow" (cf) information *can* be used to predict some important features of chemically reacting flows. This is illustrated below (Section 6.5.5.5) for diffusion flame lengths and shapes in the "flame-sheet" (diffusion controlled, or rapid kinetics-) limit.

6.5.5.4 Use of "Cold-Flow" Information

When there is *no* chemical reaction ("cold-flow") and $c_p \cong$ const, then $T(\mathbf{x}, t)$ and each chemical species mass fraction $\omega_i(\mathbf{x}, t)$ is a conserved variable; frequently, for a commonly encountered set of BC/ICs, we know (from measurements or computations) the functions $\Omega_{T,\mathrm{cf}}(\mathbf{x}, t)$, $\Omega_{i,\mathrm{cf}}(\mathbf{x}, t)$ (see, e.g., Figure 5.7-1) and, when $\alpha_{\mathrm{eff}} \approx D_{i,\mathrm{eff}}$, these functions are nearly identical. However, in the presence of homogeneous chemical reaction (i.e., upon "ignition") in general:

$$\Omega_I(\mathbf{x}, t) \neq \Omega_{i,\mathrm{cf}}(\mathbf{x}, t) \tag{6.5-47}$$

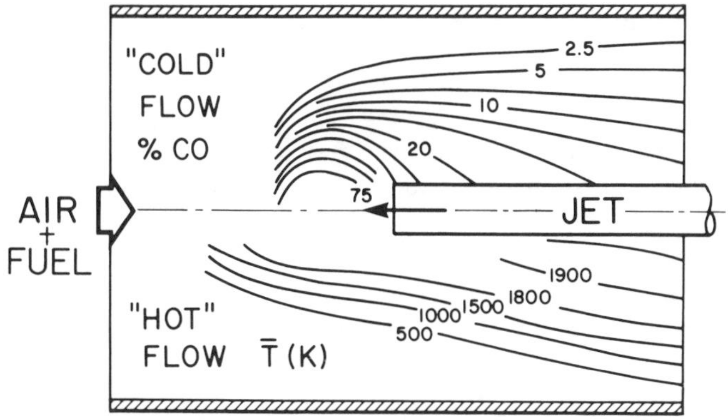

Figure 6.5-1 Density-induced differences between time-averaged "hot" (ignited CH_4/air approach) turbulent flow and corresponding "cold" (unignited CO jet) flow; opposed jet axisymmetric combustor flow field ("aerodynamic flame holder") containing a recirculation zone ($U_j = 130$ m/s, $U_{air} = 15.2$ m/s, $d_{duct} = 5.1$ cm) (after the experiments of Peck and Samuelson (1976)).

and, even if $\alpha_{eff} \approx D_{O,eff}$:

$$\Omega_{TO}\{\mathbf{x}, t\} \neq \Omega_{T,cf}\{\mathbf{x}, t\}. \tag{6.5-48}$$

These disparities will exist because, even though the IC/BC are the same and the PDEs are of the same (source-free) *form*, the PDEs involve different *coefficient functions* $\mathbf{v}\{\mathbf{x}, t\}$, $\rho v_{eff}\{\mathbf{x}, t\}$, $\rho\{\mathbf{x}, t\}$, etc. This will be especially true in nondilute laminar-flow systems with appreciable heat release in confined (e.g., ducted) situations, owing to the very different consequences of **grad** p and/or body forces in the two distinct (hf *vs.* cf) cases, and (to a lesser extent) the temperature dependence of v_{eff}, α_{eff}, and $D_{i,eff}$. An interesting example of these differences is shown in Figure 6.5-1, based on cold-flow and hot-flow measurements in a ducted[45] turbulent flow combustor.

Nevertheless, there are important applications in which these coefficient-induced differences are of secondary importance, so that, if $D_{i,eff} \approx \alpha_{eff}$, approximations like:

$$\Omega_{OF}\{\mathbf{x}, t\} \approx \Omega_{T,cf}\{\mathbf{x}, t\}, \tag{6.5-49}$$

$$\Omega_{TO}\{\mathbf{x}, t\} \approx \Omega_{T,cf}\{\mathbf{x}, t\}, \tag{6.5-50}$$

[45] For nondilute combustion systems in which the reaction product is a *condensed* phase (e.g., Al_2O_3(c)-fume, etc.), even larger differences can exist between the hot- and cold-flow fields.

and, hence:

$$(\Omega_{OF})_f \approx \Omega_{T,\text{cf}}\{\mathbf{x}_f, t\}, \tag{6.5-51}$$

provide useful information on temperature and composition profiles, flame locations, etc. As an example, consider the steady-flow, axisymmetric turbulent gaseous jet discussed in Section 5.7.1, for which experimental $\Omega_{T,\text{cf}}\{\mathbf{x}\}$ contours are plotted in Figure 5.7-1. Now, if this primary jet were pure $CH_4(g)$ and the secondary flow air (with the density ratio ρ_s/ρ_j and velocity ratio U_s/U_j approximately that specified), *upon ignition* a useful first approximation to the reaction-zone position would be provided by the contour $\Omega_{T,\text{cf}} = (\Omega_{OT})_f = 0.013$ (only a portion of which is actually shown on Figure 5.7-1). Probably the first example of the use of such predictions (for ducted, gaseous-diffusion flame shapes in *laminar* "plug flow") was provided by Burke and Schumann in 1928 (Figure 6.5-2). For *unconfined*, high Re_j axisymmetric jets, in Section 6.5.5.5 below we extract useful *flame-length* information from the $\Omega_{v,\text{cf}}\{\mathbf{x}\}$ (streamwise momentum) profiles presented in Section 4.6.

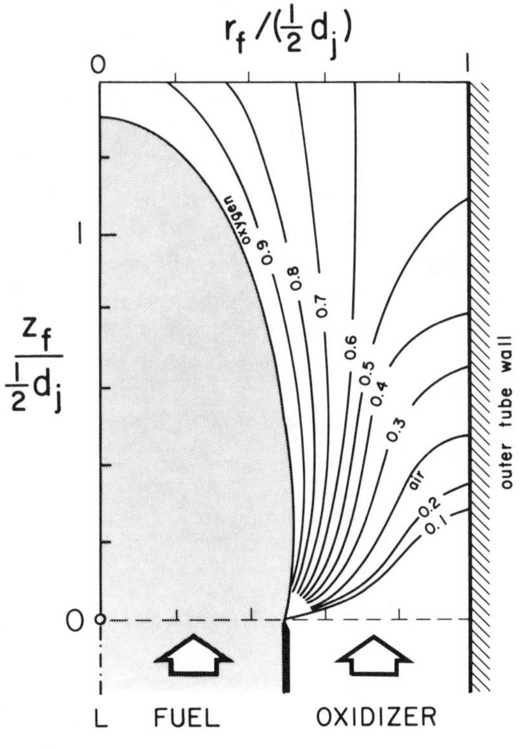

Figure 6.5-2 Possible diffusion flame shapes for a coaxial fuel jet discharging into a duct surrounded by an equal-velocity, uniform oxidizer stream (adapted from Burke and Schumann (1928)).

6.5.5.5 *Flame Lengths for Laminar or Turbulent Gaseous-Fuel Jets Discharging into Quiescent "Air" (When Buoyancy Effects are Negligible)*

"Far"-field, streamwise, momentum-density profiles are well known for *nonreactive* round jets issuing into an otherwise quiescent ambient fluid (see Section 4.6). Since this is a situation for which the streamwise momentum density in a chemically *reactive* (ignited fuel) jet would be a "conserved variable" (Section 6.5.5.3), if we are further able to make the approximations:

$$\nu_{\text{eff}} \approx D_{i,\,\text{eff}} \qquad (i = \text{O}, \text{F}) \tag{6.5-52}$$

and

$$\Omega_v(r, z) \approx \Omega_{v,\text{cf}}(r, z), \tag{6.5-53}$$

then we can use the results of Section 4.6 to extract valuable information on laminar- or turbulent-flow flame shapes $r_\text{f}(z_\text{f})$ and lengths, L_f. The validity of Eq. (6.5-52) and the flame-sheet approximation will be discussed further in Section 6.5.5.6. Here, in accordance with Section 6.5.5.4, we remark that Eq. (6.5-53) implies the negligibility of "buoyancy" effects (associated with heat-release-produced density differences in a body force field; see Sections 5.2.1 and 7.2.3.2c).

Despite our restriction to the region far from the jet mouth, we imagine that the jet emerges with uniform velocity U_j from a circular tube of diameter d_j, so that the injected axial-momentum flow rate is explicitly:

$$\dot{J} = (\rho_\text{j} U_\text{j} A_\text{j}) U_\text{j} = \rho_\text{j} U_\text{j}^2 \left(\frac{\pi d_\text{j}^2}{4}\right). \tag{6.5-54}$$

(If the fuel-vapor supply tube is of *non*circular cross section, then Eq. (6.5-54) defines its "effective" diameter, $d_{\text{j},\,\text{eff}}$.)

Moreover, since, for the cases considered below, $U_s = 0$, we see that:

$$\Omega_v(r, z) = \frac{v_z(r, z)}{U_\text{j}}. \tag{6.5-55}$$

Therefore (see Sections 6.5.5.2 and 6.5.5.4) the axisymmetric flame shape $r_\text{f}(z_\text{f})$, will be approximated by:

$$\Omega_{v,\,\text{cf}}(r_\text{f}, z_\text{f}) = (\Omega_{\text{FO}})_\text{f} \tag{6.5-56}$$

and, accordingly, the flame *length*, L_f (measured along the axis $r = 0$) will be the solution of the equation:

$$\Omega_{v,\,\text{cf}}(0, L_\text{f}) = (\Omega_{\text{FO}})_\text{f}. \tag{6.5-57}$$

The explicit implications of these conclusions, i.e., the dependence of L_f on U_j, d_j, pressure level, $\omega_{O,s}$, $\omega_{F,j}$, and fuel type for both laminar and turbulent fuel vapor jets, are stated below:

a. Laminar Fuel Jet. Using the self-similar $v_z(r, z)$ results of Section 4.6 and Eqs. (6.5-55, -57), we readily find:

$$\left(\frac{L_f}{d_j}\right)_{\text{laminar}} \approx \frac{3}{32}\left(\frac{U_j d_j}{\nu_j}\right)(1 + \Phi) \tag{6.5-58}$$

where

$$\Phi \equiv (\omega_{F,j}/\omega_{O,s})/\text{f}. \tag{6.5-40}$$

Thus, when buoyancy is negligible, the laminar flame length, L_f, should be proportional to $U_j d_j^2 p(1 + \Phi)$. Experimentally, it is found that strict linearity with respect to jet velocity is only valid over a very restricted portion of the laminar range (see Figure 6.5-3).

b. Turbulent Gaseous Fuel Jet. Using the self-similar $\bar{v}_z(r, z)$ results of Section 4.6 and Eqs. (6.5-55, -57), we readily find that, for the unconfined fully *turbulent* reacting jet:

$$(L_f/d_j)_{\text{turb}} \approx 6.57(1 + \Phi), \tag{6.5-59}$$

when use has been made of the Reichardt constant (0.0161) relating ν_{eff} to $(\dot{J}/\rho)^{1/2}$ (Section 4.6). Interestingly enough, while the dependence of the flame length on the composition parameter Φ remains the same, we see that $L_{f,\text{turb}}$ depends *linearly* on d_j and should be *independent* of both[46] U_j and p-level. These expectations are in accordance with measurements of time-averaged flame lengths in the absence of appreciable buoyancy effects (see, e.g., Sections 1.1.2 and 7.2.3.2c). The appearance of such flames as a function of jet mouth velocity is sketched in Figure 6.5-3.

It should be cautioned that these results have been derived for *gaseous* fuel jets. Fuel jets initially comprised of polydispersed sprays of liquid droplets will generally behave differently because of the preliminary requirements of phase change (see Sections 6.4.3.3 and 6.5.5.7). While phase change processes are usually *not* controlling under conditions of large industrial furnaces, nevertheless buoyancy effects are often important under such conditions.

While attention has been focused above on unconfined, nonswirling, fuel-vapor jets issuing into a quiescent medium, clearly the same strategy can be used to estimate the effects of coaxial surrounding flows, swirl, etc., on flame behavior based on corresponding "cold-flow" data. However, these more general situations are also

---

[46] In effect, the tendency of the flame to be "stretched out" by an increase in U_j is cancelled by the increase in the eddy diffusivities associated with the U_j-increase.

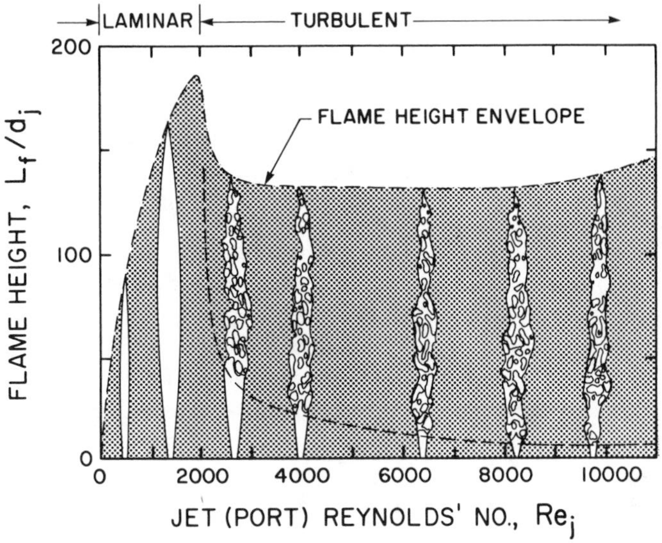

Figure 6.5-3 Cylindrical fuel-jet flame heights as a function of jet port Reynolds' number (adapted from Hottel and Hawthorne (1949)).

often more sensitive to the fluid dynamic effects of chemical heat release (Section 6.5.5.4) so that such cold-flow jet mixing information, even if quantitative, may be of only qualitative value in predicting flame behavior.

6.5.5.6 Validity of Underlying Assumptions for Laminar- or Turbulent-Flow Systems

Recapitulating, the "conserved-variable" approach to predicting the behavior of single-phase, chemically reacting flows, outlined above, is based on the following important assumptions:

1. single, stoichiometric, irreversible chemical reaction (Eq. (6.5-27));
2. simple "gradient" diffusion;
3. equality of effective diffusivities ($\nu_{\text{eff}} \approx \alpha_{\text{eff}} \approx D_{i,\text{eff}}$);
4. constancy (with respect to temperature and composition) of mixture heat capacity, c_p.

Beyond this, in Sections 6.5.5.2 and 6.5.5.4 we introduced the approximations:

5. negligible diffusion flame-zone thickness;
6. cold flow field $\approx$ hot flow field (i.e., $\Omega_I \approx \Omega_{I,\text{cf}}$ etc.).

It is prudent to briefly examine these restrictive assumptions so that results of this approach are not applied under conditions for which they are likely to fail. We shall take these briefly in turn.

Single "Global" Reaction (1). While even simple fuel-oxidizer combustion reactions are known to occur *via* a complex network ("chain") of elementary reactions, the overall effects of such reactions are often adequately represented by Eq. (6.5-27). However, this approach should *not* be adopted to describe systems limited by chemical reactions whose kinetics cannot be described by the well-known quasi-stationary-state approximation (QSSA) (linking all reaction intermediate concentrations to those of the stable reactants and/or products).

Simple Gradient Diffusion (2). In nonturbulent systems, this implies the negligibility of phoresis and coupling phenomena among the various diffusional phenomena (Chapter 3). Thus when pseudo-binary diffusion laws are valid, while the Soret and Dufour effects are negligible, assumption (2) should be useful in *laminar* fluids. Local "countergradient" transport is known to occur in some variable-density *turbulent* flows. This phenomenon would probably invalidate "conserved-variable" estimates in the corresponding regions of chemically reacting turbulent flows, although it remains possible that α_{eff} and $D_{i,\text{eff}}$ would still "track one another," even through a change in sign.

Equality of Effective Diffusivities (3). In *nonturbulent* systems, this assumption usually limits the conserved-variable approach to *gaseous* systems, in which the (molecular) *mechanisms* of momentum energy and mass diffusion are identical (Section 3.4.4). However, while the diffusivities ν, α, and D_i are of the same order-of-magnitude for gases, their near-equality may be invalidated (e.g., in systems containing very heavy and/or light molecules A, B,... for which $D_{\text{A-mix}} \neq D_{\text{B-mix}} \neq D_{\text{C-mix}} \neq \cdots$ etc.).[47]

Since the time-averaged effect of *turbulent* convection is to augment the effective diffusivities ($\nu_{\text{eff}} = \nu + \nu_t$, $\alpha_{\text{eff}} = \alpha + \alpha_t$, $D_{i,\text{eff}} = D_i + D_{i,t}$), even if the mole*cular* diffusivities were not equal, in a highly *turbulent* flow ($\nu_t \gg \nu$, etc), the assumption $\nu_{\text{eff}} \approx \alpha_{\text{eff}} \approx D_{i,\text{eff}}$ should become acceptable. However, as remarked in Section 5.7.1, fully turbulent free jet mixing appears to occur as if:

$$\text{Pr}_t \equiv \frac{\nu_t}{\alpha_t} \approx 0.5, \qquad \text{Sc}_t \equiv \frac{\nu_t}{D_{i,t}} \approx 0.5. \tag{6.5-60}$$

Thus, in fully turbulent flows the assumptions $\alpha_{\text{eff}} \approx D_{i,\text{eff}}$ are likely to be better than those involving the effective *momentum* diffusivity.

Constancy of Mixture Heat Capacity, c_p (4). Those parts of the reacting flow analogies involving a temperature variable, e.g., Ω_{TO}, require not only that $\alpha_{\text{eff}} \approx D_{\text{O,eff}}$ but also that c_p be constant (compare Eqs. (5.3-12), (6.5-29) and

[47] For example, consider the species CH_4(g), O_2(g), H_2O(g), and CO_2(g) dilute in N_2(g), at 1000 K, 1 atm. Then, while $\nu \cong 1.23$ and $\alpha \cong 1.74$, we find (Table 3.4-1: $D_{CH_4-N_2} \cong 1.78$, $D_{O_2-N_2} \cong 1.68$, $D_{H_2O-N_2} \cong 2.16$, and $D_{CO_2-N_2} \cong 1.48$, where all values are given in the units cm^2/s.

(6.5-36)). Now for an ideal gas mixture:

$$c_p\{T, \omega_i\} = \sum_{i=1}^{N} \omega_i c_{p,i}\{T\}; \tag{6.5-61}$$

hence, c_p will vary appreciably when either the mixture is comprised of dominant species whose individual heat capacities are very disparate[48] or the heat capacities of each of the dominant species is very temperature-dependent. The "$c_p \approx$ const" assumption is thus likely to be most valid in combustion systems containing an excess of air, for which $c_p\{300\,\text{K}\} = 0.240$ cal/(g-K), $c_p\{1000\,\text{K}\} = 0.273$ cal/(g-K), and $c_p\{2000\ \text{K}\} = 0.316$ cal/(g-K) (Table 8.1-1).

Negligible Diffusion Flame-Zone Thickness (5). Those predictions based on the so-called "flame-sheet" approximations require that oxidizer and fuel-vapor coexist in a zone that occupies a negligible volume fraction of the flow field. In *nonturbulent systems* this is a matter of the rapidity of the intrinsic chemical reaction compared to that of molecular-diffusional mixing at the prevailing pressures and local temperatures.

An order-of-magnitude estimate of the reaction-zone thickness, δ_{rxn}, can be obtained by considering this zone well mixed (that is, $\delta_{\text{rxn}} = V''_{\text{WSR}}$; see Section 6.1.3) and "straddling" the locus $\mathbf{x}_f$ (predicted using the "flame-sheet" (infinite reaction rate) assumption). For example, consider an intrinsic kinetic law of the simple "bimolecular" Arrhenius form:

$$-\dot{r}'''_{\text{F}} = m_{\text{F}}\left[A \exp\left\{-\frac{E}{RT}\right\}\right] \cdot n_{\text{O}} n_{\text{F}} \tag{6.5-62}$$

or, (introducing mass fractions in favor of number densities):

$$-\dot{r}'''_{\text{F}} = \frac{A}{M_{\text{O}}}\left(\frac{pM}{RT}\right)^2 \cdot \exp\left\{-\frac{E}{RT}\right\} \cdot \omega_{\text{O}}\omega_{\text{F}} \tag{6.5-63}$$

Then for this "WSR" we can write the fuel-vapor macroscopic mass balance:

$$-\left[D_{\text{F,eff}}\rho\left(\frac{\partial \omega_{\text{F}}}{\partial n}\right)\right]_{\text{f}} = [-\dot{r}'''_{\text{F}}]_{\text{WSR}} \cdot \delta_{\text{rxn}}, \tag{6.5-64}$$

where $-\dot{r}'''_{\text{F}}$ is evaluated at the *prevailing* local values of $\omega_{\text{O,rxn}}$, $\omega_{\text{F,rxn}}$ and T_{rxn}. If these are, respectively, estimated as:

$$\omega_{\text{i,rxn}} \approx \tfrac{1}{4}|(\partial\omega_{\text{i}}/\partial n)_f| \cdot \delta_{\text{rxn}} \quad (\text{i} = \text{O,F}) \tag{6.5-65}$$

[48] If this were the case, the term $-\mathbf{grad}\,T \cdot \sum_{i=1}^{N} c_{p,i}\mathbf{j}''_i$ appearing on the RHS of the $\rho c_p DT/Dt$ equation (5.3-12) might also be nonnegligible.

and

$$T_r \approx T_f - \left[\frac{1}{|(\partial T/\partial n)_-|} + \frac{1}{|(\partial T/\partial n)_+|} \right]_f^{-1} \cdot \delta_{rxn}, \qquad (6.5\text{-}66)$$

where all quantities marked f are evaluated in the flame "sheet" (asymptotic) limit), then Eq. (6.5-64) becomes a nonlinear equation for δ_{rxn}, a first approximation to which yields:

$$\delta_{rxn}^{(0)} \approx \left\{ \frac{(D_{F,\,eff}\,\rho)_f}{\frac{1}{16}\left(\frac{pM}{RT_f}\right)^2 \frac{A}{M_O} \left|\left(\frac{\partial \omega_O}{\partial n}\right)_f\right| \exp\left(-\frac{E}{RT_f}\right)} \right\}^{1/3}. \qquad (6.5\text{-}67)$$

Note that the direction normal to the flame is locally colinear with the vector $(\mathbf{grad}\,\Omega_{OF})_f$, and the directional derivative $(\partial\omega_O/\partial n)_f$ can readily be evaluated in terms of the "conserved variable," $\Omega_{OF}(\mathbf{x}, t)$.

For both laminar-[49] or turbulent-flow systems δ_{rxn}/d_j is found to diminish at high pressures and large jet diameters; however, in *turbulent* situations, local reactant unmixedness (Section 5.7.3) and $D_{F,\,eff}$ are typically large enough to cause the flame "sheet" approximation (in terms of *time-averaged* ω_i and T-variables) to *fail*. Thus, while fully turbulent systems tend to be those for which the effective diffusivities are nearly equal [cf. assumption(3)], $-\dot{r}_F'''(\bar{\omega}_O, \bar{\omega}_F, \bar{T})$ overestimates $-\bar{\dot{r}}_F'''$ and these effective diffusivities are large enough to cause the time-averaged flame zone to resemble a "brush" rather than a mathematical "sheet" (see below). Nevertheless, the *trends* derived in Section 6.5.5.5 for turbulent flame lengths, L_f, remain valid, however one chooses to define L_f.

Laminar Flamelet Approach to Turbulent Diffusion Flame "Structure" (A5, Continued). The time-averaged structure of "thick" turbulent diffusion flames can be rationalized and predicted in the limit when the characteristic chemical kinetic time, t_{chem}, is small compared to the local turbulent eddy mixing (break-up) time, κ_t/ε. In this limit the apparent thickness of the flame can be viewed as due to the passage of a succession of thin laminar diffusion "flamelets" at each observation point $\mathbf{x}$ (Toor (1962), Bilger (1980)). To understand and implement this idea consider the spatial distribution *and temporal statistics* of a conserved scalar variable Ω defined, as in Eq. (6.5-35) to have the value unity in the "fuel" jet and zero in the surrounding "oxidizer" jet. In a steady *laminar* diffusion flame there would be a one-to-one correspondence between any local scalar property (temperature, species composition, etc.) and the local value of this "mixture fraction" variable, Ω. According to the "flamelet" picture, these same *interrelationships* would prevail within the turbulent reacting flow—i.e., the spatial and turbulence characteristics of $\Omega(\mathbf{x}, t)$ contain all information needed to compute $\bar{T}(\mathbf{x})$, $\bar{\omega}_O(\mathbf{x})$, $\bar{\omega}_F(\mathbf{x})$, etc. Now the spatial distribution of a Reynolds'-averaged conserved scalar $\bar{\Omega}$ and its mean

[49] For laminar jets, low U_j-values also favor small relative reaction-zone thicknesses.

square fluctuation intensity $\overline{\Omega'\Omega'} \equiv g\{\mathbf{x}\}$ (or their Favre-averaged counterparts) can be obtained from experiments) or estimated, using, say, a 3-PDE turbulence model (e.g., κ_t-ε-g (Spalding (1971)) as outlined in Section 5.7.3). This information, combined with a simple statistical assumption about the nature of Ω-fluctuations, would then be sufficient to describe the observable features of such a turbulent reacting flow. For example, suppose we assume a simple (clipped Gaussian) *shape* for the function describing the fraction of time the scalar Ω is locally achieved—the so-called "probability density function," $p\{\Omega\}$. If the two parameters in this pdf are chosen such that:

$$\int_0^1 \Omega\, p\{\Omega\}\, d\Omega = \bar{\Omega} \tag{6.5-67a}$$

and

$$\int_0^1 (\Omega - \bar{\Omega})^2 \cdot p\{\Omega\}\, d\Omega = g \tag{6.5-67b}$$

then the time average of any scalar quantity $\phi\{\Omega\}$ of interest (temperature, species concentration, etc.) that depends upon the local mixture fraction, Ω, could be obtained from:

$$\int_0^1 \phi\{\Omega\} \cdot p\{\Omega\}\, d\Omega = \bar{\phi}. \tag{6.5-67c}$$

This ability of laminar diffusion flame "state relationships," $\phi\{\Omega\}$, to describe approximately the observed properties of turbulent diffusion flames, has recently been experimentally verified for the following gaseous fuels burning in air: H_2, CH_4, C_2H_4, C_3H_8, C_7H_{16}, and CO. Since ϕ can be applied to the local mixture temperature, each of the major polyatomic species concentrations, and, perhaps, even to the local soot volume fraction, the "wrinkled laminar flamelet" model briefly outlined above also opens the door to the rational prediction of IR-radiation emission from luminous turbulent diffusion flames (Faeth, Jen, and Gore (1985)), such as industrial furnaces, flares, and fires.

Under conditions for which the *thin* "flamelet" picture breaks down, a practical method for incorporating and predicting the effects of arbitrary homogeneous chemical kinetics is necessary. For one algorithm in widespread engineering use (Spalding (1976)), the time-averaged rate of fuel consumption $-\overline{\dot{r}_F'''}$ is limited to either $-\dot{r}_F'''\{\bar{\omega}_O, \bar{\omega}_F, \bar{T}\}$ or the local eddy break-up rate (proportional to $(\overline{\Omega'\Omega'})^{\frac{1}{2}} \cdot (\kappa_t/\varepsilon)^{-1}$), whichever is smaller. In closing this section it should also be remarked that the "laminar flamelet" idea can be carried over to describe turbulent *premixed* situations (Bray, Libby, and Moss (1984)), with the required information coming from a combination of (a) the properties of nonplanar premixed laminar flames (cf. Section 6.5.5.8) and (b) the statistics of premixed laminar flame movement past each point. For details on the implementation of these interesting ideas the reader is directed to the references cited above. Their applicability to complex multistep chemical reactions of interest for large-scale chemical synthesis should also be noted.

Cold-Flow Field $\overset{?}{\approx}$ *Hot Flow Field (6).* Homogeneous exoergic chemical reaction in a flowing fluid can alter the flow pattern *via* associated (a) density reduction, (b) changes in intrinsic fluid-transport properties (e.g., viscosity), and/or (c) changes induced in the "effective" transport properties (*via* increase or reduction in turbulence, or, in extreme cases, transition/relaminarization). Density change effects (a) can themselves be viewed as the results of:

1. reduced local "inertia" of the reaction-heated fluid and, hence, greater dynamic response to prevailing pressure gradient and viscous forces;
2. "displacement" effects associated with a local increase in velocity required for the same local mass flux; and/or
3. local "buoyancy" forces in a body force (e.g., gravity) field, with (1) and (2) usually being especially important in "confined" (e.g., duct-like) flows (see, e.g., Figure 6.5-1). Moreover, in steady-flow gas reactors, the density ratio ρ_{max}/ρ_{min} will be of the order of T_f/T_∞ (where T_f is the flame temperature (Section 6.5.5.2)) and, hence, often as large as 5 or 10 to 1. In view of these considerations, "cold-flow" experiments, even if performed on full-scale equipment (see Chapter 7), might appear to offer little chance of providing *quantitative* information about "corresponding" chemically reacting gas flows.

Nevertheless, Johnstone and Thring (1957) report: "Most of the important features of gas flow in (combustion) furnaces can be simulated on the small scale without the application of heat." These authors have reviewed the use of "seeded" (for flow visualization) cold water[50] or air models of open-hearth furnaces, and gas turbine combustion chambers, usually constructed from a transparent material, e.g., Lucite or Perspex. Of course, the economic incentives to fully exploit "cold-flow" data are enormous, and have spawned very clever examples, despite the obvious pitfalls.

In single-phase gaseous combustion applications, the dominant cause of hot/cold-flow disparities is localized density reduction (a), especially at the very high Reynolds' numbers (frequently *ca.* 10^5) characterizing industrial combustors. Even so, it is difficult to state a sufficiently general criterion to ensure that these disparities would be negligible in the face of mechanisms (1), (2), and (3) acting together. Regarding (1) and (3), respectively, it is noteworthy that even though $[(T_f/T_\infty) - 1]$ may not itself be small, "cold" forced-convection data can still be useful if:

1. $|\Delta C_p|_{cf} \cdot [(T_f/T_\infty) - 1] \ll 1$, where the premultiplier is a characteristic change in *pressure coefficient* in the cold-flow (see Section 4.4.1).
3. The group $U_j^2/(gd_j)$ (sometimes called the Froude number) is sufficiently *large* (see, also, Section 7.2.3.2c).

[50] Cold-flow models can also approximate variable-density effects, as in the use of two distinct liquids, each with its own density.

6.5.5.7 Application to Isolated Fuel Droplet ("Envelope"-Flame) Combustion

Our former restriction to single-phase systems (e.g., a chemically reacting gas mixture) does *not* preclude applications in which the chemically reacting gas is "coupled" to a nonreacting condensed phase (e.g., a fuel droplet supplying the vapor for a gaseous diffusion flame, as depicted in Section 1.1.3).

For simplicity and definiteness, consider the chemically reacting analog of the single spherical droplet quasi-steady *vaporization* problem, treated quantitatively in Section 6.4.3.3. Thus, we retain all of the stated assumptions except (7), which is now altered to read:

7′ Species A (now fuel F) reacts chemically in the vapor phase with an oxidizer species (O) supplied from the ambient gas mixture, in accord with the overall irreversible stoichiometric reaction (6.5.5.3).

Moreover, to exploit the "conserved variable" approach, we add the assumptions (1) through (4) of Section 6.5.5.6; in particular, the assumption:

$$\alpha_{\text{mix}} \approx D_{i-\text{mix}} \qquad (i = \text{O, F,P}), \tag{6.5-68}$$

where these are the *diffusivities* appearing in the ω_i and T-field equations (6.5-28, -29) governing the nonturbulent vapor phase surrounding the spherical droplet.

Chemical reaction surrounding the fuel droplet does *not* alter the boundary conditions (BC), which remain $\omega_i = \omega_{i,\infty}$ $(i = \text{O, F, P})$ and $T = T_\infty$ *at* $r = \infty$ and, at $r = d_p/2$:

$$\omega_F = \omega_{F,\text{eq}}\{T_w\},$$

$$\dot{m}'' = \begin{cases} \dfrac{1}{1-\omega_{F,w}}(D_{F-\text{mix}}\rho)\left(-\dfrac{d\omega_F}{dr}\right)_w, & (6.5\text{-}69) \\[2ex] \dfrac{\rho c_p}{L_F}\cdot(\alpha_{\text{mix}})\left(\dfrac{dT}{dr}\right)_w. & (6.5\text{-}70) \end{cases}$$

However, the effect of exoergic chemical reaction within the diffusion layer about the sphere is to steepen the gradients of ω_F and T, and, hence, increase the droplet vaporization rate (cf. discussion of $F\{\text{reaction}\}$ in Section 6.3.4).

Oxidizer and reaction-product concentrations at $r = d_p/2$ are also *not* known in advance; however, these species are assumed to have zero mass flux at the droplet surface. Thus, in view of Eq. (6.2-5):

$$0 = \dot{m}''\omega_{i,w} - (D_{i-\text{mix}}\rho)\left(\frac{d\omega_i}{dr}\right)_w \qquad (i = \text{O, P}). \tag{6.5-71}$$

Only if the homogeneous reaction rate is sufficiently rapid is it possible to assume $\omega_{O,w} \approx 0$, in which case $(d\omega_O/dr)_w = 0$ as well.

An exact mathematical correspondence between this (vaporization + reaction) problem and the corresponding vaporization problem can be established (see Table 6.5-2) by introducing normalized versions, Ω_{FO} and Ω_{TO}, of the "conserved" variables

$$\omega_{FO} \equiv \omega_F - f\omega_O \qquad \text{and} \qquad \theta_{TO} \equiv T + fQ\omega_O/c_p$$

and examining the resulting dimensionless ODEs and BCs.

In view of the vaporization-rate result (Eq. (6.4-25)), this correspondence reveals that the fuel-vaporization rate *in the presence of vapor-phase combustion* must be given by:

$$\dot{m}'' \cong \frac{2}{d_p}\left(\frac{k}{c_p}\right) \cdot \ln[1 + B_{h,\mathrm{comb}}], \tag{6.5-72}$$

where, as quoted earlier:

$$B_{h,\mathrm{comb}} = \frac{c_p(T_\infty - T_w)}{L_F} + \frac{fQ\omega_{O,\infty}}{L_F}. \tag{6.5-73}$$

In principle, the QS droplet temperature, T_w, follows from the nonlinear equation[51] $B_{m,\mathrm{comb}} = B_{h,\mathrm{comb}}$ or (cf. Table 6.5-2):

$$\frac{\omega_{F,\mathrm{eq}}\{T_w\} + f\omega_{O,\infty}}{1 - \omega_{F,\mathrm{eq}}\{T_w\}} = \frac{c_p(T_\infty - T_w) + fQ\omega_{O,\infty}}{L_F\{T_w\}}; \tag{6.5-74}$$

however, in the presence of vapor-phase combustion, it is usually sufficiently accurate to assume $T_w \approx T_{bp}\{p\}$ in Eqs. (6.5-72, -73).

Finally, the combustion analog of the droplet QS vaporization time t_{life} (cf. Eq. 6.4-29) is:

$$t_{\mathrm{life,comb}} = \frac{d_{p,0}^2}{8\alpha \dfrac{\rho_g}{\rho_\ell} \ln[1 + B_{h,\mathrm{comb}}]}. \tag{6.5-75}$$

Note that, as previously remarked (Section 1.1.3), in this "diffusion-controlled" limit the fuel-droplet lifetime depends *quadratically* on the initial diameter, and is relatively insensitive to the detailed fuel chemistry (cf. Table 6.5-1).

It is also interesting to note that, while these simple results are obtained on the assumption that the homogeneous reaction rate is "sufficiently" rapid (to ensure

---

[51] The strongest nonlinearity arises because of the Clausius'–Clapeyron vapor-pressure relation underlying $\omega_{F,\mathrm{eq}}\{T_w\}$; however, near the fuel thermodynamic critical temperature, the T_w-dependence of the latent heat L_F cannot be neglected (see Figure 6.4-2b and Rosner and Chang (1973)).

Table 6.5-2 Correspondence between Treatments of Droplet Evaporation without[a] and with[b] Homogeneous Chemical Reaction

Droplet Evaporation[a]		*Droplet Evaporation with Combustion*[b]	
Variable or Parameter	*Meaning*	*Variable or Parameter*	*Meaning*
$\omega_A \equiv \rho_A/\rho$	Vapor mass fraction	$\omega_{FO} \equiv \omega_F - f\omega_O$	"Conserved"[c] ω_F, ω_O combination
$\Omega_A \equiv \dfrac{\omega_A - \omega_{A,\infty}}{\omega_{A,w} - \omega_{A,\infty}}$	Normalized vapor mass fraction	$\Omega_{FO} \equiv \dfrac{\omega_{FO} - \omega_{FO,\infty}}{\omega_{FO,w} - \omega_{FO,\infty}}$	Normalized ω_{FO} variable
T	Vapor mixture temp	$\theta_{TO} \equiv T + fQ\omega_O/c_p$	"Conserved"[c] T, ω_O combination
$\Omega_T \equiv \dfrac{T - T_\infty}{T_w - T_\infty}$	Normalized vapor mixture temperature	$\Omega_{TO} \equiv \dfrac{\theta_{TO} - \theta_{TO,\infty}}{\theta_{TO,w} - \theta_{TO,\infty}}$	Normalized θ_{TO} variable
$B_m = \dfrac{\omega_{A,w} - \omega_{A,\infty}}{1 - \omega_{A,w}}$	Stefan flow[d] parameter $\dfrac{v_w \delta_m}{D}$	$B_m = \dfrac{\omega_{FO,w} - \omega_{FO,\infty}}{1 - \omega_{F,w}}$	Stefan flow[d] parameter $\dfrac{v_w \delta_m}{D}$
$B_h = \dfrac{c_p(T_\infty - T_w)}{L_A}$	*Stefan flow*[d] parameter $\dfrac{v_w \delta_h}{\alpha}$	$B_h = \dfrac{c_p(T_\infty - T_w) + fQ\omega_{O,\infty}}{L_F}$	Stefan flow[d] parameter $\dfrac{v_w \delta_h}{\alpha}$

[a] See Section 6.4.3.3 for outline of QS isolated droplet case.
[b] Rapid single-step stoichiometric reaction (Eq. (6.5-27)) with $\omega_{O,w} \ll \omega_{F,w}/f$ and $D_f \approx D_O \approx \alpha$ (Section 6.5.5.7).
[c] See Section 6.5.5.1.
[d] See Section 6.4.3.1.

$\omega_{O,w} \ll \omega_{F,w}/f$), no other restrictions have been placed on the form or magnitude of reaction rate, $-\dot{r}_F'''\{\omega_O, \omega_F, T\}$, and it is certainly *not* necessary that the chemical reaction be confined to a mathematical "sheet" at some radius $r_f > d_p/2$ (cf. Section 6.5.5.2). However, important questions (such as when an ignited droplet will *extinguish*, or when a fuel droplet will *ignite*) cannot be answered without considering the kinetics of the fuel + oxidizer reaction.

When the parameter $(\rho_g/\rho_\ell)\ln[1 + B_{h,\text{comb}}]$ is *not* small and yet the surface temperature T_w predicted from Eq. (6.5-74) is below the thermodynamic critical temperature of the fuel, then the QS approximation fails but the more general *transient* isolated-sphere combustion problem can still be solved (using the "conserved-variable" approach (Rosner and Chang (1973)) in terms of the corresponding transient-sphere phase change (without reaction) problem.

6.5.5.8 Further Generalizations of the "Conserved-Variables" Approach: Applications to Premixed, Chemically Reacting Systems

The fact that, if $\alpha_{\text{eff}} \cong D_{i,\text{eff}}$, variables such as: $\omega_F - f\omega_O$, $c_pT + fQ\omega_O$, $\omega_O + (1 + f)^{-1}\omega_p$ etc., satisfy the *source-free* convective-diffusion equation, (6.5-36) can also be used to simplify the prediction of *premixed* combustion systems. An important, simple illustration is the simultaneous prediction of laminar flame *structure* and *propagation speed*, S_u, when the exoergic reaction (6.5-27) occurs at some local rate $-\dot{r}_F'''\{\omega_O, \omega_F, T\}$.

In premixed systems there is usually no "diffusion-controlled limit"; that is, the chemical kinetics always play an essential role in determining the observed rate. However, as in the case of initially unmixed systems with arbitrary rates, the use of conserved variables can reduce the number of coupled field equations that must be simultaneously integrated.

Consider a laminar, planar, nearly constant pressure deflagration wave (cf. Section 4.3.2). Relative to a translating coordinate system *fixed to the "wave,"* unburned gas flows in with a mass flux ($\rho_u S_u$, as yet undetermined) which remains constant throughout the wave. The unburned gas conditions ($\omega_{O,u}, \omega_{F,u}, T_u$) are presumed known and, for a "fuel-lean" condition, the adiabatic burned gas temperature will necessarily be: $T_u + \omega_{F,u}Q/c_p$, since $\omega_{F,b} = 0$. From an overall oxygen mass balance, we will be left with: $\omega_{O,b} = \omega_{O,u} \cdot [1 - (\omega_{F,u}/f)]$.

Subject to these overall "jump" conditions, it remains to also find the *wave's internal structure*, that is, $\omega_i\{z\}$, ($i = $ O, F, P) and $T\{z\}$, which are assumed to be governed by the coupled set of ODEs (6.5-28, -29).

This formidable problem can, however, be reduced to the consideration of a *single* nonlinear ODE, say for $T\{z\}$, using "conserved variables" to eliminate the composition variables ω_O and ω_F in favor of the local temperature, T.

In general, $T\{z\}$ will satisfy an ODE of the now-familiar steady-state form:

$$\rho_u S_u c_p \frac{dT}{dz} = \frac{d}{dz}\left(k\frac{dT}{dz}\right) - \dot{r}_F'''\{\omega_O, \omega_F, T\}Q \qquad (6.5\text{-}76)$$

and, in what follows, we will neglect the variability of k and c_p. To eliminate ω_O, ω_F in favor of T, consider the "conserved variables": $c_p T + fQ\omega_O$ and $c_p T + \omega_F Q$. The latter satisfies the *source-free* ODE and the above-mentioned BC, the solution of which is simply:

$$c_p T + \omega_F Q = \text{const} = c_p T_u + \omega_{F,u} Q = c_p T_b. \quad (6.5\text{-}77)$$

Moreover, the variable $c_p T + fQ(\omega_O - \omega_{O,b})$ must be the same constant. Thus, despite the fact that T, ω_O and ω_F all change *within* the flame, these *combinations* remain constant. Returning to the ODE (6.5-76), no matter how complex the kinetic law, clearly these "first integrals" can now be used to eliminate $\omega_O(z)$ and $\omega_F(z)$ and write it as a nonlinear differential equation involving only $T(z)$.

As indicated in Section 7.2.2.2, the solution $T(z)$ to this nonlinear two-point BVP can be found only for a particular ("eigen-") value of a *parameter* involving the flame speed, S_u. In this way the laminar-flame speed, and *associated flame structure*, are found.

6.6 REMARKS ON TWO-PHASE FLOW: MASS-TRANSFER EFFECTS OF INERTIAL "SLIP" AND "ISOKINETIC" SAMPLING

When the dynamical coupling between suspended particles (or heavy "solute" molecules) and the "carrier" fluid is weak, it is best to consider these particles as comprising a distinct phase, albeit "coexisting" in the same space with the carrier fluid. This, loosely, is the distinction between "two-phase" flow and the flow of ordinary "mixtures." Quantitatively, the distinction becomes clearest in terms of the *Stokes' number*, Stk (Section 6.2.6), for above some critical value of Stk (cf. Table 6.6-1) the "second phase" can inertially impact on a target (with finite velocities), while the host *fluid* is brought to rest by the same target.

Table 6.6-1 Critical Stokes' numbers for "pure" inertial impaction.

Target Geometry	L	$\text{Re} \equiv U(2L)/\nu$	Stk(crit)	*Source*[b]
Cylinder	$d/2$	$\gg 1$	1/8	Taylor [1940]
Cylinder	$d/2$	0.1 (Oseen)	4.3	Natanson [1957]
Sphere	$d/2$	$\gg 1$	1/12	Levin [1961]
Sphere	$d/2$	$\ll 1$	1.21194	Michael and Norey [1970]
Ribbon	$w/2$	$\gg 1$	1/4	Levin [1961]
Jet normal to plate	$d_j/2$	$\gg 1$	0.637	Levin [1961]

[a] "Point" particles (cf. target dimension L).
[b] For a summary of this research, see Fuchs (1964) and Friedlander (1977).

6.6.1 Pure Inertial Impaction at Supercritical Stokes' Numbers: Cylinder in Cross-Flow

It is instructive to consider a particle-laden steady carrier flow of mainstream velocity, U, past an object of macroscopic dimension L. If the suspended particles can be assumed to be:

1. spherical, with diameter $d_p \ll L$;
2. present with a mass loading and volume fraction that are negligible;
3. large enough to neglect their Brownian diffusivity, but small enough to neglect their gravitational sedimentation (that is, $\text{Pe} \to \infty, gt_p/U \ll 1$);
4. captured (rather than "bouncing") upon impact with the target, irrespective of angle of incidence or velocity upon impact;

then each particle will move along a trajectory determined by the unperturbed host-fluid velocity field and its drag law at the prevailing particle Reynolds' number (based on the local "slip" velocity). In this extreme, and with Assumption 4 above, one can, in principle, calculate the capture efficiency function:

$$\eta_{\text{capture}}\{\text{Stk}, \text{Re}, \text{shape}, \text{orientation}\}, \tag{6.6-1}$$

by determining the "limiting-particle trajectories," that is, the upstream locations of particles whose trajectories will become tangent to the target (Langmuir, Brun, *et al.* (1958)).

An interesting feature of this idealized model of particle capture from a two-phase flow is that $\eta_{\text{capture}} = 0$ for $\text{Stk} < \text{Stk}_{\text{crit}}$; that is, capture occurs only above a *critical Stokes' number*.[52] This behavior is shown in Figure 6.6-1 for a cylindrical target in cross-flow (Rosner and Israel (1983) correlation[53] of the calculations of Brun (1958)).

In practice, some deposition will always occur at $\text{Stk} < \text{Stk}_{\text{crit}}$, owing to non-zero Brownian diffusivity, thermophoresis, and/or other mechanisms of transport. Deposition rates below Stk_{crit} are still influenced by Stk since, while the host carrier fluid may be subsonic, the particle fluid is "hypersonic," and, hence, compressible. This means there is an inertial "enrichment" (pile-up) of particles in the forward stagnation region, and a "centrifugal" depletion downstream along the cylinder (Fernandez de la Mora and Rosner (1981)).[54] Interestingly enough, the net effect on *total* deposition rate can be to *reduce* the deposition on a bluff body below that expected based only on convective diffusion of particles in the "mixture."

[52] Stk_{crit} is 1/8 for inviscid ($\text{Re} \to \infty$) host flow past a transverse circular cylinder (cf. Table 6.6-1).
[53] By redefining Stk such that the effects of non-Stokes' drag are included in t_p, the residual dependence of η_{capture} on an additional parameter has been suppressed (see Israel, R., and Rosner, D.E. (1983)).
[54] At very small Stk, these effects are recovered from a single-phase "mixture" flow analysis by including "pressure diffusion."

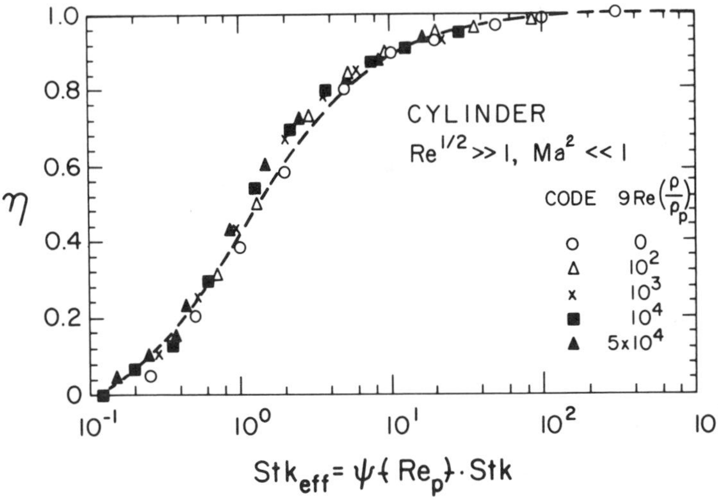

Figure 6.6-1 Particle capture fraction correlation for ideal ($\mathrm{Re} \to \infty$) flow past a transverse circular cylinder (Israel and Rosner (1983)). Here $t_{\mathrm{flow}} \equiv (d/2)/U$.

A very important combustion application of two-phase fluid mechanics/mass transfer is in the experimental *sampling* of particle-laden (e.g., sooty) combustion gases, using a small "suction" probe. If the probe sampling rate is too great (Figure 6.6-2a) the "capture efficiency" for the host gas will exceed that of the prevailing particles, leading the *sampled* stream to be relatively particle-deficient. Conversely, if the probe sampling rate is too small, (Figure 6.6-2b) the capture efficiency of the *particles* will exceed that of the gas, leading the sampled stream to be relatively particle-rich. At some particular probe sampling rate (unfortunately different for each size of particle contained in the mainstream), the effective capture efficiencies

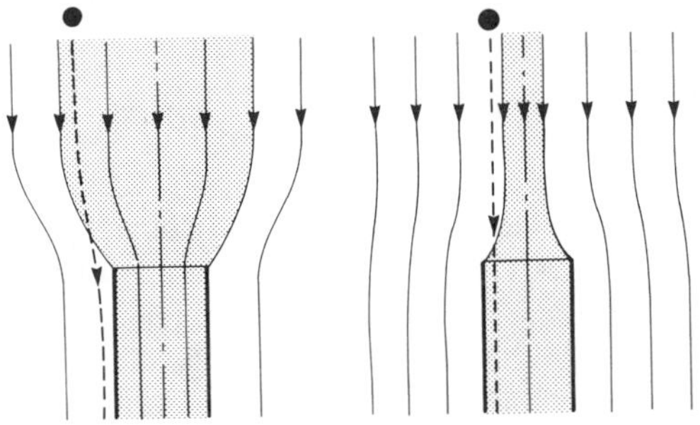

Figure 6.6-2 Effect of probe sampling rate on capture of particles and their carrier fluid.

for particles and carrier gas become equal. This is sometimes (loosely) called the "isokinetic" condition.

From the above discussion it is clear that in inertial (or diffusional) sampling from a gas containing a "polydispersed" (multisize) aerosol, there is a *biasing* associated with the fact that, in a given gas-dynamic situation, each size of particle will have its own capture efficiency. This must be corrected for, if one is interested in reporting the corresponding *mainstream* particle-size distribution (PSD).

6.6.2 Effective Diffusivity of Particles in Turbulent Flow

The ability of small suspended particles to "follow" the local turbulence (despite their "inertia") is likewise governed by a Stokes' number, written Stk_t. Here:

$$\mathrm{Stk}_t \equiv \frac{t_\mathrm{p}}{l_t/(\overline{v' \cdot v'})^{1/2}}, \tag{6.6-2}$$

where the relevant local flow time is the ratio of the *scale* of turbulence, l_t, to the *rms* turbulent velocity. A convenient alternative form of the characteristic turbulent eddy time, especially if one is characterizing the local host-fluid turbulence *via* the two scalars κ_t (turbulence kinetic energy per unit mass) and ε (turbulent viscous dissipation rate per unit mass) (cf. Section 5.7.3) is $t_{\mathrm{eddy}} \equiv k_t/\varepsilon$, in which case one would use $\mathrm{Stk}_t \equiv t_\mathrm{p}/(k_t/\varepsilon)$ (cf. Eq. (6.6-2)). In either case one expects:

$$D_{\mathrm{p,eff}}/\nu_t = (\mathrm{Sc}_t)^{-1} \cdot \mathrm{fct}\{\mathrm{Stk}_t\}, \tag{6.6-3}$$

for particles suspended in "fully" turbulent flow regions (i.e., where $\nu_t \gg \nu$). Limited data on particle dispersion in turbulent jets suggest that $\mathrm{fct}\{\mathrm{Stk}_t\}$ *increases* appreciably above unity for Stk_t above about 10^{-1} (Goldschmidt, *et al.* (1966)). Clearly, if a reasonably general expression for $D_{\mathrm{p,eff}}/\nu_t$ can be established (Eq. (6.6-3)) then, at least for fully turbulent flows with low particle mass loading and particles small enough to be closely coupled to the *time-averaged* flow, particle-concentration distributions can be economically predicted using the same Eulerian convective "eddy"-diffusion calculation procedures as for vapors.

A quite different (non-Eulerian) approach to turbulent particle dispersion, and one which appears to be particularly well suited for situations in which the particles cannot even follow the *time-averaged* motion of the carrier fluid, is the *stochastic particle-tracking technique* (see, e.g., Boysan, Ayers, and Swithenbank (1982)). For the simplest cases (in which the particle mass fraction is small enough to neglect particle–particle interactions and particle modification of the carrier-fluid motion), one first computes the host-fluid flow field, including the distribution of turbulence quantities (say κ_t and ε) using an Eulerian PDE method. One then calculates, using the (Lagrangian) equations of motion for each particle, the trajectories of an *ensemble* of particles "injected" from each point across the inlet of the particle-laden stream. To account for turbulent "dispersion," each particle is assumed to reside in a succession of eddies which modify the local free-stream velocity it "sees." These fluctuation-velocities are chosen by random sampling (such that the average local fluctuation velocity components would pertain to the prevailing root-mean-square (rms)-values) and each set of such velocities persists for a time equal to the mean local eddy lifetime (proportional to κ_t/ε). While obviously demanding of computer time and memory, this "Monte Carlo" approach to turbulent particle dispersion appears to have the versatility required to account for all particle inertial effects—with respect to both the local turbulence and the time-averaged flow. Accordingly, it should prove useful in the engineering design and optimization of two-phase equipment involving both types of inertial phenomena and, perhaps, wall impingement phenomena, such as cyclones (for gas/solid separation; *loc. cit.*) and liquid fuel spray combustors (Ewan, B.C.R., Boysan, F., and Swithenbank, J. (1985)). However, an alternative "two-coexisting-fluid" Eulerian approach (see, e.g., Section 2.6.3, and Fernandez de la Mora and Rosner (1981, 1982)) may still prove to be a more efficient way to deal with many such problems, especially when the particle *mass* loading is high enough to modify the "host" flow field, including its turbulence characteristics. This objective motivates recent activity toward developing two-equation (κ_t-ε type) models for each coexisting phase of a two-phase turbulent flow (see, e.g., Elghobashi, S., and Abou-Arab, T.W. (1983)).

6.6.3 Eddy Impaction

Most particle-containing eddies do not actually reach a deposition surface; hence, their particles must ordinarily "fight their way" to the surface by Brownian diffusion across the viscous sublayer. However, when Stk_t is sufficiently large, some eddies can "project" their particles *through* the viscous sublayer, increasing the particle deposition rate above that expected from, say, Eq. (6.5-11)—(based on the existence of a rate-limiting "laminar" sublayer). While not fully understood theoretically, such deposition-rate enhancements have been empirically correlated with a modified Stokes' number of the type (cf., e.g., Hidy and Heisler (1978));

$$\mathrm{Stk}_{t,\mathrm{eff}} \equiv \frac{t_\mathrm{p}}{\nu/(\bar{\tau}_\mathrm{w}/\rho)} \equiv t_\mathrm{p}^+ . \qquad (6.6\text{-}4)$$

Since $\bar{\tau}_\mathrm{w}/\rho$ is of the same order of magnitude as the square of the mainstream fluctuation-velocities (that is, $\overline{v'_x v'_y}$), this is equivalent to using $\nu/(\overline{v'_x v'_y})$ as the characteristic flow time in the definition of the local particle Stokes' number. The "eddy impaction" augmentation of turbulent boundary-layer mass-transfer coefficients, St_m, is found to be negligible for $\mathrm{Stk}_{t,\mathrm{eff}}$-values less than about 10^{-1}; that is, below this value a turbulent particle-containing boundary layer behaves like a "single-phase" fluid.

6.7 RESIDENCE-TIME DISTRIBUTIONS: TRACER "DIAGNOSTICS" WITH APPLICATION TO THE MATHEMATICAL MODELING OF NONIDEAL-FLOW REACTORS

6.7.1 Introduction

It is simple to apply *overall* material, energy, and momentum balances to a single chemical reactor; however, the information obtained from this "black box" approach is very far from sufficient to predict or understand its important performance characteristics (e.g., ignitability and stability limits, overall efficiency, pressure drop, selectivity, pollutant emission, noise ("roar"), etc.). On the other hand, it is usually impractical to divide the chemical reactor into subregions small enough such that we approximate (with negligible truncation error everywhere) the PDEs governing mass, momentum, and energy conservation within the entire space (cf. Figure 1.3-3). A practical "intermediate" solution is to model the real reactor by a discrete *network* of a small number of interconnected *ideal "reactor" (vessel) types.* Not only is the performance of each such vessel easy to analyze (see Section 6.1.3), but so is the performance of the resulting network.

While we have emphasized the power of an Eulerian formulation of these conservation equations, ironically it is useful to adopt here a Lagrangian viewpoint to *define* the two most important *ideal* reactor (vessel) types (Sections 6.7.2 and 6.7.3); *viz.*:

1. "Plug-Flow" Reactor (PFR), and
2. "Well-Stirred" Reactor (WSR),

both operating under fluid-dynamically steady-state conditions (feed mass flow rate, $\dot{m}$). The distinction will be made in terms of their *exit-age distribution functions* or *residence-time distribution functions* (RTDFs), $E\{t\}$, defined such that $E\{t\}\,dt$ is the *fraction of material at the vessel outlet stream that has been in the vessel for times between t and t* $\pm$ *dt.*

Thus, a PFR is a reactor whose $E\{t\}$ is a Dirac function centered at the residence time $V/(\dot{m}/\rho)$, where V is the vessel *volume.* Physically, this means that *all* fluid parcels in a PFR reside in the reactor for precisely the same time, $V/(\dot{m}/\rho)$. One realization of a PFR (from which the name is, of course, derived) is an unpacked, straight tube through which an incompressible fluid flows with a uniform ("plug flow") velocity profile.

On the other hand, a "well-stirred reactor" (WSR)[55] or vessel is one for which the entering fluid (feed) is instantaneously mixed so well that (a) no spatial gradients

[55] Also called a *stirred-tank reactor* (STR).

exist within the entire vessel volume, V, and (b) all properties of the exit stream are identical to the corresponding properties *within* the vessel (Section 6.1.3). The RTDF for a WSR will be discussed in Section 6.7.3.

An important question, to which there is no unique answer, is: What "equivalent" vessel network (combination of vessel types, volumes, and interconnections) will perform like the real reactor of interest? This question is raised here because key clues are provided by its overall RTDF—obtained by suddenly injecting a tracer chemical pulse into the reactor *inlet*, and then measuring the *output*-time history[56] (see Eq. (6.7-10)). As will be seen, RTD-analysis represents a straightforward and important application of mass-transfer theory to individual macroscopic vessels and networks of such vessels.

It is conceptually useful to think of the equivalent network as a discrete ("modular" or "lumped") approximation to the real chemical reactor. As the number of interconnected vessels increases without limit, the behavior of the discrete network approaches the behavior of the "continuum" within the entire reactor space.

6.7.2 Ideal Plug-Flow Reactor (PFR) and Its RTDF

Any vessel for which $E\{t\}$ is $\delta\{t - t_{\text{flow}}\}$, where $t_{\text{flow}} \equiv V/(\dot{m}/\rho)$ and $\delta\{\quad\}$ is the Dirac ("impulse") function, will be called a PFR (see Column 1 of Figure 6.7-1, in which $\bar{t} \equiv t_{\text{flow}}$). However, the nature of the "piping" can also influence the RTDF even for a single vessel, as can be seen by considering a PFR with *partial recycle* (Figure 6.7-2). Here, for every gram of feed, a fraction R is taken from the PFR effluent and diverted back into the inlet. This gives a portion of the effluent "another chance," increases the flow rate through the reactor itself to $(1 + R)\dot{m}$, and alters its RTDF.

6.7.3 Well-Stirred Reactor (WSR) and Its RTDF

A simple transient material balance, based on Eq. (2.1-7) with negligible reaction (for the tracer) and applied to the vessel as a whole, leads to (see Exercise 2.12, with $E = dF/dt$):

$$E(\text{WSR}) = (t_{\text{flow}})^{-1} \exp[-t/t_{\text{flow}}], \tag{6.7-1}$$

where, again, t_{flow} is the characteristic flow time: $V/(\dot{m}/\rho)$ (cf. Figure 6.7-1, Column 2). Equation (6.7-1) reveals that the most likely (probable) residence time in a

---

[56] Response to a step-function or sinusoidal tracer input can also be used to determine the RTDF. The response to a step function, suitably rescaled, immediately gives the *cumulative* residence-time distribution function (CRTDF): $F \equiv \int_0^t E\{t\}\,dt$ (Figure 6.7-1); see, e.g., Ex. 2.12.

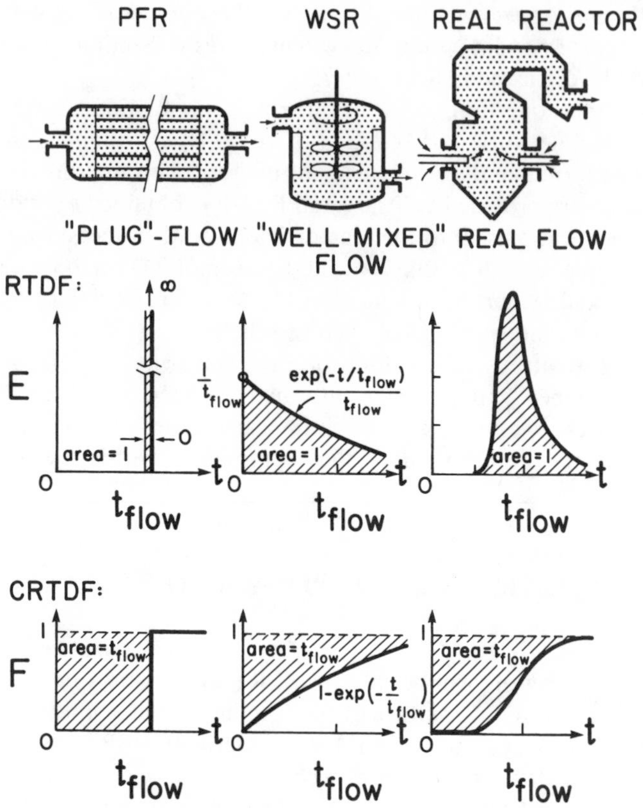

Figure 6.7-1 Tracer residence-time distribution functions for ideal and real vessels (e.g., reactors) (adapted from Levenspiel (1972)).

WSR is zero(!), and the *mean* residence time is $V/(\dot{m}/\rho)$. In contrast to the PFR, not all fluid parcels in a WSR have the same residence time.

A quantitative (but, of course, incomplete) indicator of the *spread* of residence times on a vessel is provided by the dimensionless *variance* about the mean residence time. Whereas the mean residence time is clearly related to the first moment of $E(t)$,

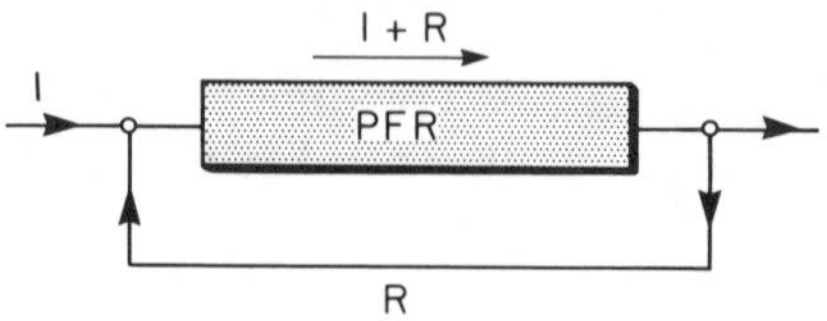

Figure 6.7-2 Ideal plug-flow reactor (PFR) with partial "recycle" (recycle introduces a distribution of residence times, and reduces the residence time per pass within the PFR).

that is,

$$t_{\text{flow}} = \int_0^\infty t \cdot E\{t\}\, dt, \tag{6.7-2}$$

the dimensionless variance σ^2 is related to the second moment of $E\{t\}$, that is,

$$\sigma^2 \equiv \frac{1}{t_{\text{flow}}^2} \cdot \int_0^\infty (t - t_{\text{flow}})^2 \cdot E\{t\}\, dt = \frac{1}{t_{\text{flow}}^2}\left[\int_0^\infty t^2 E\{t\}\, dt - t_{\text{flow}}^2\right]. \tag{6.7-3}$$

Obviously, for a plug-flow vessel $\sigma^2 = 0$; that is, there is no spread of residence times. Using Eq. (6.7-1) and the definition of σ^2, one finds that for a "well-stirred" vessel $\sigma^2 = 1$.

It is interesting to note that $E\{t\}$ for a PFR with recycle approaches Eq. (6.7-1) when $R \to \infty$ (i.e., from the RTDF viewpoint, a plug-flow reactor with infinite recycle is equivalent to a WSR).

6.7.4 RTDF for Composite Systems

For an arbitrary, nonideal vessel, the RTDF will be none of the above (cf. Figure 6.7-1, Column 3), but a "dictionary" of RTDFs for various *networks of ideal vessels* will facilitate the rational selection of an "equivalent network" for any given flow reactor.

In this connection it is easy to show that if the RTDF for vessel 1 is $E_1\{t\}$ and for vessel 2 is $E_2\{t\}$, then the RTDF for a *series* combination of vessels (1) and (2) is:

$$E_{1+2}\{t\} = \int_0^t E_1\{t'\} \cdot E_2\{t - t'\}\, dt', \tag{6.7-4}$$

etc. From this so-called *convolution formula*, one readily finds that if vessel (1) is characterized by $t_{\text{flow},1}$ and σ_1^2 and vessel (2) by $t_{\text{flow},2}$ and σ_2^2 then, for the *series* combination of vessels (1) and (2), in whatever order:

$$t_{\text{flow},1+2} = t_{\text{flow},1} + t_{\text{flow},2}, \tag{6.7-5}$$

$$\sigma_{1+2}^2 = \sigma_1^2 + \sigma_2^2; \tag{6.7-6}$$

i.e., the mean residence times and variances (about the mean residence time) are simply additive.

The theorem expressed by Eq. (6.7-4) can be used to construct $E\{t\}$ for many series networks, such as a network of n-WSRs of equal volume, for which:

$$E(n\text{-WSRs}) = \frac{t_{\text{flow}}^{-1}}{(n-1)!} \cdot \left(\frac{t}{t_{\text{flow}}}\right)^{n-1} \cdot \exp\left[-\frac{t}{t_{\text{flow}}}\right]. \tag{6.7-7}$$

(Here t_{flow} is the value of $V/(\dot{m}/\rho)$ for *each* of the vessels in series.)

For vessels 1, 2, 3, ..., *n in parallel*, receiving the fractions $f_1, f_2, f_3, \ldots, f_n$ of the total flow, it is easy to prove that:

$$E = f_1 E_1\{t\} + f_2 E_2\{t\} + \cdots + f_n E_n\{t\}, \tag{6.7-8}$$

where $\sum_i f_i = 1$ and, for each vessel:

$$\int_0^\infty E_i\{t\}\, dt = 1 \qquad (1 = 1, 2, \ldots, n). \tag{6.7-9}$$

6.7.5 Real Reactors as a Network of Ideal Reactors: Modular Modeling

On the basis of such relations it is possible to construct networks of ideal vessels, which will approximate any *experimental* reactor RTDF:

$$E_{\text{exp}}\{t\} = \left[\frac{\omega_{\text{tracer}}\{t\}}{\int_0^\infty \omega_{\text{tracer}}\{t\}\, dt}\right]_{\text{reactor exit}}, \tag{6.7-10}$$

where the tracer is put in as (nearly) a Dirac impulse function.

Figure 6.7-3 shows a network developed to approximate the performance of a particular gas turbine can combustor (Swithenbank, *et al.* (1973)). In practice, information is obtained not only from *tracer diagnostics*, but also from physical features of the combustor geometry, cold-flow observations, etc. (cf. Figure 1.2-8). This additional information is important since, on the basis of RTD-data alone, it is often difficult to *discriminate* between alternative networks (with identical values of the RTD-"moments":

$$t_{\text{flow}} = \int_0^\infty tE\{t\}\, dt, \qquad \int_0^\infty t^2 E\{t\}\, dt, \qquad \cdots, \qquad \text{etc.}).$$

Put another way, the "equivalent" vessel network is *nonunique*.[57] Whereas several rather different alternatives may describe the reactor efficiency or yield equally well, they may differ substantially in their ability to describe, say, the domain of stable operation (cf. Section 6.7.7), or the reactor selectivity for the production of particular reaction (by-)products (e.g., pollutants).

Apart from guiding the development of useful mathematical "modular" models, tracer methods can also be used to *diagnose* operating problems with existing chemical reactors or physical contactors; e.g., the experimental RTD can be used to betray the internal presence of "dead-volumes," flow "channeling," or

[57] For example, $E\{t\}$ is invariant under an interchange of the *order* of two devices in series. Yet it is easy to calculate that, if a homogeneous, irreversible chemical reaction of nonunity order is occurring, the isothermal performance of these two alternative sequences will be different. If the main reactor is first-order and a side-reaction is of nonunity order, then the yield of both reactors for the main product will be identical, but not the relative yield of by-product.

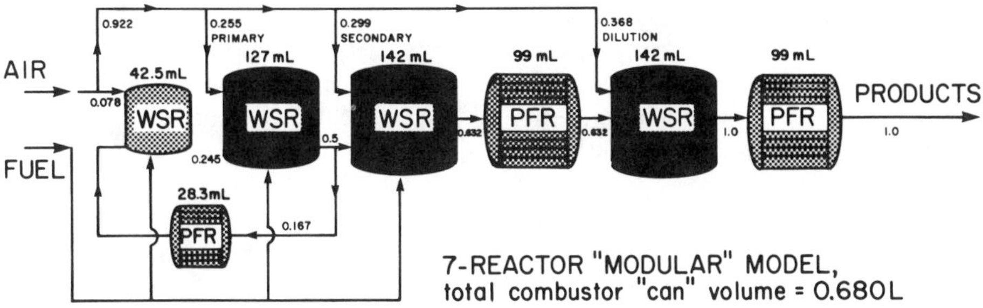

Figure 6.7-3 GT combustor; proposed interconnection of reactors comprising "modular" model (adapted from Swithenbank, *et al.* (1973)).

"bypassing," often associated with inefficient operation. This information supplied to a design engineer (or an on-line control system!) can be used to guide corrective geometric and/or fluid-dynamic changes to improve the performance of the system. Indeed, the "tracers" used for this purpose could be inevitable "disturbances" in, say, the feed composition.

It should also be noted that the RTD-function $E(t)$ described above does not convey important information about the role of concentration fluctuations associated with turbulence and/or incomplete reactant mixing at the molecular level (called "micromixing"). Indeed, in the presence of tracer concentration fluctuations at the reactor exit, we have really been discussing only the arithmetic average, $\langle E(t)\rangle$, of an ensemble of N-tracer "shots," each yielding the RTDs, $E_j(t)$ where $j = 1, 2, \ldots, N$. Two networks with identical ensemble-averaged $\langle E(t)\rangle$-relations, but with different "shot-to-shot variations" (e.g., as measured by the variance:

$$\lim_{N\to\infty} \frac{1}{N} \cdot \sum_{j=1}^{N} \int_0^\infty [E_j(t) - \langle E(t)\rangle]^2\, dt \Big), \tag{6.7-11}$$

will also, generally, perform differently as chemical reactors (see, e.g., Pratt, D.T. (1979), Villermaux, J. (1983), and Nauman and Buffham (1983)).

6.7.6 Statistical Microflow (Random Eddy Surface-Renewal) Models of Interfacial Mass Transport in Turbulent-Flow Systems (Higbie [1935], Danckwerts [1951] *et al.*)[58]

Residence-time distribution functions also play an important role in the statistical microflow models of interfacial transport in turbulent-flow systems mentioned in Section 5.7.3. In such models, mass or energy transport is visualized as occurring

[58] For a comprehensive, critical review, see, e.g., Sideman, S., and M. Pinczewski (1975); for more recent developments, see Thomas (1980).

during the intervals that eddies from the turbulent core remain in contact with the surface.[59] "Stale" eddies are presumed to be displaced, in turn, by fresh eddies, and the effective transport coefficient is calculated by appropriately time-averaging the RTDF-weighted instantaneous transfer coefficients, $\mathrm{St}(t)$. Quantitatively, if the function $E(t)$ is defined such that:

$$E(t)\,dt \equiv \begin{cases} \text{Relative portion of each unit interfacial area} \\ \text{covered by fluid eddies having "ages" between} \\ t \text{ and } t + dt, \end{cases} \tag{6.7-12}$$

then we expect:

$$\overline{\mathrm{St}} = \int_0^\infty \mathrm{St}(t) \cdot E(t)\,dt, \tag{6.7-13}$$

where $St(t)$ is calculated from a transient microfluid-dynamical analysis of the postulated individual "eddy" flow, and $\overline{\mathrm{St}}$ is the time-averaged transfer coefficient. It is interesting to note that *in all such statistical microflow models the interfacial region is, in effect, viewed as a thin "vessel" with respect to eddy residence time.*

In the earliest and simplest of such models each eddy is considered to behave like a translating "solid" body that is large compared to the transient diffusion BL ("penetration") thickness (cf. Section 5.4.5). For this microflow model one readily finds that the *dimensional* time-averaged mass-transfer coefficient is given by:

$$\frac{-\overline{j''_{\mathrm{A,w}}}}{\rho(\bar{\omega}_{\mathrm{A},b} - \bar{\omega}_{\mathrm{A,w}})} = \begin{cases} \left(\dfrac{4D_{\mathrm{A}}}{\pi t_m}\right)^{1/2} & \text{for } E(\mathrm{PFR}) \quad \text{(Higbie [1935]),} \quad (6.7\text{-}14a) \\[2ex] \left(\dfrac{D_{\mathrm{A}}}{t_m}\right)^{1/2} & \text{for } E(\mathrm{WSR}) \quad \text{(Danckwerts [1951]),} \quad (6.7\text{-}14b) \end{cases}$$

where t_m, the *mean* eddy contact time (reciprocal of the average "renewal" frequency)[60] must, of course, still be related to the prevailing geometry and bulk-flow velocity. To "repair" the currently incorrectly predicted Schmidt number dependence of $\overline{\mathrm{St}}_m$ (cf. Section 6.5.2) and obtain agreement with other experiments, both the eddy microstructure and RTDF assumptions have subsequently been generalized (see, e.g., Sideman and Pinczewski (1975), and Thomas (1980)); however, these simple examples (6.7-14a,b) capture the essence of this approach, primarily used by chemical engineers as a versatile alternative to the less detailed Prandtl–Taylor *et al.* eddy diffusivity approach.

---

[59] The same approach has been applied to *momentum* transfer in an attempt to "explain" observed $u^+(y^+)$ profile data (Section 5.7.2).

[60] It is noteworthy that these two extreme RTDFs lead to time-averaged transport coefficients that differ by only 13 percent. Thus, for the same mean eddy contact time, the predicted transfer coefficient is insensitive to the shape of the eddy RTDF.

6.7.7 Extinction, Ignition, and the "Parametric Sensitivity" of Chemical Reactors

Perhaps the simplest "modular"-model that has been used to successfully represent the steady-flow behavior of gas turbine, ramjet, and rocket-engine combustors is the series: WSR + PFR combination (Figure 6.7-4). Such devices, as well as their simple two-"module" representations, exhibit the following remarkable features:

1. There is an upper limit $\dot{m}_{max}$, to the total mass flow rate, $\dot{m}$, at each upstream condition (T_u, p_u, mixture ratio Φ) above which *extinction* of the exoergic reaction ("flame-out") abruptly occurs.
2. At any *lower* flow rate, $\dot{m} < \dot{m}_{max}$, there are *two* distinct possible steady-state conditions (each stable with respect to inevitable small disturbances), one corresponding to extensive (but not complete) fuel consumption and high temperature in the WSR, the second corresponding to negligible fuel consumption and temperature rise in the WSR.

An immediate corollary of these features is the reactor characteristic called "parametric sensitivity"; in other words, a small change in a parameter (like $\dot{m}$ or T_u) need not correspond to a small change in reactor performance, as in the case of flow-rate perturbations about $\dot{m}_{max}$.[61]

This important behavior is readily demonstrated for a WSR module (Figure 6.7-4) within which the following overall stoichiometric combustion reaction (cf. Section 6.5.5) occurs:

$$1 \text{ gram O} + f \text{ gram F} \rightarrow (1 + f)\text{gram P} + fQ \text{ cal(heat)} \qquad (6.5\text{-}27)$$

The analysis is the same as that of Section 6.1.3.2, except that we now allow for the presence of a *second* reactant (oxidant), and the associated heat-generation rate, which governs the WSR-operating *temperature*, $T_{②}$. The WSR species mass balances are, therefore:

$$\dot{m} \cdot (\omega_{i②} - \omega_{i①}) = \dot{r}_i'''\{\omega_{O②}, \omega_{F②}, T_{②}\} \cdot V_{WSR} \qquad (6.7\text{-}15)$$

($i = \text{O, F, P}$), and an overall *energy* balance (Section 3.2.3) yields:

$$\dot{m}c_p \cdot (T_{②} - T_{①}) = -\dot{r}_F'''\{\omega_{O②}, \omega_{F②}, T\ \}Q \cdot V_{WSR}. \qquad (6.7\text{-}16)$$

Note that, regardless of the functional forms of $\dot{r}_i'''$, the "source" terms for oxidizer and fuel are necessarily related by $-\dot{r}_O''' = -\dot{r}_F'''/f$, so that a method identical to that

[61] Similarly, a small *reduction* in $\dot{m}$ could bring about reactor "ignition" if the initial state is one of virtually no reaction (see Figure 6.7-5).

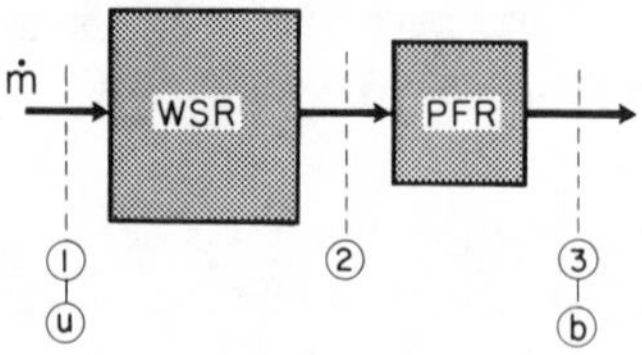

Figure 6.7-4 Simple, two-ideal reactor "modular" model of gas turbine, ramjet, or rocket engine combustor.

exploited in Section 6.5.5[62] can be used to express $\omega_{O②}$ and $\omega_{F②}$ in terms of $T_②$. These substitutions render the energy balance a single, nonlinear, transcendental equation for $T_②$, where, say, the overall kinetics are represented by the following Arrhenius-type/mass-action rate law:

$$-\dot{r}_F''' = \left[A \cdot \exp\left(-\frac{E}{RT}\right)\right] \cdot \frac{1}{M_O^{\nu_O} M_F^{\nu_F - 1}} \cdot \left(\frac{pM}{RT}\right)^n \cdot \omega_O^{\nu_O} \omega_F^{\nu_F}. \qquad (6.7\text{-}17)$$

(Note that, since Reaction (6.5-27) is really the overall result of a chain of many "elementary" chemical reactions, we now allow for nonunity individual reaction orders, ν_O and ν_F, and the "overall order" $n\,(\equiv \nu_O + \nu_F)$ is, therefore, not necessarily 2; cf. Eq. (6.5-63).) The qualitative features of the resulting transcendental equation (6.7-16) are clear from Figure 6.7-5, which shows both sides (RHS and LHS) of Eq. (6.7-16), and a typical set of intersections. It is easy to appreciate why the RHS must peak at some temperature between T_u and T_b (nearer T_b) because, while the mass fractions ω_O and ω_F are highest near T_u, the Arrhenius factor $\exp(-E/RT_u)$ is quite small there; however, while $\exp(-E/RT_b)$ is very much larger, this condition is achieved where either ω_F (for fuel-lean ($\Phi < 1$) mixtures) or ω_O (for oxidizer-lean ($\Phi > 1$) mixtures) *vanishes*.

For a given $\dot{m}$ ($<\dot{m}_{max}$) the LHS is a generally straight line that intersects the RHS ($-\dot{r}_F''' Q V_{WSR}$) at as many as *three* distinct $T_②$-values, the middle one of which can be shown to be unstable[63] (and, hence, maintainable only with the addition of some kind of control system). The upper intersection represents an "ignited" WSR steady state, whereas the lower intersection represents an "extinguished" condition (virtually no chemical reaction). Note that, if $\dot{m}$ were gradually increased, the "ignited" $T_②$-value would gradually diminish until a tangency condition, at $T_{\dot{m}-max}$,

[62] In the present case, however, it is clear that there are no equality restrictions placed on effective "diffusivities" ($D_{i,eff}$, α_{eff}); indeed such diffusivities do not even explicitly enter the overall WSR balance equations.

[63] A quasi-steady argument reveals that if for some reason $T_②$ dropped, the convective heat removal rate (LHS) would exceed the heat generation rate (RHS), causing a further drop in $T_②$, until the lowest (stable) $T_②$-value would be achieved. Conversely, a $T_②$-increase causes $T_②$ to converge on the upper (stable) $T_②$-value. A dynamical (transient) analysis of the unsteady conservation equations leads to the same conclusion.

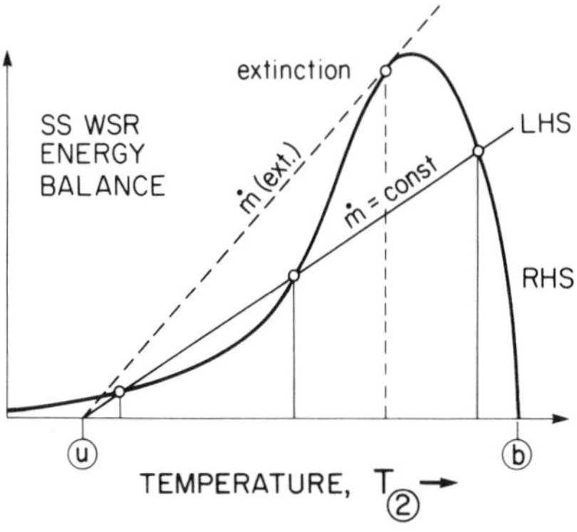

Figure 6.7-5 Influence of feed mass flow rate on WSR operating temperature and space (volumetric) heating rate (SHR); (straight line is the LHS of the energy balance equation (6.7-16)).

is achieved, corresponding to the above-mentioned maximum "throughput" rate, $\dot{m}_{\max}$. Any attempt to further increase $\dot{m}$ produces "extinction," i.e., a dramatic drop in $T_{②}$ to the lower (stable) intersection.

For the simple mathematical model considered above the maximum volumetric rate of fuel consumption (and hence maximum chemical heating rate) can be shown (Avery (1955)) to occur at a WSR-temperature:

$$T_{\dot{r}'''-\max} \approx \frac{T_b}{1 + n(RT_b/E)} \tag{6.7-18}$$

only slightly larger than the "extinction" temperature (see Figure 6.7-5). Here T_b is the adiabatic, complete combustion temperature. Laboratory experiments on devices which approximate WSRs (e.g., Longwell and Weiss (1955)) have been used to estimate the kinetic parameters E and n appearing above, with typical values (see, e.g., octane (n), Table 6.7-1) being, say, $E \approx 40$ kcal/mole and $n = 1.8$, respectively. Thus, for typical T_b-values (*ca.* 2250 K) we find $E/(RT_b) \approx 9$ and $-\dot{r}'''_{F,\max}$ occurs at a WSR temperature of about 83 percent of T_b. Note (Figure 6.7-5) that this is nearly the condition of extinction, an observation which leads to the conclusion that *steady-state combustor operation near the condition of maximum volumetric reaction rate necessarily puts one "on the brink of extinction."* Even so, a follow-on reactor (PFR above) would be needed to consume the unburned fuel emerging from the WSR stage, thereby contributing to the overall volume of the combustor configuration (see Exercise 6.12).

Of course, real reactors are not "perfectly well-stirred" (especially those also involving phase change) and their performance is also influenced by parameters controlling local mixing rates, such as geometric details and the characteristic Reynolds' number (see Section 7.2.3). For these reasons, and because of the goal of

Table 6.7-1 Some Estimates[a] of Overall Combustion Kinetics Parameters

Fuel (F)	*Oxidizer* (O)	p(atm)	T_u(K)	n	A^b	$E\left(\frac{\text{kcal}}{\text{g-mole}}\right)$	T_b(K)	$(SHR^c)_{max}$
H_2	O_2	1	310	2.17	0.16	15	2380	1.10
CH_4	O_2	1	300	1.6		29	2236	0.39
C_2H_4	O_2	1	300	1.6			2362	1.93
C_3H_8	O_2	1	380	1.56	2	31	2250	
C_4H_{10}	O_2	1	300	2.0	0.54	21	2256	
n-C_8H_{18}	O_2	1	400	1.8		40.0	2251	1.0
n-C_8H_{18}	O_2	15	400	1.8		40.0	2316	73
NH_3	O_2[d]	1	400	1.7		49.5	2782	15
NH_3	O_2[d]	15	400	1.7		49.5	3041	3000

[a] Supplemented, rounded (and selected) values based on Table 4.4 of Kanury (1975).
[b] Units are: $10^{14} s^{-1}$ (g-moles/cm^3)$^{-(n-1)}$, where n is the overall reaction order.
[c] Units are: 10^9 BTU/ft^3/hr.
[d] Stoichiometric mixture, no diluent ("diluent" is N_2) unless otherwise specified.

nearly complete fuel consumption, actual combustor "space (volumetric) heating rates" (SHR) fall well below those reported in Table 6.7-1.[64]

In closing, it should be realized that while these black-box "modular-models" obviously capture many important features of such chemical reactors, and can be useful for correlating performance data on full-scale and small-scale models (Section 7.2.3), their *ab initio* predictive value will necessarily be limited compared to more detailed pseudo-continuum mathematical models that describe the coexisting fields (of ω_i, $T, \ldots$) everywhere within the device (cf. Chapters 1 and 2). From a conceptual viewpoint, however, it is noteworthy that all of the above mathematical models have, as their basis, the macroscopic conservation conditions of Chapter 2.

SUMMARY

- Body-force-induced drift ("forced diffusion" or "phoresis") can contribute to mass transport of particular substances, in addition to convection concentration diffusion.
- In the absence of "forced diffusion" most nonreactive mass-transfer problems are "analogous" to heat-transfer problems, implying useful quantitative relations between these processes.

[64] Such high SHR-values *are*, however, achieved *within* premixed flat laminar flames (see, e.g., Section 7.2.2.2).

- Even in the presence of "analogy-breaking" phenomena, one can establish rational correction factors which account for the transport effects of, say:

  phoresis,
  chemical reaction (homogeneous).

- A *mass*-transfer coefficient correction factor can also be introduced to account for the Stefan flow (blowing or suction) necessarily associated with net interphase mass transfer in nondilute systems. This effect is, however, *not* analogy-breaking, since this convection (normal to the phase boundary) influences *energy* and tangential *momentum*-transfer coefficients in a manner analogous to the mass-transfer coefficient (Section 6.4.3.2). Many important problems involve *simultaneous* transport of mass, energy and momentum (e.g., volatile fuel-droplet vaporization) and can be solved using these algebraic transfer-rate correction factors (Section 6.4.3.3).
- When a single, stoichiometric homogeneous chemical reaction (e.g., gaseous fuel + oxidizer combustion) occurs at an arbitrary rate within a fluid, certain linear *combinations* of local species mass fractions and/or temperature, here called "conserved variables," can be shown to approximately satisfy "source-free" (nonreactive) conservation equations (Section 6.5.5.1). Accordingly, these variables can be used to establish valuable interrelations between reactive ("hot") flows and their nonreactive ("cold"-flow) counterparts (e.g., Section 6.5.5.2), as in the case of fuel-droplet evaporation with and without "envelope flame" combustion (Section 6.5.5.7), and fuel jet mixing (Section 6.5.5.5). The existence of "conserved variables" also provides the basis for relating the local time-averaged chemical reaction rate in a *turbulent* fluid mixture to the corresponding local decay rate of tracer concentration fluctuations in the same (or corresponding) flow field.
- Many multiphase flows can be treated as single-phase flows if the suspended ("dispersed") phase can be treated as a diffusing "species," closely coupled to the host flow ("continuous" phase). This approach breaks down for sufficiently large *Stokes*' numbers, corresponding to appreciable *dynamical nonequilibrium* between the two phases (Section 6.6).
- Tracer diagnostics can provide clues as to what network of ideal vessels (Section 6.1.3) will simulate the behavior of, say, a real chemical reactor (Section 6.7.3). Even the simplest such network, that of a well-stirred reactor (WSR) followed by a plug-flow reactor (PFR), exhibits many interesting properties of real chemical reactors—e.g., combustors—including a marked sensitivity to certain parameters (e.g., feed flow rate) near the condition of maximum space (volumetric) heating rate. In a deeper sense, such multi-vessel networks can be regarded as coarse discretizations, equivalent to deterministic modeling using a "finite element" approach to the governing conservation equations (see Appendices 8.2 and 8.3).

TRUE/FALSE QUESTIONS

6.1 T F Natural convection does not occur in *mass* transfer situations since there is no temperature change to cause the necessary density differences.

6.2 T F When the Schmidt number, $Sc \equiv \nu/D$, is very large, the momentum (or vorticity) boundary layer is embedded well within the mass-transfer boundary layer.

6.3 T F For two adjacent phases in local thermochemical equilibrium, each chemical-species mass fraction must be continuous across the phase boundary.

6.4 T F In ordinary chemically reacting mixtures, the equations governing chemical-*element* mass fractions are always simpler than those governing individual chemical *species* mass fractions, since ordinary chemical reactions cannot produce or destroy chemical elements.

6.5 T F In the absence of appreciable "phoresis" and/or homogeneous chemical reactions, there is a quantitatively useful "analogy" between mass- and heat-transfer coefficients, valid even in the presence of appreciable streamwise pressure gradients.

6.6 T F Multiphase flows can often be treated as single-phase flows if the suspended ("dispersed") phase can be treated as a diffusing "species," closely coupled to the host flow ("continuous" phase).

6.7 T F At sufficiently low gas pressures, even a short straight tube with *axial* flow will behave like a *well-stirred* vessel due to molecular back-mixing.

6.8 T F There is no such thing as a steady-flow chemical reactor since, in chemically nonequilibrium systems, there is always a time-rate of change of chemical composition.

6.9 T F The *minimum* allowable bed height for a desired reactant conversion in a fixed-bed chemical reactor is set by the catalytic activity of the pellets.

6.10 T F "Cold-flow" composition measurements bear no useful relations to the behavior of geometrically and dynamically similar chemically reacting fluid systems.

6.11 T F A fluidized bed chemical reactor operating close to the condition of minimum fluidization (mf) behaves in a manner similar to a plug-flow (somewhat "expanded") fixed-bed chemical reactor; indeed, fixed-bed interphase energy and mass-transfer correlations remain approximately valid, albeit with a Re-dependent void fraction, ε.

6.12 T F WSRs, while a valuable teaching tool, cannot be used in practice because the most probable residence time is zero, not the mean residence time $V/(\dot{m}/\rho)$.

6.13 T F RTDs with long "tails" usually indicate the presence of relatively "dead spaces" within a vessel.

6.14 T F For noninteracting series configurations of contactors, the overall RTD (and each of its complete set of moments) is invariant under an interchange of the *order* of vessels in the sequence.

6.15 T F When a chemical reaction of nonunity reaction order is carried out in a *series* of noninteracting contactors, the total reactant conversion is invariant under an interchange of the order of vessels in the sequence.

6.16 T F If two reactors with RTDs having the same first and second moments are used to carry out the same first-order irreversible reaction, the observed reactant conversion will necessarily be indistinguishable.

6.17 T F Reactor appearance is an accurate indicator of whether reactor performance will more closely approximate the PFR or WSR-limit.

6.18 T F To approximate the WSR-limit, the feed must be mixed with the reactor contents on a time scale comparable to the mean residence (holding-) time in the reactor.

6.19 T F The WSR limit cannot be approximated in the absence of mechanical propellers and baffles.

6.20 T F The WSR-limit ordinarily leads to a minimum-volume reactor for chemical reactions whose kinetics are of negative order, or when the reaction products (including "heat") accelerate the reaction rate ("autocatalysis").

6.21 T F Extinction conditions for a WSR are the same as ignition conditions. That is, there is no hysteresis in jumping between eligible distinct steady states.

6.22 T F When conditions (feed, parametric settings) lead to the existence of two stable operating points straddling an intermediate unstable operating point, it is impossible to obtain and maintain such an operating point, even if a control system is added to the chemical reactor.

6.23 T F Nonlinear analysis of the governing ODEs, including the use of a "phase plane," can be used to determine (a) what initial states will lead to what WSR operating points, (b) whether each approach will be monotonic or exhibit (possibly damaging) temperature "overshoots," and (c) the total time needed to achieve each steady state for each possible initial state.

EXERCISES

6.1 Consider the design of an envelope of cellophane to keep food moist at 38°C. If the maximum tolerable loss rate of water vapor is 0.17 g/d at steady state for a wrapping covering an area of 0.20 m^2, what is the required thickness of the cellophane? Assume that the vapor pressure of water is 10 Torr within the package and the air outside contains water vapor at a nominal pressure of 5 Torr. The *permeability* of H_2O in cellophane has been reported to be about 1.8×10^{-10} m^3 H_2O(STP)/[s · m^2 · atm/m] under these conditions. (1 Torr = 1 mm Hg = 133.3 Pa.)

6.2 A polyethylene film 0.15 mm thick is being considered for use in packaging a pharmaceutical product to be maintained at 30°C. If the partial pressure of O_2 outside is 0.21 atm and inside the package it is about 0.01 atm, calculate the rate of diffusional influx of O_2 at the quasi-steady state. Assume that the resistances to diffusion outside and inside the film are negligible compared to the resistance of the film itself. The O_2/polyethylene permeability is estimated to be 4.17×10^{-12} m^3 O_2(STP)/[s · m^2 · atm/m] under the prevailing conditions (after C. Geankoplis, Ex. 6.5-2).

6.3 In the preliminary design of a fermenter, it is desired to predict the rate of absorption of O_2 from agitator-produced air bubbles (at 1 atm abs. pressure) having diameters of *ca.* 100 μm into water at 37°C having a negligible concentration of dissolved O_2. At 37°C the solubility of O_2 from air in water is 2.26×10^{-4} kg mol O_2/m^3 and the diffusivity of O_2 in water is experimentally found to be $3.25 \times 10^{-9} m^2/s$. Consider only the limiting case: bubble at rest relative to the surrounding fluid (after C. Geankoplis, Ex. 7.4-1). Estimate the maximum possible rate of molecular-oxygen uptake of each microorganism, having a diameter of 0.667 μm, suspended in an agitated, 37°C, aqueous solution. Assume that (a) the surrounding liquid is saturated with O_2 from air at 1 atm abs. pressure, (b) the microorganism can utilize the oxygen much faster than it can diffuse to it, and (c) the microorganism has a negligible relative velocity since it has a density close to that of water. (*Hint:* Since the rate of O_2 uptake is assumed to be diffusion-controlled, the steady-state oxygen concentration "at" the surface will be negligible compared to that prevailing in the saturated solution.) *Ans.* 2.20×10^{-6} kg-mol O_2/s-m^2 (after C. Geankoplis, Ex. 7.4-1).

6.4 Consider the preliminary design of a steel-making process using as its starting material molten "pig" iron containing 4.0 wt.% dissolved carbon. In the "decarbonization" process, a spray of molten iron droplets falls through a pure oxygen atmosphere. During the residence time of each drop, the dissolved carbon diffuses through the molten iron to the surface, where it is assumed to react instantly (because of the high temperature) with O_2 in accord with the overall chemical reaction: $C + \frac{1}{2}O_2 \rightarrow CO(g)$.

a. If the prevailing diffusivity of carbon in iron is estimated to be 7.5×10^{-9} m^2/s, calculate the maximum drop size allowable for the final drop after a 2.0-s fall to contain on an average no more than 0.1 wt.% carbon. For this purpose use the graph in Figure 6.4E (after Rosner (1972)), summarizing the results of computations based on the Fourier method of "separation-of-variables" for transient diffusion in a *sphere.* The ordinate is the fraction of the *initial* solute content which has escaped.

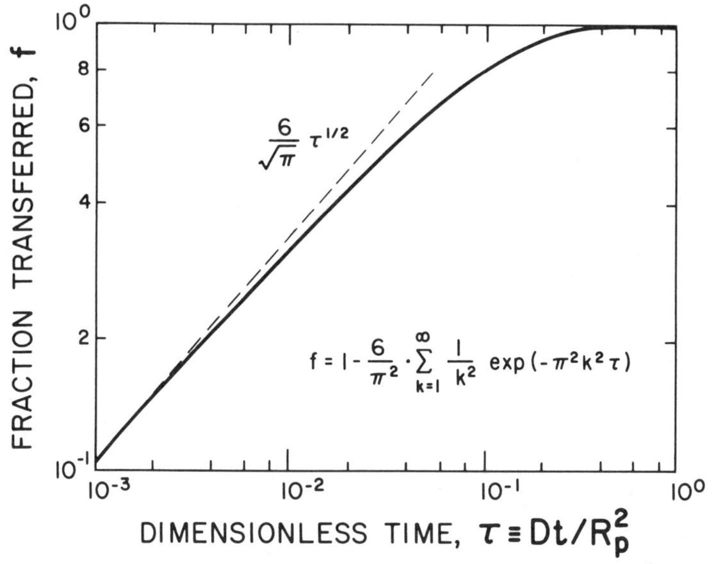

Figure 6.4E

The abscissa is the dimensionless time Dt/R^2. Assume: (i) the mass transfer rate of gases to/from the drop surface is limiting (i.e., negligible external resistance); (ii) negligible circulation within each liquid drop, and (iii) nearly constant temperature conditions for each droplet. *Ans.* radius = 0.22 mm (after C. Geankoplis, Ex. 7.5-10).

b. Can you verify the small time asymptote shown in Figure 6.4E, using our error-function solution to the related semi-infinite *slab* problem? Justify each step in your argument. Is this (small penetration depth) asymptote applicable in the above-mentioned decarbonization problem?

6.5 Consider a 2-mm diameter spherical particle of NaCl(s) salt falling through pure 20°C water (Newtonian solvent), under the influence of gravity. Given the data assembled in Table 6.5E, and the information contained in Chapters 4 and 6 of this textbook, *estimate*[65] the:

a. Terminal fall velocity (cm/s) of this isolated particle (the fall speed at which the drag is equal to the buoyancy-modified gravity force on the sphere) and the corresponding Reynolds' number, $\mathrm{Re} \equiv U_{\mathrm{fall}} d/\nu$;
b. *Forced*-convection augmented salt dissolution rate (g/s) at this fall velocity, neglecting the Stefan-flow effect (see Parts (f) and (g) below);
c. Dissolution rate (g/s) if the solvent surrounding the spherical particle were completely quiescent (compare to your result of Part (b));
d. Grashof number for *mass* transfer, Gr_m, and the corresponding Rayleigh number, $\mathrm{Ra}_m \equiv \mathrm{Gr}_m \cdot \mathrm{Sc}$;

[65] Itemize and briefly discuss important underlying assumptions, and document your sources of information (for future reference).

Table 6.5E Property Data[a] (at 20°C)

Quantity	*Symbol*	*Value*	*Units*
Density of NaCl(s)	$(\rho_A)_{solid}$	2.165	g/cm^3
Fick diffusivity of salt in solvent	$D_{A\text{-solvent}}$	1.4×10^{-5}	cm^2/s
Saturation mass fraction	$\omega_{A,sat}$	0.26	—
Density of salt-saturated water	$\rho\{\omega_{A,sat}\}$	1.1951	g/cm^3
Density of pure water	$\rho\{0\}$	0.99823	g/cm^3
Viscosity of pure water	μ	0.8904×10^{-2}	poise

[a] *Handbook of Chemistry and Physics*, CRC Press (55th Ed.), 1974–1975.

e. Parameter $Gr_m^{1/4}/Re^{1/2}$ governing the relative importance of *natural* convection to *forced* convection (is the parameter small enough to defend your estimate of part (b)?);
f. Stefan-flow parameter $B_m = (\omega_{A,w} - \omega_{A,\infty})/(1 - \omega_{A,w})$, where $\omega_{A,w}$ is taken as the *saturation* mass fraction of salt in 20°C, H_2O + salt solution;
g. Stefan-flow dissolution rate correction factor, $B_m^{-1} \ln\{1 + B_m\}$;
h. Best estimate of the actual salt-particle dissolution rate (g/s) under these conditions, taking all operative phenomena into account;
i. Reduced dissolution rate corresponding to particle free-fall through an unsaturated solvent which already contains salt in the amount $\omega_{A,\infty} = 0.16$;
j. Mean interparticle spacing at which neighboring salt particles in a "cloud" would influence your predicted dissolution rates (Part (h)).

6.6 Consider the catalytic oxidation of CO(g) to CO_2(g) in an automotive catalytic afterburner containing catalyst pellets of 5 mm radius each. Suppose that under the conditions: $T = 800$ K, $p = 1.05$ atm, $y_{CO} = 2 \times 10^{-2}$ (CO mole fraction in local combustion products "wetting" pellet), it is experimentally found that the overall rate of CO consumption by a *single pellet* is 0.463 m-moles/s. Using the effective diffusion coefficient ($\varepsilon = 0.45$, $\tau_{pore} = 2$, $d_{pore} = 10^{-2}$ μm (see Ex. 3.12)) and assuming that the oxidation reaction is taking place in the region of first-order kinetics, estimate the prevailing catalyst "effectiveness factor" η_{cat}. How much smaller could the packed-bed reactor be if the effectiveness factor in each pellet could be raised to 100%?

6.7 Absorption of Cl_2(g) (solute "A", below) into the solvent $H_2O(\ell)$ occurs more rapidly than expected based on "physical" dissolution alone, owing to the homogeneous pseudo-first-order hydration reaction:[66]

$$Cl_2(ab.) + H_2O(\ell) \xrightarrow{k_1} Cl^-(ab.) + H^+(ab.) + HOCl(ab.),$$

forming hypochlorous acid. At 76°F, under conditions where the back reac-

[66] This reaction is also responsible for systematic departures from Henry's solubility law for total dissolved chlorine in water.

tion can be neglected, Brian, *et al.* (1962) have estimated:

$$D_A = 1.59 \times 10^{-8}\ \text{ft}^2/\text{s},$$
$$k_1 = 13.6\ \text{s}^{-1} \qquad (\dot{r}'''_A/m_A = -k_1 C_A),$$

and, at $p_{Cl_2} = 0.028$ atm, $C_{A,i} \approx 1.09 \times 10^{-4}$ lb-moles Cl_2(ab.)/ft^3 solution. Consider a quiescent, "deep" layer of initially pure $H_2O(\ell)$ at 76°F, suddenly confronted with a Cl_2-gas pressure of 0.028 atm.

a. Estimate the Cl_2 diffusion rate (at $y = 0$) and the corresponding instantaneous "reaction factor":

$$F_A \equiv \frac{\text{actual transport rate at time } t}{\text{“physical” transport rate at time } t \text{ for same } C_{A,i}}$$

at an exposure time of 0.25 s by each of the following two methods:

1. An explicit (Schmidt) finite-difference method for $D\Delta t/(\Delta y)^2 = \frac{1}{2}$; that is,

$$(C_A)_{j,n+1} \cong \tfrac{1}{2}[(C_A)_{j-1,n} + (C_A)_{j+1,n}] + [(\dot{r}'''_A/m_A)]_{j,n} \cdot \Delta t;$$

(cf. Appendix 8.2).

2. The exact solution:

$$-D_A\left(\frac{\partial C_A}{\partial y}\right)_{y=0} = \dot{N}''_A(t)$$
$$= C_{A,i}(Dk_1)^{1/2} \cdot \left\{\text{erf}((k_1 t)^{1/2}) + \frac{\exp(-k_1 t)}{(\pi k_1 t)^{1/2}}\right\}.$$

b. Compare the expected Cl_2(ab.) concentration profile with reaction[67] at 0.25 s to the hypothetical profile if this homogeneous reaction did not occur.

c. An exact solution is not available for cases of *nonunity* chemical reaction order. Could the explicit finite-difference method above be used in such a case?

6.8 Convective-diffusion mass transfer plays an important role in the fabrication of small solid-state electronic devices, which often start with the "chemical vapor deposition" (CVD) of micron-thickness films of crystalline Si(s) grown on a sapphire (Al_2O_3) substrate ("wafer").

a. Use the material of Chapters 3 and 6 to make a preliminary quantitative estimate of the average silicon deposition flux (μg/min/cm^2) to a flat,

---

[67] Interestingly enough, for large values of $k_1 t$ this profile is known to approach $C_A/C_{A,i} = \exp[-y(D/k_1)^{-1/2}]$.

growing wafer exposed to a parallel flow of $SiH_4(g)/H_2$ gas mixture under the following conditions:

$$U = 50 \text{ cm/s}, \qquad T_\infty = 300 \text{ K},$$

$$L = 5 \text{ cm} \qquad T_w = 1273 \text{ K},$$

$$p = 1 \text{ atm} \qquad (y_{\mathrm{SiH_4}})_\infty = 0.005 \text{ (mole fraction)}.$$

Assume that the Si-deposition flux is determined by $SiH_4(g)$ ("silane"-) vapor transport to the wafer surface, *at* which:

$$\mathrm{SiH_4(g)} \rightarrow \mathrm{Si(s)}\!\downarrow + \mathrm{2H_2(g)}\!\uparrow.$$

That is, assume $y_{\mathrm{SiH_4,w}} \ll y_{\mathrm{SiH_4,\infty}}$ but neglect the possible complication of SiH_4(g)-decomposition in the vapor phase near the hot wafer. List and quantitatively defend each of your important assumptions and property estimates. Identify the "weakest links" in your estimates (for possible future investigation). *Ans.* 405 μg/min/cm^2.

b. If the estimated equilibrium Si(g) vapor pressure over Si(s) at this surface temperature is about 1.1×10^{-8} atm, can the "physical" sublimation rate of silicon be neglected under these CVD-reactor operating conditions?

c. Evaluate the natural convection/forced convection ratio parameter $\mathrm{Gr}_h^{1/4}/\mathrm{Re}_L^{1/2}$ (see Section 5.5.3) and comment on the possible importance of *natural* convection (caused by heat transfer) in augmenting (or suppressing?) the *mass*-transfer coefficient $\overline{\mathrm{Nu}}_{m,\mathrm{SiH_4}}$ if the heated wafer is horizontal, and the gas mixture flows past *above* it.

Note: While the Stefan-mass flow ("suction") associated with Si-deposition is negligible in this example (since $\omega_{\mathrm{Si/SiH_4}}\,\omega_{\mathrm{SiH_4},\infty} \ll 1$; see Sections 6.4.1 and 6.4.3.4), the "blowing" effect associated with thermal (Soret) diffusion is appreciable (over 20% reduction in $\overline{\mathrm{Nu}}_{m,\mathrm{SiH_4}}$) and should be taken into account (Rosner (1980)).

6.9 The length requirement for a honeycomb-type automotive exhaust catalytic converter is set by the need to reduce the CO concentration in the exhaust to about 5% of the inlet concentration (i.e., 95% conversion). Consider the basic conditions:

Inlet gas temperature	700 K
Inlet gas pressure	1 atm
Inlet gas composition (mole fractions)[68]	$y(N_2) = 0.93$, $y(CO) = 0.02$, $y(O_2) = 0.05$

[68] We explicitly neglect the inevitable presence of $H_2O(g)$ and $CO_2(g)$, lumping them into an "effective" N_2 composition.

Inlet gas velocity	10^3 cm/s
Channel cross-section dimensions	1.5 mm by 1.5 mm (each channel)
Assumed channel *wall* temperature	500 K

Assume that the Pt-based catalyst used on the walls of each channel is active enough to cause the surface-catalyzed CO oxidation reaction to be diffusion-controlled[69]; that is, the steady-state value of the CO-mass fraction established at (1 mean-free-path away from) the wall, $\omega_{CO,w}$, is negligible compared to $\omega_{CO,b}(z)$ within each channel. Also assume that the *gas-phase* kinetics of CO oxidation under these conditions preclude appreciable (uncatalyzed) homogeneous CO-consumption in the available residence times. Answer the following questions:

a. By what mechanism is CO(g) *mass* transported to the channel wall, where chemical consumption (to produce CO_2) occurs? What is the relevant transport coefficient, and to what energy-transfer process and transport property coefficient is this "analogous"?
b. Are the *m*ass-*h*eat transfer *a*nalogy *c*onditions (MAC, HAC) discussed in Section 6.3.3 approximately met in this application? What is the inlet *mass* fraction of CO gas?[70]
c. Estimate the Schmidt number $\mathrm{Sc} \equiv \nu/D_{\text{CO-mix}}$ for CO Fick diffusion through the prevailing combustion gas mixture, using the experimental observation that (Table 3.4–1):

$$D_{\text{CO-N}_2} \cong \frac{0.216}{p} \cdot \left(\frac{T}{300}\right)^{1.73} \quad \frac{\text{cm}^2}{\text{s}},$$

where p is the prevailing pressure (expressed in atmospheres) and T the mixture temperature (expressed in kelvins).
d. Under the flow rate, temperature, and pressure conditions given above and using the mass-transfer analog of Eq. (5.5-21a), estimate the catalytic duct *length* required to consume 95% of the inlet CO concentration, and the mixing cup (bulk) stream *temperature* at this length.
e. List and defend the principal assumptions made in arriving at the length estimate (of Part (d)) (see Appendix 8.1).
f. If the catalyst were "poisoned" (e.g., by lead compounds), what would

---

[69] See Section 6.5.3.

[70] Mass fraction ω_i and mole fractions y_i are related by the equation:

$$\omega_i = M_i y_i \left(\sum_{j=1}^{N} M_j y_j \right)^{-1},$$

where M_j is the molecular weight of species j and the indicated sum is taken over all species present in the single-phase mixture.

happen to the CO exit concentration? Which of the assumptions used in predicting the required converter length (Part (d)) would be violated?

g. If the heat of combustion of CO(g) is about 67.8 kcal/g-mole CO consumed, calculate how much heat must be removed to maintain the channel-wall temperature constant at 500 K?

h. Automobile operating conditions are never strictly steady, so that in practice the mass-flow rate, temperature, and gas composition entering the catalytic afterburner will be time-dependent. Under what circumstances (be quantitative) can the design equations you used be defended if used to predict the conditions exiting the duct *at each instant*? (Quasi-steady approximation.)

i. At the design condition, estimate the *fractional pressure drop*, $-\Delta p/p_0$, in the honeycomb-type catalytic afterburner. If, instead of the honeycomb-type converter, a *packed-bed* device were used to achieve the same reduction in CO-concentration, would you expect $-\Delta p/p_0$ to be larger or smaller then the honeycomb device of your preliminary design?

6.10 Consider the preliminary design of a steady-flow chemical reactor for removing trace sulfur compounds from "naphtha" (petroleum or coal-tar distillates with normal boiling points (mbp) in the range 311–422 K). This can be done by passing a mixture of H_2(g) and naphtha vapors over a solid which catalyzes the formation of the more volatile sulfur compound H_2S(g), readily separated from the product vapor stream downstream of the "hydrodesulfurization" reactor.

Heptane (C_7H_{16}, nbp = 371.7 K) can be taken to be the representative hydrocarbon in naphtha, and thiophene[71] (C_4H_4S with $M = 84.13$, mp = 233 K, nbp = 357 K) the representative trace sulfur-containing contaminant.

The feed-stream composition is to be 32.8 mole % H_2 and 17.2 mole = naphtha, and the reactor is to be operated at 660 K and 30 atm inlet pressure. We wish to find the *minimum* reactor *length* required to remove 99% of the initial sulfur content, while continuously processing 100 gm/sec of feed mixture. This presumes the available catalyst is sufficiently active that the *rate determining step* in the heterogeneous reaction is thiophene *diffusional mass transfer* to the vapor/solid catalyst interface. (Less active catalysts would, of course, lead to the need for larger reactors.)

[71] Also called thiofuran, thiophene is a heterocyclic compound

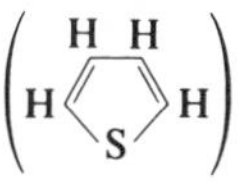

which behaves like a reactive benzene derivative. In liquid form its density is 1.064 g cm^{-3} near 293 K. Such compounds in naphtha cause the deactivation of catalysts used to "reform" naphtha into CH_4 (using steam).

Case A: Packed (Fixed)-Bed Catalytic Reactor

A 25 cm diameter duct is packed with catalyst pellets in the form of short (3.2 mm long) × 3.2 mm diameter cylinders. Estimate the required minimum reactor length, and the corresponding fractional pressure drop $((-\Delta p)/p_{\text{inlet}})$ across the fixed catalyst bed. Assume $\varepsilon \approx 0.50$.

Case B: Unpacked (Empty) Tubular Reactor

Estimate the minimum required length of an empty "multi-tubular" reactor with the active catalyst deposited on the *inner* walls of a bundle comprised of 50, 2.54 cm I.D. tubes. What would the fractional pressure drop be in *this* case?

Discuss the comparison[72] between Cases A and B. How sensitive would your results be to a change in reactor operating temperature to, say, 700 K? If the intrinsic rate of the heterogeneous catalytic desulfurization reaction were "controlling," would you expect these length requirements to be more (or less) sensitive to this temperature change? In each case, is the momentum-energy-mass transfer analogy ($C_f/2 \approx \text{St}_h\text{Pr}^{2/3} \approx \text{St}_m\text{Sc}^{2/3}$) valid? Why or why not?

General Notes

Clearly state and defend all of your important assumptions. Identify sources of basic property data and momentum/energy/mass transfer correlations. Document your reasoning and pinpoint any particularly questionable assumptions you were obliged to make. Carry out the preliminary design of the two types of naphtha vapor hydrodesulfurization reactors (transport-controlled) defined above. Note that some of the required vapor mixture (momentum) transport properties were calculated as an illustration (Exercise 3.9). For each type of reactor, include the preliminary design of a *heat exchanger* of the same type (see Figure 6.10E, p. 398) to raise the feed stream from 460 K to 660 K by direct contact with surfaces at 680 K, 30 atm. Estimate the lengths and relative fractional pressure drops in each heat exchanger and reactor section. Use the quantitative information on unpacked and packed exchangers given in this textbook, Sections 4.7, 5.5.4, 5.5.5, 5.6.2, and 6.5.1, and itemize/defend all important assumptions you make in the course of your calculations.

6.11 Consider the evaporation of an isolated 80μm diam *n*-decane droplet in 600 K, 1 atm air (Table 8.1-1) given the property data in Table 6.11E. In the absence of combustion and appreciable forced or natural convection effects, estimate:

a. Le
b. T_w
c. B_h
d. B_m

[72] Similar considerations govern the comparison between pellet-type and honeycomb-type ("monolith") catalytic converters for automotive use. In the latter case, smallness of $-\Delta p/p$ (inlet) is more important in overall system performance.

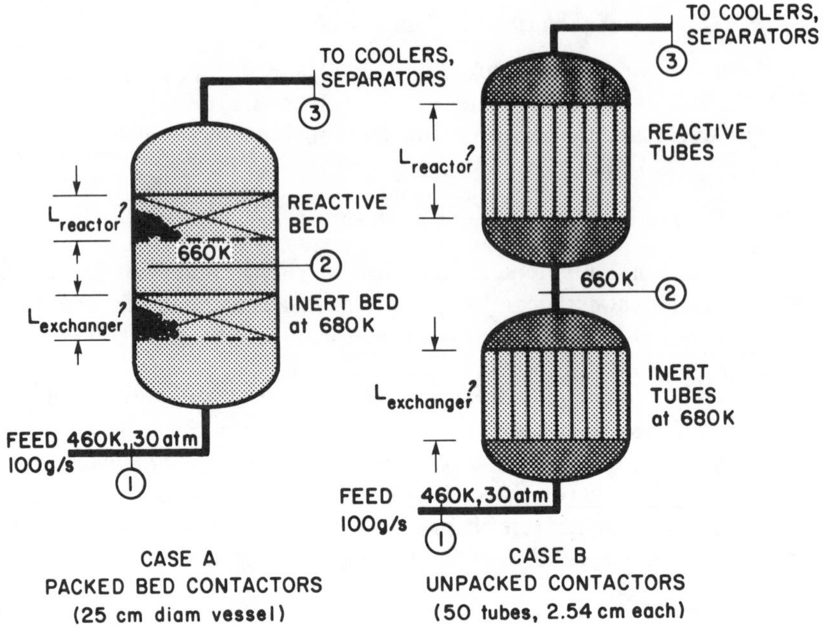

Figure 6.10E

Exercise 6.11 (*continued*)

e. $\dot{m}''$

f. $\dot{m}_p$

g. $t_{\text{life, vap}}$

h. $(\rho_g/\rho_\ell)\ln\{1 + B_h\}$

i. $-\dfrac{d}{dt}\left(\dfrac{d_p}{2}\right)\Big/ v_w$

j. $\ell/d_{p,0} \equiv \text{Kn}$

k. If envelope flame combustion is initiated, estimate $B_{h,\text{comb}}$ and the factor by which $\dot{m}''$ and $\dot{m}_p$ would be increased, and t_{life} decreased.

l. In view of the above results, assess the validity of each of the important approximations used, and the importance of the Stefan-flow effect in this example.

Table 6.11E

Substance (A)	M_A	$\rho_A^{(\ell)}\left(\dfrac{g}{cm^3}\right)$	T_{nbp}(K)	$L_A\left(\dfrac{cal}{g}\right)$	$Q\left(\dfrac{kcal}{g}\right)$	f
n-decane	142.3	0.734	447	86.0	10.56	0.0666

6.12 Consider a steady-flow single-phase combustor and its two-module idealized representation (Figure 6.7-4):

a. Verify Eq. (6.7-18) for the WSR *temperature*, $T_{②}$, at maximum $-\dot{r}_F'''$.

b. Calculate the corresponding fuel mass fraction $\omega_{F②}$ (which then enters the PFR stage).

c. Derive an expression for the PFR-volume required to achieve an overall fuel conversion (hence combustion efficiency) of 99 percent (cf. Section 6.1.1.1).

d. Using the estimated parameters ($E \approx 30$ kcal/mole, $n \approx 1.6$, $A \approx 10^{14}\text{s}^{-1}$ (g moles/cm^3)$^{-0.6}$, $Q \approx 12$ kcal/g, $f = 0.0581$; cf. Tables 6.5-1 and 6.7-1), and assuming:

$$\dot{m}_F = 0.50 \text{ kg/s} \qquad p = 20 \text{ atm}$$
$$T_u = 300 \text{ K} \qquad c_p \cong 0.28 \text{ cal/g-K (Table 8.1-1)}$$
$$\omega_{O,u} = 0.232 \qquad \Phi = 0.5,$$

estimate the required module volumes V_{WSR}, V_{PFR}, the *total* combustor volume, V(m^3), and the outlet temperature $T_{③}$.

e. Would such a chemical reactor exhibit "parametric sensitivity"? Discuss, together with any steps necessary to avoid such sensitivity.

f. Calculate the overall combustor *space-heating rate*, SHR (kcal/s-m^3) and compare your result to the (SHR$_{max}$) figures quoted in Table 6.7-1. How do these figures compare to SHR-values of common industrial gas-fired combustors? (Discuss the reasons for any disparities.)

6.13 Reconsider the situation described in Exercise 5.19, *viz.*, the estimation of the natural convective energy loss from a horizontal 3,000 K filament in the presence of chemical reaction ($H_2 \rightleftarrows 2H$). Recalculate the energy loss to the surrounding hydrogen gas on the basis of the alternate assumption that, whereas LTCE is maintained at Ⓦ and at ⓞⓞ, *no chemical reaction* (atom recombination, molecule dissociation) *occurs within the thermal boundary layer* (the so-called "chemically frozen" BL limit (Section 5.8(b)), less likely to obtain under the presently described conditions). Carefully consider the proper estimation of $\overline{\text{Nu}}_m$ for H-atom *mass* transfer away from the filament, given the fact that in the present case the *temperature* difference is the primary cause of the density change that drives the gas flow past the filament. Compare your CFBL estimate of $\dot{q}'_w$ (natural convection) (W/cm) with the corresponding value obtained using the alternative LTCE-BL assumption (Exercise 5.19). What can be concluded from this comparison (see Rosner (1961, 1965, 1975))?

6.14 Incandescent tungsten-filament bulbs filled with gaseous krypton rather than argon have recently become commercially available. Despite its higher cost, the "noble" gas krypton is claimed to have the following advantages over argon:

1. Reduced heat loss from the filament for a given filament temperature (hence, bulb brightness);
2. Reduced net rate of tungsten sublimation loss from the hot filament.
   a. To obtain a preliminary assessment of the magnitude of the natural *convection* heat-loss effect, calculate the expected heat loss (W) per centimeter length of horizontal, 0.2 mm diameter tungsten filament at 2800 K in 300 K argon, and repeat the calculation for 300 K krypton. Assume both the argon and krypton are at 2 atm total pressure, and, for this preliminary estimate, consider the gas to be

totally unconfined (i.e., neglect the "confining" effects of the outer glass envelope).

b. Predict the corresponding net rates of diffusion-controlled tungsten sublimation in argon and krypton.

c. Defend your method for estimating $\overline{\mathrm{Nu}}_m$ (what is the primary "cause" of the relative fluid-density change in this case?) and clearly outline how all necessary input data were estimated from available thermophysical-property data and theory.

d. How does the net diffusion-controlled tungsten-sublimation rate predicted in Part (b) compare with the maximum expected sublimation rate that would occur at 2800 K if the envelope were "completely evacuated"? From low-density gas-kinetic theory, the maximum molecular mass *flux* is given by $m_v \cdot [\frac{1}{4} n_v \bar{c}_v]$, where m_v, n_v, and $\bar{c}_v$ are, respectively, the vapor molecular mass, number density $(p_{v,\mathrm{eq}}/(k_B T_w))$ and mean molecular (thermal) speed $((8k_B T_w/(\pi m_v))^{1/2})$.

e. Imagine that the resistance to heat loss posed by the inert gas fill is the result of a hypothetical quiescent (stagnant) annular "sheath" (I. Langmuir [1912]) of (radial) thickness δ_h surrounding the filament of radius $d_w/2$. Use Table 5.4-1 and your results from Part (a) to calculate δ_h(argon) and δ_h(krypton). Discuss the significance of the comparison of δ_h with $d_w/2$ and the prevailing gas mean-free-path. Also, above what actual outer-envelope radius would you expect your assumption about negligibility of the "confining" effect (Part (a)) of the envelope on the gas flow to be valid?

Note: Experimental data for natural-convection heat transfer from/to an isolated horizontal cylinder in a Newtonian fluid are summarized (correlated) in Figure 7.2-1 for $10^{-4} \leqslant \mathrm{Ra}_h \leqslant 10^9$. To complete Parts (b), (c), (d) use the experimental result that the equilibrium vapor pressure of tungsten at 2800 K is 0.984×10^{-2} dyne/cm^2. *Ans.* (a) argon: 2.0 W/cm; krypton 1.3 W/cm; (b) argon: 19 pg/cm-s; krypton 14 pg/cm-s; (d) vacuum sublimation rate: 6900 pg/cm-s; (e) δ_h(argon) = 6.7 mm; δ_h(krypton) = 4.5 mm.

REFERENCES

Bilger, R.W., Chapter 3 in *Turbulent Reacting Flows* (P.A. Libby, F.A. Williams, eds.), Vol. 44 of *Topics in Applied Physics*. Berlin: Springer-Verlag, pp. 65–113 (1980).

Boysan, F., W.H. Ayers, and J. Swithenbank, *Trans. Inst. Chem. Eng.* **60**, 222–230 (1982).

Bray, K.N.C., P. Libby, and J.B. Moss, *Comb. Sci. Technol.* **41**, pp. 143–172 (1984).

Brian, P.L.T., J.E. Vivian, and A.G. Habib, *AIChE J.* **8**, pp. 205–209 (1962).

Brun, R.J., *et al.*, *NACA Report 1215* (1955).

Burke, S.P., and T.E.W. Schumann, *Ind. Eng. Chem.* **20**, No. 10, pp. 998–1004 (1928).

Elghobashi, S., and T.W. Abou-Arab, *Phys. of Fluids* **26**, p. 931 (1983).

Ewan, B.C.R., F. Boysan, and J. Swithenbank, Paper No. 62 in *Proc. 20th Symposium (Int.) in Combustion*. Pittsburgh, PA: The Combustion Institute (1985).

Faeth, G.M., S.M. Jen, and J. Gore, in *Heat Transfer in Fire and Combustion Systems*, ASME HTD–Vol. 45 (C.K. Law, *et al.*, eds.). New York: Amer. Soc. Mech. Engineers, pp. 137–151 (1985).

Fernandez de la Mora, J., and D.E. Rosner, *J. PhysicoChem. Hydrodynamics* **2**, pp. 1–21 (1981).

Fernandez de la Mora, J., and D.E. Rosner, *J. Fluid Mech.* **125**, pp. 379–395 (1982).

Friend, W.L., A.B. Metzner, *AIChE J.* **4**, pp. 393–402 (1958).

Goldschmidt, V.W., *et al.*, *Prog. in Heat and Mass Transfer*, Vol. 6, Proceedings of Int. Symposium on Two-Phase Systems (G. Hetsroni, S. Sideman, and J.P. Hartnett, Eds.), Oxford: Pergamon Press (1972); see also, Goldschmidt, V.W., and S. Eskinazi, *J. Appl. Mech.* **33**, pp. 735–747 (1966).

Hidy, G.M., and S.L. Heisler, Chapter 7 in *Recent Developments in Aerosol Science* (D.T. Shaw, ed.) New York: J. Wiley, (1978), pp. 135–165.

Hottel, H.C., and W.R. Hawthorne, in *Third Symposium on Combustion, Flame and Explosion Phenomena*, Baltimore, MD: Williams and Wilkins, pp. 254–266 (1949).

Johnstone, R.E., and M.W. Thring, *Pilot Plants, Models and Scale-up Methods in Chemical Engineering*. New York: McGraw-Hill (1957).

Labowsky, M., and D.E. Rosner, Paper 4 in *Evaporation-Combustion of Fuels*, Amer. Chem. Soc. Advances in Chem. Series, No. 166, pp. 63–79 (1978).

Laurent, A., and J.C. Charpentier, *Chem. Eng. Journal* **8**, pp. 85–101 (1974).

Longwell, J., and M.A. Weiss, *Ind./Engrg. Chemistry* **47**, 1634 (1955).

Nauman, E.B., *Chem. Engrg. Commun.* **8**, pp. 53–131 (1981); see, also: Nauman, E.B., and B.A. Buffham, *Mixing in Continuous Flow Systems*, New York: J. Wiley (1983).

Ostrach, S., *PCH-PhysicoChem. Hydrodynamics* **1**, pp. 233–247 (1980).

Peck, R.E., and G.S. Samuelson, *Sixteenth Symp. (Int.) on Combustion*, Pittsburgh, PA: The Combustion Institute, pp.1675–1687 (1976).

Petukhov, B.S., in *Advances in Heat Transfer*, Vol. 6. New York: Academic Press, pp. 504–564 (1970).

Pratt, D.T., in *Energy and Combustion Science* (N.A. Chigier, Ed.), Oxford: Pergamon Press, pp. 75–88 (1979).

Rosner, D.E., *AIAA J.* **2**, pp. 593–610 (1964).

Rosner, D.E., *J. Phys. Chem.* **75**, pp. 2969–2976 (1971).

Rosner, D.E., *High Temp. Sci.* **4**, No. 6, pp. 357–371 (1972).

Rosner, D.E., *Comb. Sci. and Technol.* **10**, pp. 97–108 (1975).

Rosner, D.E., *J. PhysicoChem. Hydrodynamics* **1**, pp. 159–185 (1980).

Rosner, D.E., and W.S. Chang, *Comb. Sci. and Technol.* **7**, pp. 145–158 (1973).

Rosner, D.E., and J. Fernandez de la Mora, *ASME Trans.-J. Engineering for Power* **104**, pp. 885–894 (1982).

Rosner, D.E., and R. Israel, *J. Aerosol Sci. and Technol.* **2**, pp. 45–51 (1983).

Rosner, D.E., and R. Nagarajan, *Chem. Eng. Sci.* **40**, pp. 177–186 (1985).

Shaw, D.A., and T.J. Hanratty, *AIChE J.* **23**, pp. 28–37 (1977).

Sherwood, T.K., and R.L. Pigford, *Absorption and Extraction*. New York: McGraw-Hill, (1952).

Spalding, D.B., *Chem. Eng. Sci.* **26**, pp. 95–107 (1971).

Spalding, D.B., *Comb. Sci. Technol.* **13**, p. 13 (1976).

Swithenbank, J., *et al.*, in *14th Symposium (Int.) on Combustion*. Pittsburgh, PA: The Combustion Institute, pp. 627–638 (1973); see, also, *20th Symposium (Int.) on Combustion*, pp. 541–547 (1984).

Thomas, L.C., *Int. J. Heat Mass Transfer* **23**, pp. 1099–1104 (1980).

Toor, H.L., *AIChE J.* **8**, 70–78 (1962).

Truesdell, C., *J. Chem. Phys.* **37**, pp. 2336–2344 (1962).
Villermaux, J., Paper No. 6 in *Chemical Reaction Engineering—Plenary Lectures*, Symposium Series No. 226 (J. Wei, C. Georgakis, Eds.), Washington, D.C.: Amer. Chem. Soc., pp. 135–185 (1983); see also *Chem. Eng. Commun.* **21**, pp. 105–122 (1983).
Weisz, P.B., and J.S. Hicks, *Chem. Eng. Sci.* **17**, pp. 265–275 (1962).
Zeldovich, Y.B., *NACA TM 1296* (1951); see, also, *The Mathematical Theory of Combustion and Explosions*. New York: Consultants Bureau (1985).

BIBLIOGRAPHY

Elementary

Edwards, D.K., V.E. Denny, and A.F. Mills, *Transfer Processes*. New York: Holt, Rinehart and Winston (1979).
Levenspiel, O., *Chemical Reaction Engineering*. New York: J. Wiley (1972).
Welty, J.R., C.E. Wicks, and R.E. Wilson, *Fundamentals of Momentum, Heat, and Mass Transfer* (Third ed.). New York: J. Wiley (1984).

Intermediate

Astarita, G., *Mass Transfer with Chemical Reaction*. Amsterdam: Elsevier (1967).
Bennett, C.O., and J.E. Myers, *Momentum, Heat, and Mass Transfer*. New York: McGraw-Hill (1974), Third ed. (1982).
Bird, R.B., W. Stewart, and E.N. Lightfoot, *Transport Phenomena*. New York: J. Wiley (1960).
Csanady, G.T., *Turbulent Diffusion in the Environment*, Dordrecht, Holland: D. Reidel Publishing Co. (1973).
Cussler, E.L., *Diffusion-Mass Transfer in Fluid Systems*. Cambridge Univ. Press (1984).
Danckwerts, P.V., *Gas-Liquid Reactions*. New York: McGraw-Hill (1970).
Denbigh, K.G., and J.C.R. Turner, *Chemical Reactor Theory*. Cambridge Univ. Press (1984).
Eckert, E.R.G., and R.M., Jr. Drake, *Analysis of Heat and Mass Transfer*. New York: McGraw-Hill (1972).
Frank-Kamenetskii, D.A., *Diffusion and Heat Transfer in Chemical Kinetics* (Second ed.) (J.P. Appleton, ed.). New York: Plenum Press (1969).
Friedlander, S.K., *Smoke, Dust and Haze—Fundamentals of Aerosol Behavior*. New York: J. Wiley (1977).
Fristrom, R., and A.A. Westenberg, *Flame Structure*. New York: McGraw-Hill (1965).
Froment, G.F., and K.B. Bischoff, *Chemical Reactor Analysis and Design*. New York: J. Wiley (1979).
Fuchs, N.A., *Mechanics of Aerosols*. New York: Pergamon Press (1964).
Geankoplis, C.J., *Transport Processes and Unit Operations*. Boston: Allyn & Bacon (1978).
Kanury, A.M., *Introduction to Combustion Phenomena*. New York: Gordon and Breach (1975).
Kays, W.M., *Convective Heat and Mass Transfer*. New York: McGraw-Hill (1966).
Lee, H.H., *Heterogeneous Reactor Design*. Boston: Butterworth (1985).
Pasquill, F., *Atmospheric Diffusion*. Toronto: D. Van Nostrand (1962).
Rudinger, G., *Fundamentals of Gas-Particle Flow*, Vol. 2, *Handbook of Powder Technology*. Holland: Elsevier (1980).

Sherwood, T.K., R.L. Pigford, and C.R. Wilke, *Mass Transfer*. New York: McGraw-Hill (1975); see also *Absorption and Extraction*. New York: McGraw-Hill (1952).

Spalding, D.B., *Convective Mass Transfer—An Introduction*. New York: McGraw-Hill (1963).

Treybal, R.E., *Mass Transfer Operations* (Second ed.). New York: McGraw-Hill (1968).

Advanced

Aris, R., *The Mathematical Theory of Diffusion and Reaction in Permeable Catalysts*, Vols. 1, 2. Oxford: Clarendon Press (1975).

Borghi, R., "On the Structure and Morphology of Turbulent Premixed Flames," Paper 7 in *Recent Advances in the Aerospace Sciences* (C. Casci and C. Bruno, eds.) New York: Plenum Press, pp. 113–138 (1985).

Crank, J., *The Mathematics of Diffusion* (Second ed.). New York: Clarendon (Oxford Univ.) Press (1975); reprinted 1979.

Hirschfelder, J.O., C. F. Curtiss, and R. B. Bird, *Molecular Theory of Gases and Liquids*. New York: J. Wiley (1954).

Krishna, R., and R. Taylor, "Multicomponent Mass Transfer—Theory and Applications," in *Handbook of Heat and Mass Transfer Operations*, Gulf Publ. Corp. (Chapter 7, (1985)).

Lapidus, L., and N. Amundson, eds., *Chemical Reactor Theory: A Review*. Englewood Cliffs, NJ: Prentice-Hall (1977).

Levich, V.G., *Physicochemical Hydrodynamics* (Second ed.). Englewood Cliffs, NJ: Prentice-Hall (1962).

Libby, P.A., and F.A. Williams, eds., *Turbulent Reacting Flows* (Vol. 44 of *Topics in Applied Physics*. Berlin: Springer-Verlag (1980)).

Peters, N., "Structure of Turbulent Jet Diffusion Flames," *Int. Chem. Engrg*. New York: AIChE (July 1985); **25**, no. 3, pp. 406–417. (trans of *ChE Technik* (1983); **55**, no. 10, pp. 743–751).

Pope, S.B. "PDF Methods for Turbulent Reactive Flows," *Progress Energy Combustion Sci.* Oxford, England: Pergamon Press (1985); **11**, pp. 119–192.

Slattery, J.C., *Momentum, Energy and Mass Transfer in Continua*. New York: McGraw-Hill (1972).

Sideman, S., and W.V. Pinczewski, "Turbulent Heat and Mass Transfer at Interfaces: Transport Models and Mechanisms," Ch. 2 in *Topics in Transport Phenomena* (C. Gutfinger, ed.). New York: J. Wiley (1975); pp. 47–207.

Williams, F.A., *Combustion Theory*. Reading, MA: Addison-Wesley (1964); Second Ed. Menlo Park, CA: Benjamin Cummings (1985).

Additional Sources of Information

The following periodicals also regularly contain important contributions to the field of *mass transfer*:

Int. J. Heat and Mass Transfer
J. Aerosol Science
Amer. Inst. Chem. Engrs. (AIChE) J.
J. Physicochemical Hydrodynamics
Advances in Transport Processes (Wiley Eastern, Ltd., New Delhi)
Aerosol Science and Technology (AAAR)
Chemical Engineering Science
J. Multiphase Flow
J. Colloid and Interface Sci.
Environmental Sci. and Tech.

7

Similitude Analysis with Application to Chemically Reactive Systems—Overview of the Role of Experiment and Theory

7.1 INTRODUCTION AND OBJECTIVES

7.1.1 Possibility of Scale-Model Testing and the Quantitative Exploitation of Similarity in Generalizing the Results of Experiments or Computations

The cost of running full-scale, long-duration experiments to develop, say, a new industrial chemical synthesis reactor, a furnace, or a rocket engine, is very great. For this reason there is a considerable incentive to obtain much of the required information using *small-scale models* and/or short-duration ("accelerated") tests. The applied chemist L.H. Baeckeland [1863–1944] summarized this philosophy well when he said: "Commit your blunders on a small scale and make your profits on a large scale."

Of course, scale-model tests can always be run—the real question is: *Under what circumstances (if any) can one quantitatively predict full-scale ("prototype") behavior from the results of such scale-model experiments?* Clearly, it is necessary that the small-scale model be similar in all important respects to the prototype—i.e., dynamically, thermally, chemically, as well as geometrically. We consider below the following two questions: (a) What conditions must be met to ensure these levels of "similarity," at least approximately? and (b) How will prototype (p) and model (m) performance variables be quantitatively interrelated?

Note that here we are seeking simple quantitative interrelations between (measured) model performance and (anticipated) prototype performance; i.e., we wish to avoid, if possible, complex interrelations requiring a significant invest-

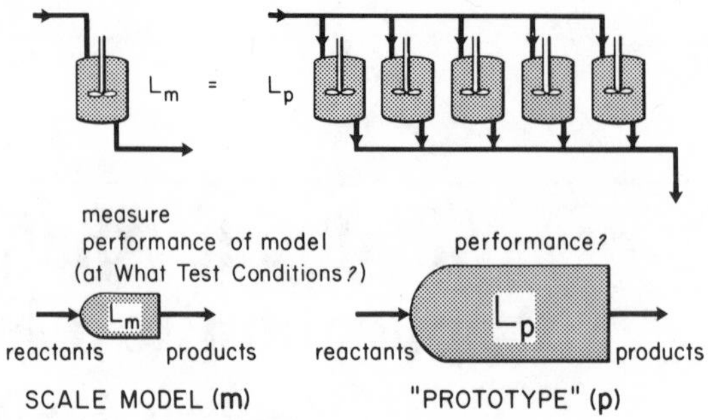

Figure 7.1-1 Schematic of two alternative approaches to "scale-up"; illustration of the cost-*vs.*-risk trade-off.

ment in mathematical modeling and computer time. We demonstrate here that, by a proper choice of test conditions, it *is* often possible to establish such simple interrelations. However, in the presence of many interacting physico-chemical phenomena a judicious blend of scale-model testing and mathematical modeling is usually needed to design large-scale equipment (Sections 7.3, 7.4.2). Note also that, for reasons of capital cost, a single large-scale device is usually much more attractive than simply obtaining the desired capacity by "stringing together" many smaller-scale units (e.g., in parallel) (see Figure 7.1-1).

Usually the geometrical *scale factor* L_p/L_m is much *greater* than unity (to minimize the cost of the model tests); however, there are occasions when tests are done using "models" that are actually *larger* than full-scale(!)—such results must be "scaled down" to apply to "full"-scale. Examples would be: (a) the use of experiments on a supported 1 mm ($=1000\ \mu$m) diameter fuel droplet to predict the behavior of a 10-μm fuel droplet in a furnace (a size which is less accessible, experimentally), and (b) the probing of expanded (low pressure) premixed flames to infer the internal structure of (higher pressure) flames which are too thin for conventional probing (see Figure 1.2-5 and Section 7.2.2.2).

It will be seen that maximum use of all experiments or computations can be made by reporting the results in terms of certain *dimensionless ratios*. The reader has probably noticed that most of the results quoted in Chapters 4, 5, and 6 have already been cast in this form, using suitable "internal" (rather than arbitrary) reference quantities (i.e., *relevant* lengths, times, temperature differences, energy fluxes, etc.). This process is sometimes called "eigen-measure," and the resulting ratios: "eigen-ratios." Just as Pythagoras could judge the *geometric* similarity of triangles *via length* eigen-ratios, the similarity of more complex physico-chemical systems will be seen to be determined by eigen-ratios that go beyond length alone.

7.1.2 Types of Similarity

To be quantitatively useful, the model and prototype must usually be similar in the following respects:

1. *Geometrically*; i.e., corresponding distances in the prototype and model must all stand in the same ratio: L_p/L_m.
2. *Dynamically*; i.e., force or momentum flux ratios responsible for fluid motion must be the same for prototype and model.
3. *Thermally*; i.e., corresponding ratios of temperature differences between any two points in the prototype and model must be equal.
4. *Compositionally*; i.e., corresponding ratios of key species composition differences between any two points in the prototype and model must be equal.

We show below that, by making certain "compensating changes" in other system parameters, *all* of these types of similarity can be simultaneously and readily attained in nonreactive flow situations over wide *nonunity* ranges of the scale factor L_p/L_m. Strict similarity is far more difficult to achieve in systems with exoergic *chemical* change, especially if chemical nonequilibrium prevails with respect to homogeneous or heterogeneous reactions. Specific examples of both types (nonreactive/reactive) will be given in Section 7.2, together with the methods of establishing similarities (2) through (4) and the quantitative implications of strict or approximate similarity ("partial modeling").

7.1.3 Nondimensional Presentations in Science and Engineering

The use of appropriate reference quantities, or "eigen-measures," to generalize the results of experiments or theoretical calculations is probably already familiar to the reader. Consider the following simple examples:

7.1.3.1 Law of "Corresponding States"

Equation-of-State (EOS) thermodynamic *data* for the rare gases He, Ar, Ne, Kr, and Xe are rather different, owing in part to large differences in molecular size and weight. Thus, p, V, T data for each gas would plot differently. However, if the *critical* quantities p_c, V_c, T_c are used as reference values; *viz.*:

$$p_{\text{ref}} \equiv p_c, \qquad V_{\text{ref}} \equiv V_c, \qquad T_{\text{ref}} \equiv T_c, \tag{7.1-1a,b,c}$$

then, on a "reduced" plot of p/p_{ref} *vs.* V/V_{ref} and T/T_{ref}, these gases become indistinguishable—i.e., they all have the same EOS in terms of:

$$\frac{p}{p_c} = \text{fct}\left(\frac{T}{T_c}, \frac{V}{V_c}\right). \tag{7.1-2}$$

This implies that, if you went to the expense of determining $p(V, T)$ experimentally for, say, argon, this simple replotting technique (introduction of $p_{\text{ref}} \equiv p_c$, $V_{\text{ref}} \equiv V_c$, $T_{\text{ref}} \equiv T_c$) would allow these same data to be used for predictions of the behavior of He, Ne, Kr, and Xe *at their corresponding states.* It is remarkable that

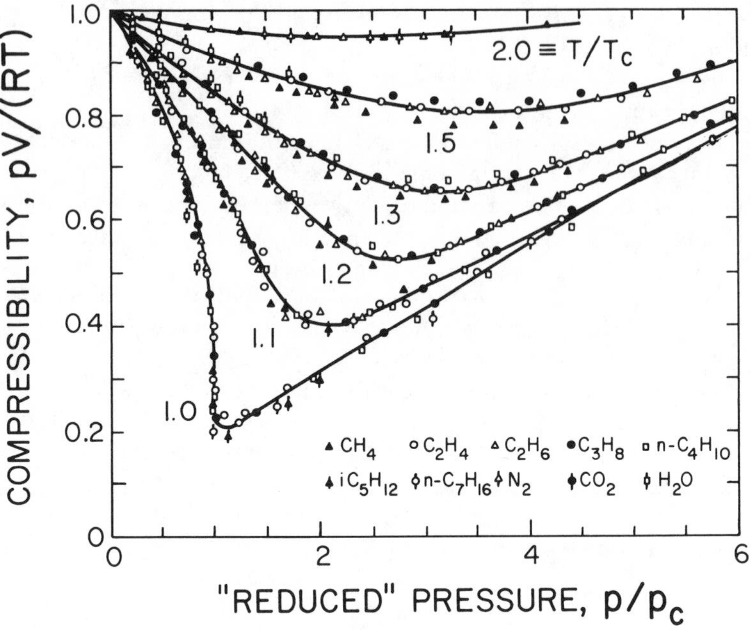

Figure 7.1-2 "Corresponding states" correlation for the compressibility $pV/(RT)$ of ten vapors (after G.-J. Su (1946)).

this procedure also works well for many other vapors, chemically unlike the noble gases. Figure 7.1-2 shows such a correlation of "compressibility" ($pV/(RT)$) data[1] for ten gases that are "superficially" rather dissimilar. Correlations of this type are also widely used for other equilibrium properties (e.g., surface tension) as well as *transport properties* (see Sections 3.3.4 and 3.4.4).

7.1.3.2 Newtonian Viscosity of a Vapor

Consider a pure vapor comprised of molecules of mass m which interact with each other according to any spherically symmetric potential $\phi\{r\}$ with $\phi\{\sigma\} = 0$ and having the "well depth" ε (see Figure 7.1-3). Without actually solving any equations, the viscosity, μ, of such a vapor can be shown to depend on:

$$\begin{aligned} k_B T &\equiv \text{product of Boltzmann const and local temperature} \\ m &\equiv \text{mass/molecule} \\ \sigma &\equiv \text{size parameter defined by } \phi\{\sigma\} = 0 \\ \varepsilon &\equiv \text{energy-well parameter (cf. Figure 7.1-3)} \\ v &\equiv n^{-1} \text{ (average volume/molecule)} \end{aligned} \tag{7.1-3}$$

[1] Here R is the *universal gas constant*. The ratio $pV/(RT)$ would be near unity under conditions such that the perfect-gas EOS is an adequate approximation.

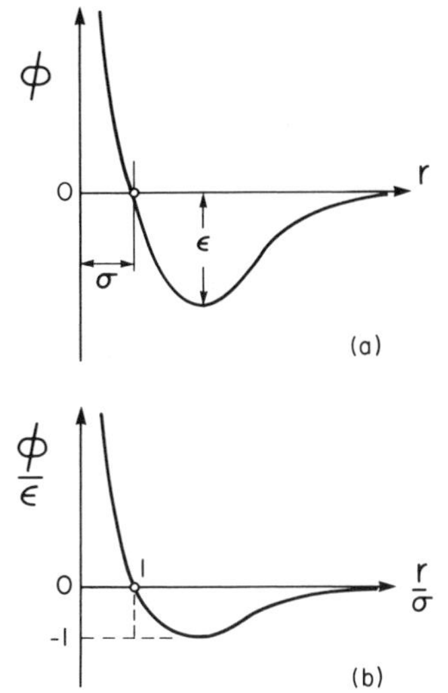

Figure 7.1-3 Two-parameter (σ, ε) spherically symmetric intermolecular potential: (a) dimensional; (b) non-dimensional (scaled).

i.e., we expect:

$$\mu = \mu\{k_B T, m, \sigma, \varepsilon, v\}.$$

However, using a method to be discussed in Section 7.2.1, we can show that since this relation must be *dimensionally homogeneous*, it can be rewritten in the more useful form:

$$\mu = \frac{(mk_B T)^{1/2}}{\sigma^2} \cdot \frac{1}{\text{fct}\left(\dfrac{k_B T}{\varepsilon}, \dfrac{\sigma}{v^{1/3}}\right)}. \tag{7.1-4}$$

Thus, the viscosity coefficient of *all* such vapors should be correlated in this particular way.[2]

As a corollary, data using any particular vapor could be used to obtain the indicated function of $k_B T/\varepsilon$, σ^3/v *once and for all*.

Note that, in the perfect-gas limit ($\sigma^3/v \to 0$), the well-known viscosity law of Chapman and Enskog, *viz.*:

$$\mu = \frac{5(\pi m k_B T)^{1/2}}{16\pi\sigma^2} \cdot \frac{1}{\Omega_\mu\{k_B T/\varepsilon\}}, \tag{7.1-5}$$

[2] Note that while the dependence of μ on T and σ can be rather complex, all such gases must have $\mu \sim m^{1/2}$.

is indeed of this form, showing that, for such a gas, μ is *independent* of p, and the momentum diffusivity $\nu(\equiv \mu/\rho)$ varies like p^{-1}.

7.1.3.3 Similitude in Biology: Mammal "Invariants" (Stahl (1962))

In the section above, we exploited the intrinsic similarity of "all" vapors (or certain families of vapors, with quasi-spherical interaction potentials). In other fields of study one can establish and exploit other intrinsic similarities.

Consider the intrinsic similarities that have been found between all mammals (a rather complex class of chemical reactors!). While each mammal, depending on its type and size (say, mass) has a different:

$$\begin{aligned} m_{(\text{heart})} &\equiv \text{mass of heart}, \\ m_{(\text{blood})} &\equiv \text{mass of blood (liquid "inventory")}, \\ m_{(\text{kidney})} &\equiv \text{mass of kidney}, \\ m_{(\text{liver})} &\equiv \text{mass of liver}, \qquad \text{etc.}, \end{aligned} \tag{7.1-6}$$

certain dimensionless *ratios* of such quantities are approximately *the same for all mammals, irrespective of their size* (*or total mass*). These ratios, called mammal "invariants" (or "allometric ratios") include (Stahl (1962)):

$$\frac{m_{(\text{blood})}}{m_{(\text{heart})}}, \qquad \frac{m_{(\text{kidney})}}{m_{(\text{liver})}}, \qquad \text{etc.} \tag{7.1-7}$$

Moreover, in the lifetimes of mammals, corresponding biological events (e.g., puberty, menopause, etc.) take place at approximately corresponding times.

These are quantitative statements that can be used to make rational *predictions* based on extensive data on some mammals and fragmentary information on others. In a certain sense, all mammals are "models" of one another, and this similarity is being exploited in biological studies.

7.1.3.4 Fluid Dynamics of Incompressible Fluid Flow Past a Sphere: The "Universal" Drag Law

The steady drag, D, on a smooth sphere of diameter d_w in a uniform but nonturbulent mainstream of velocity U also depends on the fluid density, ρ, and Newtonian viscosity, μ—that is,

$$D = \text{fct}\{d_w, U, \rho, \mu\}. \tag{7.1-8}$$

However, as is well known, this *dimensional* relation can be cast in the *non-*

dimensional form:

$$\frac{D}{\frac{1}{2}\rho U^2\left(\frac{\pi d_w^2}{4}\right)} = C_D\left\{\frac{\rho U d_w}{\mu}\right\} \quad (7.1\text{-}9)$$

involving an undetermined function (the so-called "drag coefficient") of only *one* parameter, *viz.*:

$$\mathrm{Re} \equiv \frac{\rho U d_w}{\mu} = \frac{\rho U U}{(\mu U/d_w)} = \frac{\text{char. momentum flux by convection}}{\text{char. momentum flux by diffusion}}. \quad (7.1\text{-}10)$$

Thus, Eq. (7.1-9) will correlate drag-data and/or calculations for all Newtonian fluids, all sphere sizes, and all imposed fluid velocities (subject to the underlying assumptions: continuum, low mainstream Mach number, constant properties, etc.). Moreover, while one may be interested in the drag on a sphere larger than 1 meter in diameter, in principle it is sufficient to experiment on a sphere of, say, diameter 1 cm ($L_p/L_m = 10^2$) to determine the function $C_D\{\mathrm{Re}\}$ once and for all, or even C_D at only one Re (see Figure 4.4-4).

In this case we have exploited the fact that all spheres are *geometrically* similar; moreover, if we arrange the scale-model experiment such that the above-mentioned momentum flux ratio is the same as that for the prototype; i.e.,:

$$(\mathrm{Re})_m = (\mathrm{Re})_p, \quad (7.1\text{-}11)$$

then clearly,

$$(C_D)_m = (C_D)_p; \quad (7.1\text{-}12)$$

i.e., *dynamic* similarity has also been achieved, and we conclude that:

$$\left[\frac{D}{\frac{1}{2}\rho U^2(\pi d_w^2/4)}\right]_p = \left[\frac{D}{\frac{1}{2}\rho U^2(\pi d_w^2/4)}\right]_m. \quad (7.1\text{-}13)$$

This is the *quantitative* relation between D_p and D_m that we sought.

The Reynolds'-number equivalence condition (7.1-11) ensures that the model flow is also *dynamically* similar to the full-scale flow. Usually, only in that case (equal Re), is it possible to establish the simple quantitative relation (7.1-13) between the *geometrically* similar model and prototype drags.

An exception to the above statement is possible when the Re-dependence of C_D is known. For example, by an asymptotic *similitude analysis* of the momentum equations (Section 7.1.1), one can show, even without solving[3] any PDEs, that, when

[3] By *solving* these equations, G. Stokes showed that the proportionality coefficient in Eq. (7.1-14) is 24, a result accurate up to about $\mathrm{Re} = 10^{-1}$ (see Figure 4.4-4).

Re ≪ 1, then:

$$C_D \sim (\mathrm{Re})^{-1}. \tag{7.1-14}$$

If *both* Re_p and Re_m satisfied this Re ≪ 1 condition, it would *not* be necessary to test at precisely the same Re-value to establish a quantitative relation between D_p and the measured D_m.

A corollary of the above discussion is that any drag measurements or calculations that are made should be cast in the form (7.1-9) rather than (7.1-8), for in *dimensionless* form the results are far more widely applicable.

We now turn to the methods available for establishing the remaining types of similarity (thermal, compositional) and the possibility of scale-model testing of systems with *chemical reaction.*

7.2 DIMENSIONAL ANALYSIS, SIMILITUDE ANALYSIS, ANALOGIES, AND SCALE-MODEL THEORY

7.2.1 Dimensional Analysis and Its Advantages/Limitations

Vaschy [1892] and, independently, Buckingham [1914], showed that any *dimensional* interrelation involving N_v variables can be rewritten in terms of a smaller number, N_π, of independent *dimensionless* variables, where N_π is usually smaller than N_v by precisely the number of fundamental dimensions (e.g., five in a problem involving length, time, mass, heat, and temperature). Thus, the above-mentioned drag relation, Eq. (7.1-8), involving five *dimensional* physical variables can be rewritten as Eq. (7.1-9), involving only $5 - 3 = 2$ *nondimensional* variables or "groups."[4]

Called Buckingham's "Pi Theorem," this result can be used in any branch of science/technology, even without a detailed knowledge of the underlying constitutive equations, field equations, and/or boundary conditions (see, e.g., Murphy (1950), Bridgman (1963), and Schuring (1977)). It is sufficient that the variables influencing the *quantity of interest* be listed in their entirety (e.g., Eq. (7.1-8) for the sphere drag). This procedure will be seen to provide the relevant similarity criteria for problems that go beyond *geometrical* and *dynamical* similarity. Indeed, the resulting dimensionless groups (the Π's) *are* themselves the quantities that must be kept invariant in model testing.

Now consider the somewhat more general problem of steady *heat flow* from

[4] In this purely dynamical problem heat and temperature are irrelevant, hence, there are only 3 fundamental dimensions used to describe each of the participating quantities ($\rho, U, \ldots$ etc.).

an isothermal sphere in a steady uniform (forced) fluid flow. If we expect:

$$\frac{\dot{q}_{\mathrm{w}}}{A_{\mathrm{w}}} = \mathrm{fct}\{d_{\mathrm{w}}, \mu, \rho, U, k, c_{\mathrm{p}}, (T_{\mathrm{w}} - T_{\infty})\}, \tag{7.2-1}$$

then this interrelation between eight *dimensional* quantities can be rewritten in terms of three dimensionless groups:[5]

$$\frac{(\dot{q}_{\mathrm{w}}/A_{\mathrm{w}})}{[k(T_{\mathrm{w}} - T_{\infty})/d_{\mathrm{w}}]} = \overline{\mathrm{Nu}}_h\left\{\frac{\rho U d_{\mathrm{w}}}{\mu}, \frac{(\mu/\rho)}{[k/(\rho c_{\mathrm{p}})]}\right\}. \tag{7.2-2}$$

This is in the form of Eqs. (5.5-10) and (5.6-10) stated earlier, and one notes that, while Re establishes *dynamic similarity*, equality of the fluid Prandtl number,

$$\mathrm{Pr} \equiv \frac{\mu/\rho}{[k/(\rho c_{\mathrm{p}})]} \equiv \frac{\nu}{\alpha} \qquad \left(\frac{\text{momentum diffusivity}}{\text{energy diffusivity}}\right), \tag{7.2-3}$$

in model and prototype is evidently necessary to ensure simultaneous *thermal similarity*. Thus, *prototype* heat fluxes could readily be determined from *model* heat fluxes, provided the model test is run under conditions such that, not only:

$$\mathrm{Re}_{\mathrm{m}} = \mathrm{Re}_{\mathrm{p}} \tag{7.2-4}$$

but also

$$\mathrm{Pr}_{\mathrm{m}} = \mathrm{Pr}_{\mathrm{p}}. \tag{7.2-5}$$

More generally, *geometric* similarity for bodies of complex shape requires similarity with respect to *shape* and *orientation* (relative to the oncoming stream and/or gravity, etc.). Thus, for *fluid convection* in a constant-property, low-Mach-number Newtonian fluid flow, Eq. (7.2-2) takes the form:

$$\overline{\mathrm{Nu}}_h\{\mathrm{Re}, \mathrm{Pr}, \text{shape}, \text{orientation}\}, \tag{7.2-6}$$

defining a dimensionless function which, in principle, can be determined using model objects of "any" scale.

As the relevant variable list grows, reflecting the importance of additional physical and/or chemical phenomena, the number of similarity parameters likewise grows, making scale-model similarity more difficult to achieve. Thus, in the

[5] Note that in such problems energy can be considered a fundamental "dimension," bringing this number to 5 (Rayleigh [1915]). There are several systematic procedures for determining an independent set of the N_π dimensionless groups. Usually, however, they are easy to write down from inspection, and chosen in accord with established practice for closely related problems.

simultaneous presence of variable thermophysical properties, forced convection, natural convection, and free-stream turbulence, Eq. (7.2-6) becomes:

$$\overline{\mathrm{Nu}_h}\{\mathrm{Re}, \mathrm{Ra}_h, \mathrm{Pr}, I_{t,\infty}, L_{t,\infty}/L, T_w/T_\infty, \text{shape, orientation}\}. \qquad (7.2\text{-}7)$$

For high-speed (compressible) gas flows, one must also add:

$$\mathrm{Ma} \equiv \frac{U}{a_\infty} \quad \text{(Mach number)}, \qquad (7.2\text{-}8)$$

$$\gamma \equiv \frac{c_p}{c_v} \quad \text{(specific heat ratio)}, \qquad (7.2\text{-}9)$$

etc. The consequences of additional phenomena in systems with phase change and/or chemical reaction will be considered in Section 7.2.3.

It is often stated that the greatest advantage of this method ("dimensional analysis") of establishing similarity is that one does not need to know (much less solve) the underlying constitutive equations, and field equations subject to their appropriate initial and boundary conditions. However, this fact is at the same time the method's greatest *weakness*, since:

a. one can never be sure that the initial variable list is *sufficient* to adequately describe the phenomenon in question;
b. there is information *contained* in the constitutive laws, governing field equations/boundary conditions/initial conditions that can be exploited to further reduce the required number of dimensionless parameters, even without having to solve these PDEs + BCs + ICs (Section 7.2.2);
c. the physical significance of the relevant dimensionless groups is obscured;
d. no insight is provided concerning the possible use of *analogs*—i.e., models based on other phenomena obeying the same laws.

7.2.2 Similitude Analysis

In contrast to dimensional analysis, *similitude analysis*[6] is based on the governing constitutive equations, conservation (macro- or microscopic CV) equations and their initial/boundary conditions (Chapter 2). Here, by inspection and the exploitation of justifiable approximations, one can extract the desired similitude

[6] What we call "similitude" analysis has been called "inspectional" analysis by several earlier authors (see, e.g., Becker (1976)). Moreover, whereas in the next examples we identify the relevant dimensionless groups using the governing conservation equations in their *microscopic* control volume (CV) form (i.e., as PDEs), the underlying *macroscopic* CV conservation equations (cf. Chapter 2) can, of course, also be used for this purpose (see, e.g., Becker (1976)). Thus, for *multiphase* flows (Section 2.6.4), one could identify relevant dimensionless groups either from the governing macroscopic CV conservation principles (integral equations) or from the "volume-averaged" pseudo-continuum *field* equations (PDEs) derived therefrom.

conclusions without actually solving the resulting set of dimensionless equations. The method is more powerful than formal dimensional analysis because it:

a. removes the guesswork or intuition concerning what are the relevant variables.
b. clearly demonstrates the physical significance of each dimensionless group, and suggests when certain groups will be irrelevant based on the dominance of competing effects.
c. often enables a significant reduction in the number of relevant dimensionless groups actually needed to adequately describe a problem.
d. suggests the existence and use of *analogies*, i.e., other phenomena (governed by *identical* dimensionless PDEs + BCs + ICs) in a system of the same geometry.

7.2.2.1 Similitude Analysis of Convective Heat Flow

As a simple physical example, prior to our discussion of chemical reactor similitude, consider the problem of steady *heat flow* from an isothermal horizontal cylinder of characteristic transverse dimension, L, in a Newtonian fluid that is motionless except for the "natural convection" induced by the simultaneous presence of the body force field $\mathbf{g}$.

From the viewpoint of *dimensional analysis* we might postulate a *dimensional* interrelation of the form:

$$\frac{\dot{q}'_{\text{w}}}{\pi L} = \text{fct}_1(L, g, \beta_T, T_{\text{w}}, T_\infty, k, \rho, c_{\text{p}}, \mu, \text{shape, orientation}), \qquad (7.2\text{-}10)$$

where $\dot{q}'_{\text{w}}$ is the total rate of heat loss per unit axial *length* of cylinder, πL (where L is a characteristic *transverse* dimension of the cylinder cross-section) is proportional to the cylinder surface area per unit axial length, and β_T is the thermal expansion coefficient of the fluid.[7] According to dimensional analysis (Π-theorem), this interrelation between ten dimensional variables can be rewritten in terms of "only" six independent dimensionless groups, e.g.:

$$\frac{[\dot{q}'_{\text{w}}/(\pi L)]}{[k(T_{\text{w}} - T_\infty)/L]} = \text{fct}_2\left\{\frac{gL^3}{\nu^2}, \beta_T(T_{\text{w}} - T_\infty), \frac{\nu}{\alpha}, \frac{T_{\text{w}}}{T_\infty}, \frac{(\nu/L)^2}{c_{\text{p}}(T_{\text{w}} - T_\infty)}, \text{shape, orientation}\right\}. \qquad (7.2\text{-}11)$$

Thus, even if the cross-sectional shape and orientation (with respect to gravity) were preserved in the model and prototype, it would also appear to be necessary to preserve five additional dimensionless groups to allow $(\dot{q}'_{\text{w}})_{\text{m}}$ measurements to be usable in predicting $(\dot{q}'_{\text{w}})_{\text{p}}$-values.

---

[7] The local buoyancy force per unit volume of fluid, $-\mathbf{grad}\, p + \rho\mathbf{g}$ (Section 5.3.1), can then be approximated by $\rho_\infty g \beta_T (T - T_\infty)$, where

$$\beta_T \equiv v^{-1}(\partial v/\partial T)_{\text{p}} = -\rho^{-1}(\partial \rho/\partial T)_{\text{p}}.$$

However, a *similitude analysis* (see Appendix 7.1 and the discussion below) of the basic constitutive equations, PDEs, and BCs, leads us to conclude that, in the present case:

$$\frac{(\dot{q}'_{\mathrm{w}}/(\pi L))}{[k(T_{\mathrm{w}} - T_{\infty})/L]} \cong \mathrm{const}\{\mathrm{shape}\} \cdot \overline{\mathrm{Nu}}_h\{\mathrm{Ra}_h, \mathrm{Pr}, \mathrm{shape}, \mathrm{orientation}\}, \quad (7.2\text{-}12)$$

showing that the key "set-up rules" for dynamic and thermal similarity actually involve *only* Pr and the so-called *Rayleigh number for heat transfer*:

$$\mathrm{Ra}_h \equiv \frac{g\beta_T(T_{\mathrm{w}} - T_{\infty})L^3}{\nu^2} \cdot \frac{\nu}{\alpha} \equiv \mathrm{Gr}_h \cdot \mathrm{Pr} \quad (7.2\text{-}13)$$

Note that Ra_h is the result of "merging" the first three arguments appearing in Eq. (7.2-11). Note also that the temperature *ratio* $T_{\mathrm{w}}/T_{\infty}$ is itself irrelevant (unless variable thermophysical fluid properties become significant), and the next group in Eq. (7.2-11) can be shown to be unimportant for a viscous continuum flow, being of the order of only $(\mathrm{Kn})^2/[(T_{\mathrm{w}}/T_{\infty}) - 1]$ for a perfect gas. This group turns out to be the multiplier of the *viscous dissipation term* appearing in the PDE governing the dimensionless temperature field, and viscous dissipation plays a minor role in most natural convection problems (compared to Fourier conduction and convection).

Briefly, the equations and boundary conditions governing the velocity[8] and temperature fields are first made dimensionless (cf. Appendix 7.1) using:

$$L_{\mathrm{ref}} \equiv L, \quad (7.2\text{-}14)$$

$$(T - T_{\infty})_{\mathrm{ref}} \equiv (T_{\mathrm{w}} - T_{\infty}), \quad (7.2\text{-}15)$$

$$U_{\mathrm{ref}} \equiv \nu_{\infty}/L. \quad (7.2\text{-}16)$$

Then, in the coupled PDEs governing $\mathbf{v}^* \equiv \mathbf{v}/U_{\mathrm{ref}}(=\mathbf{v}L/\nu_{\infty})$ and $T^* \equiv (T - T_{\infty})/(T_{\mathrm{w}} - T_{\infty})$ appear the parameters Gr_h, Pr and $(\nu_{\infty}/L)^2/[c_p(T_{\mathrm{w}} - T_{\infty})]$, with the latter multiplying dimensionless products quadratic in the local velocity gradients (collectively the viscous dissipation term). When the latter is negligible, $\mathbf{v}^*$ and T^* are the solutions of the PDE-system:

$$\mathrm{div}^* \mathbf{v}^* = 0 \quad \text{(mass)}, \quad (7.2\text{-}17)$$

$$\mathbf{v}^* \cdot \mathbf{grad}^* \mathbf{v}^* = \mathrm{div}^*(\mathbf{grad}\, \mathbf{v}^*) - \mathrm{Gr}_h \cdot (\mathbf{g}/g) \cdot T^* \quad \text{(momentum)}, \quad (7.2\text{-}18)$$

$$\mathbf{v}^* \cdot \mathbf{grad}^* T^* = (\mathrm{Pr})^{-1} \mathrm{div}^*(\mathbf{grad}^* T^*) \quad \text{(energy)}, \quad (7.2\text{-}19)$$

[8] Recall that, in natural convection problems, there is no obvious choice of reference velocity (cf. Section 7.1.3.4). For the purposes of this similitude ("inspectional") analysis, however, U_{ref} could equally well have been chosen as $(2\beta_T \Delta T gL)^{1/2}$ (Section 5.4.2). How would this have altered the location of Gr_h in Eq. (7.2-18)?

and in this case no additional parameters enter conditions at the boundaries, along which $\mathbf{v}^*$ and T^* are either zero or unity.

The quantity of interest, $\dot{q}'_w$, the rate of heat loss per unit depth of cylinder, can be computed from $\int -k(\partial T/\partial n)\, dA'_w$. Clearly, even without carrying out the actual solution analytically or numerically, Nu_h can, at most, be a function of Ra_h and Pr (or Gr_h and Pr) for a given cylinder cross-sectional *shape* and *orientation.*

This method also shows that the dimensionless groups that appear as parameters in our basic equations have an obvious physical significance. Note, e.g., that:

$$\left|\frac{\text{local buoyancy force/mass}}{\text{local viscous force/mass}}\right| = \mathrm{Gr}_h \cdot \frac{|T^*|}{|\mathrm{div}^*(\mathbf{grad}^*\,\mathbf{v}^*)|}. \tag{7.2-20}$$

Thus, if the local quantity which Gr_h multiplies is never very large or very small, the coefficient Gr_h itself *serves as a measure of the relative magnitudes of buoyancy and viscous forces in such a flow.*

This method also reveals the existence of a potentially useful *mass-transfer* "*analog*" of the heat-transfer problem. Consider the seemingly different problem of a slowly subliming (or dissolving) solid cylinder of the same shape and orientation as before, with the "solute" mass fraction $\omega_{\mathrm{A,w}} = \mathrm{const}(\ll 1)$ and $\omega_{\mathrm{A},\infty}$ (also $\ll 1$) specified. Then the local buoyancy force/mass can be written $g\beta_\omega(\omega_\mathrm{A} - \omega_{\mathrm{A},\infty})$ and the composition variable ω^*, defined by:

$$\omega^* \equiv \frac{\omega_\mathrm{A} - \omega_{\mathrm{A},\infty}}{\omega_{\mathrm{A,w}} - \omega_{\mathrm{A},\infty}} \tag{7.2-21}$$

satisfies a PDE identical in form to (7.2-18), i.e.:

$$\mathbf{v}^* \cdot \mathbf{grad}^*\,\omega^* = (\mathrm{Sc})^{-1}\mathrm{div}^*\,\mathbf{grad}^*\,\omega^*, \tag{7.2-22}$$

where we neglect homogeneous chemical reaction and assume the local validity of Fick's law for dilute species A diffusion through the Newtonian fluid. Moreover, $\mathbf{v}^*$ satisfies the nonlinear PDE:

$$\mathbf{v}^* \cdot \mathbf{grad}^*\,\mathbf{v}^* = \mathrm{div}^*(\mathbf{grad}\,\mathbf{v}^*) - \mathrm{Gr}_m(\mathbf{g}/g)\omega^* \tag{7.2-23}$$

identical in form to (7.2-18). Here we have introduced the transport property (diffusivity ratio):

$$\mathrm{Sc} \equiv \frac{\nu}{D_\mathrm{A}} \qquad \text{(Schmidt number)} \tag{7.2-24}$$

and the Grashof number for mass transport:

$$\mathrm{Gr}_m \equiv \frac{g\beta_\omega(\omega_{\mathrm{A,w}} - \omega_{\mathrm{A},\infty})L^3}{\nu^2} \equiv \frac{\mathrm{Ra}_m}{\mathrm{Sc}}. \tag{7.2-25}$$

Since, in this case, no new parameters enter the BC, by inspection and comparison we can immediately conclude that

$$\frac{[j'_{A,w}/(\pi L)]}{[D_A\rho(\omega_{A,w} - \omega_{A,\infty})/L]} = \text{const}\{\text{shape}\} \cdot \overline{\text{Nu}}_m\{\text{Ra}_m, \text{Sc, shape, orientation}\}. \tag{7.2-26}$$

Furthermore, it follows that $\overline{\text{Nu}}_m$ must be the *same function* of Ra_m, Sc, ..., as $\overline{\text{Nu}}_h$ is of Ra_h, Pr. This means that this function can be established by either geometrically similar model *mass-transfer* experiments, *or* model *heat-transfer* experiments, whichever are more convenient. Whereas "dimensional analysis" might lead to the correlation (7.2-26), it would *not* predict that $\overline{\text{Nu}}_h$ and $\overline{\text{Nu}}_m$ are *the same function of their respective arguments.*

Figure 7.2-1 demonstrates the success of the $\overline{\text{Nu}}_h\{\text{Ra}_h\}$ correlation for horizontal circular cylinders ($L \equiv d_w$) of various dimensions and temperatures maintained in different Newtonian fluids (air, H_2, CO_2, CCl_4, glycerine, toluene, aniline, and water). Note that the correlation succeeds in the rather broad range:

a. $3 \times 10^{-3} \leqslant d_w \leqslant 1.6 \times 10^1$ cm,
b. $1.1 \times 10^{-1} \leqslant p \leqslant 10^3$ atm,
c. $283 \leqslant T_\infty \leqslant 376$ K,
d. $292 \leqslant T_w \leqslant 1300$ K.

It is especially remarkable since these Newtonian fluids have significantly different Prandtl numbers (e.g., at 298 K, air, glycerine, and water have the Pr-values: 0.71, 170, and 8.07, respectively). Thus, in this Ra_h-presentation the "residual" Pr-dependence is evidently small (had Gr_h been chosen as the correlating variable, this would not have been the case). This can be understood theoretically since from $\text{Gr}^{1/4} \gg 1$ laminar boundary layer (LBL) theory one can prove that $\text{Nu}_h \sim (\text{Gr}_h\text{Pr})^{1/4}$ if Pr is "large"—hence, we expect that the use of $\text{Ra}_h \equiv \text{Gr}_h \cdot \text{Pr}$ will embrace the dominant Pr-dependence, unless $\text{Pr} \ll 1$ (liquid metals) or LBL theory fails.

7.2.2.2 Similitude Analysis of Laminar Flame Speed

Perhaps the simplest problem involving transport by convection and diffusion, along with *simultaneous homogeneous chemical reaction*, is that of predicting the steady propagation speed of the "wave" of chemical reaction observed subsequent to local ignition in an *initially premixed* quiescent, nonturbulent gas. In such a case, heat and, perhaps, reaction intermediates diffusing from the initial zone of intense chemical reaction prepare the adjacent layer of gas for rapid chemical reaction. This reaction is accompanied by heat release and diffusion, which prepares the next gas layer for rapid chemical reaction, etc. An understanding of this fundamental problem has, among other things, opened the way to the creative use of relatively abundant flame-speed information (experimental (Table 6.5-1) and theoretical) in

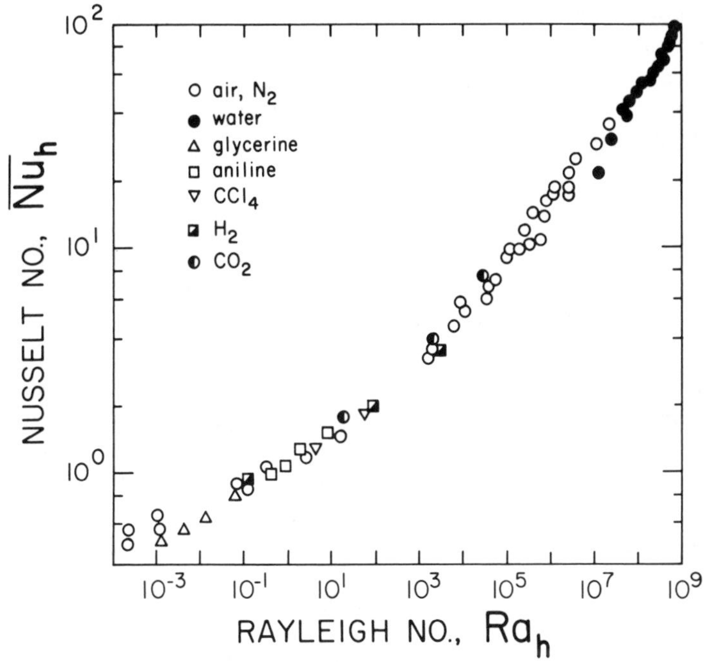

Figure 7.2-1 Correlation of perimeter-averaged "natural convection" heat transfer from/to a horizontal circular cylinder in a Newtonian fluid (adapted from McAdams (1954)).

predicting the performance of complicated combustion devices. Not surprisingly, the abundance of premixed gas-flame ("deflagration") speed data is accounted for by the fact that the steady propagation *speed*, S_u, relative to the unburned gas is quite simple to measure (but not trivial to properly interpret; Dixon-Lewis and Islam (1982)). Of interest here[9] is the dependence of this speed on the thermophysical and chemical properties and state of the unburned gas.

Even though this situation is geometrically simple and we are willing to approximate the laws of *transport* (Chapter 3), the fact that almost all combustion reactions are now known to occur *via* a complex network of many elementary reactions involving molecular fragments as intermediates (see, e.g., Figure 1.2-5) would appear to preclude simple, general S_u-results. However, in the spirit of the "physical" examples already discussed in this chapter, this physico-*chemical* problem will be shown to yield to an instructive *similitude analysis* of the conservation and constitutive equations describing a simplified "mathematical model" of the actual processes.

---

[9] Other important premixed flame characteristics (such as the existence of fuel-lean and fuel-rich composition "*limits of flammability*," and the *dynamic stability* of the flame with respect to disturbances (momentum, energy, and/or mass) in the unburned gas) are now also understood, but beyond the scope of this introductory treatment.

Thus, consider what can be learned by (a) adopting assumptions (1) through (4) of Section 6.5.5.6 and (b) treating deflagration waves that propagate slowly enough to neglect the relative change of pressure, $(p_u - p_b)/p_u$, across them. Moreover, the stoichiometric fuel + oxidizer vapor reaction of assumption (1) is assumed to explicitly occur at the local rate:

$$-\dot{r}_F''' = \frac{1}{M_O{}^{\nu_O} M_F{}^{\nu_F - 1}} \cdot \left(\frac{pM}{RT}\right)^n \cdot A \exp\left\{-\frac{E}{RT}\right\} \cdot \omega_O{}^{\nu_O} \omega_F{}^{\nu_F}, \qquad (7.2\text{-}27)$$

where $n \equiv \nu_O + \nu_F$ is the overall reaction "order." This generalization of the "bimolecular" ($n = 2$) form (6.5-63) is necessary to describe the overall effect of many "elementary steps" of different reaction orders.

Pursuing the similitude-analysis steps (Appendix 7.1), we next write the governing equations/conditions; i.e., those relations which, if completely solved, would provide the quantity of interest. This has already been done in Section 6.5.5.8, where we showed that the temperature distribution $T\{z\}$ within the wave satisfies the nonlinear ODE (6.5-76) subject to the two-point BC: $T\{-\infty\} = T_u$, $T\{\infty\} = T_b$, and the integrals:

$$c_p T + \omega_F Q = c_p T_u + \omega_{F,w} Q = c_p T_b, \qquad (6.5\text{-}77a)$$

$$c_p T + (\omega_O - \omega_{O,b}) f Q = c_p T_b. \qquad (6.5\text{-}77b)$$

Presumably, there is only one speed, namely the flame speed S_u sought, for which this boundary value problem can be solved.

We now introduce appropriate "yardsticks" for the dependent and independent variables. Thus, all temperature rises are compared to the adiabatic temperature rise—i.e., introduce the normalized temperature variable:

$$\Theta \equiv \frac{T - T_u}{\omega_{F,u} Q / c_p}. \qquad (7.2\text{-}28a)$$

All lengths (or length-increments) are compared to the characteristic length α/S_u, where α is the mixture thermal diffusivity; i.e., we introduce the dimensionless distance variable:

$$\zeta \equiv \frac{S_u z}{\alpha}. \qquad (7.2\text{-}28b)$$

Finally, all local reaction rates are compared to the *maximum* reaction rate, $-\dot{r}'''_{F,\max}$, which occurs at a calculable[10] intermediate temperature $T_{\dot{r}'''\text{-max}}$ satisfying $T_u < T_{\dot{r}'''\text{-max}} < T_b$ (usually close to the "hot boundary" temperature, T_b, since

[10] Obtained by expressing $-\dot{r}_F'''$ in terms of T alone (with the help of Eq. (6.5-77)), differentiating once with respect to temperature and finding the value $T_{\dot{r}'''\text{-max}}$ at which $-\dot{r}_F'''$ maximizes.

$E/(RT_b) \gg 1$; see below). Thus, we introduce the normalized reaction-rate function:

$$\mathscr{R} \equiv \frac{-\dot{r}_F'''\{\omega_O\{T\}, \omega_F\{T\}, T\}}{-\dot{r}_{F,\max}'''\{T_{\dot{r}'''\text{-max}}\}}. \tag{7.2-29}$$

In these terms the problem now reduces to finding the "eigen-value" $\mathfrak{w}$ corresponding to the solution of the BVP:

$$\frac{d\Theta}{d\zeta} = \frac{d^2\Theta}{d\zeta^2} + \frac{1}{\mathfrak{w}^2} \cdot \mathscr{R}\{\Theta\} \tag{7.2-30}$$

where

$$\Theta = \begin{cases} 0 & \text{at } \zeta = -\infty, \\ 1 & \text{at } \zeta = +\infty, \end{cases} \tag{7.2-31a,b}$$

$$\mathfrak{w}^2 \equiv \frac{\rho_u S_u^2 \omega_{F,u}}{\alpha(-\dot{r}_{F,\max}''')}. \tag{7.2-32}$$

It is not necessary to actually solve this highly nonlinear problem to extract valuable information about the laminar flame speed, S_u. Inspection of Eqs. (7.2-27, -29, and 6.5-77) reveals that in addition to the temperature variable Θ, the normalized rate function $\mathscr{R}$ must depend upon the dimensionless *parameters*:

$$\Phi \equiv \frac{(\omega_{F,u}/\omega_{O,u})}{f}, \qquad \begin{pmatrix}\text{mixture}\\ \text{ratio}\end{pmatrix}, \tag{7.2-33}$$

$$\text{Arr} \equiv \frac{E}{RT_b}, \qquad (\text{Arrhenius}) \tag{7.2-34}$$

$$\mathscr{Q} \equiv \frac{\omega_{F,u}Q}{c_p T_b} = \left(1 + \left(\frac{\omega_{F,u}Q}{c_p T_u}\right)\right)^{-1} \qquad \begin{pmatrix}\text{chemical}\\ \text{energy}\\ \text{release}\end{pmatrix} \tag{7.2-35}$$

and, of course, the individual reaction orders, ν_O, ν_F, appearing in the kinetic law (7.2-27).

It therefore follows that, at most:

$$\mathfrak{w} = \text{fct}\{\text{Arr}, \mathscr{Q}, \Phi, \nu_O, \nu_F\}, \tag{7.2-36}$$

or, in view of the definition of $\mathfrak{w}$ (Eq. (7.2-32)), the flame speed must be given by:

$$S_u = \left\{\frac{\alpha(-\dot{r}_{F,\max}''')}{\rho_u \omega_{F,u}} \cdot \text{fct}\{\text{Arr}, \mathscr{Q}, \Phi, \nu_O, \nu_F\}\right\}^{1/2}, \tag{7.2-37}$$

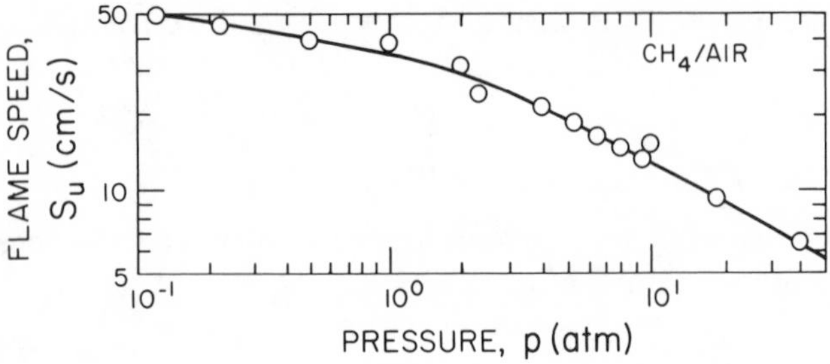

Figure 7.2-2 Pressure dependence of methane/air laminar flame speed (adapted from Diedrichsen and Wolfhard (1956)).

where the undetermined function fct{ } can only be evaluated by further numerical or analytical methods; they will not be pursued here.

Without going any further, the similitude result Eq. (7.2-37) already contains the *pressure* dependence of S_u since, subject to our underlying assumptions, none of the arguments of fct{ } is itself pressure-dependent. Therefore, since $\alpha \sim p^{-1}$, $-\dot{r}'''_{F,max} \sim p^n$ and $\rho_u \sim p^{+1}$, it can be concluded that:

$$S_u \sim p^{(n/2)-1}. \tag{7.2-38}$$

Since the overall reaction order which represents a complex chain reaction is not easy to estimate (i.e., S_u would be pressure independent only if $n = 2$), an effective overall reaction order can be determined from experimental $S_u\{p\}$-data in accord with:

$$n_{eff} = 2\left(1 + \frac{d \ln S_u}{d \ln p}\right). \tag{7.2-39}$$

Experimentally, it has been found that most flames with speeds less than about 0.5 m/s (see Table 6.5-1) exhibit an effective overall reaction order *less* than 2; i.e., the flame speed *decreases* somewhat at higher pressures (see, e.g., Lewis and von Elbe (1951) and Figure 7.2-2).

As a corollary, since $(d\Theta/d\zeta)^{-1}_{max}$ is a measure of the dimensionless flame *thickness*, the pressure dependence of the *dimensional premixed gas-flame thickness* can be extracted from the equations above, with the result $\delta_f \sim p^{-n/2}$ (cf. Eq. (6.5-68), derived for a "diffusion" flame). For this reason, low pressure, premixed flames are easier to probe (to determine structure, extract kinetics/mechanism, etc.) than their higher pressure counterparts.

Further study of this problem (see, e.g., the summaries of Frank-Kamenetskii (1969), Williams (1985), Zeldovich *et al.* (1984), and Buckmaster and Ludford (1982)) has revealed that, since, typically, Arr $\gg$ 1,

1. the dominant *temperature* dependence of S_u is contained in the factor

$\{-\dot{r}'''_{F,\max}(T_{\dot{r}'''\text{-max}})\}^{1/2}$ and $T_{\dot{r}'''\text{-max}}$ is very near to the burned gas temperature T_b.

2. the function $\mathfrak{w}$ depends only algebraically on the *product* Arr · $\mathcal{Q}$.

Thus, in the important asymptotic limit Arr ≫ 1, to a first approximation the laminar flame speed has the temperature dependence (see, e.g., Figure 7.2-3):

$$S_u \sim \exp\left(-\frac{E}{2RT_b}\right), \tag{7.2-40}$$

so that an *effective overall activation energy* can be determined from flame speed data *via*:

$$E_{\text{eff}} \approx -2R \cdot \frac{d \ln S_u}{d(1/T_b)}. \tag{7.2-41}$$

It will be seen below that the new dimensionless parameters (7.2-33, -35) that entered this very simple, but representative, transport problem with *simultaneous homogeneous chemical reaction* also appear in more complex chemical-reactor examples. Moreover, since the parameter $\mathfrak{w}^2$ (Eq. (7.2-32)) can obviously be regarded as the ratio of a characteristic *flow time* δ_f/S_u to a characteristic *chemical time*

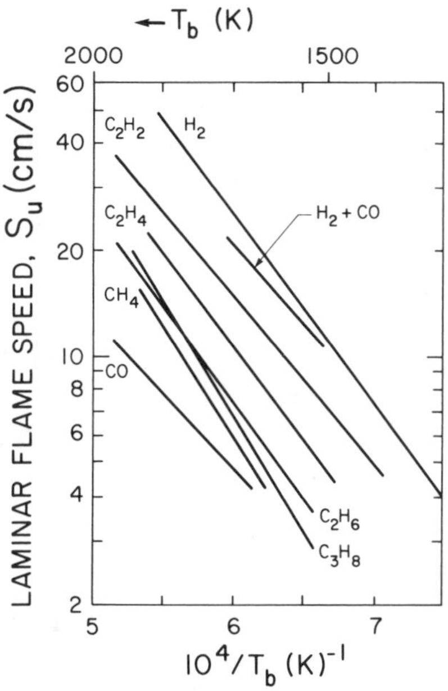

Figure 7.2-3 Dependence of laminar flame speed on burned gas temperature for several ($\Phi = 0.8$) fuel/air mixtures (adapted from Kaskan (1951)).

$\rho_u \omega_{F,u}/(-\dot{r}'''_{F,\max})$, in view of (2) above, α/S_u^2 is often taken as a measure of the *characteristic chemical time* in more complex flow systems involving the same oxidizer/fuel combination.

7.2.3 "Partial Modeling" of Chemically Reacting Systems (see, e.g., Spalding (1963))

Chemically reacting systems introduce many additional parameters; this becomes clear when the complete set of basic equations of Chapter 2 is made nondimensional using any particular set of reference quantities. For example, we can select:

$$\begin{aligned} L_{\text{ref}} &\equiv L, \\ U_{\text{ref}} &\equiv U_\infty, \qquad \text{(forced convection)} \\ t_{\text{ref}} &\equiv \frac{L}{U_\infty}, \\ p_{\text{ref}} &\equiv \frac{1}{2}\rho_\infty U_\infty{}^2, \end{aligned} \tag{7.2-42,a-d}$$

and, for a combustor,

$$(T - T_\infty)_{\text{ref}} \equiv (T_{\text{adiab}} - T_\infty), \text{ etc.} \tag{7.2-42e}$$

An important question then is the following: Are the additional dimensionless thermochemical and chemical kinetic parameters of a number and type such that true similarity cannot be achieved except in the "trivial" (and expensive!) case $L_p = L_m$? Generally this *is* found to be the case; however, since not all of the similarity groups are of equal importance, often some can be dispensed with to allow *approximate similarity*, or "partial modeling."

We illustrate below several simple examples of combustion similitude relations. These are approximate similitudes, in the sense that each necessarily has a restricted domain of validity. However, within this domain scale-model tests can be (and have been) used to establish very valuable information on large-scale combustor performance, and the effects of altering fuel type and operating parameters can be anticipated.

7.2.3.1 Gas-Turbine Combustor Efficiency (after Way (1956))

The number of parameters potentially influencing the combustion efficiency of an aircraft gas-turbine combustor "can" (Figure 1.2-8) is very great—especially since the geometry is rather complex, and the *liquid* fuel is introduced into the enclosure as a spray. Each such spray can be characterized by a spray angle, spray momentum flux, droplet size distribution, etc. Rather than include these *two-phase* effects, let us

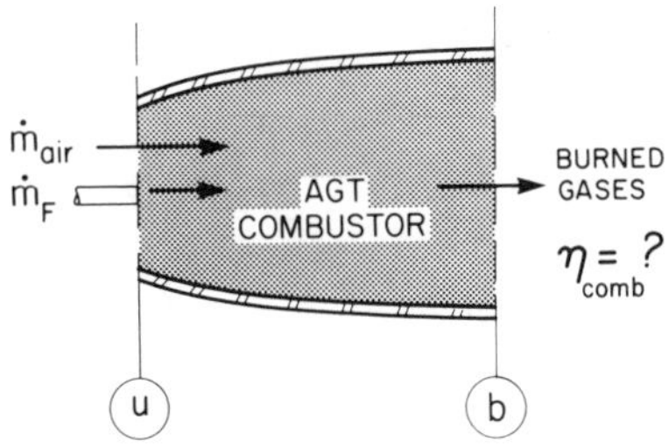

Figure 7.2-4 Aircraft gas turbine GT combustor (schematic).

consider the simpler limiting case in which the fuel droplets are sufficiently small that their "penetration" is small, and their vaporization is rapid enough to not limit the overall chemical heat release rate.

Consider the simple combustor type schematically shown in Figure 7.2-4, in which we specify mean quantities at the upstream (u) and burned gas (b) stations. As a relevant *performance criterion* we select the *combustion efficiency*:

$$\eta_{\text{comb}} \equiv \frac{T_{0,\text{b}} - T_{0,\text{u}}}{T_{0,\text{b;adiab}} - T_{0,\text{u}}} \tag{7.2-43}$$

(where the subscript 0 implies "stagnation" quantities (cf. Section 4.3.1), and the denominator is the gas total temperature rise, ΔT_0, that would occur if *all* of the heat of combustion were delivered to the prevailing gas mixture).

In addition to the following now-familiar similarity criteria for *physical* flow systems such as:

$$\begin{aligned} &\text{shape} \\ \text{Re} &\equiv U_{\text{u}} L / \nu_{\text{u}}, \\ \text{Pr} &\equiv \nu_{\text{u}} / \alpha_{\text{u}}, \\ \text{Sc} &\equiv \nu_{\text{u}} / D_{\text{F,u}}, \\ \gamma &\equiv (c_p / c_v)_{\text{u}}, \qquad \text{and} \\ \text{Ma} &\equiv (U/a)_{\text{u}}, \end{aligned} \tag{7.2-44,a-e}$$

we now expect η_{comb} to depend additionally upon:

$$\Phi \equiv \frac{(\text{fuel/air mass flow ratio})}{(\text{fuel/air})_{\text{stoich}}}, \tag{7.2-45}$$

$$\mathcal{Q} \equiv \frac{\omega_{\text{F,u}} Q}{c_{\text{p,u}} T_{\text{b, adiab}}}, \tag{7.2-46}$$

as well as on two parameters arising from the overall fuel-vapor homogeneous chemical *kinetics*, *viz.*:

$$\mathrm{Arr} \equiv \frac{E}{RT_{\mathrm{b,ad}}} \quad \text{(dimensionless Arrhenius activation energy)}, \tag{7.2-47}$$

$$\mathrm{Dam} \equiv \frac{t_{\mathrm{flow}}}{t_{\mathrm{chem,ref}}} \quad \text{(Damköhler-ratio of characteristic flow time to chemical oxidation time)}. \tag{7.2-48}$$

This would appear to constitute a minimum list[11] necessary to achieve geometric, dynamic, thermal, and chemical similarity between the model (m) and prototype (p) AGT-type combustors.

If, indeed, the combustion efficiency η_{comb} exhibits the functional dependencies:

$$\eta_{\mathrm{comb}} = \eta\{\mathrm{Re}, \mathrm{Pr}, \mathrm{Sc}, \gamma, \mathrm{Ma}, \Phi, \mathscr{Q}, \mathrm{Arr}, \mathrm{Dam}, \mathrm{shape}\}, \tag{7.2-49}$$

we can certainly conclude $\eta_{\mathrm{m}} = \eta_{\mathrm{p}}$ if each of the indicated nondimensional parameters is the same for both model and prototype.

If the scale model is run with the same fuel and at the same inlet temperature (T_{u}) and mixture ratio, Φ, as the prototype, most of these arguments will indeed be the same for both model and prototype. It remains for us to require that:

$$(\mathrm{Re})_{\mathrm{m}} = (\mathrm{Re})_{\mathrm{p}}, \tag{7.2-50}$$

$$(\mathrm{Ma})_{\mathrm{m}} = (\mathrm{Ma})_{\mathrm{p}}, \tag{7.2-51}$$

$$(\mathrm{Dam})_{\mathrm{m}} = (\mathrm{Dam})_{\mathrm{p}}. \tag{7.2-52}$$

We now ask: Is there a combination of model pressure, velocity, and scale ($p_{\mathrm{m}}, U_{\mathrm{m}}$, and L_{m}) such that these remaining similarity conditions can be met?

To answer this question requires specification of the p, U, and L dependence of each of these parameters. For a perfect gas, Re-equivalence implies:

$$(pUL)_{\mathrm{m}} = (pUL)_{\mathrm{p}}. \tag{7.2-53}$$

Since Ma-equivalence implies:

$$U_{\mathrm{m}} = U_{\mathrm{p}} \tag{7.2-54}$$

[11] Indeed, we have not explicitly included a "micro-mixing" Damköhler number (Section 6.5.5.6), i.e., one based upon the turbulent (eddy-breakup) time κ_t/ε (or, alternatively, the characteristic time: $(D \ln g/Dt)^{-1}$, where $g \equiv \overline{\Omega_I' \Omega_I'}$) rather than the macroscopic mean fluid residence time. If departures from strict geometric- and dynamic (Re-) similarity are considered (see below), this omission will inevitably limit the domain of validity of the resulting "partial similitude."

and $L_m \neq L_p$, we will therefore have to test at the model pressure:

$$p_m = p_p\left(\frac{L_p}{L_m}\right). \tag{7.2-55}$$

In general, however, this conflicts with Dam-equivalence. This is easiest to see for a simple nth-order homogeneous fuel-consumption reaction. In such a case:

$$t_{\text{chem}} \equiv \frac{\rho_{F,\text{ref}}}{(-\dot{r}_F''')_{\text{ref}}} \sim \frac{p}{p^n} \sim p^{-(n-1)}. \tag{7.2-56}$$

Since $t_{\text{flow}} \equiv L/U_u$, Dam-equivalence therefore requires:

$$\left(\frac{Lp^{n-1}}{U}\right)_m = \left(\frac{Lp^{n-1}}{U}\right)_p. \tag{7.2-57}$$

But, in the light of the Ma-equivalence requirement, this becomes:

$$p_m = p_p\left(\frac{L_p}{L_m}\right)^{1/(n-1)}, \tag{7.2-58}$$

which is seen to *differ* from Eq. (7.2-55) whenever $n \neq 2$ (as is usually the case; cf. Table 6.7-1 and Section 6.7.6).

Note that, even in this "simple" combustor application, strict scale-model similarity appears to be unattainable. That is, in principle, no combination of p_m, U_m, and L_m allows us to confidently use scale model η_{comb}-measurements to quantitatively predict full-scale combustor performance.

However, if we consider available combustor efficiency measurements, it appears that η_{comb} is much more sensitive to the loading parameter, Dam, than to Re, especially at high (fully turbulent) Reynolds' numbers. The nature of this behavior is sketched in Figure 7.2-5. Apparently, then, if Re is sufficiently large, the dependence of η_{comb} on Re can be neglected. (This behavior is reminiscent of many "fully" turbulent nonreactive flows, for which important performance criteria and characteristics become insensitive to Re, provided Re is sufficiently large.)

If we adopt this "Re-dispensation," then an *approximate similitude* can be achieved, and the scale-model combustor tests should be run with:

$$U_m = U_p \tag{7.2-59}$$

and

$$p_m = p_p\left(\frac{L_p}{L_m}\right)^{1/(n-1)} \tag{7.2-60}$$

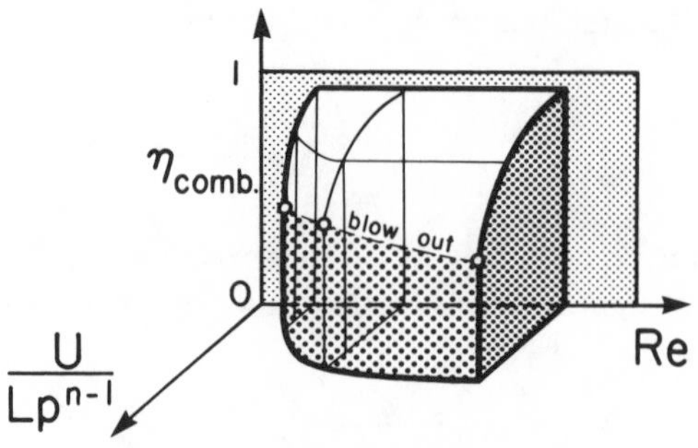

Figure 7.2-5 Dependence of GT combustor efficiency on Re at constant (inverse) Damköhler number (schematic, adapted from S. Way (1956)).

Experimentally, the apparent reaction order, n, appearing in Eq. (7.2-4) is often near 1.3–1.6, depending upon the fuel (cf. Table 6.7-1). If, for example, $n \approx 1.5$ then the smaller combustor model should be tested at a pressure higher than that of the (full-scale) prototype by the factor $(L_p/L_m)^2$.

Equivalently, efficiency and, perhaps, stability data on combustors of various sizes, air-flow rates, and pressures should approximately correlate with a parameter proportional to Dam or, if one prefers, $(\mathrm{Dam})^{-1}$, *viz.*:

$$\frac{U}{p^{n-1}L} \quad \text{or} \quad \frac{\dot{m}_{\text{air}}}{p^n L^3}. \tag{7.2-61}$$

Figures 7.2-6 and -7 show two examples of the success of this type of performance correlation, one dealing with *efficiency* (of a family of rather "inefficient"combustors!), and the second dealing with fuel-lean and -rich *stability limits*.

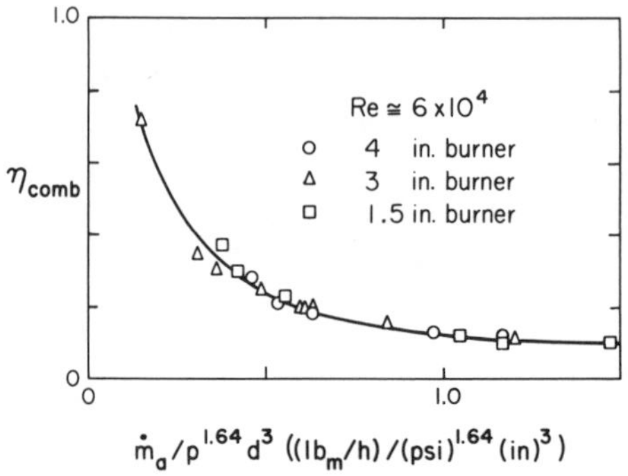

Figure 7.2-6 Correlation for the GT combustor efficiency *vs.* parameter proportional to (inverse) Damköhler number (adapted from S. Way (1956)).

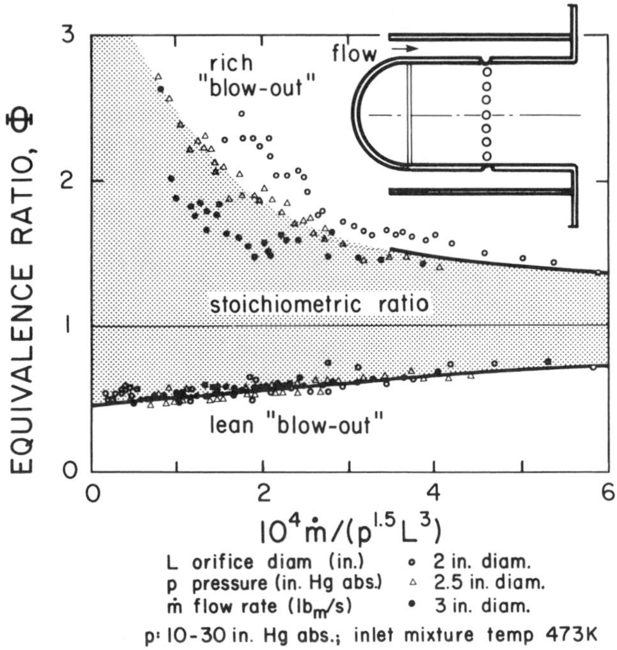

Figure 7.2-7 Correlation of GT combustor stability limits *vs.* parameter proportional to (inverse) Damköhler number (after D. Stewart (1956)).

7.2.3.2 "Similitudes" in Combustion—Some Other Examples

We complete this discussion with several further, instructive examples of approximate "similitudes" in combustion. They deal, respectively, with (a) the "flashback" of gaseous fuel/air mixtures in ducts, (b) "blow-off" (extinction) of flames anchored by flame "holders" in high-speed streams, (c) the length of laminar fuel-jet diffusion flames influenced by buoyancy, and (d) the burning time of an isolated fuel droplet at elevated pressures. Further examples are introduced in Exercises 7.5 and 7.6.

a. Flashback of a Flame in a Duct. The ability of any flame to propagate back upstream against reactant flow in a duct will depend upon the existence of a region near the duct wall where the local streamwise velocity is *below* the prevailing laminar flame speed, S_u. However, since no flame can propagate closer to the wall than some "quenching distance" δ_q, approximately given by:[12]

$$\delta_q \approx \text{const} \cdot \alpha_u / S_u, \tag{7.2-62}$$

[12] The *thickness* of a laminar flame is also found to be proportional to α_u/S_u, both experimentally and theoretically (see Section 7.2.2.2).

the *critical condition for flashback* should, therefore, be of the "gradient" form: $U/d \sim S_u/\delta_q$ or:

$$\left(\frac{U}{d}\right)_{fb} = \text{const} \cdot \left(\frac{S_u{}^2}{\alpha_u}\right). \tag{7.2-63}$$

Multiplying both sides of this equation by d^2/α_u leads to the expectation that flashback data for such ducted premixed systems should yield a correlation law of the Peclet-number form:

$$\left(\frac{Ud}{\alpha_u}\right)_{fb} = \text{const} \cdot \left(\frac{S_u d}{\alpha_u}\right)^2. \tag{7.2-64}$$

Figure 7.2-8 (after Putnam and Jensen (1949)) reveals that flashback data for different fuels, tube diameters, and pressures indeed "correlate" in this way. Again, such information provides the basis for accurate (flashback) "predictions" in new systems of a physicochemically "similar" nature.

b. "Blow-Off" from Premixed Gas Flame "Holders." It is experimentally observed that "oblique" premixed gas flames can be "anchored" or stabilized in ducts even at feed-flow velocities much greater than S_u (see Sections 6.5.5.8, 7.2.2.2). In effect, the "anchor" is a relatively well-mixed zone of recirculating reaction products, as found immediately downstream of "bluff" objects (rods, disks, gutters (Figures 1.2-4, 7.2-9), etc.) or (backward-facing) steps in the duct itself. However, in any particular system there is a sharp upper limit to the feed-flow velocity above which "blow-out" (BO) (or "extinction") occurs. This blow-out velocity, U_{bo}, of course, depends upon many physico-chemical parameters, and, in the spirit of our earlier examples, we ask if some useful "similitude" or "correlation" can be established, albeit only approximately.

The abundance of S_u-data, and the similitude results of Section 7.2.2.2 (for the "canonical" laminar flame-speed problem) make it attractive to invoke S_u as a measure of the reaction kinetics. As is common, several alternative approaches can be followed, each leading to self-consistent tentative conclusions, which can then be experimentally tested.

Pursuing first the approach of the gas-turbine combustor efficiency example, we postulate a plausible dimensionless relation of the form:

$$\text{Re}_{bo} \approx \text{fct}_1\{\text{Dam}, \text{Arr}, \mathcal{Q}, \Phi, \text{Pr}, \text{Sc}, \text{geometry}\} \tag{7.2-65}$$

where

$$\text{Re}_{bo} \equiv \frac{U_{bo} L}{\nu_u} \tag{7.2-66}$$

is the blow-off Reynolds', number based on stabilizer transverse dimension. But, in

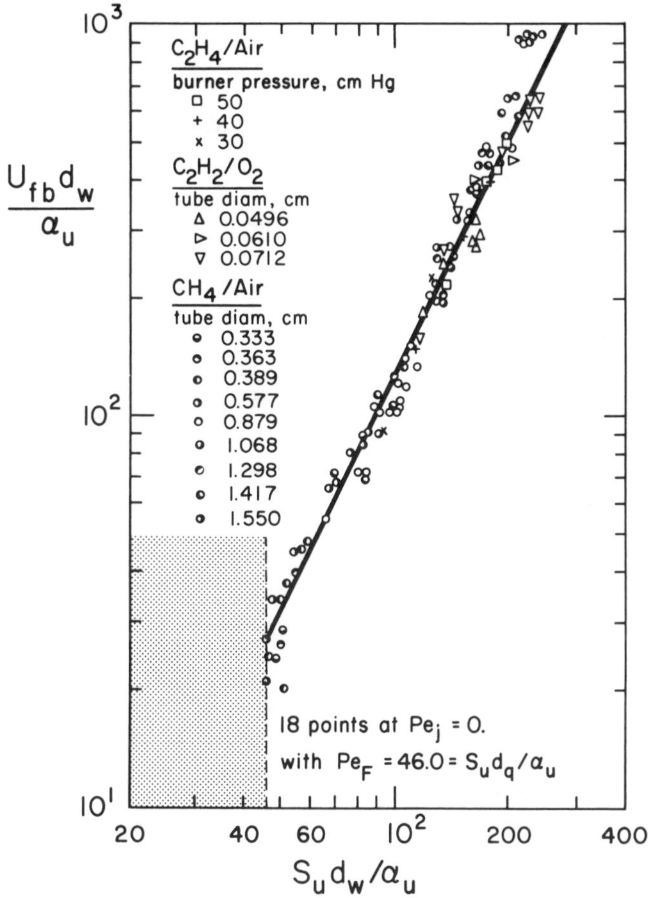

Figure 7.2-8 Correlation of flashback limits for premixed combustible gases in tubes (after Putnam and Jensen (1949)).

the light of the results of our similitude analysis of S_u data, we can write:

$$\mathrm{Dam} = \frac{(L/U_{bo})}{(\alpha_u/S_u^{\ 2})} \cdot \mathrm{fct}_2\{\mathrm{Arr}, \mathcal{Q}, \Phi, \nu_O, \nu_F\}. \tag{7.2-67}$$

Combining these results, and "solving" for the Peclet number at blow-off: $U_{bo}L/\alpha_u$, we conclude:

$$\frac{U_{bo}L}{\alpha_u} \approx \mathrm{fct}_3\left\{\frac{S_u L}{\alpha_u}, \mathrm{Arr}, \mathcal{Q}, \Phi, \nu_O, \nu_F, \mathrm{Pr}, \mathrm{Sc}, \text{geometry}\right\}. \tag{7.2-68}$$

Experimental flame stabilization data (examined by Spalding and Tall (1954), with

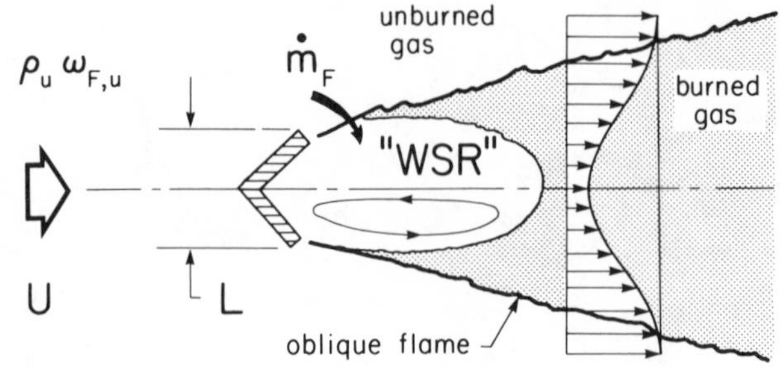

Figure 7.2-9 Schematic of an oblique flame "anchored" to a gutter-type (2 dimensional) flame-holder (stabilizer) in a uniform stream of premixed combustible gas ($S_u < U < U_{bo}$) (see, also, Figure 1.2-4).

results reported by Spalding (1955)) for methane, propane, or naphtha vapors in air, reveal that the observed blow-off velocities indeed correlate in this way. Fortunately, however, the *explicit*[13] dependence on the parameters Arr, $\mathcal{Q}$, Φ, ν_O, ν_F, Pr, and Sc is evidently weak; i.e., experimentally:

$$\frac{U_{bo}L}{\alpha_u} \approx \text{fct}_4\left(\frac{S_u L}{\alpha_u}, \text{geometry}\right). \tag{7.2-69}$$

Figure 7.2-10 shows this useful correlation for several two-dimensional and three-dimensional flame-holder geometries.

It is instructive to briefly examine a second approach, which postulates a specific physico-chemical *model*, and, accordingly, extracts more information about the likely dependence of $U_{bo}L/\alpha_u$ on $S_u L/\alpha_u$. Suppose, following Longwell *et al.* (1953), that the recirculation zone (Figure 7.2-9) can be likened to a WSR, with "blow-off" corresponding to the *extinction* process (due to excessive fuel-feed rate) discussed in Section 6.7.6. Now, a three-dimensional stabilizer of characteristic transverse dimension L will exhibit a recirculation zone with an effective *volume* given by:

$$(V_{WSR})_{eff} \approx L^3 \cdot \text{fct}_5\{\text{Re}, \text{Pr}, \text{Sc}, \text{geometry}\}. \tag{7.2-70}$$

The fuel-flow rate into this recirculation–reaction zone can likewise be written:

$$\dot{m}_F \approx \rho_u U \omega_{F,u} \cdot [L^2 \cdot \text{fct}_6\{\text{Re}, \text{Pr}, \text{Sc}, \text{geometry}\}], \tag{7.2-71}$$

and "blow-out" should occur when the corresponding volumetric fuel consumption

[13] $U_{bo}L/\alpha_u$ *does* depend upon the parameters Arr, $\mathcal{Q}$, Φ, ν_O, and ν_F, but mainly *via* the first argument, $S_u L/\alpha_u$ of $\text{fct}_3\{\ldots\}$. Not only is the dependence of $\text{fct}_3\{\ldots\}$ on Pr and Sc probably weak, but, also, these transport property parameters do not vary much from system to system (see Chapter 3).

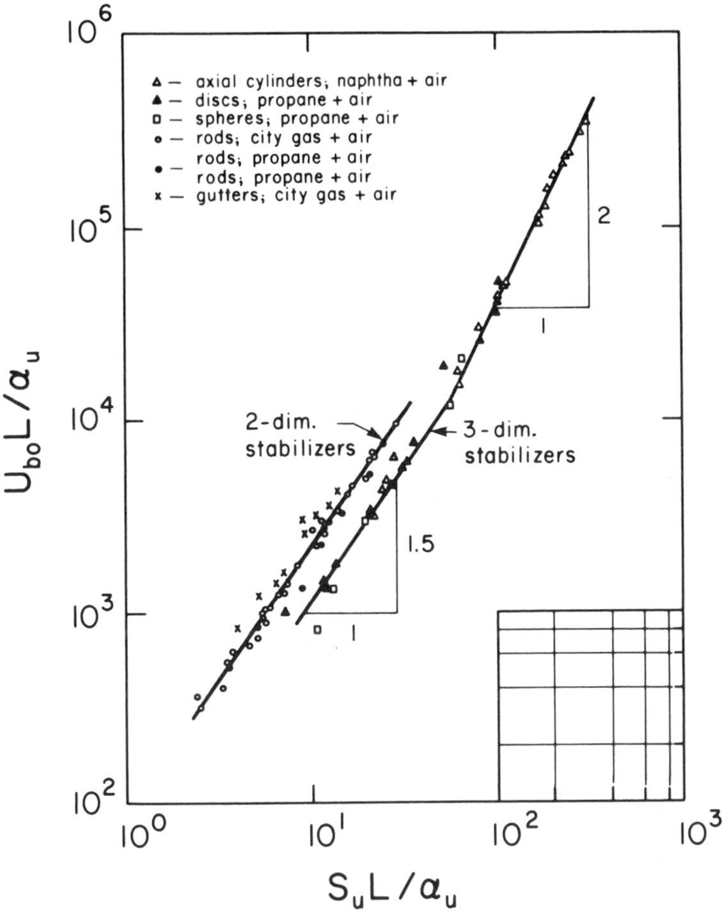

Figure 7.2-10 Test of proposed correlation of the dimensionless "blow-off" velocity, $U_{bo}L/\alpha_u$ *vs.* S_uL/α_u for a flame stabilized by a bluff body of transverse dimension L in a uniform, premixed gas stream (adapted from Spalding (1955)).

rate is near $\dot{r}'''_{F,max}$ (see Section 6.7.6). That is,

$$(\dot{m}_F)_{bo} \approx (-\dot{r}'''_{F,max}) \cdot (V_{WSR})_{eff}. \qquad (7.2\text{-}72)$$

Invoking the relation between the laminar flame speed, S_u, and $-\dot{r}'''_{F,max}$ (see Eq. (7.2-37)) and rearranging, Eq. (7.2-72) immediately leads to the expectation:

$$\frac{U_{bo}L}{\alpha_u} = \left(\frac{S_uL}{\alpha_u}\right)^2 \cdot \mathrm{fct}_7\{\mathrm{Arr}, \mathscr{Q}, \Phi, \nu_O, \nu_F, \mathrm{Re}, \mathrm{Pr}, \mathrm{Sc}, \mathrm{geometry}\}, \qquad (7.2\text{-}73)$$

where $\mathrm{fct}_7\{\ldots\}$ is, hopefully, a "weak" function of its indicated arguments, especially at high Reynolds' numbers. Note (Figure 7.2-10) that the quadratic dependence on $(S_u L/\alpha_u)$ is, indeed, observed at Re_{bo}-values above about 10^4. Thus, by invoking a simple, yet plausible, physico-chemical model of the process, we have been rewarded with information beyond that contained in Eq. (7.2-68); e.g., the blow-off velocity should scale linearly with the transverse *dimension* of the flame stabilizer, at sufficiently high Reynolds' numbers.

The same line of reasoning leads to an identical result, Eq. (7.2-74), for the two-dimensional (e.g., "gutter"-type) stabilizer, including such stabilizers when "bent" into a large circle for use in, say, turbo-jet "afterburner" ducts.

c. Laminar Diffusion Flame Height. Both "buoyancy" and fuel-jet momentum contribute to the observed height, L_f, of a fuel-jet diffusion flame. In deciding how to present ("correlate") such data, it is useful to introduce a simple model that suggests *relevant groupings* of the variables—again, something beyond the realm of ordinary "dimensional analysis." For example, by treating the hot "flame sheet" region as the cause of a natural convective inflow of ambient oxidizer, one can show that, for any fuel/oxidizer pair, when buoyancy dominates:

$$\left(\frac{L_f}{R_j}\right) \cdot (\mathrm{Re}^2 \mathrm{Fr})^{-1/3} \rightarrow \text{const}, \tag{7.2-74}$$

whereas, if fuel-jet momentum dominates then, at constant $\mathrm{Re} \equiv (UR_j/\nu)_j$:

$$\left(\frac{L_f}{R_j}\right) \cdot (\mathrm{Re}^2 \mathrm{Fr})^{-1/3} \rightarrow \text{const}' \cdot (\mathrm{Fr})^{-1/3}. \tag{7.2-75}$$

Here $R_j = \frac{1}{2} d_j$ and Fr is the *Froude number*:

$$\mathrm{Fr} \equiv U_j^2/(gR_j). \tag{7.2-76}$$

Accordingly, gaseous fuel-jet flame height measurements, made using a centrifuge capable of up to $31 \cdot g(\text{Earth})$, have been plotted in this way (cf. Figure 7.2-11, after Altenkirch, *et al.* (1976)). In addition to indicating attainment of the "buoyancy-controlled" condition for C_3H_8 and $C_2H_6(g)$ flames at $\mathrm{Fr} < 1$, results presented in this *dimensionless* form should apply to other fuel/jet situations which are "similar" in the *geometric, dynamical, thermal, and compositional sense.*

d. Fuel Droplet Combustion at High Pressures (cf. Section 6.5.5.7). The pressure dependence of the burning time, t_{comb}, of an individual fuel droplet is of great fundamental and practical interest. While many laboratory studies have been carried out at or near $p = 1$ atm, in actual rockets, diesel engines, and modern aircraft gas turbines, pressure levels in excess of 20 atm are attained. If new but fragmentary high-pressure data were plotted against the pressure level, the results would differ widely according to fuel type and initial droplet diameter, even at a fixed ambient temperature, T_∞, and oxidizer concentration $(\omega_{O_2\infty})$. However:

i. Diffusion-limited burning-rate theory and available $p \gtrsim 1$ atm experiments

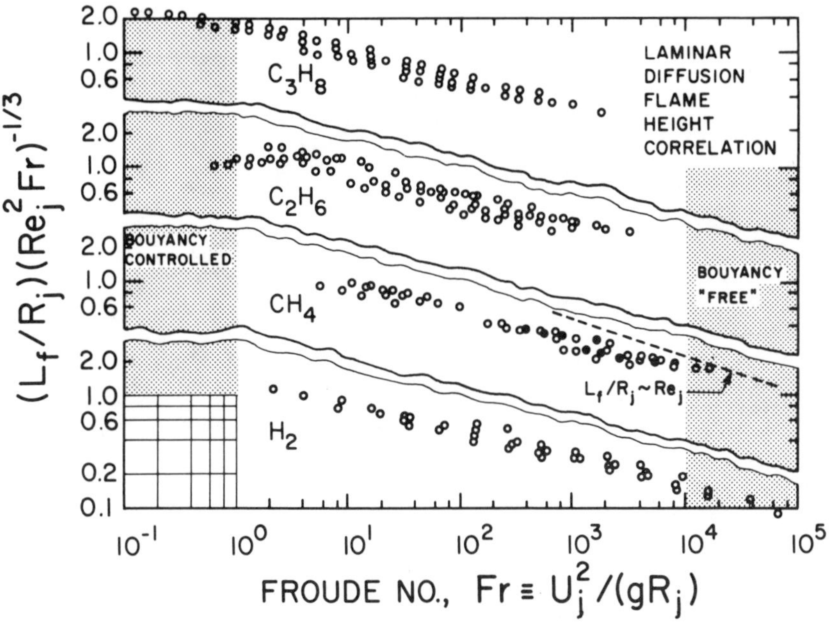

Figure 7.2-11 Correlation of laminar jet diffusion flame lengths (adapted from Altenkirch *et al.* (1977).

reveal that

$$t_{\text{comb}} = d_{p,0}^2/K \tag{7.2-77}$$

where K, a "burning rate constant," is dependent on the fuel type and environmental conditions but independent of droplet diameter.[14]

ii. Each fuel possesses its own thermodynamic critical pressure, p_c, to which the prevailing pressure p should be compared (cf. our discussion of the "law of corresponding states," in Section 7.1.3.1).

These observations suggest that a useful *correlation* may result from replotting available droplet combustion data in the form:

$$\frac{K^{-1}\{p, \text{fuel type}, \ldots\}}{K^{-1}\{p_c, \text{same fuel}, \ldots\}} \quad vs. \quad \frac{p}{p_c\{\text{same fuel}\}}. \tag{7.2-78}$$

Figure 7.2-12 (data of Kadota and Hiroyasu (1981)) suggests that such a "*corresponding states*" *correlation for high-pressure droplet combustion* is reasonably

[14] This remarkable proportionality holds even when the quasi-steady approximation fails (Rosner and Chang (1973)). Thus, even though d^2 does *not instantaneously* decrease linearly with time, the *overall* combustion time remains proportional to d_0^2.

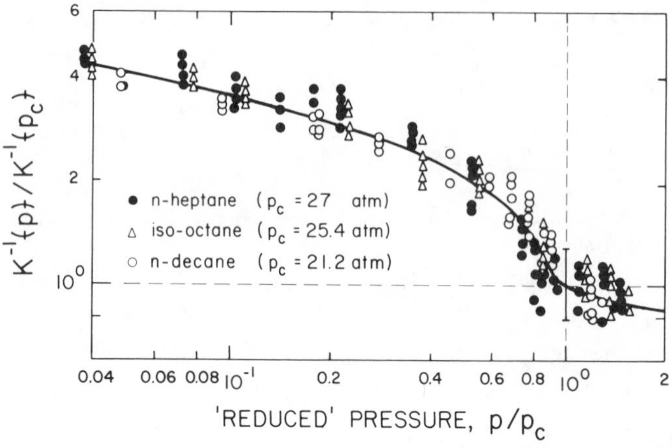

Figure 7.2-12 Approximate correlation of fuel-droplet burning rate constants at elevated pressures (based on data of Kadota and Hiroyasu (1981)).

successful. This correlation will allow rational estimates for the high-pressure burning time of fuels for which data are not available.

7.2.4 Alternative Interpretations of Dimensionless Groups ("Eigen-Ratios")

The generalized "Pythagorean method" of establishing the conditions of similarity by forming a sufficient set of eigen-ratios (a process called "configurational analysis" by Becker (1976)) clearly leads to similitude criteria in the form of dimensionless ratios of inventories, source strengths, and fluxes as well as lengths and times. This is true whether one starts from the governing conservation equations in their macroscopic or "microscopic" CV forms (Chapter 2). But the *interpretation* of each of the resulting groups is really not unique, and it may be useful to regard them all as ratios of the same type, e.g., *characteristic times*. As examples, a length ratio in a forced convection system can be regarded as a fluid *transit-time* ratio; the momentum flux ratio, which we call the Reynolds' number, Re (Eq. (7.1-10)), could equally well be regarded as the ratio of times required for momentum diffusion and convection through a common area L^2. Even the Prandtl number, $\mathrm{Pr} \equiv \nu/\alpha$, can be regarded as the ratio of two times, *viz.*, L^2/α and L^2/ν, governing, respectively, the diffusive decay of nonuniformities of energy and momentum; and so on. Note that such characteristic *times* and the ratios found therefrom are relevant even to *steady-state* problems.

We have already observed that fluid-phase ("homogeneous" reaction) Damköhler numbers can be viewed as the ratio of the characteristic flow time to the characteristic chemical times of the system (see, e.g., Section 7.2.3.1). But *surface* ("heterogeneous") reaction Damköhler numbers can also be viewed in this way

(Rosner (1959))—an example being the "catalytic parameter" $C \equiv k_w\delta_m/D$ appearing in Section 6.5.3. Rewriting this dimensionless parameter in the form:

$$C \equiv \frac{k_w\delta_m}{D} = \frac{(\delta_m^2/D)}{\left(\dfrac{\rho\omega_e\delta_m}{\rho\omega_e k_w}\right)} \equiv \frac{t_{\text{diff}}}{t_{\text{chem,w}}} \tag{7.2-79}$$

makes it clear that C can equally well be regarded as the ratio of the characteristic time, t_{diff}, for the reactants to diffuse across the boundary layer of thickness δ_m, to the characteristic time, $t_{\text{chem,w}}$, that would be required by the surface to chemically consume those reactant molecules which "initially" filled the boundary layer. Finally, we recall that the dimensionless *Stokes' number* governing *dynamical* non-equilibrium in two-phase flows (Sections 6.2.6, 6.6.1) could also be regarded as the ratio of two characteristic times; *viz.*, a particle "stopping" time, t_p, and a fluid transit time, $t_{\text{flow}} \equiv L/U$.

The establishment of approximate similarity criteria by identifying and forming ratios using an independent set of relevant *characteristic times* is an attractive way of dealing with many complex physico-chemical problems, including those for which detailed "computer modeling" (see Section 7.3, below) is still unrewarding (due to excessive uncertainties), or totally beyond reach (see, e.g., the treatment of gas turbine-spray combustor performance (Mellor, A., *et al.* (1979)) and coal-particle devolatilization (Kalson, P.A. (1981))). However, we have already seen that even an "inspired" application of this "configurational" approach may mask the fact that, in some cases, further simplifications are possible through rational parameter groupings (see, e.g., Sections 7.2.2.1, 7.2.2.2, and Rosner, Gökoĝlu, and Israel (1983)), or by dropping parameters to which the quantity of interest is insensitive (Spalding (1963)). In summary, it seems clear that the most powerful similitudes will always result from making maximum use of the available (experimental and analytical) information; in this sense we will "get what we pay for."

7.3 CONCLUDING REMARKS: MATHEMATICAL MODELING IN CHEMICALLY REACTING SYSTEMS—AN INTERACTIVE, BALANCED APPROACH

The history of all technologies, including chemical reactor technology, amply demonstrates that maximum progress requires parallel interactive efforts in the areas of:

a. scale-model testing,
b. full-scale testing,
c. mathematical modeling, incorporating the results of relevant *basic research*, and "validated" based on the above-mentioned "model" experiments.

From an engineering point of view, isolated, intensive efforts in any one of these areas will inevitably be sterile compared to what can be achieved by a judicious blend of all three.

In this chapter we have seen that the same conservation- and constitutive-equations that provide the basis for mathematical modeling (with the help of electronic computers and the numerical methods outlined in Appendix 8.2) also provide the basis for establishing simple, quantitative "similitudes" between model and full-scale systems.

SUMMARY

Similitude Analysis: Overview of the Role of Experiment and Theory in Reactive Fluid Mechanics

- The governing balance + constitutive laws reveal when:

  two systems of different scale ($L_p/L_m \neq 1$) are *quantitatively* similar (geometric, dynamic, thermal, compositional);

  two apparently different *processes* at the same scale are quantitatively similar (*analogy*);

  by a process called *similitude*-(or "inspectional"-)analysis (Appendix 7.1 and Section 7.2.2).

- Dimensionless groups ("eigen-ratios" such as shape factors, Re, Ra, Pr, ...) comprise the similarity parameters which yield "set-up rules" for designing *scale-model* experiments amenable to *quantitative* use. Appearing as coefficients in the dimensionless conservation equations (macro- or micro-control volume) and side conditions (boundary, initial), they betray the relative importance of competing transport phenomena. Inspection of these dimensionless equations, after the introduction of defensible approximations, usually leads to a smaller set of necessary parameters than would be obtained by formal application of "dimensional analysis" (Vaschy-Buckingham) to the same physico-chemical problem.
- While for systems with many participating phenomena (such as occur in chemical reactors) strict similarity for $L_p/L_m \neq 1$ is usually precluded, such systems are often *approximately* similar, because of the insensitivity of the quantity of interest to one or more parameters. For example, the efficiency of a turbulent flow combustor may be sufficiently insensitive to the Reynolds' number to allow quantitatively useful small-scale testing at other Re-values, provided, however, that Damköhler similarity is strictly preserved (Section 7.2.3). Such a test combustor would be a "partial" (or approximate) model of the full-scale prototype.

Overview

Before embarking on the detailed discussion of an interrelated set of illustrative numerical exercises (Chapter 8) covering the sometimes abstract material included in Chapters 2 through 7, it is perhaps timely to pause here and "take stock" of what has been presented in this text. It will be remembered that our goals included developing an understanding of the role of transport phenomena in chemically reacting systems, the underlying laws of which are set forth in Chapters 2 and 3. In Chapters 4, 5, and 6 we have first developed the important laws of momentum, energy, and mass transport and their consequences in simpler (usually nonreactive) situations. "Armed" with this perspective and the topics we have chosen to include on transport *with* simultaneous chemical reaction, each student should now be able to profit from more specialized courses and participate in future research in which this comprehensive background is absolutely essential. Any residual doubts the student may have about the relevance of this material to practical chemical-reactor or heat/mass-exchanger problems will be removed in subsequent courses and work, if not in studying the interrelated exercises of Chapter 8.

- Chapter 2 provided a concise account of the basic laws governing the behavior of chemically reacting fluids treated as continua. These laws are seen to be of two distinct types, *viz.*; (a) *conservation* ("balance") laws, leading to equations applicable to all fluids, which embrace all of the laws of classical dynamics and thermodynamics, and (b) *constitutive* laws which describe the behavior of particular classes of fluids (e.g., with respect to momentum, energy, and mass *diffusion* fluxes and chemical *kinetics*). When applied to each representative subregion of fixed space we are led to partial differential equations (PDEs) interrelating the local field densities; however, these same laws are also applied to (partially) macroscopic control volumes and even control volumes straddling phase boundaries or gas-dynamic discontinuities (leading to boundary conditions or "jump" conditions). While it is true that these laws provide the foundation for the quantitative understanding of all chemical-reaction engineering problems, we first illustrated in Chapters 4 through 6 their application to nonreactive systems in which the number of interacting phenomena is smaller and the consequences of these laws are, correspondingly, easier to extract. In addition to these basic insights, some of the methods and results discussed in Chapters 4 through 7 are shown to carry over to chemical-reaction engineering applications.
- In Chapter 3 we discussed those equations, specific to individual classes of fluids, which must be added to the basic conservation laws in order to solve practical transport problems. Attention was focused on these "constitutive" equations governing the *diffusion fluxes* of momentum, energy, and/or species mass, as well as the coefficients (viscosity, heat conductivity, diffusion coefficients) that appear therein. It was shown that while these fluid property values can simply be regarded as measurable coefficients, valuable information about them can be extracted from a molecular "model" of the substance of interest (e.g.,

low-density gas, ideal crystal, etc.). (Equations of state, and chemical-reaction rate equations were also mentioned, but are more appropriately considered in detail in courses devoted, respectively, to equilibrium thermodynamics and chemical kinetics.)

- In Chapter 4 we illustrated the consequences of total *mass* conservation and *linear-momentum* conservation for a variable density (but constant composition) fluid in the absence of momentum *diffusion* (the so-called inviscid-limit). Emphasis was first placed on steady, "one-dimensional" flow as influenced by area change and/or heat addition (without inquiring into the mechanism or "details" of heat addition).
- In Chapter 4 we also discussed the consequences of total mass and linear momentum "conservation" in the presence of momentum *diffusion* (i.e., viscous effects). The simplest multidimensional steady-flow examples were considered, involving *friction drag* at solid walls (boundary-layer flow) and fluid *entrainment* by jets for both laminar and turbulent flow.
- In Chapter 5 we illustrated the consequences of energy conservation in moving fluids (with quiescent media, including stationary solids, being a simple special case!). Emphasis was placed on energy transport by *convection* and/or *diffusion* (conduction) in nonreactive systems and the exploitation of *analogies* between energy and momentum transfer in quantitatively understanding *turbulent* flows (Sections 5.6 and 5.7). However, we included brief accounts of the treatment of energy transport effects of chemical reaction (Section 5.8) and *radiative* transport (Section 5.9). Fortunately, in many engineering applications radiation can be uncoupled from energy transport by the mechanism of convection and/or diffusion.
- In Chapter 6 we illustrated some of the interesting consequences of individual *species mass conservation* in moving fluids (again including quiescent media as a special case). In general, both diffusion and convection participate in species mass transport, and special attention is given to:
  a. analogies to momentum and/or energy transport (Section 6.3.3), including (Section 6.5.5) those that hold in some simple but important chemically reacting systems;
  b. quantitative inclusion of "analogy-breaking" phenomena (e.g., homogeneous chemical reaction, or *solute* "phoresis") *via* correction factors (Section 6.3.4);
  c. the application of tracer stimulus–response methods to "diagnose" mixing in nonideal vessels (Section 6.7).
- Last, in Chapter 7 we illustrated how our understanding of the consequences of momentum, energy, and mass conservation in nonreactive flow systems could be used to extract information about their combined influences. While we have avoided detailed discussions of prototype "nonideal" and/or multiphase chemical-reactor problems, we introduced and exploited the notion of "similarity" (geometric, thermal, compositional) in nonreacting and chemically reacting systems, and demonstrated the interactive role of experiment and theory in advancing our quantitative understanding of chemical-reactor problems.

Review of the Uses of Conservation and Constitutive Laws

We have illustrated or touched upon many different *uses* of the fundamental laws introduced and discussed in Chapter 2. Accordingly, it is instructive to briefly return to Figure 1.6-1, this time citing more specific examples—many of which have by now been explicitly or implicitly included in Chapters 2 through 7.

Design of Scale Experiments/Generalizations of Experimental Data or Theoretical Predictions

The use of nondimensionalized forms of the basic laws of Chapter 2 to design quantitatively useful small-scale experiments or generalize new (experimental or theoretical) results has been discussed earlier (Chapters 4, 5, and 6) and in this chapter for both nonreacting and reactive systems. Specific examples are to be found in the determination and use of such dimensionless transfer coefficients as:

$\mathrm{Nu}_h \equiv$ dimensionless heat-transfer coefficient	(Nusselt number)	(Ch. 5)
$\mathrm{Nu}_m \equiv$ dimensionless mass-transfer coefficient	(Nusselt number)	(Ch. 6)

in static and/or (natural or forced) flow systems (see Section 8.3 below) or

$C_f \equiv$ skin friction (momentum transfer) coefficient		(Ch. 4)
$\mathrm{St}_h \equiv$ dimensionless heat-transfer coefficient	(Stanton number)	(Ch. 5)
$\mathrm{St}_m \equiv$ dimensionless mass-transfer coefficient	(Stanton number)	(Ch. 6)

in forced-flow systems.

A specific set of interrelated numerical examples will be provided and discussed in Chapter 8.

Determination of Phenomenological Coefficients via Theory and Experiments on Canonical Configurations

Many of the simple situations considered above (constant fluid properties, simple geometry) provide the basis for necessary "property" measurements. Thus, the theory of capillary-tube viscous flow discussed in Chapter 4, combined with flow-rate and pressure-drop *measurements*, allows the determination of the Newtonian *viscosity* μ, for many fluids, including gases. Similarly, the theory of diffusion from a continuous point source in a uniform laminar stream, discussed in Sections 5.5.6 and 6.5.1, combined with the corresponding temperature or composition *measurements*, have been used to determine *molecular* thermal *conductivity*, k, and binary (Fick) diffusion coefficients (D_{ij}), often at combustion temperatures (by using a stream of combustion-heated gases, such as that above a flat flame burner). Even the theory of mainstream kinetic energy "recovery" by a surface in a high-speed stream (Section 6.5.4) has been used to determine gaseous *Prandtl numbers* and could likewise be used (in the presence of mainstream chemical energy) to determine gaseous *Lewis numbers*. Both of these *molecular-diffusivity ratios* play an important role in reaction engineering applications. Moreover, we have also seen how time-averaged velocity profile data (Section 5.7.2) can be used to determine an "effective" transport property—*viz.*, the "eddy" diffusivity distribution in a turbulent wall boundary layer. Without being exhaustive, it is clear that the principles discussed in Chapter 2, as applied to simple momentum, energy, and mass-transport situations in Chapters 4, 5, and 6, respectively, provide us with the phenomenological coefficients needed to analyze more complex reaction engineering configurations.[15]

Solution of the Prototypical "Unit Processes" of Chemically Reacting Systems

Many of the simple configurations we have considered, using the conservation equations and constitutive laws of Chapter 2, lead to results needed to analyze more comprehensive, including chemical reaction, problems. For example, Eq. (6.4-27), equivalent to the "local" law:[16]

$$\dot{m} = 2\pi d_{\mathrm{p}}(D_{\mathrm{F\text{-}mix}}\rho)_{\mathrm{gas}} \cdot \tfrac{1}{2}\overline{\mathrm{Nu}}_{m,0}\{\mathrm{Re}, \mathrm{Sc}\} \cdot \ln\{1 + B_m\} \qquad (7.4\text{-}1)$$

for the quasi-steady vaporization rate of an isolated (fuel) droplet, of instantaneous diameter, *d*, is the basis for many two-phase flow (vaporizing spray) analyses (see Sections 2.6.3 and 8.2). Equation (7.4-1) also leads to the definition of a "vaporization Damköhler number" $t_{\mathrm{flow}}/t_{\mathrm{vap}}$, important in modeling such flows. Thus, if the environmental conditions are constant, Eq. (7.4-1) implies:

$$t_{\mathrm{vap}} = d_{p,0}^2/K_{\mathrm{vap}} \qquad (7.4\text{-}2)$$

where t_{vap} is the time to completely vaporize a droplet of initial diameter $d_{p,0}$ (much larger than the prevailing gaseous mean-free-path) and

$$K_{\mathrm{vap}} = \left\{\frac{8(D_{\mathrm{F\text{-}mix}}\rho)_{\mathrm{gas}}}{\rho_{\mathrm{F}}(\ell)} \cdot \frac{\overline{\mathrm{Nu}}_{m,0}}{2} \cdot \ln\{1 + B_m\}\right\}. \qquad (7.4\text{-}3)$$

By adopting a simple chemical model, and invoking mass- and energy-conservation principles, we have also seen (Section 6.5.5.7) that these results can be generalized to account for "envelope flame" combustion in the immediate vicinity of each droplet (Figure 1.1-3). (The effect is to increase $\dot{m}$ and increase K by an amount close to

---

[15] For the molecular properties, microscopic theories are, however, also needed to carry out inevitable interpolations and, especially, extrapolations, using a necessarily limited experimental data base.

[16] In general, the appropriate value of the mass-transfer driving force parameter B_m (Eq. (6.4-4)) cannot be evaluated without considering the droplet *energy* balance. When Le = 1 this additional equation is simply $B_m = c_p(T_\infty - T_w)/L_{\mathrm{vap}}$, where L_{vap} is the latent heat of fuel evaporation (per unit mass). (See the discussion of Section 6.4.3.3.)

that expected if the droplet were simply evaporating into the products of adiabatic combustion between fuel vapor and ambient oxidizer.)

The equations of Chapters 2 and 3 have also been used to analyze the structure and speed, S_u, of a flame propagating into a premixed, nonturbulent, combustible gas mixture. The results for simple kinetic models reveal how S_u depends on *both* kinetic and transport parameters (cf. Section 7.2.2.2). Indeed, since S_u-data are often readily available, Damköhler numbers explicit in S_u (e.g., $S_u L/\alpha_u$) are in widespread use to correlate performance results in premixed systems (e.g., flame-holder blow-off limits, etc.) as well as diffusion flames situations. In effect, $S_u{}^2$, for which data are comparatively readily available, is used as a quantitative indicator of the global chemical kinetics (cf. Section 7.2.3.2b).

Computer Modeling of Complex Chemical Reactors

Instead of relying exclusively on scale-model experiments on separators, combustors, etc., the results of which may not be fully transferrable to full scale (Chapter 7), the fundamental principles and equations of Chapter 2 provide the foundation for the economically attractive method of computer "simulation" or computer "modeling." Here one seeks to predict, based on a comprehensive mathematical model "alone," the behavior of a full-scale device (e.g., synthesis reactor, furnace, stationary power plant) including its response to changes in various parameters at the control of the designer. Ordinarily, this requires large-scale digital computation, and, for complex devices (three-dimensional, multiphase), this promising approach is just becoming technically and economically feasible.

It should be recognized that the rational development of such "computer models" still requires repetitive comparison with *experiments* (perhaps on simpler, smaller, configurations) and the proper quantitative *interpretation* of, say, *probe data*, also involves the use of the same laws of momentum, energy, and mass transport (cf. Section 1.6.5). An example is the role of surface reactions in "falsifying" the reading of a "catalytic" thermocouple probe, discussed in Section 6.5.4. Another closely related example is the need to correct temperature-probe readings for the simultaneous effects of thermal radiation and possible conduction losses (down the leads or probe support, etc.). Our final example is the use of transport principles to "deconvolute" observed sampling tube data. This enables the experimenter to infer the particle-size distributions (e.g., of soot) which must have existed in the mainstream (Section 6.6.1).

Concluding Remarks

Obviously, it is quite impossible, in an introductory text, to fully illustrate the diverse purposes to which the above-mentioned laws of momentum, energy, and mass conservation can be fruitfully put. However, it is hoped that this discussion, combined with the interrelated numerical examples to be discussed in Chapter 8, will convince the skeptic that an understanding of the seemingly "abstract" laws stated in Chapter 2 will pay "concrete" dividends.

In one's preparation for engineering courses, emphasis is usually placed on the underlying laws of thermochemistry and chemical kinetics, irrespective of fluid motion and energy/mass transport. Here we have shown how this information can be incorporated into useful mathematical models of chemically reacting fluid flows, based on fundamental conservation (balance) equations, and a continuum ("field"-theory) approach to momentum, energy, and mass transfer for a wide variety of "fluids" (constitutive relationships). If, on this basis, rational mathematical models can be developed to the point where they can adequately predict the results of simpler "canonical" experiments, scale-model experiments and full-scale reactor or separator experiments, then these comprise powerful and economical tools for the optimal design of new devices, and the modification or control of existing ones (e.g., subject to new conditions (e.g., fuels) or constraints). The laws of transport and the conceptual framework developed here can be applied in subsequent courses to each of the important *types* of chemical reactors. Moreover, the techniques required to obtain reliable and sufficiently detailed *experimental* information from operating systems must be critically studied. In both cases, frequent recourse will be necessary to the material here, the understanding of which will inevitably pay handsome dividends.

TRUE/FALSE QUESTIONS

7.1 T F When made dimensionless, the underlying equations of conservation, combined with their boundary conditions, contain no more information about a transport problem than that already contained in the Buckingham "π-Theorem."

7.2 T F Dimensionless groups ("eigen-ratios," such as shape factors, Re, Ra, Pr, etc.) comprise the similarity parameters which yield "set-up rules" for designing *scale-model* experiments amenable to quantitative use.

7.3 T F Similitude analysis may be regarded as a procedure for discovering quantitative similarities ("co-relations") for a class of superficially dissimilar systems having some intrinsic similarity.

7.4 T F There are no quantitatively useful scale-up or scale-down principles applicable to systems with chemical reaction. For this reason, scale-model tests of chemical reactors are useless.

7.5 T F If an equation is not "dimensionally homogeneous," it is either wrong or valid only in a particular unit system.

7.6 T F Conservation (balance) laws for *macro*scopic control volumes cannot be used for purposes of similitude analysis; rather, *differential* equations are necessary.

7.7 T F Nondimensional groups and variables are useful even in cases amenable to complete mathematical (analytical or numerical) solution since they give the predicted results maximum generality.

7.8 T F Dimensional analysis using the formal Buckingham "π-Theorem" has the merit that it can be accomplished with no understanding of the fundamentals of the problem.

7.9 T F While dimensional analysis and similitude analysis provide useful results in the area of "pure" fluid mechanics (momentum transfer), these techniques cannot be applied to systems with simultaneous heat and/or mass transfer.

7.10 T F Two phenomena are said to be mathematically *analogous* if they are governed by identical nondimensional field equations (PDEs).

7.11 T F The relative importance of any two terms in a nondimensional PDE can be determined by simply evaluating and comparing their nondimensional coefficients (parameters).

7.12 T F Dimensionless parameters do not have a unique physical interpretation; rather, several alternative interpretations are common and instructive.

EXERCISES

7.1 For ease of manufacture, compactness, and maintenance, the convective heat exchanger tubes in a proposed fossil-fuel combustion power station will contain clusters of four two-inch (O.D.) tubes (see Figure 7.1E) with $s/d = 1.5$, inclined at the angle $\psi = 15°$ from the horizontal, exposed to the vertical flow of 1600 K combustion products at $U_\infty = 10$ m/s, $p = 1$ atm. However, information in the heat-transfer literature on the performance of such clusters is deemed inadequate, so that the project engineer authorizes you to perform such calculations or experiments as necessary to accurately predict the *forced-convection* heat-transfer performance of such inclined clusters of long, straight isothermal tubes. The quantity of maximum interest is the total heat-transfer rate (gas-to-wall) per unit length of the tube cluster. Because of the geometric and fluid-dynamic complexity, theoretical calculations (e.g., finite differences,

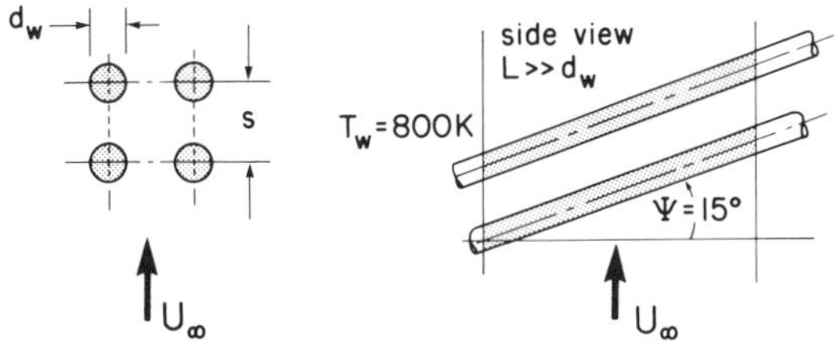

Figure 7.1E

or finite elements) are rejected in favor of *scale-model experiments*, if the latter are feasible using available laboratory facilities.

a. Based on the similarity principles in this chapter, can you select a working fluid, and practical scale-model experimental *conditions* such that heat-transfer measurements made on the model (m) configuration can be used to make reliable quantitative predictions of the performance of the above mentioned prototype (p)? (Or, as one skeptical official of the corporation has pessimistically concluded, will it be necessary to do expensive and time-consuming *full-scale* experiments in a prototype "laboratory" furnace?)

b. Demonstrate that it is reasonable to expect that additional (new) phenomena will not enter your scale-model experiments, complicating the relationship between *m*odel and *p*rototype performance.

c. Under the model experimental conditions you choose, what would the relation be between observed model (m) and predicted prototype (p) heat-transfer rates?

d. On what "coordinates" should your scale-model heat transport data be plotted so that the results are apt to apply to similar configurations at "any" scale?

e. The corrosion engineers are concerned about the condensation of certain salts on the surface of each such cooled tube bundle. What information and assumptions would allow you to predict the maximum possible rate of vapor transport and condensation onto the tube bundle, based on your small-scale *heat*-transfer measurements?

7.2 Application of the Mass/Energy Transfer "Analogy" and Scale-Model Theory. Carefully study Exercise 7.1 and the accompanying description of one possible solution in which the scale-model test involves *heat*-transfer measurements on a cluster of 2-mm diameter rods in a "water tunnel." Consider the following *alternative* solution, based on the feasibility of making *dissolution* (convective *mass* transfer)-*rate measurements* on a cluster of 5-mm diameter rods exposed to the flow of liquid *water* at 25°C.

a. Given the physical data below, systematically work out how you would use benzoic-acid *dissolution*-rate data[17] to predict the *heat*-transfer performance of the full-scale tube bundles, also circumventing the need for expensive full-scale furnace tests. Defend the validity of each of your underlying assumptions and itemize the experimental precautions you would have to take. In your opinion, does the present *mass*-transfer analogy approach have any decisive advantages or drawbacks?

Basic data: Pure benzene carboxylic acid[18] (commonly called

---

[17] Obtained either from measurements of the total mass lost by the benzoic-acid-coated rods in the course of each experimental run, or from direct measurements of the mixing-cup averaged ("bulk") composition of the aqueous solution downstream from the cluster of slowly dissolving rods.

[18] Written -COOH, or C_6H_5COOH.

"benzoic acid") is a solid (mp 122°C, density = 1.266 g/cm^3) which is sparingly soluble in 25°C water (0.0278 g-moles/liter H_2O). Its diffusion coefficient has been measured to be 1.00×10^{-5} cm^2/s in 25°C water (Table 2.6 of Sherwood, Pigford, and Wilke (1975)).

b. If the simultaneous role of *natural* convection cannot conveniently be "simulated" in your mass-transfer model experiment, is it possible to dispense with Gr_m-similarity altogether? Examine this possibility on the basis of available experimental data *for an isolated cylinder under the simultaneous influence of forced convection and* ("*opposed*") *natural convection.* (In that case how much is $\overline{Nu}_h$ decreased below its "pure" forced convection counterpart) when $Gr_h^{1/4}/Re^{1/2}$ is as large as expected in our "prototype"?)

c. Why in Exercise 7.1, and in Part (a) above, is it reasonable to test with $Re_m = Re_p$, assuming the dependence of $\overline{Nu}$ on Pr (or Sc) is "known" (from isolated cylinder data)? Would it be reasonable to also dispense with the "set-up rule": $Re_m = Re_p$ by assuming that the Re-dependence of $\overline{Nu}$ (tube bundle) is the same as the Re-dependence of $\overline{Nu}$ (isolated cylinder)? Are any experimental data available which would allow you to compare the Re-dependence of $\overline{Nu}$ for both an isolated cylinder and, say, two cylinders "in tandem"?

d. In the solution of Exercise 7.1(a-e), no mention was made of the possible importance of simultaneous *natural* ("free") convection in the full-scale furnace application. Investigate this possibility by first calculating the magnitude of Gr_h and then forming the important *ratio* $Gr_h^{1/4}/Re^{1/2}$ (governing the relative importance of free and forced convection). Is this ratio small enough to consider natural convection negligible, or will it be necessary to *also* do model testing at the same Grashof number as for the full-scale prototype? In the benzoic-acid dissolution *model* (*mass*-transfer analog) experiment considered in this exercise, what is the value of $(Gr_m)_m$, given the fact that the density difference between benzoic-acid-saturated water and pure water is only 1.42×10^{-3} g/cm^3 at 25°C. To force $(Gr_m)_m = (Gr_h)_p$ what value of d_m would be needed, and what water velocity, U_m would now be needed to preserve $Re_m = Re_p$? Would the Pr and/or Sc-dependence of $\overline{Nu}$ be known under these more complicated (combined forced convection and natural convection) conditions?

7.3 A six-bladed turbine-agitated pilot plant, stirred-tank, jacketed chemical reactor (STR), a small-scale model of a proposed production unit, is of such size that 1 g charged to the pilot-plant (model) reactor will be the equivalent of 0.5 kg of the same material charged to the production unit. The production unit (STR) will be 2 m in diameter and 2 m deep, and will contain a six-bladed turbine agitator of 0.6 m diameter. In the pilot scale STR, the optimum agitator speed is found (by experiment) to be 330 rpm.

a. What are the other significant dimensions of the pilot plant (scale-model) reactor? What will the linear scale-up factor be?

b. If the reaction mass has the properties of water at 70°C, and the power input per unit volume is kept constant[19] (in model (m) and prototype (p)), at what speed should the impeller turn in the full-scale (prototype) production reactor?

c. At what speed would the production-unit impeller have to turn to hold the *mixing* (blending) *time* constant?[20]

d. At what speed should the production-unit impeller turn if the Reynolds', number (based on impeller tip speed) is held constant? [*Note*: Re $\equiv \Omega d_a^2/\nu$.] Would this ensure equal nondimensional heat-transfer coefficient (charge/vessel wall)?

e. Which of the above operational "choices" would you recommend for actual STR scale-up? Why? (After Ex. 9-5, McCabe and Smith (1976).)

7.4 a. Use the correlation shown in Figure 7.2-10 to "predict" the blow-off velocity, U_{bo}, in a jet-engine afterburner which will employ a gutter-type flame stabilizer of transverse dimension 2.5 cm in a premixed kerosene vapor/"air" stream with the estimated local properties:

$$S_u = 50 \text{ cm/s}, \qquad T_u = 600 \text{ K}, \qquad p_u = 0.5 \text{ atm}.$$

b. Under these conditions, over what range of gas velocities, U, in the duct could this "afterburner" be operated?

c. If the afterburner pressure level was 10^{-1} atm (rather than 0.5 atm), what would be the effect on U_{bo}?

d. Briefly discuss each of the important assumptions you have made to "predict" U_{bo} under these conditions.

e. Since the experimental data underlying Figure 7.2-10 were obtained using *other* gaseous fuels at higher pressures, how "safe" is your prediction? Would you recommend performing costly new experiments under precisely the above conditions?

7.5 Besides laminar flame-speed data (Section 7.2.2.2 and Table 6.5-1), measurements have been made of the minimum energy (e.g., spark) needed to initiate ("ignite") combustion in premixed gases (Lewis and von Elbe (1951)). But, since steady flame propagation is, in effect, "continuous ignition," there should be a correlation between the *minimum ignition energy* (MIE) and the flame speed (S_u) for the same gaseous fuel/oxidizer mixture. To pursue this, consider the tentative postulate that MIE should be proportional to the energy required to raise a volume of gas $\pi\,\delta_f^3/6$ from T_u to T_b, where δ_f is the steady-flow flame thickness (Section 7.2.2.2).

a. Show that this postulate immediately leads to the conclusion that the

---

[19] The turbulent mixing effectiveness of such a device is approximately dependent on the impeller power dissipated per unit volume of liquid (e.g., hp per 1000 gal). At high Re we expect $P/d_a^3 \sim \Omega^3 d_a^2$, where Ω is the impeller rotational speed and d_a the agitator impeller tip diameter.

[20] See, e.g., the miscible Newtonian fluid blending time correlation of Norwood and Metzner, *AIChE J.* 6, A32 (1960), which, at high Re ($>10^3$) predicts that the blending time should be proportional to $\Omega^{-2/3} d_a^{1/6}$.

dimensionless quantities:

$$\underbrace{\frac{\text{MIE}}{(\alpha_u/S_u)^3(\rho_u\omega_{F,u}Q)}}_{\Phi \leqslant 1} \quad \text{and} \quad \underbrace{\frac{\text{MIE}}{(\alpha_u/S_u)^3(\rho_u\omega_{O,u}fQ)}}_{\Phi \geqslant 1}$$

should be approximately the same for all gaseous fuel/oxidizer combinations.

b. Test this proposed correlation using the data assembled in Table 6.5-1, and estimate the "best" (universal) value of this *dimensionless MIE*.
c. Is the initial postulate plausible? If so, how can it be justified?
d. How could such a correlation be used?
e. What are the likely limitations of this correlation? (That is, what additional parameters might influence the magnitude of the dimensionless MIE?)

7.6 Of interest in predicting metal temperatures (e.g., cylinder walls, piston face; see Figures 1.1-1 and 5.4-2) in internal combustion-(IC-) engines is the heat-flux-*vs.*-time "history," $-\dot{q}''_w(t)$, when a premixed flame propagating toward the wall "collides" with that surface. Suppose that several such heat-flux-*vs.*-time records are available in the literature for specific fuel-lean systems, and these are all *qualitatively* similar. Outline a procedure whereby these available data could be used to construct and test a "universal" "reduced" heat flux *vs.* "reduced" time plot, from which *quantitative* $-\dot{q}''_w(t)$ curves for new systems of special interest to you could easily be obtained in terms of readily available laminar flame-speed data. Motivate and defend your choices of an appropriate reference *energy flux* and reference *time*. Under what conditions might this "correlation" break down? [Motivated by paper of Vosen *et al.* (1985).]

Appendix 7.1

Outline of Procedure for Similitude Analysis[21]

1. Write the necessary and sufficient equations which would determine the quantity of interest (if they could be completely solved) in the most appropriate coordinate system. These will usually be partial differential equations (PDEs), boundary conditions, initial conditions, and auxiliary laws (e.g., constitutive equations, equations of state, etc.).
2. Introduce nondimensional variables using appropriate *reference* lengths, times, temperature differences, etc. (the natural "yardsticks") of the problem selected from the known geometry and boundary conditions:
   a. Wherever possible variables should be "normalized" (defined so that they range from zero to unity).
   b. When no obvious choice is available, insert a temporarily unspecified reference quantity, and select it later such that the number of dimensionless parameters appearing in the resulting set of equations is minimized.
3. From the expected magnitude of the parameters and terms in the resulting equations make suitable (defensible) approximations (i.e., drop negligible terms).
4. Formally express the dimensionless quantity of interest in terms of the dimensionless variables and parameters of the problem.
5. Inspect this result for its implied parametric dependence. It constitutes the "similitude relation" sought. Especially as a result of Step 3, this result will usually constitute a much "stronger" similitude than that obtainable (for the same problem) from conventional dimensional analysis (Vaschy, Buckingham-type), i.e., fewer similitude criteria ("set-up rules") will be found to be essential in determining the value of the dimensionless quantity of interest.[22]

REMARKS

1. Apparently dissimilar physico-chemical problems may lead to identical dimensionless equations and BCs/ICs, thereby establishing useful "analogies" (e.g., between heat and mass transfer).

---

[21] A procedure to discover quantitative similarities ("co-relations") for a class of superficially dissimilar systems (problems) having some intrinsic similarity.
[22] Cf. Section 7.2.2.

2. Suitably chosen dimensionless parameters can be simply interpreted as ratios between typical or "characteristic" terms in the governing equations (e.g., the Reynolds' number $\mathrm{Re} \equiv UL/\nu$ can be regarded as the ratio of the characteristic convective momentum flux ρUU to the characteristic momentum flux $\mu U/L$ due to momentum transport on a "microscopic" level). The problem and resulting similitude may simplify considerably when such parameters become very large or negligibly small compared to unity.
3. Approximate similitudes ("partial modeling") may be possible for complicated problems; i.e., because of the nature of the quantity of interest, not all parameters are equally important (some conditions may be "escapable").
4. As a result of Step 3 and Remark 2, the resulting simplified (or asymptotic) equations may have *invariance properties* which allow a further reduction in the number of governing dimensionless parameters, and the extraction of functional dependencies (e.g., $\overline{\mathrm{Nu}}_h = \mathrm{const} \cdot (\mathrm{Re})^{1/2} (\mathrm{Pr})^{1/3}$ for laminar boundary-layer flow at high Prandtl numbers; This implies that meaningful model heat-transfer experiments need not be done at the *same* Reynolds' number and/or Prandtl number as the full-scale "prototype").

REFERENCES

Altenkirch, R.A., *et al.*, *Sixteenth Symposium (Internat.) on Combustion.* Pittsburgh, PA: The Combustion Inst. 1165–1174 (1976).

Buckmaster, J.D., and G.S.S. Ludford, *Theory of Laminar Flames.* Cambridge: Cambridge Univ. Press (1982).

Churchill, S.W., and M. Bernstein, *ASME J. Heat Transfer* **99**, pp. 300 (1977).

Diederichsen, J. and H.G., Wolfhard, Proc. Roy. Soc. A., **236**, 89 (1956).

Dixon-Lewis, G., and S.M. Islam, *Nineteenth Symposium (Internat.) on Combustion.* Pittsburgh, PA: The Combustion Inst., 283–291 (1982).

Fristrom, R.M., and A.A. Westenberg, *Flame Structure.* New York: McGraw-Hill (1965).

Hlavacek, V., M. Kubicek, and M. Marek, *J. Catalysis* **15**, 17, 31 (1969).

Kadota, T., and H. Hiroyasu, *Eighteenth Symposium (Int.) on Combustion* Pittsburgh, PA: The Combustion Institute (1981).

Kalson, P. Ph.D. Dissertation, Ann Arbor, MI: University of Michigan (1981).

Kaskan, W., *Proc. Sixth Symp. (Int.) on Combustion.* New York: Reinhold (1946), 134-143.

Lewis, B., and G. von Elbe, *Combustion, Flames, and Explosions of Gases.* New York: Academic Press, Inc. (1951).

Longwell, J.P., *et al.*, *Ind. Eng. Chem.* **45**, p. 1629 (1953); see, also, *Ind. Eng. Chem.* **47**, p. 1634 (1955).

McAdams, W.H., *Heat Transmission* (Third Ed.). New York: McGraw-Hill (1954).

McCabe, W.L., and J.C. Smith, *Unit Operations of Chemical Engineering* (Third Ed.). New York: McGraw-Hill (1976); (ChE Series), Appendix 4: Dimensional Analysis.

Norwood, K.W. and A.B. Metzner, *AIChE J.* **6**, pp. 432–437 (1960).

Putnam, A.A., and R.A. Jensen, in *Third Symposium on Combustion.* Baltimore: Williams and Wilkins, pp. 89–98 (1949).

Rosner, D.E., *J. Aero/Space Sci* (now *AIAA J*) **26**, pp. 281–286 (1959).

Rosner, D.E., *Amer Rocket Soc J.* (now *AIAA J*) **31**, 816–818 (1961).

Rosner, D.E., *Comb. Sci. and Technol.* **10**, 97–108 (1975).

Rosner, D.E., *Comb. and Flame* **9**, 199–201 (1965).

Rosner, D.E., and W.S. Chang, *Comb. Sci. and Technol.* **7**, pp. 145–158 (1973).
Sherwood, T.K., R.L. Pigford, and C.R. Wilke, *Mass Transfer*. New York: McGraw-Hill (1975).
Spalding, D.B., *Some Fundamentals of Combustion*. Gas Turbine Series, Vol. 2. New York: Academic Press (1955).
Stewart, D.G., in *Selected Combustion Problems* II, AGARD/NATO. London: Butterworths (1956), pp. 384–413.
Su, G.-J., *Ind. Eng. Chem.* **38**, p. 803 (1946).
Vosen, S.R., R. Grief, and C. Westbrook, Paper No. 8, Session A2, *Proc. 20th Symposium (Int.) on Combustion*. Pittsburgh, PA: The Combustion Institute, pp. 75–83 (1984).
Zeldovich, Y.B., *et al.*, *The Mathematical Theory of Combustion and Explosions*. New York: Consultants Bureau (1985).

BIBLIOGRAPHY: SIMILITUDE/DIMENSIONAL ANALYSIS

Historical

Bridgman, P., *Dimensional Analysis*. New Haven, CT: Yale University Press Reprint (1963).
Damköhler, G.I., "Influence of Flow, Diffusion, and Heat Transfer on Reactor Performance" (in German), *Z. Elektrochemie* **42**, 846 (1936).
Ehrenfest-Afanassjewa, T., "Dimensional Analysis Viewed from the Standpoint of Similarities," *Phil. Mag.* **1**, 257–272 (1926).
Macagno, E.O., "Historico-Critical Review of Dimensional Analysis," *J. Franklin Inst.* **292**, No. 6, 391–402 (1971).
Rayleigh (Lord), "The Principle of Similitude," *Nature* (London) **95**, 66–68 (1915).

Elementary

Beer, J.M., and N.A. Chigier, *Combustion Aerodynamics* (Chap. 7). New York: Halsted Press, Div. J. Wiley & Sons, 1972.
Johnstone, R.E., and M.W. Thring, *Pilot Plants, Models and Scale-up Methods in Chemical Engineering*, New York: McGraw-Hill, 1957.
Schmidt-Nielsen K., *How Animals Work*. Cambridge Univ. Press (London (1972)); Chap. 6.
Stahl, W.R., "Similarity and Dimensional Methods in Biology," *Science* **137**, 205–212 (1962).
Weller, A.E., "Similarities in Combustion: A Review," *Selected Combustion Problems, II*, London: Butterworths (1956), pp. 371–383.

Intermediate

Bisio, A., and R.L. Kabel, *Scale-up of Chemical Processes: Conversion from Laboratory Scale Tests to Successful Commercial Size Design*, New York: J. Wiley (1985).
Churchill, S.W., *The Interpretation and Use of Rate Data: The Rate Concept* Scripta (Wash. DC)/McGraw-Hill (New York) (1974); Part IV.
Hellums, J.D., and S.W. Churchill, "Simplification of the Mathematical Description of Boundary and Initial-Value Problems," *AIChE J.* **10**, 110–114 (1964).

Isaacson, E. de S.Q., and M. de S.Q. Isaacson, *Dimensional Methods in Physics.* New York: Halsted Press (1975).

Kline, S.J., *Similitude and Approximation Theory.* New York: McGraw-Hill (1965).

Lefebvre, A.H., *Gas Turbine Combustion.* New York: McGraw-Hill (1983).

Mellor, A.M., in *Progress in Energy and Combustion Sci.* **6**, 347–358 (1980).

Murphy, G., *Similitude in Engineering.* New York: Ronald Press (1950).

National Research Council: *The Use of Models in Fire Research.* NAS-NRC Publication No. 786 (1961).

Schuring, D.J., *Scale Models in Engineering.* New York: Pergamon (1977).

Spalding, D.B., "The Art of Partial Modeling," *Ninth Symposium (International) on Combustion.* New York: Academic Press, 883–843 (1963).

Stewart, D.G., "Scaling of Gas Turbine Combustion Systems," *Selected Combustion Problems, II*, London: Butterworths (1956), pp. 384–413.

Szücs, E., *Similitude and Modelling.* Amsterdam: Elsevier Science Publishers (1980).

Way, S., "Combustion in the Turbojet Engine," *Selected Combustion Problems, II*, London: Butterworths (1956), pp. 296–327 (especially Section 4).

Woodward, E.C., "Similitude Study of Idealized Combustors," *Sixth International Combustion Symposium* (1957).

Zierep, J., *Similarity Laws and Modeling.* New York: M. Dekker (1971).

Advanced

Barenblatt, G.I., *Similarity, Self-Similarity, and Intermediate Asymptotics*, New York: Consultants Bureau (1979).

Becker, H.A., *Dimensionless Parameters—Theory and Methodology.* New York: J. Wiley (Halsted Press) (1976).

Hanson, A.G., *Similarity Analysis of Boundary-Value Problems in Engineering.* Englewood Cliffs, NJ: Prentice-Hall (1964).

Hayes, W.D., and R.F. Probstein, *Hypersonic Flow Theory.* Vol. 1 Inviscid Flows (Sections 2.2–2.4 Hypersonic Similitudes), Appl. Math. and Mechanics Series Vol. 5A. New York: Academic Press (1966); second edition, pp. 39–50.

Lin, C.C., and L.A. Segel, *Mathematics Applied to Deterministic Problems in the Natural Sciences.* New York: MacMillan (1974).

Ostrach, S., "Role of Analysis in the Solution of Complex Physical Problems," *Proc. Third International Heat Transfer Conference*, ASME (1966).

Penner, S.S., "Similarity Analysis for Chemical Reactors and the Scaling of Liquid-Fuel Rocket Engines," Chapter 12 in *Combustion Researches and Reviews.* London: Butterworths (1956), pp. 140–162.

Rosner, D.E., S. Gökoĝlu, and R. Israel, "Rational Engineering Correlations of Diffusional and Inertial Particle Deposition Behavior in Non-Isothermal Forced-Convection Environments," in *Fouling of Heat Exchange Surfaces* (R. Bryers, ed.). New York: Engrg. Foundation (1983); 235–256.

Sedov, L., *Similarity and Dimensional Methods in Mechanics.* New York: Academic Press (1959).

Zeldovich, Y.B., *et al.*, *The Mathematical Theory of Combustion and Explosions.* New York: Consultants Bureau (1985).

8

Problem-Solving Techniques, Aids, Philosophy: Forced Convective Heat and Mass Transfer to a Tube in Cross-Flow

INTRODUCTION

The following quantitative interrelated exercises illustrate how the material of Chapters 5, 6, and 7 can be applied to the preliminary design of a heat exchanger for extracting power from the *products of combustion* in a stationary (fossil-fuel-fired) power plant. As is typical of most such design problems, a blend of fundamental *principles* of "transport phenomena" and experimental *data* (suitably presented and generalized (Chapter 7)) is required for its solution.[1]

Consider the preliminary design of a representative tube in a convective heat exchanger exposed to the products of pulverized coal (PC) combustion. Our preliminary design requires that we quantitatively estimate:

the rate at which *energy* is transferred from the hot combustion products to the tube, per unit tube length (e.g., kW/m);

the rate at which each such tube will accumulate ash (present as particulate *matter* in the PC combustion products).

Figure 8-1 shows the proposed overall PC furnace + heat exchanger section. We will focus attention on energy and mass transfer to one representative tube of the heat exchanger, of circular cross section and diameter 5 cm (Figure 8-2) exposed to a "mainstream" gas estimated to have the following properties:

$$T_\infty = 1200 \text{ K}, \qquad p = 1 \text{ atm}, \qquad U_\infty = 10 \text{ m/s}.$$

[1] We also seek to transmit certain problem-solving "habits" (see Appendix 8.1).

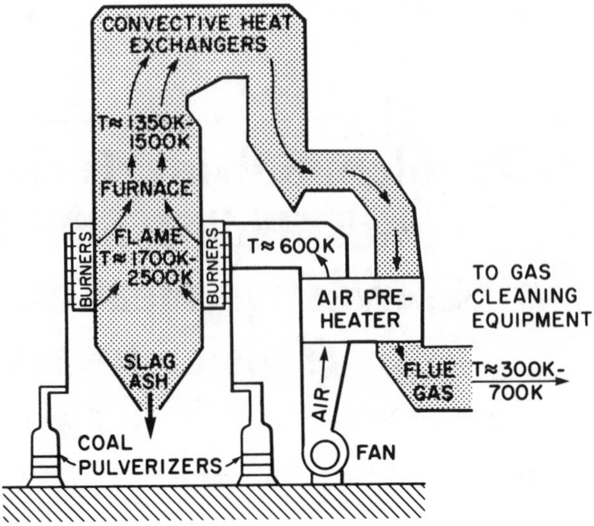

Figure 8-1 Pulverized coal combustion stationary power plant (schematic, after Flagan and Friedlander (1978)).

This gas contains ash particles, which for preliminary design purposes will be considered SiO_2 and grouped into two categories:

Submicron:	$d_p = 10^{-1}$ μm at $\omega_p = 2 \times 10^{-4}$	(mass fraction),
Supermicron:	$d_p = 10$ μm at $\omega_p = 1 \times 10^{-2}$	(mass fraction).

Problem 8.1. Estimate the rate of energy gain (kW) per meter of tube length if the outer surface of the tube is maintained at its design value $T_w = 800$ K.

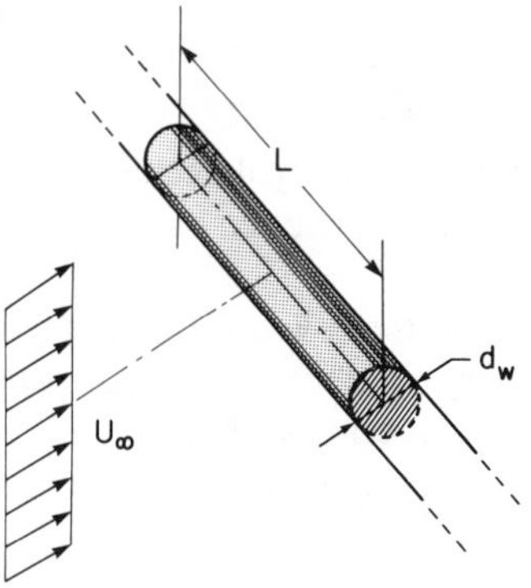

Figure 8-2 Configuration and nomenclature: heat-exchanger tube in a cross-flow of combustion production gas.

Problem 8.2. Under the conditions of Problem 8.1, estimate the rate at which both the submicron and supermicron ash will be accumulated per unit length of cylinder (such "fouling" will ultimately be associated with a heat-transfer barrier ("resistance" (Chapter 5)) requiring periodic steam-jet cleaning ("soot-blowing") or shutdown and cleaning, or tube replacement).

Tentative Simplifying Assumptions

As in all engineering problems, to proceed it is necessary to make certain simplifying assumptions, the accuracy of which can hopefully be assessed *a posteriori*. Assumptions which cannot be defended must be relaxed either immediately or upon obtaining the requisite missing information. In the latter case, suspicious assumptions should be "flagged" so that undue credence is not given to your preliminary results, and the critical missing information is pinpointed for further study.

Based on our experience with somewhat similar problems, we *tentatively* assume that:

1. The gas flowing around the cylinder can be treated as a *continuum* (on the scale of the cylinder diameter).
2. The heat-exchanger tube may be treated as an *isolated* cylinder in cross-flow, with *forced*-convection dominating *natural* convection.
3. Viscous dissipation can be neglected in determining the gas temperature profile and hence heat-transfer rate from the gas to the solid.
4. The mainstream turbulence level is not so high as to noticeably alter the time-averaged energy or mass-transfer rates.
5. At least for the *submicron* particle transport (by Brownian motion and convection), the analogy between mass and energy-transfer holds.

The validity of these tentative assumptions will be reconsidered in the exercises.

8.1 ENERGY-TRANSFER RATE FROM THE COMBUSTION PRODUCTS TO THE TUBE

Note first that we have been asked to calculate the *total* energy-flow rate to a meter of tube, not the detailed velocity or temperature *profiles* in the gas surrounding the tube, or the angular distribution of heat flux around the cylinder perimeter. Thus, information on the *average heat-transfer coefficient*:

$$\frac{-\dot{q}_{w}/(\pi d_{w} L)}{[k(T_{\infty} - T_{w})/d_{w}]} \equiv \overline{\mathrm{Nu}}_{h}\{\mathrm{Re}, \mathrm{Pr}\} \tag{8.1-1}$$

for this particular flow configuration (circular cylinder shape ($L_{ref} = d_{w}$), transverse

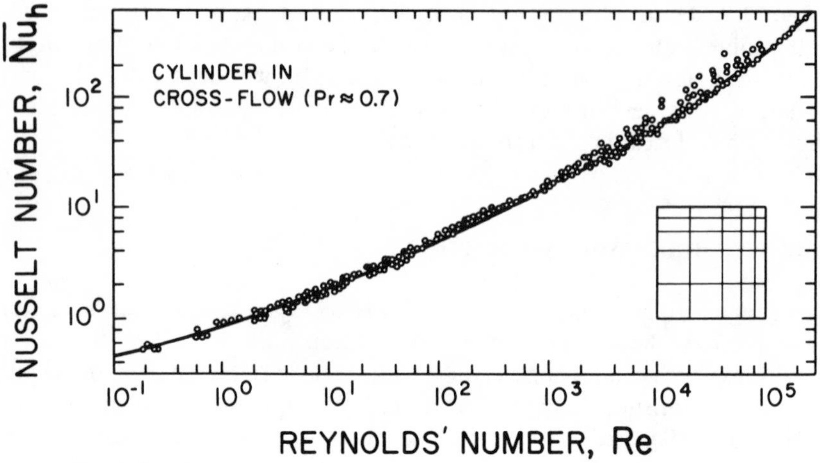

Figure 8.1-1 Correlation of heat loss/gain by circular cylinder in a steady cross-flow of air (after McAdams (1954)).

to flow) will be *sufficient* to solve this problem, provided this information is available in the appropriate range of the important dimensionless parameters:

$$\mathrm{Re} \equiv \frac{U_\infty d_w}{\nu} \qquad \text{(Reynolds number)}, \tag{8.1-2}$$

$$\mathrm{Pr} \equiv \frac{\nu}{(k/\rho c_p)} \qquad \text{(Prandtl number)}. \tag{8.1-3}$$

Whether the information has been obtained theoretically or *via* experiments (say, on a circular cylinder of nearly 5 cm diameter) in combustion products is *not* important since only these *groups* are important (subject to the above-mentioned assumptions) in determining the prevailing value of the dimensionless average heat flux, $\overline{\mathrm{Nu}}_h$ (perimeter-averaged Nusselt number).

In the present case we can exploit the rather extensive *experimental*[2] data acquired for heat *loss* from small cylindrical wires heated in *air*, when such data are cast in the form of a "universal" $\overline{\mathrm{Nu}}_h\{\mathrm{Re}, \mathrm{Pr(air)}\}$-relation (Figure 8.1-1), or equivalent formulae which "fit" these data).[3] In effect, by invoking Figure 8.1-1 we are using experimental *data* obtained on a scale model of our situation (Chapter 7) to make the desired "prediction" of $-\dot{q}_w/L$. Because of the large amount of

[2] Even for this simple geometry, the flow pattern is sufficiently complex to render the actual prediction of $\mathrm{Nu}_h\{\mathrm{Re}, \mathrm{Pr}\}$ impractical over much of the important Re-range.

[3] For repetitive calculations using a digital computer, it is useful to replace Figure 8.1-1 by a suitable "curve-fit" expression, as represented by Eqs. (5.5-10) and (5.6-10) for flow past a sphere.

nonreactive N_2 (MW = 28) in the air used for PC combustion[4] we have:

$$(\mathrm{Pr})_{\text{comb. products}} \approx (\mathrm{Pr})_{\text{air}}, \tag{8.1-4}$$

so that scale-model experimental data obtained using *air* (Figure 8.1-1) should be quite applicable to combustion products.

Inspection of Figure 8.1-1 reveals that if we evaluate the Re expected to prevail for the heat-exchanger tube, then $\overline{\mathrm{Nu}}_h$ can be read off[5] immediately. Knowledge of k, T_∞, T_w, d_w and the definition of $\overline{\mathrm{Nu}}_h$ then complete the calculation of $-\dot{q}_w$ *via* Eq. (8.1-1), i.e.,

$$\frac{-\dot{q}_w}{(\pi d_w L)} = \frac{k(T_\infty - T_w)}{d_w} \cdot \overline{\mathrm{Nu}}_h\{\mathrm{Re}, \mathrm{Pr}\} \tag{8.1-5}$$

from which the rate of energy flow $-\dot{q}_w$ will be calculated for $L = 1$ meter $= 10^2$ cm.

8.1.1 Reynolds' Number and Dimensionless Heat-Transfer Coefficient

Inspection of Eq. (8.1-2) and the basic data of this problem reveals that the momentum diffusivity $\nu \equiv \mu/\rho$ must be evaluated to evaluate Re. Again, it will probably be sufficient to obtain ν from μ(air) and ρ(air) at some appropriate state. Here we immediately note that temperatures range between 800 K (near the wall) and 1200 K in the burned gases "far" away (upstream) from the wall. However, a corresponding variation in properties also existed in the model experiments leading to Figure 8.1-1 and these data points were in fact plotted using properties evaluated at the so-called "film" (mean) temperature:

$$T_f \equiv \frac{1}{2}(T_w + T_\infty). \tag{8.1-6}$$

Therefore, in the present problem we should also evaluate[6] the "fluid" properties μ and ρ at $T_f = \frac{1}{2}(800 + 1200) = 1000$ K and 1 atm. There is no point in searching the literature for the gas density under these conditions, since it is well approximated by the perfect gas law:

$$\rho = \frac{pM}{RT} \cong \frac{(1)(28)}{(82.06)(1000)} = 3.4 \times 10^{-4} \frac{\mathrm{g}}{\mathrm{cm}^3}$$

---

[4] And the almost compensating effects of CO_2 and H_2O (MW = 44 and 18, respectively) produced in the combustion process.

[5] For repetitive calculations using a digital computer, it is useful to replace Figure 8.1-1 by a suitable "curve fit" expression, as represented by Eqs. (5.5-10) and (5.6-10) for flow past a sphere.

[6] It is useful to adopt the habit of writing all numbers as the product of a number near unity and 10 raised to the appropriate (integer) power. (See notes on problem Solving, Appendix 8.1.)

Table 8.1-1 Thermodynamic and Transport Properties of Air at 20 atm[a]

Temperature, *T*	*Isentropic Exponent*, γ	*Molecular Weight*, *M*	*Viscosity*, $10^6\ \mu$	*Specific Heat* c_p	*Thermal Conductivity*, $10^6\ k$	*Prandtl Number*, Pr	*Enthalpy*, *h*	*Temperature*, *T*
300	1.4000		184	.2401	63	.706	−.5	300
400	1.3951		227	.2422	78		23.6	400
500	1.3865		265	.2461	92		48.0	500
600	1.3759		299	.2511	106		72.8	600
700	1.3646		331	.2568	121		98.2	700
800	1.3537		362	.2626	135		124.2	800
900	1.3439	28.964	391	.2681	148	.706	150.7	900
1000	1.3356		419	.2730	162	.705	177.8	1000
1100	1.3288		445	.2773	175	.705	205.3	1100
1200	1.3224		470	.2814	188	.705	233.2	1200
1300	1.3162		494	.2855	200	.704	261.6	1300
1400	1.3103		517	.2897	213	.704	290.3	1400
1500	1.3045		540	.2939	226	.703	319.5	1500
1600	1.2989		563	.2981	239	.703	349.1	1600
1700	1.2933		585	.3025	252	.702	379.2	1700
1800	1.2879		607	.3070	266	.702	409.6	1800

1900	1.2825	28.965	629	.3115	280	.701	440.6	1900
2000	1.2772	28.964	651	.3163	294	.700	471.9	2000
2100	1.2719	28.963	672	.3214	309	.699	503.8	2100
2200	1.2666	28.961	694	.3270	325	.698	536.2	2200
2300	1.2612	28.958	715	.3332	343	.696	569.3	2300
2400	1.2556	28.953	736	.3404	362	.692	602.9	2400
2500	1.2498	28.945	758	.3488	384	.688	637.4	2500
2600	1.2437	28.933	779	.3588	409	.683	672.7	2600
2700	1.2374	28.915	800	.3707	439	.676	709.2	2700
2800	1.2309	28.890	821	0.3850	473	0.669	747.0	2800
K	—	g/g-mole	g/cm-s	cal/(g · K)	cal/(cm · s · K)	—	cal/g	K

[a] While originally tabulated for application at 20 atm (e.g., gas-turbine operation), at these temperature levels all values tabulated apply with acceptable accuracy to air at *lower* pressures, provided oxygen dissociation is not important. It is also interesting to note that the dry "air" for which this table was constructed contained traces of CO_2 and Ar (with carbon and argon *element* mole fractions of 150.7 and 4681 ppm, respectively). The CO_2 causes the slightly negative absolute enthalpy tabulated at 300 K (cf. Figure 2.5-1). [From Poferl, D.J., and R. Svehla, NASA TN D-7488 (1973).]

(or 0.34 kg/m^3). Many convenient tabulations of the dynamic *viscosity* of air are available, although not all in the same units. A convenient set of values for *air*, but at 20 atm. (based on calculations of Poferl and Svehla (1973)), is reproduced here as Table 8.1-1. Since, in the perfect gas range, μ is *pressure independent* (cf. Section 3.2.5) we can use these values here even though our interest is in an application at $p = 1$ atm. Thus, we take:

$$\mu\{1000 \text{ K}, 1 \text{ atm.}\} \approx 4.19 \times 10^{-4} \text{ g cm}^{-1}\text{s}^{-1},$$

so that the momentum diffusivity may be estimated as:

$$\nu\{1000 \text{ K}, 1 \text{ atm.}\} = \frac{4.17 \times 10^{-4}}{3.41 \times 10^{-4}} = 1.23 \text{ cm}^2\text{s}^{-1}.$$

We can then evaluate:

$$\text{Re} \equiv \frac{U_\infty d_w}{\nu} = \frac{(1000)(5)}{(1.23)} = 4.06 \times 10^3 \qquad \text{(dimensionless)}.$$

For air flowing over a cylinder at this Reynolds' number, Figure 8.1-1 reveals that:

$$\overline{\text{Nu}}_h \approx 36 \qquad \begin{pmatrix}\text{perimeter-averaged dimensionless} \\ \text{heat-transfer coefficient}\end{pmatrix}$$

8.1.2 Heat-Transfer Rate

Reference to Eq. (8.1-5) reveals that only an estimate of the relevant air *thermal conductivity* k (at $T_f \equiv \frac{1}{2}(T_\infty + T_w)$) is needed to complete our prediction of the energy-flow rate for each meter length of cylinder. Table 8.1-1 is useful again since k is also p-independent in this "perfect gas" regime ($Z \approx 1$). Thus:

$$k\{1000 \text{ K}, 1 \text{ atm}\} \approx 1.62 \times 10^{-4} \text{ cal cm}^{-1}\text{s}^{-1}\text{K}^{-1}.$$

We can then evaluate $-\dot{q}_w$ for $L = 10^2$ cm using Eq. (8.1-5) (from which d_w is seen to explicitly cancel). Thus:

$$-\dot{q}_w = \pi L[k(T_\infty - T_w) \cdot \overline{\text{Nu}}_h\{\text{Re}, \text{Pr}\}] \tag{8.1-7}$$

or

$$-\dot{q}_w \cong (3.142)(10^2)[(1.62 \times 10^{-4})(1200 - 800) \cdot (36)] = 0.733 \times 10^3 \text{ cal s}^{-1}.$$

Since 1 Watt $\equiv$ 1 Joule/s and 4.184 Joule = 1 cal, it is easy to convert this figure to

Watts (or kilowatts); thus,

$$0.733 \times 10^3 \frac{\text{cal}}{\text{s}} \times \frac{4.184\ \text{J}}{1\ \text{cal}} \times \frac{1\ \text{kW}}{10^3\ \text{J/s}} = 3.07\ \text{kW}.$$

We conclude that, under the specified conditions, each meter of heat-exchanger tube would extract[7] energy from the combustion gases at the rate of about 3.1 kW (answer to Problem 8.1).

8.2 HEAT-EXCHANGER TUBE-FOULING RATE PREDICTIONS DUE TO ASH ACCUMULATION

The principles expounded in Chapters 6 and 7 allow us to make a corresponding prediction of the *mass-transfer rate*, provided we can assume that the laws of particle transport are analogous to those of energy transport. This will be true only if the particles can be treated like a "heavy gas" constituent governed by a Brownian diffusion law analogous to Fourier's heat-flux law. As usual, let us first proceed on this basis, and then quantitatively investigate these assumptions *a posteriori.*

For each of the two particle sizes the above-mentioned assumptions lead us to expect the functional relationship:

$$\frac{-j_{\text{p,w}}/(\pi d_{\text{w}} L)}{[D_{\text{p}} \rho(\omega_{\text{p},\infty} - \omega_{\text{p,w}})/d_{\text{w}}]} \equiv \overline{\text{Nu}}_m\{\text{Re}, \text{Sc}\}, \tag{8.2-1}$$

where $\overline{\text{Nu}}_m$ is the mass-transfer Nusselt number at the prevailing Re and *Schmidt number* $\text{Sc} \equiv \nu/D_{\text{p}}$, again *for a circular cylinder in cross-flow*. Relevant estimates of Sc are readily obtained from Figure 3.4-1, albeit at 1500 K rather than 1000 K. For $d_{\text{p}} = 10^{-1}$ μm we see that $\text{Sc} \approx 3 \times 10^4$, a value which should also prove sufficiently accurate at 1000 K.[8] The situation is somewhat more complicated for $d_{\text{p}} =$ 20 μm, for which $\text{Sc} \approx 10^8$ at 1500 K. For such particles, however, $d_{\text{p}} > \ell$, and we can extrapolate to 1000 K using the Stokes-Einstein law (Eq. (3.4-14)) which implies that Sc will have the same T-dependence as μ^2 (or about $T^{1.4}$ for gases). Thus, we estimate that for $d_{\text{p}} = 20$ μm and $T = 1000$ K:

$$\text{Sc} \approx 10^8 \cdot \left(\frac{1000}{1500}\right)^{1.4} \approx 0.567 \times 10^8.$$

In the absence of $\overline{\text{Nu}}_m\{\text{Re}, \text{Sc}\}$ data in this large Sc-range (3×10^4 and 0.567×10^8), we ask whether this can be estimated from $\overline{\text{Nu}}_h\{\text{Re}, \text{Pr}\}$ for $\text{Pr} = \text{Pr}_{\text{air}} \approx 0.7$ shown

---

Our sign convention is that $\dot{q}_{\text{w}}$ is positive when heat is *added* to the fluid (that is, $\dot{q}''_{\text{w}}$ is in the direction of the outward surface normal). Heat transfer *to* the cylinder corresponds to a *negative* $\dot{q}_{\text{w}}$.

[8] When particles act like heavy molecules ($d_{\text{p}} \ll \ell$) the diffusivity *ratio* Sc is virtually temperature-independent.

in Figure 8.1-1. This is not unreasonable, since the analogy between energy and mass transfer was shown in Chapters 6 and 7 to imply that $\overline{\text{Nu}}_h\{\text{Re}, \text{Pr}\}$ and $\overline{\text{Nu}}_m\{\text{Re}, \text{Sc}\}$ are really the same function! However, only $\overline{\text{Nu}}\{\text{Re}, 0.7\}$ is available (Figure 8.1-1), whereas we need $\overline{\text{Nu}}$ at values of the second argument in excess of 10^4. But this can be accomplished by noting that the dependence of $\overline{\text{Nu}}_h$ on Pr is known to be well-approximated by the proportionality $\text{Pr}^{1/3}$ for $\text{Pr} \geqslant 0.7$ at sufficiently high Re values. Thus, for the submicron ash we estimate:

$$\overline{\text{Nu}}_m \approx 36\left(\frac{3 \times 10^4}{0.7}\right)^{1/3} = 1.26 \times 10^3.$$

Similarly, for the supermicron ash we formally estimate:

$$\overline{\text{Nu}}_m \approx 36\left(\frac{0.567 \times 10^8}{0.7}\right)^{1/3} = 1.56 \times 10^4.$$

To complete the calculations of $-j_{\text{p,w}}$ we evidently need $D_{\text{p}}\{1000\ \text{K}, 1\ \text{atm}\}$ and the ω_{p} values in the mainstream (given) and at the gas/tube surface interface. Regarding the latter, an argument identical to that given in Section 6.5.3 reveals that, if nearly each incident particle sticks upon diffusing "up to" the wall, then we can assume $\omega_{\text{p,w}} \ll \omega_{\text{p},\infty}$. The relevant values of D_{p} are computed from:

$$D_{\text{p}} = \frac{\nu}{\text{Sc}} = \begin{cases} \dfrac{1.23}{3 \times 10^4} = 4.1 \times 10^{-5}\,\dfrac{\text{cm}^2}{\text{s}} & (\text{for } d_{\text{p}} = 10^{-1}\ \mu\text{m}), \\[2ex] \dfrac{1.23}{0.567 \times 10^8} = 2.17 \times 10^{-8}\,\dfrac{\text{cm}^2}{\text{s}} & (\text{for } d_{\text{p}} = 20\ \mu\text{m}), \end{cases}$$

so that we can formally calculate $-j_{\text{p,w}}$ from:

$$-j_{\text{p,w}} = \pi L[D_{\text{p}}\rho\omega_{\text{p},\infty} \cdot \overline{\text{Nu}}_m\{\text{Re}, \text{Sc}\}]. \tag{8.2-2}$$

For $L = 10^2$ cm and the parameters corresponding to $d_{\text{p}} = 10^{-1}\ \mu$m, we obtain:

$$-j_{\text{p,w}} = (3.14)(10^2)(4.1 \times 10^{-5})(3.41 \times 10^{-4})(2 \times 10^{-4})(1.26 \times 10^3)\ \text{g} \cdot \text{s}^{-1}$$

or 1.1 μg/s (0.035 kg/year).[9] For $L = 10^2$ cm and the parameters corresponding to $d_{\text{p}} = 20\ \mu$m we formally obtain:

$$-j_{\text{p,w}} = (3.14)(10^2)(2.08 \times 10^{-8})(2.41 \times 10^{-4})(1 \times 10^{-2})(1.56 \times 10^4)$$

or 0.35 μg/s (or 0.011 kg/yr).

[9] These fouling rates can be converted to nominal deposit *thicknesses* if one estimates the deposit density and further assumes the deposit to be *uni*formly distributed around the perimeter. More accurately, the angular distribution of deposit thickness could be estimated using $\text{Nu}_h\{\theta; \text{Re}, \text{Pr}\}$ or $\text{Nu}_m\{\theta; \text{Re}, \text{Sc}\}$ data (see, e.g., Figure 5.3-3).

This does *not* complete the solution to Problem 8.2 since it remains for us to defend our basic assumptions. Here we sequentially examine the assumptions that the particles (a) are not appreciably influenced by thermal forces within the gaseous boundary layer, and (b) behave like a constituent of the gas mixture—at most "diffusing" relative to the gas motion.

8.2.1 Evaluation and Inclusion of the Role of Particle Thermophoresis

Despite the simultaneous presence of a *temperature* gradient, we have thus far assumed that convection and *Brownian diffusion* are the primary mechanisms of particle transport to the heat-exchanger tube. However, as pointed out in Section 6.2.5.3, due to the force experienced by a small particle in a nonisothermal gas, there will inevitably be a thermophoretic contribution $\rho\omega_p(\mathbf{c}_p)_T$ to the particle mass flux, where:

$$(\mathbf{c}_p)_T = (\alpha_T D)_p \cdot [-(\mathbf{grad}\ T)/T]. \tag{8.2-3}$$

A remarkable finding from the kinetic theory of gas/particle mixtures (Waldmann (1966)) is that even though the ash particles considered here have diameters large enough to cause the Brownian diffusivity, D_p, to fall well below the gas momentum diffusivity ν_g (see Figure 3.4-1), the *thermophoretic diffusivity* $(\alpha_T D)_p$ appearing in Eq. (8.2-3) for the thermophoretic drift speed remains of the order of $(3/4)\nu_g$, independent of particle size (as long as the particle size remains negligible compared to the prevailing mean-free-path). Using selected results from Section 6.3.4, we show below that, far from being negligible, thermophoresis is expected to increase the previously estimated submicron particle capture rate by over a factor of 200. A similar calculation could be made for the supermicron 20 μm particle fraction; however, in the next section we show that this fraction cannot be self-consistently treated using a single-phase "diffusion" treatment—i.e., *inertial* impaction, rather than thermophoretically modified diffusion, will dictate the capture fraction for ash particles in that size range.

To proceed, we note that in Section 6.3.4 we estimated that phoresis (due to any cause) toward the collector would increase convective-diffusion rates by a factor of the order of:

$$F\{\text{suction}\} = \frac{\text{Pe}_{\text{suction}}}{1 - \exp\{-\text{Pe}_{\text{suction}}\}} \tag{6.3-19}$$

where, in the present case:

$$\text{Pe}_{\text{suction}} \equiv \frac{(-c_p)_T d_w}{D_p \overline{\text{Nu}}_{m,0}}. \tag{6.3-17}$$

Using Eq. (8.2-3) and the definition of the dimensionless surface-averaged heat-transfer coefficient $\overline{\mathrm{Nu}}_h$, we can now estimate the thermophoretic drift speed toward the heat-exchanger tube as follows:

$$(-c_{\mathrm{p}})_T = (\alpha_T D)_{\mathrm{p}} \cdot \left|\frac{\overline{\mathbf{grad}\, T}}{T}\right| = (\alpha_T D)_{\mathrm{p}} \cdot \frac{(-\bar{\dot{q}}''_{\mathrm{w}}/k)}{T_{\mathrm{w}}} = (\alpha_T D)_{\mathrm{p}} \frac{\overline{\mathrm{Nu}}_h}{d_{\mathrm{w}}} \cdot \frac{T_\infty - T_{\mathrm{w}}}{T_{\mathrm{w}}}.$$

Inserting this result into Eq. (6.3-17), we find that:

$$(\mathrm{Pe}_{\mathrm{suction}})_T = (\alpha_T)_{\mathrm{p}} \cdot \frac{\overline{\mathrm{Nu}}_h}{\overline{\mathrm{Nu}}_{m,0}} \cdot \frac{T_\infty - T_{\mathrm{w}}}{T_{\mathrm{w}}}. \tag{8.2-4}$$

However, since we estimated the mass-transfer coefficient $\overline{\mathrm{Nu}}_{m,0}$ from $\overline{\mathrm{Nu}}_h$ data (Figure 8.1-1) in the first place, this expression can be rewritten in the following equivalent form:

$$(\mathrm{Pe}_{\mathrm{suction}})_T = (\alpha_T \mathrm{Le})_{\mathrm{p}} \cdot \frac{\mathrm{Sc_p}^{2/3}}{\mathrm{Pr}^{2/3}} \cdot \left(\frac{T_\infty - T_{\mathrm{w}}}{T_{\mathrm{w}}}\right) \tag{8.2-5}$$

seen to be Reynolds'-number insensitive. Here $(\alpha_T \mathrm{Le})_{\mathrm{p}}$ is the ratio of the thermophoretic particle diffusivity to the heat diffusivity of the combustion products, a ratio estimated to be about 0.42 for 10^{-1} μm diameter SiO_2-particles in 1500 K, 1 atm gases (Rosner and Fernandez de la Mora (1982). Since this ratio is unlikely to change much in the temperature interval between 1500 K and 1000 K, we now evaluate:

$$(\mathrm{Pe}_{\mathrm{suction}})_T = (0.42) \cdot \frac{(3 \times 10^4)^{2/3}}{(0.705)^{2/3}} \cdot \left(\frac{1200 - 800}{800}\right) = 256,$$

which, *via* Eq. (6.3-19) leads immediately to the estimate $F\{$thermophoretic suction$\}$ ≈ 256, hardly negligible! While correct as to order-of-magnitude, for diffusing species with such large $\mathrm{Sc_p}$-values, we have recently shown that this augmentation factor should be reduced by a factor of the order of:

$$\exp\left\{-(\alpha_T \mathrm{Le})_{\mathrm{p}} \cdot \frac{T_\infty - T_{\mathrm{w}}}{T_{\mathrm{w}}}\right\} = \exp\left\{-0.42\left(\frac{1200 - 800}{800}\right)\right\} = 0.81$$

(Gökoĝlu and Rosner (1984)), making our improved estimate of $F\{$thermophoresis$\}$ about 207.[10] Thus, instead of an expected Brownian deposition rate of 10^{-1} μm diameter particles of *ca.* 0.035 kg/yr, we now expect *ca.* 7.2 kg/yr. These figures correspond to capture efficiencies (100 η_{cap}) of about 0.004 (without thermophoresis)

[10] If $\mathrm{Pe}_{\mathrm{suction}}$ has been evaluated using $\frac{1}{2}(T_\infty + T_{\mathrm{w}})$ in the denominator (rather than T_{w}), a similar result would have been obtained.

and 0.8% (with thermophoresis), respectively. Interestingly enough, because of the dominance of *thermophoresis* in determining the deposition rate of the submicron ash fraction, the expected angular distribution of deposition rate should be closer to that of $\mathrm{Nu}_h\{\theta; \mathrm{Re}, \mathrm{Pr}\}$ (cf. Figure 5.3-3) than that of $\mathrm{Nu}_m\{\theta; \mathrm{Re}, \mathrm{Sc}_p\}$.

We now turn to our evaluation of the assumption of "single-phase" (multi-component) flow, and show that, while indeed valid for the submicron (10^{-1} μm) diameter ash fraction considered in detail here, the capture of the 20 μm diameter fraction will instead be determined by *inertial* effects (cf. Figure 8.2-1) quite unrelated to those which determine the above-mentioned convective diffusion mass- (and heat-transfer) coefficients. The overall situation is therefore somewhat similar to that sketched in Figure 8.2-2 (after Rosner and Atkins (1982)), which displays how the particle capture fraction varies with particle diameter for a typical heat-exchanger tube in the cross-flow of pulverized coal combustion products. Note that, in contrast to our simple numerical example, real combustors produce a broad *distribution* of mainstream ash particle sizes (see second panel, Figure 8.2-2). Figure 8.2-2 clearly shows the over two-decade thermophoretic enhancement of submicron particle capture (albeit for the case $T_w/T_\infty = 0.6$ rather than our present $800/1200 = 0.667$), and the dramatic effects of particle *inertia* on η_{cap}, which set in for mainstream particles above about 10 μm diam.

8.2.2 Defense of "Single-Phase" Flow Assumption

In accord with the discussion of Chapter 6 we must evaluate whether the particle Stokes' number, Stk, is small enough to justify our "single-phase" treatment of

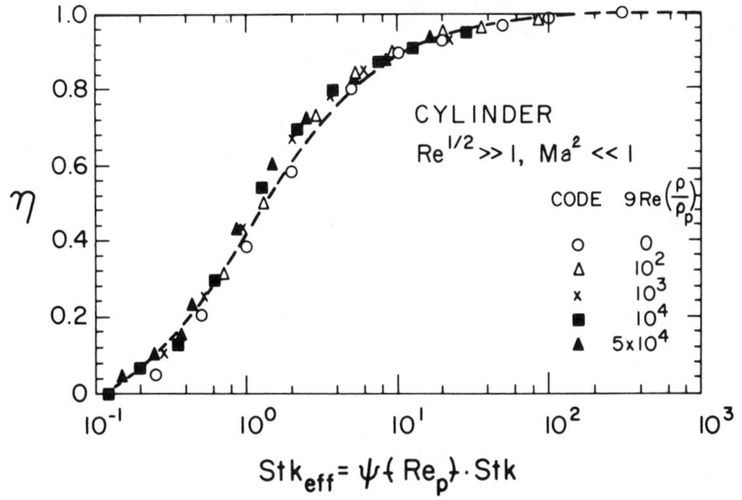

Figure 8.2-1 Correlation of inertial capture of particles by a circular cylinder in cross-flow (Israel and Rosner (1983)).

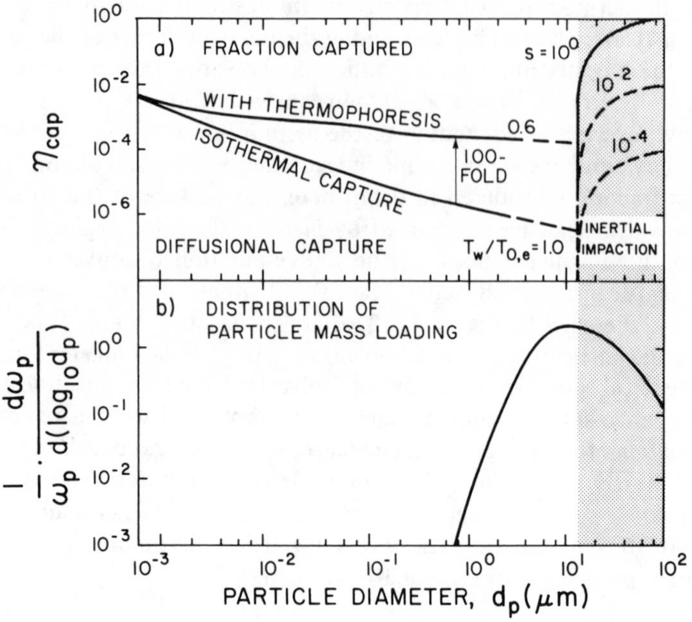

Figure 8.2-2 Representative heat-exchanger tube fouling-rate conditions; (a) particle size dependence of the capture fraction, η_{cap}; (b) size distribution of mainstream particle mass loading. Overall mass fouling rate will be proportional to the integral of the product of these two functions.

particle deposition. Here Stk $\equiv t_p/t_{flow}$, where $t_{flow} = \frac{1}{2}d_w/U_\infty$ and t_p can be calculated either from Eq. (6.2-7) (together with a slip-flow (noncontinuum) drag correction factor) or, more directly, from:

$$t_p = \frac{D_p}{(k_B T/m_p)}. \tag{8.2-6}$$

Thus, for each particle size we will evaluate the "dynamical" (inverse) Damköhler number:

$$\text{Stk} = \frac{\pi}{6}\frac{D_p}{k_B T}\cdot\frac{\rho_p d_p{}^3 U_\infty}{\frac{1}{2}d_w} \tag{8.2-7}$$

at the mainstream temperature $T_\infty = 1200$ K. For the submicron ($d_p = 10^{-1}$ μm) particles, D_p(1200 K, 1 atm) is estimated from the hard-sphere kinetic theory result (Eq. 3.4-8)):

$$D_p \approx (3.93 \times 10^{-5})\left(\frac{1200}{1000}\right)^{3/2} = 0.517 \times 10^{-4}\ \frac{\text{cm}^2}{\text{s}},$$

whereas, for the supermicron ($d_p = 20\ \mu m$) particles $D_p\{1200\ K, 1\ atm\}$ is estimated from:

$$D_p \approx (2.08 \times 10^{-8})\left(\frac{1200}{1000}\right)^{0.3} = 2.2 \times 10^{-8}\ \frac{cm^2}{s}$$

(consistent with $D_p \sim T/\mu$ (Eq. (3.4-14)) and $\mu \sim T^{0.7}$). Inserting the remaining estimate $\rho_p = 2.2\ g\ cm^{-3}$ for SiO_2 and the Boltzmann const ($k_B = 1.38 \times 10^{-6}$ erg particle^{-1} K^{-1}), we then obtain:

$$\text{Stk} = \begin{cases} 1.44 \times 10^{-4} & (d_p = 10^{-1}\ \mu m), \\ 0.462 & (d_p = 20\ \mu m). \end{cases}$$

Interestingly enough, while Stk is acceptably small for $d_p = 10^{-1}\ \mu m$, Stk for $d_p = 20\ \mu m$ is *above the critical value* (1/8) *for "inertial impaction"*[11] (see Figure 8.2-1 or 6.6-1). This means that 20 μm particles cannot "follow" gas streamlines around the cylinder and, hence, our previous formal estimate of $-j_{p,w}$ for 20 μm particles *cannot* be defended *a posteriori*. However, using Figure 8.1-2, we can roughly estimate a *capture efficiency*, η_{cap} (defined in Chapter 6), of about 20% for this size particle. This corresponds to an *impaction* rate of:

$$(-j_{p,w})_{\text{impaction}} = \eta_{cap} \cdot [\rho_\infty \omega_{p,\infty} U_\infty d_w L], \tag{8.2-8}$$

where the bracketed quantity is the ash mass-flow rate through the (upstream) projected area of the tube. Inserting the appropriate numbers:

$$(-j_{p,w})_{\text{impaction}} = (2 \times 10^{-1})(2.841 \times 10^{-4})(1 \times 10^{-2})(10^3)(5)(10^2).$$

The result is now 0.28 g s^{-1} (exceeding our previous "diffusional" estimate by a factor of 8.2×10^5!). We conclude that if all of the 20 μm particles that strike the heat-exchanger tube *stick*, the fouling rate per meter of tube could be as high as 0.28 g/s (9 metric tons/year!).[12] This potentially catastrophic fouling rate might, for example, dictate the use of still smaller velocities in the heat-exchanger section to reduce Stk (Eq. (8.2-7)) to "sub-critical" values.

EXERCISES AND DISCUSSION QUESTIONS

8.1 By stating and, where possible, numerically evaluating the relevant *dimensionless ratios*, quantitatively defend the following approximations made in solving

---

[11] It should *not* be assumed that *inertial* effects occur only above Stk_{crit}, since even in the absence of actual wall *impaction* (i.e., below Stk_{crit}), inertial effects can alter the local distribution of particle concentrations "available" for Brownian diffusion or thermophoresis (see, e.g., Fernandez de la Mora, J., and D.E. Rosner, *J. Fluid Mech.* **125**, 379–395 (1982)).

[12] One "metric ton" = 10^3 kg = 2200 lb.

Problems 8.1, 8.2:

a. continuum approximation,
b. neglect of viscous dissipation,
c. neglect of natural convection (Section 5.5.3).

With additional information it would also be possible to defend (or relax) the:

d. neglect of mainstream turbulence (Section 5.6.4),
e. neglect of effects of neighboring heat-exchanger tubes (Exercise 7.1),
f. neglect of deposition on the cylinder by "eddy impaction" (Section 6.6.3),
g. neglect of radiative energy gain (from gases and suspended ash).

8.2 Suppose, for structural-design reasons, you wanted to estimate the *drag force* per unit length of the heat-exchanger tube. Could this also be done based on the $\overline{\mathrm{Nu}}_h\{\mathrm{Re}, \mathrm{Pr}\}$-relation (Figure 8.1-1) using the so-called "analogy" between *momentum* and energy transfer?

8.3 In estimating the submicron particle (ash) mass-transfer rate we exploited the (a) mass-energy transfer "analogy" (i.e., $\overline{\mathrm{Nu}}_h\{\mathrm{Re}, \mathrm{Pr}\}$ and $\overline{\mathrm{Nu}}_m\{\mathrm{Re}, \mathrm{Sc}\}$ are the same functions). Can this analogy be derived from "dimensional analysis"? Similitude analysis? (b) proportionality $\overline{\mathrm{Nu}}_h \sim \mathrm{Pr}^{1/3}$ for large Prandtl numbers. Can this proportionality be derived from "dimensional analysis"? Similitude analysis?

8.4 Verify the effective "capture efficiency" figures quoted in the text for *submicron* particle collection by the heat-exchanger tube under the conditions treated above. How does it compare to that estimated above for inertial capture of 20 μm particles?

8.5 Show that the particle Stokes' number (Eq. (8.2-7)) can be re-expressed in terms of the prevailing values of the Mach, Reynolds', and Schmidt numbers in accordance with:

$$\mathrm{Stk} = \frac{2\gamma(\mathrm{Ma})^2}{(m_g/m_p)(\mathrm{Sc})_p(\mathrm{Re})},$$

where $\gamma \equiv c_p/c_v$ and m_g/m_p is the ratio of individual gas molecule-to-particle mass. Can analogous momentum nonequilibrium phenomena be observed for mixtures containing trace amounts of a high-molecular-weight *vapor* (e.g., $WF_6(g)$) in a low-molecular-weight carrier gas (e.g., $H_2(g)$) under continuum flow conditions (see, e.g., Fernandez de la Mora, J. (1982)?

REFERENCES

Fernandez de la Mora, J., *Phys. Review A* **25**, No. 2, pp. 1108–1122 (1982); see, also, *J. Chem. Phys.* **82** (7), 3453–3464 (1985).

Fernandez de la Mora, J., and D.E. Rosner, *J. Fluid Mechanics* **125**, pp. 379–395 (1982).

Flagan, R.C., and S.K. Friedlander, in *Recent Developments in Aerosol Science* (J. Davies, ed.). New York: J. Wiley (1978), p. 25.

Gökoĝlu, S.A., and D.E. Rosner, *Int. J. Heat Mass Transfer* **27**, pp. 639–645 (1984).

Israel, R., and D.E. Rosner, *J. Aerosol Sci. and Technol.* **2**, pp. 45–51 (1983).
McAdams, W.H., *Heat Transmission* (Third Ed.). New York: McGraw-Hill (1954).
Poferl, D.J., and R.A. Svehla, *NASA TN D-7488* (Dec., 1973).
Rosner, D.E., and R.A. Atkins, in *Fouling and Slagging Resulting from Impurities in Combustion Gases* (R. Bryers, Ed.). New York: Engrg. Foundation, pp. 469–492 (1983).
Rosner, D.E., and J. Fernandez de la Mora, *ASME Trans.-J. Engrg. for Power* **104**, pp. 885–894 (1982).
Waldmann, L., in Chapter 6 of *Aerosol Science* (C.N. Davies, Ed.). New York: Academic Press; p. 137 (1966).

Appendix 8.1

Recommendations on Problem Solving

1. Upon examining the problem statement, write the relevant conservation (balance) principle(s) in a *general* form (don't prematurely specialize), using "suggestive" and/or conventional notation.
2. Sketch the physico-chemical situation and carefully select convenient control volume(s) for imposing the relevant conservation principle(s). Control surfaces should be chosen such that terms vanish or are readily evaluated along them; don't probe unnecessarily deeply (when a black-box approach will suffice)!
3. Specialize to the problem at hand by dropping terms which are either: (a) identically zero (such as "accumulation terms" in a steady-state situation), and (b) likely to be insignificant *compared to the terms retained* (e.g., "flow" work done by normal stresses other than thermodynamic pressure; gravitational potential-energy differences *compared to* internal energy changes accompanying chemical reaction, etc.). Maintain a cumulative list of all assumptions, expressed in the form of quantitatively testable, dimensionally and tensorially homogeneous inequalities.
4. Symbolically solve the resulting equations for the quantity of interest in terms of quantities presumed to be known, checking each step for reasonableness (dimensional homogeneity, limiting behavior, symmetry, etc.). Add (to your cumulative list) assumptions made for mathematical simplicity.
5. Insert numerical values (and their units) to obtain the result sought. Document all sources of numerical data and conversion factors (don't trust memory on obscure conversions). Simplify numerical work and order-of-magnitude estimation by using a power-of-ten notation such that all numbers are written as the product of a number of order unity and the appropriate power of ten (e.g., $980.2\ \text{cm s}^{-2} = 0.9802 \times 10^3\ \text{cm s}^{-2}$, etc.).
6. *A posteriori*, numerically evaluate the validity of all physico-chemical and/or mathematical approximations and simplifications made for the situation at hand. Relax those assumptions/simplifications which cannot be defended.
7. Don't rush off to the next problem! Instead, examine your solution for reasonableness; is your result in some ways surprising? suggestive? "obvious" (in retrospect)? Can the same result be obtained from a different viewpoint? (E.g., using different control volume(s)?) If a different result is obtained (in this

way) can you rationalize this (potentially instructive) disparity? What are some potentially important *implications* of your result? Consider your problem as embedded in a class of problems with somewhat different specified conditions. Was the problem you solved "singular"? "typical"? Has the present problem "taught you a lesson" worth remembering for the future?

Appendix 8.2

Outline of the Method of Finite Differences (MFD) for the Numerical Solution of Partial Differential (Field) Equations (PDEs) and Ancillary Boundary Conditions (BCs)

In the problems discussed in Chapter 8, to obtain the engineering estimates required, it was possible to make direct use of existing correlations of the results of prior experiments or calculations. For simple boundary-value problems (BVPs) requiring the *ab initio* solution of a PDE we have seen that *analytical* methods (e.g., combination-of-variables (Section 5.4.6), or separation-of-variables (Appendix 5.1)) are often available, which allow the desired results to be written in terms of previously tabulated special functions (sine{ }, cos{ }, erfc{ }, etc.). However, in many practical situations we now require direct solutions of the coupled, nonlinear PDEs governing the fields of momentum density, energy density, and species mass densities of interest (cf. Chapter 2), often for domains (and flow inlet/outlet locations) of quite complicated geometry. Indeed, paraphrasing Cebeci and Smith (1974), the subject of transport processes has been beset by the difficulty that our ability to write down the governing PDEs and BCs (Chapter 2) far outran our ability to solve them. This situation has, however, improved markedly owing to (i) the availability of high-speed/large-memory electronic digital computers, and (ii) the development of efficient *numerical* methods based on converting the governing PDEs and BCs to a coupled system of *algebraic* equations amenable to digital computation, often using matrix methods. For completeness, one class of such methods, called the "method of finite differences" (MFD), is outlined below. For details consult the references listed at the end of this appendix.

Basic Idea. Give up the notion of attempting to determine the dependent field quantity $u\{\mathbf{x}, t\}$—say, at "every" point within the domain D of independent variables. (There is an infinite number of such field values.) Accept instead the values of u at discrete "node" points within the domain; i.e., seek a finite number of values: $u_{i,j,k,n}$, where the indices $i, j, k\ (= 0, 1, 2, \ldots)$ denote discrete spatial locations in D, and $n\ (= 0, 1, 2, \ldots)$ denotes discrete locations in the time domain.

Question. How can the numerical values of $u_{i,j,k,n}$ be found such that the original PDE + BC + IC can be acceptably approximated?

Steps

1. Write the dimensionless PDE + BC + IC for the dependent variable(s) u in the most appropriate orthogonal coordinate system. The field equation is generally of the form:

$$\frac{\partial u}{\partial t} = \mathcal{N}\{u\} \qquad \text{(sometimes written } u_t = \mathcal{N}\{u\}\text{)},$$

where, in most problems arising in transport theory, $\mathcal{N}\{u\}$ will be a nonlinear (or quasi-linear) second-order (in space variables) *differential* operator.[13] The boundary conditions (BC) are frequently of the form

$$a_{\mathrm{B}}\left(\frac{\partial u}{\partial v}\right)_{\mathrm{B}} + b_{\mathrm{B}}u_{\mathrm{B}} + c_{\mathrm{B}} = 0,$$

where v is a length coordinate normal to the boundary B and a, b, c are coefficients independent of u itself. This BC includes the commonly encountered special cases $a = 0$ (u_{B} specified) and $b = 0$ (specified "flux"). Initial conditions (usually $u\{\mathbf{x}, 0\}$ must also be specified.

2. Lay down a "mesh" in both time and physical (transformed and/or normalized) space variables and accept as a solution the discrete values $u_{i,j,k,n}$, of the field density at the "node" points (provided the conditions implied by Steps 3, 4, and 6 below are satisfied).
3. Replace the PDE by an appropriate "difference analog" which will provide *algebraic* interrelations between the $u_{i,j,k,n}$. Difference analogs (which can be obtained by truncating Taylor series, integration formulae, or variational methods) are not unique; i.e., they exhibit important differences with respect to the local magnitude (truncation) and/or growth (or damping) of inevitable random numerical errors ("stability"). According to the so-called "equivalence theorem" of PDE-numerical analysis, a numerical method which locally approximates the PDE derivatives with mesh refinement ("consistency"), and which is free from the exponential amplification of random numerical errors (i.e., a globally "stable" method) will converge on the solution to the originally posed PDE-BVP.
4. Replace the BC by a difference analog; in all, sufficient algebraic equations must now be available to calculate all of the required $u_{i,j,k,n}$, frequently by "marching" in t-space. The difference analog of the BC may be complicated if the boundary B of the domain D is not a simple shape (e.g., not a coordinate surface).
5. Solve the resulting algebraic system of equations for the $u_{i,j,k,n}$ using matrix methods/notation to organize the calculations. The solution procedure is simplest if the equations are linear and the analog expressions for $\mathcal{N}\{u\}$ at most

[13] Only in the simplest problems (linear, quiescent medium, no sources) is $\mathcal{N} = \mathrm{div}(\mathbf{grad}) \equiv \nabla^2$. For problems in radiative transfer, spatial *integrals* can appear in $\mathcal{N}\{u\}$.

link variables at three adjacent mesh points. The solution procedure is trivial if an "explicit method" is used since each $u_{i,j,k,n+1}$ is calculated directly from the "known" (previously determined) values $u_{i,j,k,n}$; however, such methods are usually *unstable* unless the mesh is sufficiently fine. "Implicit" methods, i.e., FD algorithms for which one must solve implicit algebraic equations for the nodal unknowns, are stable (with respect to the growth of cumulative errors) for any mesh size, even if the source term in the DE is extremely sensitive to the dependent variable (so-called "stiff" equation).

6. Repeat the above process with a finer or coarser mesh; accept the result only if the corresponding u-values (and/or their integrated consequences) are insensitive to mesh size. (*Remark*: The mesh size cannot be so small that inevitable roundoff errors cause large relative errors in computing first and second differences of values at adjacent mesh points.) Ordinarily computation time (and cost) increase rapidly with mesh fineness, and cost limitations set in before roundoff errors do.

Comments

1. There are now many excellent books available on numerical methods for solving PDEs (the selection might well be made based on the relevance of the illustrations to the particular interests of the reader). An excellent introductory reference for the MFD is still Von Rosenberg, D. (1969). (See reference list below.)
2. For many commonly encountered PDE-BVPs pre-programmed "subroutines" are available on call. An example is Harwell Subroutine DPØLA (Crank–Nicolson method) for quasi-linear "parabolic" problems of the form: $u_t = a(u, \ldots)u_{xx} + bu_x + cu + d$, with linear two-point boundary data. (Here, the subscripts t, xx, and x denote *derivatives* with respect to the indicated variable).
3. For problems involving irregularly shaped domains and/or steep local gradients, an alternative "discretization" procedure, called the method of finite *elements* (MFE) is also in widespread use (see, e.g., Appendix 8.3 and Huebner, K.H. (1975)); however, the ability to rapidly compute and use "body-fitted" orthogonal coordinates has also permitted the MFD to be effectively applied to irregularly shaped domains (see, e.g., Thompson *et al.* (1982)). More generally, one can use MFD to "simultaneously" solve for the fields of interest *and* an optimal (nearly orthogonal) coordinate grid, "adaptively" distributing the grid points in accordance with a criterion that accounts for the locations of maximum gradients in the dependent variable (see, e.g., Shyy, Tong, and Correa (1985)).
4. Regions of a flow field in which convection dominates diffusion (momentum, energy, and/or mass) are approximately described by PDEs *without* second-order derivative terms (called locally "hyperbolic" PDEs). Within such regions there are local directions (called characteristic directions in the space of independent variables) *along* which the dependent variables can be shown to satisfy ODEs. This property is numerically exploited in the "method-of-characteristics" (MC) (see, e.g., Abbott (1966)). In effect, of all possible local

coordinate systems, an optimal set (called "characteristic coordinates") is calculated and then used to advance the solution. When applicable, the MC is usually preferable to the MFD, especially for highly nonlinear problems. However, usually boundary layers (within which diffusion *cannot* be neglected) prevent global applicability of a MC.

REFERENCES

Abbott, M.B., *An Introduction to the Method of Characteristics.* London: Thames and Hudson (1966).

Ames, W.F., *Numerical Methods for Partial Differential Equations* (Second Ed.). New York: Academic Press (1977); London: T. Nelson & Sons (1977).

Huebner, K.H., *The Finite Element Method for Engineers.* New York: J. Wiley (1975).

Patankar, S.V., *Numerical Heat Transfer and Fluid Flows.* New York: McGraw-Hill (1980).

Shyy, W., S.S. Tong, and S.M. Correa, "Numerical Recirculating Flow Calculation Using a Body-Fitted Coordinate System," *J. Numerical Heat Transfer* **8**, 99–113 (1985); see, also, AIAA Paper AIAA-84-1381.

Thompson, J.F., Z.V.A. Warsi, and C.W. Mastin, *J. Computational Phys.* **47**, 1–108 (1982); see, also, *AIAA J.* **22**, 1505–1523 (1984).

Von Rosenberg, D., *Methods for the Numerical Solution of Partial Differential Equations.* American Elsevier (1969).

Appendix 8.3

Outline of the Method of Finite Elements (MFE) for the Numerical Solution of PDEs on Domains of Complicated Shape

Steps (Steady-State Illustration)

1. If available,[14] first write a valid *variational principle* for the governing PDE and BCs on the domain of interest. For linear problems, the corresponding Lagrange functional (whose extremum is to be found) usually involves an integrand (Lagrange "density") at most quadratic in the dependent variable u and its space derivatives (see footnote below).
2. Subdivide the domain into convenient "finite elements" (e.g., tetrahedra for three dimensions) with dependent variables u_{ijk} at the *nodes* as the basic unknowns. Use some interpolation procedure to define u *everywhere else.*
3. Use the variational principle, together with the fact that each finite element contributes additively to the Lagrange integral, to derive a set of *algebraic* equations interrelating the nodal values u_{ijk}. This leads to sparse matrix equations for the u_{ijk}.
4. The solution to these algebraic equations defines the MFE-discretized solution to the original continuum boundary-value problem.

Advantages (Relative to Finite Differences)

1. Ease with which shapes and sizes of the finite elements (defined by the positions of the node points) can be chosen. This allows elliptic PDEs within complex geometries to be handled, allowing for steep local gradients.

[14] For example, the inhomogeneous linear elliptic (Poisson) PDE: $\mathrm{div}(\mathbf{grad}\, u) = p\{\mathbf{x}\}u$ with $u\{\mathbf{x}_B\}$ specified (say) has a unique solution such that:

$$\int_V \tfrac{1}{2}\{(\mathbf{grad}\, u)^2 + pu^2\}\, dV$$

is a *minimum.* This integral is called the *Lagrange functional* or *Dirichlet integral* (associated with the Poisson PDE), and the relevant Lagrange "density" is the integrand: $(1/2)\{(\mathbf{grad}\, u)^2 + pu^2\}$. Approximation methods which exploit extremum principles (such as the above-mentioned minimum condition) are called *variational* methods; however, valid variational principles are not available for most PDEs (see Remark 1).

2. Ease of handling *boundary conditions via* the basic variational integral, and less truncation error for gradient conditions.

Remarks

1. Often a variational principle is *not* available, in which case a "weighted PDE-residual" procedure (Appendix 8.4) is used to generate similar algebraic equations (i.e., a weighted local error of the trial solution is "minimized" within each finite element). In such cases the trial function $\tilde{u}$ is defined *via* the nodal values u_{ijk} and *interpolation functions* of position ("local coordinates"); the domain residual within each finite element is made orthogonal to each interpolation formula.
2. When the mesh, domain, geometry, and trial functions are particularly simple, MFE becomes equivalent to the method of finite differences (MFD; see Appendix 8.2).
3. As in finite-difference methods, nonlinearities are handled by linearization at each stage of the matrix computations, using convergent *iterative* methods.

REFERENCES

Becker, E.B., G.F. Carey, and J.T. Oden, *Finite Elements—An Introduction*, Vol. 1. Englewood Cliffs, NJ: Prentice-Hall, Inc. (1981).

Finlayson, B., *The Method of Weighted Residuals and Variational Principles*, Vol. 87 of *Mathematics in Science and Engineering*. New York: Academic Press (1972).

Huebner, K.H., *The Finite-Element Method for Engineers*. New York: J. Wiley (1975).

Mitchell, A.R., and R. Wait, *Finite-Element Method in Partial Differential Equations*. New York: J. Wiley (1977).

Strang, G., and G. Fix, *An Analysis of the Finite-Element Method*. Englewood Cliffs, NJ: Prentice-Hall (1973).

Wilson, E.L., and R.W. Nickell, "Applications of Finite-Element Method to Heat-Conduction Analysis," *Nuclear Engineering and Design*, Vol. 4, 276–286 (1966).

Zienkiewicz, O.C., *The Finite-Element Method in Engineering Science*. New York: McGraw-Hill (1971).

Appendix 8.4

Outline of the Method of Weighted Residuals (MWR) for the Approximate Solution of Partial Differential (Field) Equations and Ancillary Conditions (BCs, ICs)

Efficient approximate methods have been developed which, in effect, reduce the problem of solving nonlinear PDE-BVPs to the solution of either ODEs or algebraic equations (without necessarily introducing the nodal values u_{ijkn} as basic unknowns). An important class of such methods, generically called "Weighted Residual" (WR) methods [Finlayson (1972)] is particularly well suited to obtain approximate semi-analytical or efficient numerical solutions to transport problems in which:

Only integrated properties are needed (as opposed to local accuracy);

The solution is not expected to exhibit steep local spatial gradients;

The spatial domain possesses a high degree of symmetry.

The WR-strategy may be summarized as follows: Consider the problem of finding $u\{\mathbf{x}, t\}$ satisfying the PDE: $u_t = \mathcal{N}\{u\}$ (where $\mathcal{N}\{u\}$ is typically a nonlinear, second-order spatial differential operator) on some spatial domain, V, with specified initial conditions (ICs) and boundary conditions (BCs). Suppose one introduces an approximate solution $\tilde{u}\{\mathbf{x}, t\}$ which satisfies the BCs, and contains N *specified* "basis-," trial-, or shape-functions of the spatial coordinates $\mathbf{x}$ (defined over the entire domain V) and N as yet *undetermined* functions of time. By satisfying the PDE (expressing local conservation) as accurately as possible (globally, not necessarily locally) it is possible to obtain N ODEs (and their ICs) allowing the N previously undetermined functions of time to be found, where N may be as small as 1. Satisfying global conservation conditions is usually discussed in terms of certain (weighted-) integrals of the local "domain-(interior) residual" or local error:

$$\varepsilon_V = \mathcal{N}\{\tilde{u}\} - \frac{\partial \tilde{u}}{\partial t}.$$

[ε_V would clearly vanish everywhere if the approximating function $\tilde{u}(\mathbf{x}, t)$ were indeed exact.] A subclass of such methods (called the "least-squares" method) simply minimizes the integral $\int_V \varepsilon_V^2 \, dV$. More generally, one can force any N "weighted-domain residuals" to vanish;[15] that is,

$$\int_V w_i \varepsilon_V(\tilde{u}) \, dV = 0 \qquad (i = 1, 2, \ldots, N),$$

where the weighting functions $w_i(\mathbf{x})$ can be smooth functions of position, unrelated to the trial functions, chosen to emphasize locations where accuracy is felt to be especially important.[16] Alternatively, the weighting functions can be the same as the trial spatial functions themselves, and members of a complete set of functions in the domain V (Galerkin method), or even Dirac delta functions centered about preselected points in the domain [equivalent to simply demanding that the domain residual (interior error ε_V) vanish only at these selected ("collocation"-) points $\mathbf{x}_i$ $(i = 1, 2, \ldots, N)$, thereby avoiding time-consuming quadratures].

While such WR-methods can be computationally efficient (especially when considerable knowledge of the spatial dependence and symmetry of the solutions is "built in" to the trial spatial functions), in general they suffer from a lack of information about their convergence properties (e.g., when N increases) and, hence, accuracy. Usually it is wise to demonstrate at least one of the following properties:

Additional terms (incrementing N) lead to only a small change in the calculated quantity of interest (QOI).

Alternative but qualitatively similar trial functions (spatial dependencies), given only a small change in the QOI.

Your method, when applied to a well-known special case (for which an exact solution may be available) gives acceptable accuracy for a similar QOI.

A pedagogically interesting simple example is provided by the transient linear (heat-)diffusion problem treated exactly in Section 5.4.6. In this case the ultra-simple approximating function:

$$\frac{T_w - T(x, t)}{T_w - T_0} = \begin{cases} x/\delta_h(t) & 0 \leqslant x \leqslant \delta_h(t), \\ 1 & x \geqslant \delta_h(t) \end{cases}$$

[15] When the trial functions do not satisfy the BCs, both weighted *domain*-residuals, ε_V, and *boundary* residuals, ε_S, are forced to vanish.

[16] In one space (-like) dimension, if the weighting functions are simply $x^m (m = 0, 1, 2, \ldots)$ chosen from the set of all integral powers of the coordinate x (measured from the boundary or symmetry axis), such a MWR is called a "moment" method since successive "moments" of the interior residual ε_V are forced to vanish. If one does not go beyond $m = 0$, then the approximate ("integral"-) method has (merely) satisfied the PDE "on the average" over the entire domain.

(involving only *one* undetermined function of time; *viz.*, the thermal penetration depth, $\delta_h\{t\}$, which vanishes at $t = 0$), combined with a macroscopic energy conservation condition over the "entire" domain $0 \leqslant x \leqslant \delta_h\{t\}$, leads to an ODE for $\delta_h\{t\}$ with the analytic solution: $\delta_h = 2(\alpha t)^{1/2}$. While local relative errors in the temperature profile are excessive, the corresponding wall *heat fluxes* are found to be (under-)predicted by only 11.4 percent (cf. exact solution)!

Comments

1. The theoretical basis for systematically choosing the weighting functions $w_i\{\mathbf{x}\}$ is as follows: If the weighting functions $w_i\{\mathbf{x}\}$ are members of a "complete" set of functions on the domain V, then the only function $\varepsilon_V\{\tilde{u}\}$ for which *every* weighted integral of $w_i\varepsilon_V\{\tilde{u}\}$ vanishes is, in fact, $\varepsilon_V\{\tilde{u}\} = 0$. Of course, in any practical realization of MWR, only the first few weighted residuals of ε_V are forced to vanish; hence the associated $\tilde{u}\{\mathbf{x}, t\}$ will be an *approximate* solution of the PDE-BVP.
2. Few general statements can be made about the relative merits of the above-mentioned variants of MWR. Indeed, there are deep interrelations between these variants, usually demonstrated for simple linear problems. Perhaps the Galerkin method [1915] has been investigated most extensively, and has the strongest theoretical foundation (since it has the property that, if the exact solution is contained in the trial (basis-)functions, the Galerkin method will find it [Finlayson (1972)]).
3. MWR is also used "locally" in implementing the method of finite elements (MFE) (see Appendix 8.3) when (a) a variational principle does not exist for the PDE-BVP under investigation (the usual situation for nonlinear transport problems) and (b) MFE is desirable because of the complexity of the overall domain V, and/or the expected presence of strong local spatial nonuniformities. In such cases, instead of being defined over the entire domain V, the trial (shape-) functions are defined only within each *finite element* (into which the entire domain has been judiciously subdivided) and can be considered to vanish elsewhere. Thus, the Galerkin method can be used to approximately satisfy the PDE "over" each finite element, and the "local solution" can be explicitly rewritten in terms of relevant adjacent nodal values $u_{i,j,k,n}$ of the dependent variable.

REFERENCES

Finlayson, B., *The Method of Weighted Residuals and Variational Principles*, Vol. 87 of *Mathematics in Science and Engineering*. New York: Academic Press (1972).

Goodman, T.R., "The Heat Balance Integral Method," in Vol. 1, *Advances in Heat Transfer* (T.F. Irvine and J.P. Hartnett, eds.). New York: Academic Press (1964); pp. 51–122.

Stewart, W.E., "Simulation and Estimation by Orthogonal Collocation," *Chemical Engineering Education* (Fall, 1984), pp. 203–212.

EXERCISES (Appendixes 8.2, 8.3, and 8.4)

1. For what sort of energy/mass/momentum transport problems would recourse be made to numerical methods (e.g., MFD)? (See Figure 4.4-5, Spectrum of Methods.)
2. What is meant by (a) truncation error? (b) round-off error? (c) convergence? (d) "stability"? How do these considerations influence the choice of algorithm and mesh spacing? What other factors govern the choice and uniformity of mesh spacing?
3. How are difference analog equations obtained? Outline *one* such method. Is there more than one "correct" difference analog of a particular PDE + BC + IC? Write two possible difference analogs for the transient heat-conduction equation in one dimension, one being "explicit" and one being "implicit," and discuss their relative merits. How is the Schmidt (graphical) method (see, e.g., MFD Exercise 6.7) related to the explicit analog given above?
4. Why are matrix methods convenient for organizing and carrying out numerical calculations of the discretized (algebraic) equations? Why is it desirable to linearize (*via* estimation methods) nonlinear terms at each stage of the calculation?
5. Should the "primitive" differential equations be attacked directly, or is it desirable to first introduce: (a) nondimensionalized, normalized variables, (b) "stretched" (distorted) coordinates to magnify regions in which gradients are expected to be large? If strategy (b) is used, does this influence the nature of the differential equation to be discretized?
6. In a one-dimensional transient problem (say y, t) is it possible to discretize only one variable (say y) and leave the other variable (t) continuous? What methods could be used to solve the resulting numerical problem?
7. What is the basic difference between problems that are mathematically "parabolic" (containing a time-like or "one-way" variable) and "elliptic" (containing all "two-way" variables)? How does this influence the numerical procedures adopted?
8. Suppose one obtains a discretized solution (so that all relevant dependent variables are known at the "mesh points"). How could these variables be estimated elsewhere? (*Optional*: What is a "spline" function, and how could it be applied for interpolation purposes?)
9. An alternative to the method of finite difference (MFD) which retains most of its attributes but offers advantages for complicated-shaped domains and boundary conditions, is the "finite-element method" discussed in Appendix 8.3. Compare MFE with MFD and indicate the essential advantages of MFE over the method of finite differences.

Appendix 8.5

Physical Constants

Quantity	*Symbol*	*Value*	[*Units*]
Universal gas constant	R	$= 1.987$	[cal/g-mole · K]
		8.3143×10^3	[joules/kg-mole · K]
		82.05	[cm^3 · atm/g-mole · K]
		1.545×10^3	[ft · lb$_f$/lb-mole · °R]
Standard gravitational acceleration	g	$= 980.665$	[cm/s^2]
		9.80665	[m/s^2]
		32.174	[ft/s^2]
Avogadro's number	N_A	$= 6.02252 \times 10^{23}$	[molecules/g-mole]
Boltzmann constant	k_B	$= 1.38054 \times 10^{-16}$	[erg/molecule · K]
		1.38054×10^{-23}	[joule/molecule · K]
Atomic mass unit	amu	$= 1.66043 \times 10^{-24}$	[g]
		1.66043×10^{-27}	[kg]
Planck's constant	h	$= 6.6256 \times 10^{-27}$	[erg · s]
		6.6256×10^{-34}	[joule · s]
Speed of light	c	$= 2.997925 \times 10^{10}$	[cm/s]
		2.997925×10^8	[m/s]

Appendix 8.6

Metric System Notes/Conversion Factors

SI units

Quantity	Name	Symbol
Base		
length	meter	m
mass	kilogram	kg
time	second	s
electric current	ampere	A
thermodynamic temperature	kelvin	K
amount of substance	mole	mole
luminous intensity	candels	cd
Supplementary		
plane angle	radian	rad
solid angle	steradian	sr

SI Derived Units with Special Names

	SI Unit		
Quantity	*Name*	*Symbol*	*Expression in terms of other units*
frequency	hertz	Hz	s^{-1}
force	newton	N	$kg \cdot m/s^2$
pressure stress	pascal	Pa	$N \cdot m^2$
energy, work quantity of heat	joule	J	$N \cdot m$
power, radiant flux	watt	W	J/s
quantity of electricity, electric charge	coulomb	C	$A \cdot s$
electric potential voltage, potential difference electro-motive force	volt	V	W/A

SI Prefixes

Factor	*Prefix*	*Symbol*	*Factor*	*Prefix*	*Symbol*
10^{18}	exa	E	10^{-1}	deci	d
10^{15}	peta	P	10^{-2}	centi	c
10^{12}	tera	T	10^{-3}	milli	m
10^{9}	giga	G	10^{-6}	micro	μ
10^{6}	mega	M	10^{-9}	nano	n
10^{3}	kilo	k	10^{-12}	pico	p
10^{2}	hecto	h	10^{-15}	femto	f
10^{1}	deka	da	10^{-18}	atto	a

Directions for Use

SI symbols are not capitalized unless the unit is derived from a proper name; e.g., Hz for H.R. Hertz. Unabbreviated units are not capitalized; e.g., hertz, newton, kelvin. Only E, P, T, G, *and* M *prefixes are capitalized.*
Except at the end of a sentence, SI units are not to be followed by periods.
In derived unit symbols, use center dot to denote multiplication and a slash for division; e.g., newton-second per square meter = $N \cdot s/m^2$.

Units in Use with the International System

Name	*Symbol*	*Value in SI Units*
minute	min	1 min = 60 s
hour	h	1 h = 60 min = 3600 s
day	d	1 d = 24 h = 86400 s
year	yr	1 yr = 365 d
degree	°	$1° = (\pi/180)$ rad
liter	L	$1\ L = 10^{-3}\ m^3$
metric ton	t	$1\ t = 10^3$ kg
angström	Å	1 Å = 0.1 nm = 10^{-10} m
bar	bar	1 bar = 0.1 MPa = 10^5 Pa
standard atmosphere	atm	1 atm = 101325 Pa

SI Derived Units Expressed by Means of Special Names

	SI Unit	
Quantity	*Name*	*Symbol*
dynamic viscosity	pascal · second	Pa · s
moment of force	meter · newton	N · m
surface tension	newton per meter	N/m
heat flux density, irradiance	watt per square meter	W/m^2
heat capacity, entropy	joule per kelvin	J/K
specific heat capacity, specific entropy	joule per kilogram · kelvin	J/(kg · K)
specific energy	joule per kilogram	J/kg
thermal conductivity	watt per meter-kelvin	W/(m · K)
energy density	joule per cubic meter	J/m^3
electric field strength	volt per meter	V/m
electric charge density	coulomb per cubic meter	C/m^3
electric flux density	coulomb per square meter	C/m^2
molar energy	joule per mole	J/mol
molar entropy, molar heat capacity	joule per mole-kelvin	J/(mol · K)
radiant intensity	watt per steradian	W/sr
radiance	watt per square meter-steradian	$W \cdot m^{-2} \cdot sr^{-1}$

SI Derived Units Expressed in Terms of Base Units

	SI Unit	
Quantity	*Name*	*Symbol*
area	square meter	m^2
volume	cubic meter	m^3
speed, velocity	meter per second	m/s
acceleration	meter per second squared	m/s^2
kinematic viscosity	square meter per second	m^2/s
wave number	1 per meter	m^{-1}
density, mass density	kilogram per cubic meter	kg/m^3
current density	ampere per square meter	A/m^2
concentration (of amount of substance)	mole per cubic meter	mol/m^3
specific volume	cubic meter per kilogram	m^3/kg
angular velocity	radian per second	rad/s
angular acceleration	radian per second squared	rad/s^2

Acceptable Metric Units

Quantity	*SI Unit*	*Accepted Alternate*
time	second	year
		day
		hours
pressure	pascal	
energy	joule	—
force	newton	—
mass	kilogram	metric ton
volume	m^3	liter
viscosity	Pa s	—

Conversion Factors to SI for Selected Quantities

To convert from	*To*	*Multiply by*	
barrel (for petroleum, 42 gal)	meter^3 (m^3)	1.5898729	E − 01
British thermal unit (Btu, International Table)	joule (J)	1.0550559	E + 03
Btu/lb_m-deg F (heat capacity)	joule/kilogram-kelvin (J/kg · K)	4.1868000	E + 03
Btu/hour	watt (W)	2.9307107	E − 01
Btu/second	watt (W)	1.0550559	E + 03
Btu/ft^2-hr-deg F (heat transfer coefficient)	joule/meter^2-second-kelvin (J/m^2 · s · K)	5.6782633	E + 00
Btu/ft^2-hour (heat flux)	joule/meter^2-second (J/m^2 · s)	3.1545907	E + 00
Btu/ft-hr-deg F (thermal conductivity)	joule/meter-second-kelvin (J/m · s · K)	1.7307347	E + 00
calorie (International Table)	joule (J)	4.1868000	E + 00
calorie (Thermochemical)	joule (J)	4.1840000	E + 00
cal/g · deg C	joule/kilogram-kelvin (J/kg · K)	4.1868000	E + 03
centimeter	meter (m)	1.0000000	E − 02
centimeter of mercury (0°C)	pascal (Pa)	1.3332237	E + 03
centimeter of water (4°C)	pascal (Pa)	9.80638	E + 01
centipoise	pascal-second (Pa · s)	1.0000000	E − 03
centistokes	meter^2/second (m^2/s)	1.0000000	E − 06
degree Celsius (°C)	kelvin (K)	$T_K = t_c + 273.2$	
degree Fahrenheit (°F)	kelvin (K)	$T_K = (t_F + 459.67)/1.8$	
degree Rankine (°R)	kelvin (K)	$T_K = t_R/1.8$	
dyne	newton (N)	1.0000000	E − 05
erg	joule (J)	1.0000000	E − 07
fluid ounce (U.S.)	meter^3 (m^3)	2.9573530	E − 05
foot	meter (m)	3.0480000	E − 01
foot of water (39.2°F)	pascal (Pa)	2.98898	E + 03

$foot^2$	$meter^2$ (m^2)	9.2903040	E − 02
foot/$second^2$	meter/$second^2$ (m/s^2)	3.0480000	E − 01
$foot^2$/hour	$meter^2$/second (m^2/s)	2.5806400	E − 05
foot-pound-force	joule (J)	1.3558179	E + 00
$foot^2$/second	$meter^2$/second (m^2/s)	9.2903040	E − 02
$foot^3$	$meter^3$ (m^3)	2.8316847	E − 02
gallon (U.S. liquid)	$meter^3$ (m^3)	3.7854118	E − 03
gram	kilogram (kg)	1.0000000	E − 03
horsepower (550 ft · lb_f/s)	watt (W)	7.4569987	E + 02
inch	meter (m)	2.5400000	E − 02
inch of mercury (60°F)	pascal (Pa)	3.37685	E + 03
inch of water (60°F)	pascal (Pa)	2.48843	E + 02
$inch^2$	$meter^2$ (m^2)	6.4516000	E − 04
$inch^3$	$meter^3$ (m^3)	1.6387064	E − 05
kilocalorie	joule (J)	4.1868000	E + 03
kilogram-force (kg_f)	newton (N)	9.8066500	E + 00
micron (micrometer)	meter (m)	1.0000000	E − 06
mil	meter (m)	2.5400000	E − 05
mile (U.S. Statute)	meter (m)	1.6093440	E + 03
mile/hour	meter/second (m/s)	4.4704000	E − 01
millimeter of mercury (0°C)	pascal (Pa)	1.3332237	E + 02
ounce-mass (avoirdupois)	kilogram (kg)	2.8349523	E − 02
ounce (U.S. fluid)	$meter^3$ (m^3)	2.9573530	E − 05
pint (U.S. liquid)	$meter^3$ (m^3)	4.7317647	E − 04
poise (absolute viscosity)	pascal-second (Pa · s)	1.0000000	E − 01
poundal	newton (N)	1.3825495	E − 01
pound-force (lb_f)	newton (N)	4.4482216	E + 00
pound-force-second/ft^2	pascal-second (Pa · s)	4.7880258	E + 01
pound-mass (lb_m avoirdupois)	kilogram (kg)	4.5359237	E − 01
pound-mass/$foot^3$	kilogram/$meter^3$ (kg/m^3)	1.6018463	E + 01

(*continued*)

Conversion Factors to SI for Selected Quantities (*continued*)

To convert from	*To*	*Multiply by*	
pound-mass/foot-second	pascal-second (Pa · s)	1.4881639	E + 00
pound/inch^2 (psi)	pascal (Pa)	6.8947573	E + 03
quart (U.S. liquid)	meter^3 (m^3)	9.4635295	E − 04
slug	kilogram (kg)	1.4593903	E + 01
stokes (kinematic viscosity)	meter^2/second (m^2/s)	1.0000000	E − 04
ton (short, 2000 lb_m)	kilogram (kg)	9.0718474	E + 02
torr (mm Hg, 0°C)	pascal (Pa)	1.3332237	E + 02
watt-hour	joule (J)	3.6000000	E + 03
yard	meter (m)	9.1440000	E − 01

Solutions to Selected Exercises

2.8 a. KE/mass for 1 g H_2O @ 1 m/s (m) → 1 m/s

$m = 1$ g

$$\frac{\text{KE}}{\text{mass}} = \frac{v^2}{2} = \frac{1}{2}\left(10^2\,\frac{\text{cm}}{s}\right)^2 = 0.5 \times 10^4\,\frac{\text{erg}}{\text{g}} \times \frac{1\text{ J}}{10^7\text{ erg}} \times \frac{1\text{ cal}}{4.184\text{ J}}$$

$$\frac{\text{KE}}{\text{mass}} = 1.19 \times 10^{-4}\text{ cal/g}$$

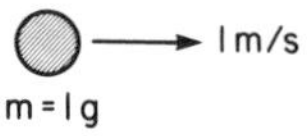

b. $\Delta\phi = g\,\Delta z$

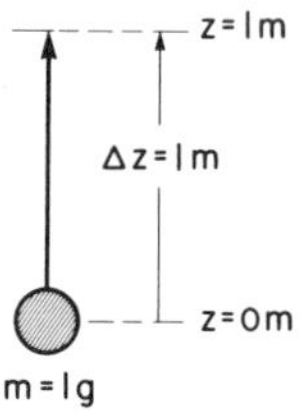

$$= \left(0.9807 \times 10^3\,\frac{\text{cm}}{\text{s}^2}\right)(10^2\text{ cm})$$

$$= 0.9807 \times 10^5\,\frac{\text{erg}}{\text{g}} \times \frac{1\text{ J}}{10^7\text{ erg}} \times \frac{1\text{ cal}}{4.184\text{ J}}$$

$$= 234 \times 10^{-3}\text{ cal/g}$$

c. $(\Delta h)_{\text{thermal}} \cong c_p\,\Delta T = (1)(100) = 1 \times 10^2\text{ cal/g}$

$H_2O(\ell)$ (points to c_p); $\frac{\text{cal}}{\text{g K}}$ (points to (1)); K (for (100))

d. $$(\Delta h)_{\text{fusion}} = 1.4363\,\frac{\text{kcal}}{\text{g-mole}} \times \frac{10^3\text{ cal}}{1\text{ kcal}} \times \frac{1\text{ g-mole}}{18.016\text{ g}} = 79.7\text{ cal/g}$$

$$= 0.797 \times 10^2\text{ cal/g}$$

e. $$(\Delta h)_{\text{vaporiz.}}\text{ @ 373 K} = 9.717\,\frac{\text{kcal}}{\text{g-mole}} \times \frac{10^3\text{ cal}}{1\text{ kcal}} \times \frac{1\text{ g-mole}}{18.016\text{ g}} \cong 540\text{ cal/g}$$

f. $(\Delta h)_{\text{comb}}$ to form 1 g $H_2O(\ell)$: 3.79×10^3 cal/g @ 298 K

Remark: $\Delta H_f^\circ(298) = -57.798\,\frac{\text{kcal}}{\text{g-mole}}$; this means that for the formation reaction: $H_2(g) + \frac{1}{2}O_2(g) \rightarrow H_2O(g)$, $\Delta H_{298} = -57.798$

Table 2.8E Summary of Energy Changes

Type of Energy		cal/g (kcal/kg)
Δ(kinetic energy) (@ 1 m/s)		1.19×10^{-4}
Δ(potential energy) (@ 1 m)		2.34×10^{-3}
Δ(thermal energy) ($\Delta T = 100$ K)		1.00×10^{2}
Δh(fusion)	Phase change	0.797×10^{2}
Δh(vaporiz.)	Phase change	0.539×10^{3}
Δh(comb)		3.79×10^{3}

Study Questions:

a. Which values are *independent* of chemical substance?
b. What are some "general" implications of these findings? (cf. relative importance of Δ(KE), Δ(PE), Δh_T, $(\Delta h)_{\text{phase change}}$, $\Delta h_{\text{chem. change}}$.)

2.9

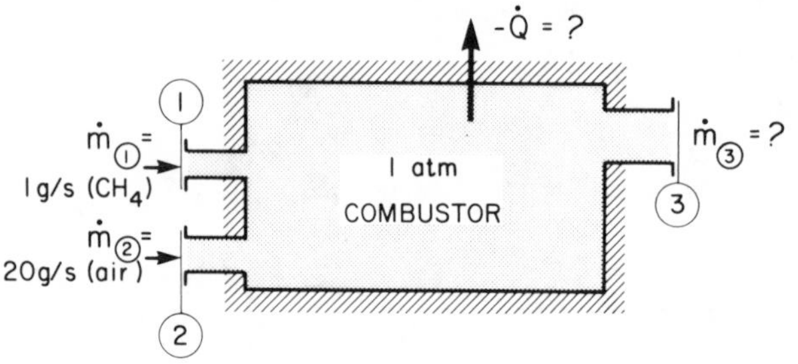

Stream ① Pure $CH_4(g)$
@ 1 atm
300 K,
$\dot{m}_{①} = 1$ g/s

Stream ③ Air + Comb Products
@ ≈ 1 atm
1000 K

Stream ② Air @ 1 atm
300 K
$\dot{m}_{②} = 20$ g/s

$\dot{m}_{③} = ?$
$\omega_{i,③} = ?\ (i = 1, 2, \ldots, 5)$
$\dot{m}_{③} = ?$

Total Mass Balance:

$$\underbrace{\frac{\partial}{\partial t}\int_V \rho\, dV}_{0 \text{ in } ss} + \underbrace{\int_S \rho \mathbf{v} \cdot \mathbf{n}\, dA}_{\dot{m}_{③} - (\dot{m}_{①} + \dot{m}_{②})} = 0$$

Conclusion

$$\dot{m}_{③} - (\dot{m}_{①} + \dot{m}_{②}) = 0,$$

or

$$\dot{m}_{③} = \dot{m}_{①} + \dot{m}_{②},$$

or

$$\dot{m}_{③} = 1 + 20 = 21 \text{ g/s};$$

i.e., exit stream (1000 K, 1 atm) has mass-flow rate of 21 g/s (*via* overall mass balance).

Chemical Composition of the Exit Stream, i.e.,
ω_i, (where $i =$ 1, 2, 3, 4, 5)
O_2 N_2 H_2O CO_2 CH_4
This can be found *via* the chemical element mass balances, i.e.,

$$\frac{\partial}{\partial t}\int_V \rho_{(k)}\, dV + \int_S \rho_{(k)}\mathbf{v}\cdot\mathbf{n}\, dA = 0;$$

for a *ss* this can be written:

$$\int_S \omega_{(k)}\rho\mathbf{v}\cdot\mathbf{n}\, dA = 0$$

or

$$\omega_{(k),③}\dot{m}_{③} - [\omega_{(k),①}\dot{m}_{①} + \omega_{(k),②}\dot{m}_{②}] = 0,$$

or

$$\omega_{(k),③} = \frac{\omega_{(k),①}\dot{m}_{①} + \omega_{(k),②}\dot{m}_{②}}{\dot{m}_{③}},$$

where

$$\omega_{(k),③} = \sum_{i=1}^{5} \omega_{k/i}\,\omega_{i,③}$$

(See matrix below) — for $\omega_{k/i}$; Sought — for $\omega_{i,③}$

Species

Here $\omega_{k/i}$:

Elements	(j)	O_2 ($i=1$)	N_2 ($i=2$)	H_2O ($i=3$)	CO_2 ($i=4$)	CH_4 ($i=5$)
H	1	0	0	0.1119	0	0.25137
C	2	0	0	0	0.2727	0.74813
N	3	0	1	0	0	0
O	4	1	0	0.8881	0.7271	0
Sum:		1	1	1	1	1

(Each column must sum to unity.)

Note that we need $\omega_{(k),\textcircled{1}}$ for k = 1,.... where
$\omega_{(k),\textcircled{2}}$ for k = 1,.... $\omega_{(k)} = \sum \omega_{k/i}\omega_i$.

We readily find:

Element	k	$\omega_{(k),\textcircled{1}}$	$\omega_{(k),\textcircled{2}}$
H	1	0.25137	0
C	2	0.74813	0
N	3	0	0.77
O	4	0	0.23
		Stream ① (pure CH_4)	Stream ② (air)

Element 1 (Hydrogen)

$$\omega_{(1),\textcircled{3}} = \frac{\omega_{(1)\textcircled{1}}\dot{m}_{\textcircled{1}} + \omega_{(1)}\ \dot{m}_{\textcircled{2}}}{\dot{m}_{\textcircled{3}}}$$

$$\omega_{(1),\textcircled{3}} = \frac{(0.25137)(1) + (0)(20)}{(21)} = 1.197 \times 10^{-2}$$

Element 2 (Carbon)

$$\omega_{(2),\textcircled{3}} = \frac{\omega_{(2)\textcircled{1}}\dot{m}_{\textcircled{1}} + \omega_{(2),\textcircled{2}}\dot{m}_{\textcircled{2}}}{\dot{m}_{\textcircled{3}}}$$

$$= \frac{(0.74813)(1) + (0)(20)}{21} = 3.5625 \times 10^{-2}$$

Similarly,

$$\omega_{(3),\textcircled{3}} = 0.7333 \qquad (N\text{-balance})$$

$$\omega_{(4),\textcircled{3}} = 0.21905 \qquad (O\text{-balance})$$

Calculation of $\omega_{i,\textcircled{3}}$ *via* $\omega_{(k),\textcircled{3}}$ $i = 1, \ldots, 5 \qquad j = 1, \ldots, 4$
In stream ③ for each k = 1,...,4 we have:

$$\omega_{(k)} = \sum_{i=1}^{S} \omega_{k/i} \quad \omega_i$$

Just calculated ($\omega_{(k)}$) — Matrix on page 495 ($\omega_{k/i}$) — Sought (ω_i)

k = 1 (*H-balance*) *gives*:

$$1.1970 \times 10^{-2} = 0.1119\,\omega_3 \rightarrow \omega_{3,\textcircled{3}} = 0.10697$$

k = 2(*C-balance*) *gives*:

$$3.5625 \times 10^{-2} = 0.2727\,\omega_4 \rightarrow \omega_{4,③} = 0.13065$$

k = 3 (*N-balance*) *gives*:

$$0.7333 = 1.000\,\omega_2 \rightarrow \omega_{2,③} = 0.7333$$

k = 4(*O-balance*) *gives*:

$$0.21905 = 1.00\,\omega_1 + 0.8881\,\underset{\text{above}}{\underset{\uparrow}{\omega_3}} + 0.7271\,\underset{\text{above}}{\underset{\uparrow}{\omega_4}}$$

Therefore: $\omega_{1,③} = 0.02905$

This completes the composition calculation for the exit stream (*Note:* $\omega_{5,③} = 0$ (no unburned methane)).
These values will be needed to calculate

$$h_③\{1000\text{ K}\} = \sum_{i=1}^{5} \omega_{i,③} \cdot h_i\{1000\text{ K}\},$$

which appears in the energy balance below:

Apply M-Scopic Energy Balance to Calculate $-\dot{Q}$
When the total stress $\mathbf{\Pi}$ is split into $-p\mathbf{I}$ and $\mathbf{T}$ Eq. (2.3-3) can be written in the useful form (see Eq. (3.2-6)):

$$\frac{\partial}{\partial t}\int_V \rho\left(e + \frac{v^2}{2}\right)dV + \int_S \rho\left(h + \frac{v^2}{2}\right)\mathbf{v}\cdot\mathbf{n}\,dA = -\int_S \dot{\mathbf{q}}''\cdot\mathbf{n}\,dA$$
$$+ \int_V \dot{q}'''\,dV + \int_S \mathbf{v}\cdot\mathbf{T}\cdot\mathbf{n}\,dA + \sum_{i=1}^{N}\int_V \dot{\mathbf{m}}_i''\cdot\mathbf{g}_i\,dV$$

in which the convective term now contains the *enthalpy* $e + (p/\rho)$. If we now neglect:

a. KE/mass ($v^2/2$) terms,
b. accum. term (ss)
c. $\dot{q}'''$ term (no volumetric heat addition; see Section 2.5.4)
d. body force work (see Exercise 2.6)

the macroscopic energy balance equation then can be written:

$$\dot{m}_③ h_③ - [\dot{m}_① h_① + \dot{m}_② h_②] = \dot{Q}_{\text{added}} - (\dot{\bar{W}}_{\text{shaft}} + \dot{\bar{W}}_\tau)$$

where

$$\dot{Q}_{\text{added}} = -\int \dot{\mathbf{q}}'' \cdot \mathbf{n}\, dA$$

and

$$\bar{\dot{W}}_\tau \equiv \left(-\int_S \mathbf{v} \cdot \mathbf{T} \cdot \mathbf{n}\, dA\right)_{\substack{\text{all surfaces except those}\\ \text{enveloping a turbine}\\ \text{or pumping device}}}$$

However, there is no "shaft work" (see Section 2.5.5); moreover,

regarding: ■ where **v** is a large (①, ②, ③) normal component, **T** is small;
$\mathbf{v} \cdot \mathbf{T} \cdot \mathbf{n}\, dA$ ■ where **T** is large (walls), $\mathbf{v} = 0$ (no "slip"; Ch. 4).

Therefore $\bar{\dot{W}}_\tau$ can be neglected and:

$$-\dot{Q} \cong (\dot{m}_{①} h_{①} + \dot{m}_{②} h_{②}) - \dot{m}_{③} h_{③}.$$

To complete *calculation of* $-\dot{Q}$, we therefore need $h_{①}\{300\text{ K}\}$
$h_{②}\{300\text{ K}\}$
$h_{③}\{1000\text{ K}\}$

where

$$h = \sum_{i=1}^{5} \omega_i \cdot h_i\{T\} \quad \text{ideal gas mixture}$$

for each stream, and:

$$h_i\{T\} = \underbrace{\frac{+\Delta H^0_{f_i}\{T_{\text{ref}}\}}{M_i}}_{\text{Tabulated}} + \frac{\overbrace{H_i\{T\}}^{\text{Tabulated}} - H_i\{T_{\text{ref}}\}}{M_i}$$

Now $T_{\text{ref}} = 298.16$ K is very close to $T_{①} = T_{②} = 300$ K;

therefore

$$h_{①} = \omega_{5,①} h_5\{300\text{ K}\} \cong (1.00)\frac{-17.895}{16.04} \cong -1.1156\frac{\text{kcal}}{\text{g}};$$

$$h_{\textcircled{2}} = \omega_{1,\textcircled{2}} h_1\{300\ \mathrm{K}\} + \omega_{2,\textcircled{2}} h_2\{300\ \mathrm{K}\} \qquad \text{(air)},$$

$$h_{\textcircled{2}} \cong 0.23(0.000) + 0.77(0.000) = 0.$$

Finally,

$$h_{\textcircled{3}} = \omega_{1,\textcircled{3}} h_1\{1000\ \mathrm{K}\} + \omega_{2,\textcircled{3}} h_2\{1000\ \mathrm{K}\} + \omega_{3,\textcircled{3}} h_3\{1000\ \mathrm{K}\}$$
$$+ \omega_{4,\textcircled{3}} h_4\{1000\ \mathrm{K}\} + \omega_{5,\textcircled{3}} h_5\{1000\ \mathrm{K}\};$$

i.e.,

$$h_{\textcircled{3}}\{1000\ \mathrm{K}\} = 0.02905\left(\frac{5.427}{32.00}\right) + 0.7333\left(\frac{5.129}{28.01}\right)$$
$$+ 0.10697\left[-\frac{57.798}{18.016} + \frac{6.209}{18.016}\right]$$
$$+ 0.13065\left[-\frac{94.054}{44.01} + \frac{7.984}{44.01}\right] + 0[\cdots]$$

or $h_{\textcircled{3}} = -0.42262$ kcal/g.

Therefore

$$\dot{Q}_{\mathrm{removal}} = (1\ \mathrm{g/s})(-1.1156) + (20\ \mathrm{g/s})(0) - (21\ \mathrm{g/s})(-0.42262)$$
$$= 7.76\ \mathrm{kcal/s}\ (\text{to have } T_{\textcircled{3}} = 1000\ \mathrm{K}).$$

Calculation of $\int_S \rho s \mathbf{v} \cdot \mathbf{n}\, dA$ (net outflow rate of entropy from M-scopic CV)

Known ↙ (over $\dot{m}_{\textcircled{3}}$); ? ↙ (over $s_{\textcircled{3}}$); ? ↙ (over $s_{\textcircled{2}}$)

$$\int_S s\rho \mathbf{v} \cdot \mathbf{n}\, dA = \dot{m}_{\textcircled{3}} s_{\textcircled{3}} - (\dot{m}_{\textcircled{1}} s_{\textcircled{1}} + \dot{m}_{\textcircled{2}} s_{\textcircled{2}})$$

Note that, whereas stream ① is *pure* CH_4—for which

$s_{CH_4}\{300\ \mathrm{K},\ 1\ \mathrm{atm}\} \Rightarrow$ obtainable from, say, JANAF Thermochemical Tables,

Streams ② and ③ are *mixtures*; hence,

"Mixing entropy contribution" ↙ (over $\frac{R}{M_i}\ln$ term)

$$s_{\textcircled{3}} = \sum_{i=1}^{5} \omega_{i,\textcircled{3}} \left\{\frac{S_i\{1000\ \mathrm{K}, 1\ \mathrm{atm}\}}{M_i} + \frac{R}{M_i}\ln\left\{\left(\frac{1}{y_{i,\textcircled{3}}}\right)\right\}\right\},$$

where

$$y_{i,③} = \frac{\omega_{i,③}/m_i}{\sum_{j=1}^{5} \omega_{j,③}/m_j} \qquad \begin{pmatrix} \textit{mole} \text{ fractions calculable} \\ \text{from the } \textit{mass} \text{ fractions} \end{pmatrix}.$$

Equivalently,

$$s_③ = \frac{\sum_{i=1}^{5} y_{i,③}[S_i\{1000\text{ K}, 1\text{ atm}\} + R\ln\{1/y_{i,③}\}]}{\sum_{i=1}^{5} y_{i,③} M_i}$$

(Similarly, stream ② is a mixture, and this affects calculation of $S_②$.)

2.13 The situation/nomenclature is displayed below. As usual we use y and x for mole fractions in the vapor and liquid phases, respectively. If we denote W ≡ water, G ≡ glycol and A ≡ air, then in the present problem note that $y_W, y_G \ll 1$, and the liquid inventory will not change much with time; i.e., $L \approx L_0 = \text{const}$. Moreover, a species material balance on water for the fixed control volume shown (dashed) must be of the transient nonreactive form:

$$\text{(Rate of accumulation within CV)} \qquad (1)$$
$$+ \text{(Net convective outflow rate from CV)} = 0,$$

since there is a batch charge of glycol being dehydrated by a steady air flow ($\dot{V}$ kg-moles air/h), which is initially dry ($y_{A,①} = 1$). Using dry air as the flow basis for the evaluation of the convective terms, conservation of H_2O (≡ W) gives:

$$\frac{d}{dt}(x_W L) + \left[\left(\frac{y_W}{y_A}\right)_② - \left(\frac{y_W}{y_A}\right)_①\right]\dot{V} = 0, \qquad (2)$$

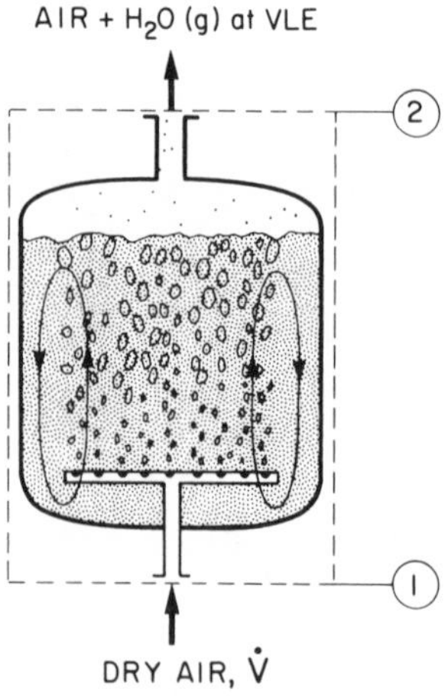

where subscripts ② and ① denote exit and inlet station conditions. In writing the accumulation term of Eq. (2), use has been made of the *well-stirred* assumption in characterizing the entire liquid by the single water composition variable $x_W(t)$. To proceed with the exploitation of Eq. (2), in the present case we note that:

$$\left| x_W \frac{dL}{dt} \right| \ll \left| L \frac{dx_W}{dt} \right|,$$

$$L = L_0 = L_{G,0} = \text{constant},$$

$$(y_W)_{①} = 0 \quad \text{(dry air input)},$$

$$(y_A)_{②} \cong 1 \quad \text{(outlet gas is predominantly air)};$$

hence Eq. (2) simplifies to the first-order ordinary differential equation:

$$L \frac{dx_W}{dt} + y_{W,②} \dot{V} = 0. \tag{3}$$

However, at each instant the outlet stream water-vapor content $y_{W,②}(t)$ is, to a first approximation, in equilibrium with the water content of the glycol; that is,

$$y_{W,②} = K_w \cdot x_W$$

(where, in view of the approximate validity of Raoult's law for this system, the partition coefficient for water is

$$K_w = 149/760 \quad \text{(at 60°C)}). \tag{4}$$

Therefore Eq. (3) can be written in terms of x_W and t alone; and "separated" to obtain:

$$dt = \frac{L_0}{K_w \dot{V}} \cdot \frac{dx_W}{x_W} \tag{5}$$

(which, incidentally, is seen to be dimensionally consistent, since $L_0/\dot{V}$ has the units of time, the remaining quantities being dimensionless). Integrating to find the time for the glycol to go from $x_{W,0}(= 0.02)$ to $x_{W,f}(= 0.001)$ gives the general expression

$$t = \frac{L_0}{\dot{V}} \cdot \frac{1}{K_w} \ln\left(\frac{x_{W,0}}{x_{W,f}}\right). \tag{6}$$

Inserting the specified data:

$$t = \left(\frac{10}{5}\right)\left(\frac{760}{149}\right)\cdot \ln\left(\frac{0.02}{0.001}\right) = 30.6\ \text{h}.$$

Thus, if our basic assumptions are realistic, the desired amount of water will have been "stripped" from the glycol charge in about 30.6 hours. Note that an *infinite* amount of time would be required to reduce the final water mole fraction $x_{W,f}$ to *zero*.

3.7 $k(NH_3(g))$ and $D_{NH_3-H_2}$ Calculation
Given the tabular data:

Table 3.7E Molecular Parameters: $NH_3 + H_2$ System

i	*Species*	M_i	σ_i	$(\varepsilon/k_B)_i$	$C_{p,i}$(306 K)
1	$NH_3(g)$	17.031	2.900	558.3	8.559
2	$H_2(g)$	2.016	2.827	59.7	6.900
Units:		g/g-mole	Å	K	cal/mole K

Calculate $k(NH_3(g))$ and $D_{NH_3-H_2}$ @ 1 atm, 306 K.

Thermal Conductivity of $NH_3(g)$
See Eq. (3.3-6):

$$k \approx \underbrace{\frac{15}{4}\frac{R\mu}{M}}_{k_{\text{trans}}}\cdot\underbrace{\left\{1 + \frac{4}{15}\left(\frac{C_p}{R} - \frac{5}{2}\right)\right\}}_{\text{Eucken factor}}$$

μ is calculated as before, to give 1.063×10^{-4} poise. Therefore:

$$k_{\text{trans}} = \frac{15}{4}\cdot\frac{R}{M}\cdot\mu = \frac{15}{4}\cdot\frac{(1.987)}{(17.031)}(1.063 \times 10^{-4}) = 4.651 \times 10^{-5}\,\frac{\text{cal/cm}^2\text{s}}{\text{K/cm}}$$

$$\text{Eucken factor} = 1 + \frac{4}{15}\left(\frac{8.559}{1.987} - \frac{5}{2}\right) = 1.482.$$

Therefore

$$k = k_{\text{trans}}\cdot(\text{Eucken factor})$$

$$k = (4.651 \times 10^{-5})(1.482) = 6.892 \times 10^{-5}\,\frac{\text{cal/cm}^2\text{s}}{\text{K/cm}}$$

$$\textit{Note: } \text{Pr} \equiv \frac{\nu}{(k/\rho c_p)} = \frac{c_p\mu}{k} = \left(\frac{C_p}{M}\right)\frac{\mu}{k} = 0.775.$$

Calculation of $D_{NH_3-H_2}$ @ 306 K, 1 atm: *and* Sc *for* $y_1 = 0.5$.
Use Eq. (3.48) with $\sigma_{ij} \cong \frac{1}{2}(\sigma_{ii} + \sigma_{ij})$, $\varepsilon_{ij} \cong (\varepsilon_{ii} \cdot \varepsilon_{jj})^{1/2}$:

$$D_{ij} = \frac{3\,k_B T}{8\,p} \cdot \left[\frac{k_B T}{2\pi} \cdot \left(\frac{m_i + m_j}{m_i m_j}\right)\right]^{1/2} \cdot \frac{1}{\sigma_{ij}{}^2 \Omega_D(kT/\varepsilon_{ij})}.$$

In the present case:

$$k_B T = 1.38054 \times 10^{-16}(306) = 4.2245 \times 10^{-14}\,\text{erg/molecule}$$

$$\frac{m_i m_j}{m_i + m_j} = 2.9929 \times 10^{-24}\ \text{g};$$

$$p = 1\ \text{atm} = 1.0133 \times 10^6\ \text{dyne/cm}^2;$$

$$\sigma_{ij} = 2.8635\ \text{Å}; \qquad \varepsilon_{ij}/k_B = [(\varepsilon_{ii}/k_B)(\varepsilon_{jj}/k_B)]^{1/2} = 182.6\,\text{K};$$

$$\sigma_{ij}{}^2 = 8.1996\ (\text{Å})^2 = 8.1996 \times 10^{-16}\ \text{cm}^2;$$

$$k_B T/\varepsilon = \frac{306}{182.6} = 1.676$$ (outside of range of curve-fit, which formally gives 1.03).

From tabular values (Hirschfelder, Curtiss, and Bird (1954)):

$$\Omega_D(1.676) \approx 1.145;$$

therefore: $$D_{NH_3-H_2} = 0.7895 \approx 0.790\,\text{cm}^2/\text{s}.$$

For an equimolar mixture of ammonia and hydrogen: $\mu \approx 123\,\mu$ poise = 1.23×10^{-4} P (*via* graph). Moreover, $M_{mix} = \sum_{i=1}^{N} y_i M_i = 0.5(17.031) + 0.5(2.016) = 9.5235$.
Assuming perfect gas behavior:

$$\rho = \frac{pM}{RT} = \frac{(1)(9.5235)}{(82.06)(306)} = 3.79 \times 10^{-4}\ \text{g/cm}^3.$$

The momentum diffusivity of the mixture is therefore:

$$\nu \equiv \mu/\rho = 0.324\ \text{cm}^2/\text{s},$$

corresponding to a diffusivity ratio of:

$$\text{Sc} \equiv \nu/D_{ij} = \frac{0.3273}{0.7892} = 0.411 \qquad \text{(dimensionless)}.$$

3.9 Property Estimation for the Hydrodesulfurization of Naphtha Vapors

a. Suppose we need viscosity of

$$\underbrace{H_2}_{y_{H_2}=0.828} \quad + \quad \underbrace{C_7H_{16}}_{y_{C_7H_{16}}\approx 0.172} \quad + \ldots \text{trace } C_4H_4S$$

@ 660 K, 30 atm

b. Further, if

$$\dot{m} = 2\,\text{g/s per tube and } d_w(\text{tube}) = 2.54\text{ cm, Re} = ?$$

The first step is the estimation of the mixture viscosity based on its composition and the properties of its constituents under the anticipated operating conditions (660 K, 30 atm). Tentatively we use:

$$\mu \approx \frac{\sum_{i=1}^{N} M_i^{1/2} y_i \mu_i}{\sum_{i=1}^{N} M_i^{1/2} y_i} = \frac{M_1{}^{1/2} y_1 \mu_1 + M_2{}^{1/2} y_2 \mu_2}{M_1{}^{1/2} y_1 + M_2{}^{1/2} y_2}$$

where

$$\mu_i = \frac{5}{16} \cdot \frac{(\pi m_i k_B T)^{1/2}}{(\pi \sigma_{ii}{}^2) \cdot \Omega_\mu \{k_B T/\varepsilon_{ii}\}}$$

($i = 1, 2$, where $1 \equiv H_2(g)$, $2 \equiv C_7H_{16}(g)$.)
(cf. Enskog–Chapman theory.)

In what follows, we sequentially consider the viscosities of each of the constituents of the vapor mixture.

Viscosity of Hydrogen(g) at 660 K, 30 atm:

$$m = \frac{M}{N_A} = \frac{2.016\text{ g/g-mole}}{6.0225 \times 10^{23}\text{ molec/mole}} = 3.347 \times 10^{-24}\text{ g};$$

$$\sigma_{ii} = 2.827 \times 10^{-8}\text{ cm};$$

$$\pi \sigma_{ii}{}^2 = 2.511 \times 10^{-15}\text{ cm}^2;$$

$$5/16 = 0.3125;$$

$$(\pi m k_B T)^{1/2} = ((3.141\ldots)(3.347 \times 10^{-24})(1.38054 \times 10^{-16})(660))^{1/2}$$
$$= 9.789 \times 10^{-19},$$

and

$$k_B T/\varepsilon_{ii} = 660/59.7 = 11.055.$$

Therefore

$$\underbrace{\Omega_\mu\{11.055\} = 0.8114}\text{[a]} \quad \text{(N.B.: } 1.22\left(\frac{k_B T}{\varepsilon}\right)^{-0.16} \text{ gives 0.831.)}$$

---

[a] Remark: @ 660 K H_2 behaves like a "hard-sphere" with a diameter:

$$\sigma \Omega_\mu^{1/2} = (2.827\text{ Å})(0.8114)^{1/2} = 2.546\text{ Å}.$$

using table for L-J 12:6 potential (Hirschfelder, Curtiss, and Bird (1954)).

Therefore

$$\mu_{H_2} = \frac{(0.3125)(9.789 \times 10^{-19})}{(2.511 \times 10^{-15})(0.8114)} = 1.50 \times 10^{-4} \frac{g}{\text{cm-s}} \quad \text{(poise)};$$

(to convert to MKS units (Pa-s), multiply by 10^{-1}).

We now consider a similar calculation for the species C_7H_{16}(g) and check whether "dense vapor" corrections are important.

Viscosity of n-Heptane at 660 K, 30 atm.

$$m = \frac{100.128}{6.023 \times 10^{23}} \text{ g} = 1.6626 \times 10^{-22} \text{ g};$$

$$\sigma = 8.88 \times 10^{-8} \text{ cm};$$

$$(\pi m k_B T) = 4.759 \times 10^{-35};$$

$$(\pi m k_B T)^{1/2} = 6.8986 \times 10^{-18};$$

$$\pi \sigma^2 = 2.4773 \times 10^{-14} \text{ cm}^2;$$

$$k_B T/\varepsilon = 660/282 = 2.34.$$

Therefore $\Omega_\mu\{2.34\} = 1.12$ *via* Hirschfelder *et al.* (1954) table;
cf. $1.22(2.34)^{-0.16} = 1.065$),

and

$$\mu^{(0)}_{C_7H_{16}}\{660 \text{ K, } 30 \text{ atm}\} = \frac{(0.3125)(6.8986 \times 10^{-18})}{(2.4773 \times 10^{-14})(1.12)}$$

$$= 0.817 \times 10^{-4} \text{ poise.}$$

Dense-Vapor Correction? This can be estimated *via* the principle of "corresponding states."

For C_7H_{16}: $\quad p_r \equiv p/p_c = 30/26.8 = 1.119,$

and $\quad T_r \equiv T/T_c = 660/540 = 1.222.$

From

$\mu/\mu_c\{T/T_c, p/p_c\}$ graph (Figure 3.2-3), $\mu/\mu_{gas} \approx 1.25$
(cf. $p/p_c \to 0$ asymptote).

Therefore $\mu_{C_7H_{16}} \cong 1.25\,(0.817 \times 10^{-4}\ \text{P}) = 1.017 \times 10^{-4}$ poise.

Viscosity of Binary Mixture @ 660 K, 30 atm
To facilitate the numerical use of Eq. (3.2-13), we now assemble the relevant data and above-mentioned estimates:

Table for calc. $\mu_{(\text{mix})}$

i	M_i	$M\ M_i^{1/2}$	y_i	$10^4\mu_i$
1	2.016	1.4199	0.828	1.502
2	100.128	10.006	0.172	1.017

Then

$$10^4\mu \approx \frac{1.4199(0.828)1.502 + 10.006(0.172)(1.017)}{1.4199(0.828) + 10.006(0.172)} = 1.21 \text{ poise.}$$

Re-Number Calculation:

$$\text{Re} \equiv \frac{Ud_w}{\nu}, \quad \text{where} \begin{cases} U \equiv \left(\dfrac{\dot{m}}{\rho A}\right) & \text{average velocity} \\ \nu \equiv (\mu/\rho) & \text{momentum diffusivity} \end{cases}$$

For each $d_w = 2.54$ cm tube

$$A = \frac{\pi d_w^2}{4} = \frac{\pi}{4}(2.54)^2 = 5.067 \text{ cm}^2$$

and

$$\rho \approx \frac{pM}{RT} = \frac{(30)(18.891)}{(82.06)(660)} = 1.046 \times 10^{-2} \frac{\text{g}}{\text{cm}^3}$$

if the perfect gas EOS is valid for the mixture, where

$$M_{\text{mix}} = \sum_{i=1}^{N} y_i M_i = 0.828(2.016) + 0.172(100.128) = 18.891 \frac{\text{g}}{\text{g-mole}}.$$

Now:

$$\nu \equiv \mu/\rho \cong \frac{1.214 \times 10^{-4} \text{ Poise}}{1.046 \times 10^{-2} \text{ g/cm}^3} = 1.16 \times 10^{-2} \frac{\text{cm}^2}{\text{s}}$$

and

$$U = \frac{2}{(1.046 \times 10^{-2})(5.067)} = 3.77 \times 10^{1} \text{ cm/s}.$$

Therefore $\quad \text{Re} \equiv \frac{U d_w}{\nu} = \frac{(3.77 \times 10^{1})(2.54)}{(1.16 \times 10^{-2})} = 8.26 \times 10^{3}.$

Conclusion: Since $\text{Re}_{\text{trans}} \cong 2.1 \times 10^{3}$ for tube, @ $\text{Re} = 8.26 \times 10^{3}$ the flow should be *turbulent* (under anticipated hydrodesulfurization conditions; see, also, Exercise 6.10).

4.4 a. Upstream (Chamber) Pressure
Isentropic gas flow from $T_0 = 2000$ K to sonic speed @ 1 atm. Assume $\gamma = 1.28$ (cf. Table 8.1-1).

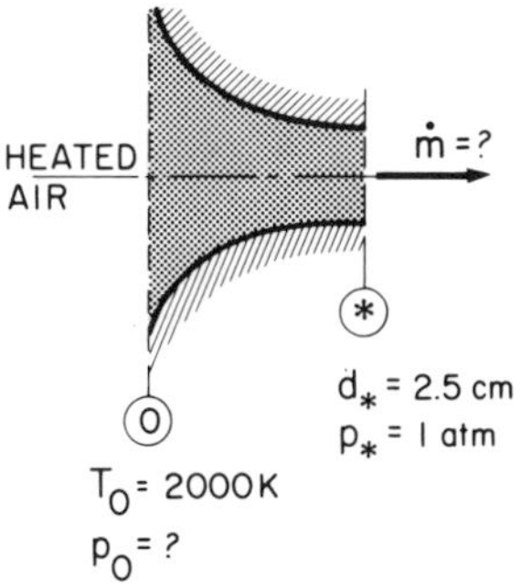

Therefore, $\quad \frac{\gamma - 1}{2} = 0.140$

and

$$\frac{\gamma}{\gamma - 1} = 4.571, \quad \frac{\gamma + 1}{2} = 1.14$$

Now,

$$\frac{p_0}{p_*} = \left[1 + \frac{\gamma - 1}{2}(\text{Ma})_*^2\right]^{\gamma/(\gamma - 1)} \qquad \text{(cf. Eq. (4.3-42))}.$$

Therefore, $\frac{p_0}{1} = [1 + 0.140(1)^2]^{4.571} \rightarrow p_0 = 1.82$ atm.

Also,

$$\frac{T_0}{T_*} = \frac{\gamma + 1}{2} = 1.14 \quad \rightarrow \quad T_* = 1754 \text{ K}.$$

For a perfect gas EOS:

$$\frac{\rho_0}{\rho_*} = \frac{p_0}{p_*} \cdot \frac{T_*}{T_0} = (1.82) \cdot \frac{1}{1.14} = 1.6 \qquad \text{(gas density } \textit{ratio} \text{ across nozzle)},$$

b. Nozzleshape (see Figure (4.4E))

d. Exit (Jet) Velocity

$$u_* = a_* = \left(\frac{\gamma R T_*}{M}\right)^{1/2} = \left[\frac{(1.28)(0.83144 \times 10^{8})(1754)}{(28.97)}\right]^{1/2}$$

Therefore $\quad u_* = a_* = 8.03 \times 10^{4}$ cm/s.

c. Mass Flow Rate

$$\dot{m} = \rho_* u_* A_*.$$

But, for a perfect gas:

$$\rho_* = \frac{p_* M}{RT_*} = \frac{(1)(28.97)}{(82.06)(1754)} = 2.01 \times 10^{-4}\ \text{g/cm}^3,$$

$$u_* = 8.03 \times 10^4\ \text{cm/s} \qquad \text{(Part d)},$$

and

$$A_* = \frac{\pi d_*^2}{4} = \frac{\pi}{4}(2.5)^2 = 4.91\ \text{cm}^2;$$

Therefore $\dot{m} = (2.01 \times 10^{-4})(8.03 \times 10^4\ \text{cm/s})(4.91) = 79.1\ \text{g/s}.$

e. Flow Rate of Axial Momentum, $\dot{J}$ at exit:

$$\dot{J} = \dot{m}u_* = (79.1\ \text{g/s})(8.03 \times 10^4\ \text{cm/s})$$
$$= 6.35 \times 10^6\ \text{g} \cdot \text{cm/s}^2.$$

f. $\rho_0/\rho_* = 1.6$ (calculated above). (See also Solution to Exercise 4.11: Turbulent Jet Mixing Downstream of Exit Plane.)

4.5 C_2H_2/Air Detonation

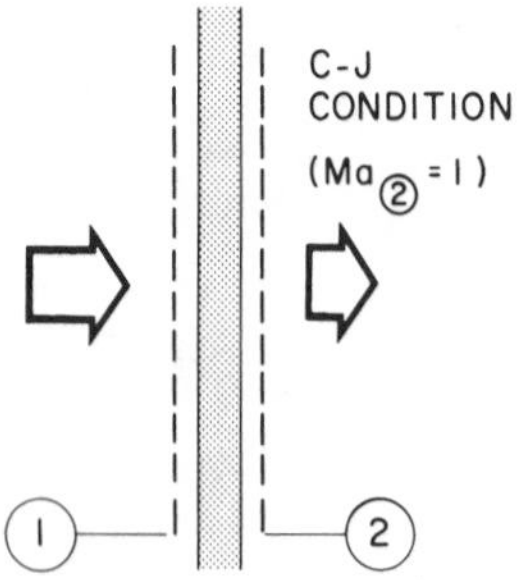

$$\omega_{C_2H_2 ①} = 0.045;$$

$$q = \left(0.045 \frac{\text{g}(C_2H_2)}{\text{g(mix)}}\right)\left(\frac{11.52\ \text{kcal}}{\text{g}(C_2H_2)}\right)\frac{10^3\ \text{cal}}{1\ \text{kcal}}$$

$$q = 5.184 \times 10^2 \frac{\text{cal}}{\text{g}}.$$

Assume: $M \approx 28.97$, $\gamma \approx 1.35$, $R = 1.987$, $T_① = 300$ K

$$\mathscr{H} = \frac{\gamma^2 - 1}{2\gamma} \cdot \frac{Mq}{RT_①}$$

$$\mathscr{H} = \frac{(1.35)^2 - 1}{2(1.35)} \cdot \frac{(28.97)(5.184 \times 10^2)}{(1.987)(300)}$$

or

$$\mathscr{H} = 7.675$$

$[(\mathrm{Ma})_{①}]_{\mathrm{CJ\,deton}} = (1 + \mathscr{H})^{1/2} + \mathscr{H}^{1/2} = 5.7156 \approx 5.716.$

$$u_{①} = a_{①} \cdot \mathrm{Ma}_{①} = \left(0.3472 \frac{\mathrm{km}}{\mathrm{s}}\right)(5.716),$$

$$u_{①} = 1.98\ \mathrm{km/s}.$$

Calculation of $T_{②}$

$$\underbrace{T_{②} + \frac{\gamma R T_{②}^2}{2c_p M}}_{\text{since } u_{②} = a_{②}} = T_{0\,②} = \overbrace{T_{0\,①} + \frac{q}{c_p}}^{\text{effect of "heat addit."}} = \left(T_{①} + \frac{u_{①}^2}{2c_p}\right) + \frac{q}{c_p}$$

therefore

$$T_{②} = \frac{T_{①} + \dfrac{\frac{1}{2}u_{①}^2 + q}{c_p}}{\left[1 + \dfrac{\gamma R/M}{2c_p}\right]}.$$

Since $u_{①} = 1.98\ \mathrm{km/s}$,

$q = 5.184 \times 10^2\ \mathrm{cal/g}$,

$c_p \approx 0.263\ \mathrm{cal/(g \cdot K)}$,

we calculate: $T_{②} \approx 3453\ \mathrm{K}$.

This fixes $\rho_{②}$ since:

$$G = \rho_{①} \cdot u_{①} = \rho_{②} u_{②} = \rho_{②} \cdot \left(\frac{\gamma R T_{②}}{M}\right)^{1/2},$$

or, equivalently:

$$\rho_{②} = \rho_{①} \cdot \frac{a_{①}\mathrm{Ma}_{①}}{a_{②}} = \rho_{①}\left[\frac{T_{①}}{T_{②}}\right]^{1/2} \cdot (\mathrm{Ma}_{①}).$$

But

$$\rho_{①} = \frac{p_{①} M_{①}}{R T_{①}} \approx \frac{(1)(28.97)}{(82.06)(300)} = 1.177 \times 10^{-3} \frac{\mathrm{g}}{\mathrm{cm}^3};$$

therefore $\rho_{②} = 1.983 \times 10^{-3}\ \mathrm{g/cm^3}$;

therefore $$p_{②} = \frac{\rho_{②} R T_{②}}{M} = 19.39 \text{ atm};$$

and, since $\text{Ma}_{②} = 1$ (Chapman-Jouguet Condition):

$$p_{0②} = p_{②} \cdot \left(\frac{\gamma + 1}{2}\right)^{\gamma/(\gamma - 1)}$$
$$= 19.39(1.863) = 36.1 \text{ atm}.$$

Conclusion: Detonation Speed = 1.98 km/s ($(\text{Ma})_{①} = 5.7$). Stagnation Pressure: $p_{0②} = 36.1$ atm.

4.6 a. $$l(\text{air}) \cong 0.065\left(\frac{T}{300}\right)^{1.2} \cdot \frac{1}{p} \ \mu\text{m},$$

$$l(\text{air}) = 0.065\left(\frac{1500}{300}\right)^{1.2}\left(\frac{1}{1}\right) \mu\text{m} = 0.448 \ \mu\text{m};$$

$$d_{\text{w}} = 2a_{\text{w}} = 2(0.5 \text{ cm}) = 1.00 \text{ cm};$$

$$\text{Kn} \equiv l/d_{\text{w}} = 0.448 \times 10^{-6} \text{ m}/1.00 \times 10^{-2} \text{ m},$$

$$\text{Kn} = 0.448 \times 10^{-4} \qquad \text{(well into the continuum regime).}$$

b. $$\rho = pM/(RT) = \frac{(1)(28.97)}{(82.06)(1500)} = 2.35 \times 10^{-4} \frac{\text{g}}{\text{cm}^3},$$

$$\dot{m} = \int_0^{a_{\text{w}}} \rho v_z(r) \cdot 2\pi r \, dr = \rho \int_0^{a_{\text{w}}} 2U\left[1 - \left(\frac{r}{a_{\text{w}}}\right)^2\right] 2\pi r \, dr,$$

$$\dot{m} = \rho U(\pi a_{\text{w}}^2) = (2.35 \times 10^{-4})(10^3)(\pi)(0.5 \text{ cm})^2;$$

therefore $\dot{m} = 0.185$ g/s.

c. $$v_{z,\text{avg}} \equiv \dot{m}/(\rho A) = U = 10^3 \text{ cm/s};$$

$$\text{Re} \equiv \frac{\rho U d_{\text{w}}}{\mu} = \frac{(2.35 \times 10^{-4})(10^3)(1.0)}{(5.40 \times 10^{-4})} = 4.36 \times 10^2 \qquad \text{(laminar range)}$$

(where $\mu = 540 \times 10^{-6}$ poise (Table 8.1-1)).

d. $$\dot{J} \equiv \int_0^{a_{\text{w}}} (\rho v_z)(v_z) \, 2\pi r \, dr = \int_0^{a} \left\{2U\left[1 - \left(\frac{r}{a_{\text{w}}}\right)^2\right]\right\}^2 2\pi r \, dr,$$

$$\dot{J} = 4\rho U^2 a_{\text{w}}^2 \int_0^1 (1 - \xi^2)^2 \, 2\pi\xi \, d\xi = \frac{4}{3}\pi\rho U^2 a_{\text{w}}^2.$$

Flename

Flename

flename

Flename

~~Flename~~

$$\begin{cases} \delta = \dfrac{[NG]_{out}}{[NG]_{in}} = e^{-\phi_0 c H} \\ H = \dfrac{\dot{W}}{[NG]_{in} \dot{V}} \end{cases}$$

But since $\dot{m} = \rho U \pi a_w^2$ we find:

$$\dot{J} = \frac{4}{3}(\dot{m}U) = \frac{4}{3}(0.185)(10^3) = 246.5\ \frac{\text{g} \cdot \text{cm}}{\text{s}^2}.$$

Note that $\dot{m}/(\rho A) \neq \dot{J}/\dot{m}$.

e. $\dot{K} \equiv \int_0^{a_w} (\rho v_z)\left(\frac{v_z^2}{2}\right) 2\pi r\, dr$

$$= \int_0^{a_w} \rho \frac{\{2U[1-(r/a)^2]\}^3}{2} \cdot 2\pi r\, dr$$

$$= 4\pi\rho U^3 a_w^2 \cdot \int_0^1 (1-\xi^2)^3 2\xi d\xi$$

$$= 2\,\dot{m}(U^2/2) = (2)(0.185)(10^6/2),$$

$\dot{K} = 1.85 \times 10^5$ g cm^2/s^3.

Note that:

$$\left(\frac{\dot{m}}{\rho A}\right) \neq \left(\frac{\dot{J}}{\dot{m}}\right) \neq \left(\frac{2\dot{K}}{\dot{m}}\right)^{1/2};$$

that is,

$$U \neq \frac{4}{3}U \neq (2)^{1/2}U \qquad \text{(see, also, Section 2.5.5).}$$

f. $v_\theta(r, z) \neq 0$ would *not* influence $\dot{m}$, $\dot{J}$, $\dot{K}$ if v_z were $2U[1-(r/a_w)^2]$. Discuss.

g. $\tau_{rz} = 0$ at $r = 0$ and $(\tau_{rz})_w = \tau_w = \dfrac{4\mu U}{a_w}$ (see Eq. (4.5-24)).

Thus

$$\tau_w = \frac{(4)(5.40 \times 10^{-4})(10^3)}{(0.5)} = 4.32\frac{\text{dyne}}{\text{cm}^2} = 0.452\ \text{Pa}$$

$$C_f = \frac{\tau_w}{\frac{1}{2}\rho U^2} = \frac{16}{\text{Re}} \qquad \text{(see Eq. (4.5-25))}$$

therefore $C_f = \dfrac{16}{4.359 \times 10^2} = 3.67 \times 10^{-2}$ (see Figure 4.5-3.)

h. Descriptors:

Continuum	Laminar
"Incompressible"	Quasi-one-dimensional
Newtonian (viscous)	Internal
Steady	Single-Phase

4.9 Momentum Transfer to (Drag on) Immersed Objects
Drag/meter of axial length = ? for objects of transverse dimension 5 cm. in $U = 10$ m/s, air @ 1 atm., 1200 K.

a. *Cylinder in Crossflow* (*cf. Section 4.4-1*)

$$\mu_\infty = 4.7 \times 10^{-4} \text{ poise} \qquad \text{(Table 8.1-1);}$$

$$\rho_\infty = \frac{pM}{RT_\infty} = \frac{(1)(28.97)}{(82.06)(1200)} = 2.94 \times 10^{-4} \frac{\text{g}}{\text{cm}^3};$$

$$\nu_\infty \equiv \mu_\infty/\rho_\infty = 1.60 \text{ cm}^2/\text{s};$$

$$\text{Re} \equiv \frac{Ud_w}{\nu} = \frac{(10^3 \text{ cm/s})(5 \text{ cm})}{1.60 \text{ (cm}^2/\text{s)}}$$

$$= 3.13 \times 10^3;$$

$$(C_D)_{\text{graph}} \cong 0.95 \equiv \frac{(\text{Drag/Length})}{(\frac{1}{2}\rho_\infty U^2)(\text{Frontal Area/Length})}.$$

(Figure 4.4-3)

Frontal area/meter = 5 cm × 100 cm = 5×10^2 cm^2/m;

$$\frac{1}{2}\rho_\infty U^2 = \left(\frac{1}{2}\right)(2.94 \times 10^{-4})(10^3)^2 = 1.47 \times 10^2 \frac{\text{dyne}}{\text{cm}^2}.$$

$$\text{Therefore Drag/Length} = ((1.47 \times 10^2)(0.95))(5 \times 10^2 \text{ cm}^2)$$

$$= 6.99 \times 10^4 \frac{\text{dyne}}{\text{meter}}$$

$$= \left(6.99 \times 10^4 \frac{\text{dyne}}{\text{m}}\right) \cdot \left(10^{-5} \frac{\text{N}}{\text{dyne}}\right)$$

$$= 0.699 \text{ N/m}.$$

Most of this drag is due to the $p(\theta)$ distribution—that is, "form" drag.

b. *Plate Normal to Flow*: $C_D(\text{Re})$ not given in Chapter 4 (check literature).

c. *Plate Aligned with Flow*:

In this case $\mathrm{Re}_L \equiv \dfrac{UL}{\nu_\infty}$

and $\mathrm{Re}_L = 3.13 \times 10^3$.

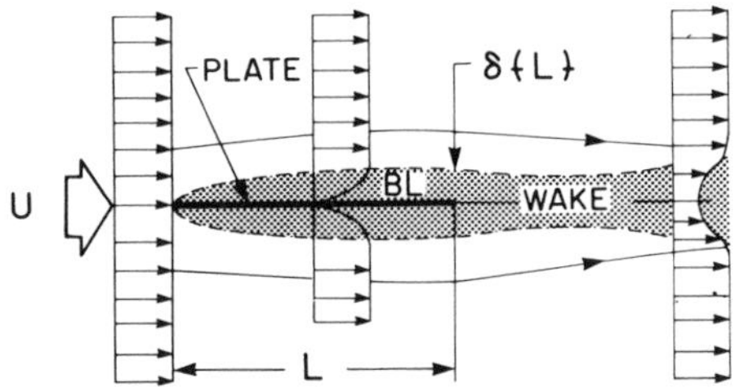

Since $\mathrm{Re}_L < 10^6$ (approx.) we expect flow in the momentum defect Boundary Layer to be *laminar*. Then (Eq. 4.5-38),

$$\bar{C}_f = \frac{1.328}{(\mathrm{Re}_L)^{1/2}} \qquad \text{(Blasius)}$$

where

$$\mathrm{Re}_L^{1/2} = (3.13 \times 10^3)^{1/2} = 5.59 \times 10^1$$

and

$$\left.\begin{aligned} \bar{C}_f &= \frac{1.328}{5.59 \times 10^1} = 2.37 \times 10^{-2} \\ \text{therefore}^{b} \quad & \frac{\text{Drag (both sides)}}{(\frac{1}{2}\rho_\infty U^2)(\text{total wetted area})} = 2.37 \times 10^{-2}. \end{aligned}\right\} \text{dimensionless}$$

But total wetted area/meter = $(2)(5 \times 10^2) = 10^3\ \mathrm{cm^2/m}$. Therefore

$$\text{Drag} = (2.37 \times 10^{-2})\left(1.47 \times 10^2 \frac{\text{dyne}}{\text{cm}^2}\right)\left(10^3 \frac{\text{cm}^2}{\text{m}}\right)\left(10^{-5} \frac{\text{N}}{\text{dyne}}\right),$$

or

$$\text{Drag} = 3.49 \times 10^{-2} \text{ newtons.}$$

---

[b] Since the *frontal* area of such a plate is negligible, this drag corresponds to $C_D \to \infty$.

This drag is entirely due to $\tau_w\{x\}$—i.e., it is "friction drag."

$$\delta\{L\} \cong \frac{5L}{\mathrm{Re}_L^{1/2}} = \frac{(5)(5)}{5.59 \times 10^1} \cong 0.45 \text{ cm}$$

therefore {2 plates could be brought to within 1 cm apart w/o interference.

4.11 *Momentum Transfer: Turbulent Round Jet*

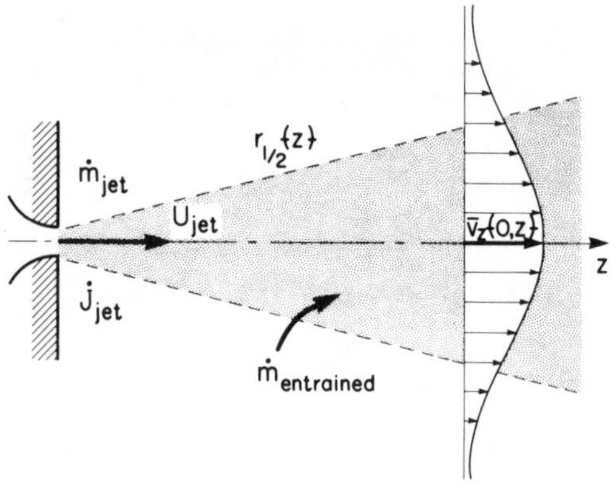

$$\dot{J} = \dot{m}U_{\text{jet}} = (79.125)\left(8.028 \times 10^4 \frac{\text{cm}}{\text{s}}\right)$$
$$= 6.35 \times 10^6 \text{ gm} \cdot \text{cm/s}^2.$$

ν_t is determined by $(\dot{J}/\rho)^{1/2}$ (see Section 4.6.1). Using $(\rho_j + \rho_\infty)/2$, we obtain:

$$\rho \approx 6.9 \times 10^{-4} \text{ g/cm}^3.$$

Therefore $\quad (\dot{J}/\rho)^{1/2} = 0.96 \times 10^5 \rightarrow \nu_t = 1.55 \times 10^3 \text{ cm}^2/\text{s},$

cf. $\quad \nu_j = 2.96 \text{ cm}^2/\text{s} \quad$ (*laminar* (fluid) momentum diffusivity);

that is, $\quad \nu_t/\nu_j = 5.2 \times 10^2.$

Where does $\bar{v}_z\{0, z\}$ drop to $U_j/10$? For a turbulent jet:

$$\underbrace{\bar{v}_z\{0, z\}}_{U_j/10} = \frac{3}{8\pi} \cdot \frac{1}{0.0161} \cdot \left(\frac{\dot{J}}{\rho}\right)^{1/2} \cdot \frac{1}{z},$$

we find: $z = 88.6 \text{ cm} = 0.89 \text{ m}.$

(This would have been 0.46 km if there had been no turbulent enhancement in momentum diffusion.)

$\dot{m}_{\text{entrained}}/\dot{m}_{\text{primary jet}}$ at this location?

$$\dot{m}_{\text{entrained}} \approx 8\pi\rho \cdot \left[0.0161\left(\frac{\dot{J}}{\rho}\right)^{1/2}\right] z \qquad \text{(turbulent case)},$$

where
$$\rho \approx 6.9 \times 10^{-4} \text{ g/cm}^3,$$
$$(\dot{J}/\rho)^{1/2} \approx 0.96 \times 10^5 \text{ cm}^2/\text{s};$$

therefore
$$\nu_t = 1.55 \times 10^3 \frac{\text{cm}^2}{\text{s}},$$

so that at
$$z \approx 88.6 \text{ cm},$$
$$\dot{m}_{\text{entrained}} \approx 4.05 \times 10^3 \text{ g/s};$$

cf.
$$\dot{m}_j = 0.792 \times 10^2 \text{ g/s}.$$

Therefore $\dfrac{\dot{m}_{\text{entrained}}}{\dot{m}_{\text{jet exit}}} \cong 51$ at location where $\bar{v}_z(0, z) = \dfrac{U_j}{10}$.

Exercises

1. Calculate the time-averaged $\bar{v}_z(r, z)$ *profile* at this location.
2. Can you estimate the centerline, time-averaged temperature and $CO_2(g)$ concentration at this point? (Itemize and discuss the underlying assumptions.)

5.10 *k(insul) via* T_{hot}, T_{cold}

The manufacturer of the insulation reports T_h, T_w-combinations for the configuration shown in Figure 5.10E. *What is the k and the "R"-value (thermal resistance) of their insulation?* We consider here the intermediate case:

$$T_{\text{hot}} \equiv T_h = 2400°\text{F} = 1589 \text{ K},$$
$$T_w \cong 670°\text{F} = 628 \text{ K},$$

and carry out all calculations in metric units.

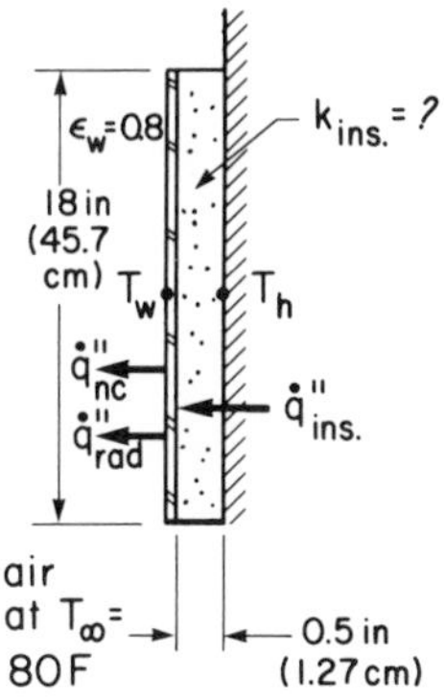

Figure 5.10E

Note:

$$\bar{\dot{q}}''_{ins} = \bar{\dot{q}}''_{rad} + \bar{\dot{q}}''_{nat\ conv.}$$

$\bar{\dot{q}}''_{rad}$: Calc. *via* ε, T_w, T_∞

$\bar{\dot{q}}''_{nat\ conv.}$: Calc. *via* $\overline{Nu_h}(Ra_h)$

Then:

$$k_{ins} = \bar{\dot{q}}''_{ins}/(T_h - T_w) \quad \text{at } \tfrac{1}{2}(T_h + T_w)$$

and

$$\text{“}R''_{ins} \equiv \frac{(\text{thickness})_{insul}}{k_{insul}}$$

Radiation Flux

$$\bar{\dot{q}}''_{rad} = \varepsilon_w \sigma_B (T_w{}^4 - T_\infty{}^4)$$

or

$$\bar{\dot{q}}''_{rad} = \varepsilon_w \left\{ 56.72 \left[\left(\frac{T_w}{1000} \right)^4 - \left(\frac{T_\infty}{1000} \right)^4 \right] \right\} \quad \frac{\text{kW}}{\text{m}^2}$$

Inserting

$$\left.\begin{aligned} \varepsilon_w &= 0.90 \\ T_w &= 628 \text{ K} \\ T_\infty &= 300 \text{ K} \end{aligned}\right\} \text{yields: } \bar{\dot{q}}''_{rad} = 7.52 \frac{\text{kW}}{\text{m}^2}$$

Natural Convection Flux: *Vertical Flat Plate*

$$Ra_h \equiv \frac{g\beta_T(\Delta T)L^3}{\nu^2} \cdot \left(\frac{\nu}{\alpha}\right) \quad \left\{ \begin{array}{l} \text{Rayleigh Number} \\ \text{for heat transfer} \\ \text{(based on } L\text{)} \end{array} \right.$$

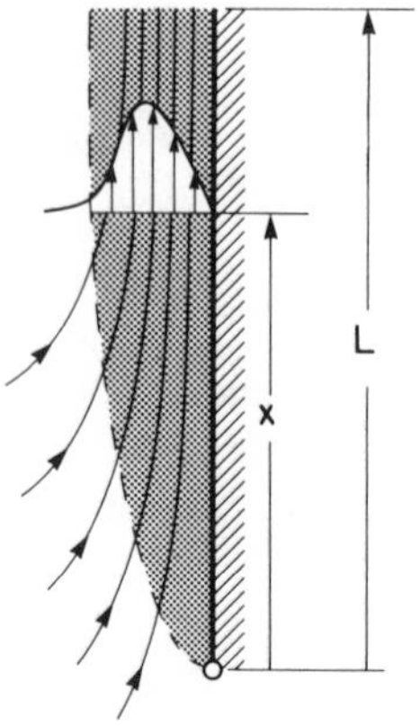

But:

$$g = 980 \text{ cm/s}^2, \; L = 45.7 \text{ cm}$$

and, for a perfect gas:

$$\beta_T = 1/T_{film} \text{ where } T_{film} = \tfrac{1}{2}(628 + 300) = 464 \text{ K}$$

Therefore $$\beta_T \Delta T = \frac{628 - 300}{464} = 0.707.$$

For air:

$$\mu\{464\} \cong \mu\{400\} \cdot \left(\frac{464}{400}\right)^{0.67}$$

$$= (2.27 \times 10^{-4})(1.105) = 2.51 \times 10^{-4}$$

and $$\rho_{\text{film}} = \frac{pM}{RT_{\text{film}}} = 7.61 \times 10^{-4} \text{ g/cm}^3$$

Therefore $$\nu = (\mu/\rho)_{\text{film}} = \frac{2.5 \times 10^{-4}}{7.61 \times 10^{-4}} = 0.3295 \frac{\text{cm}^2}{\text{s}}$$

and $$\text{Pr} \equiv (\nu/\alpha) = 0.706 \qquad \text{(Table 8.1-1).}$$

Therefore $$\text{Ra}_h = 4.3 \times 10^8 \qquad \text{based on } L = 45.7 \text{ cm.}$$

This is in the *laminar* BL range (see Figure 5.6-1).

Now,

$$\frac{\dot{q}''_{\text{w,nc}}\{x\}}{k(\Delta T/x)} \equiv \text{Nu}_{h,x} = 0.517(\text{Ra}_{h,x})^{1/4}$$

and $$\overline{\dot{q}''_{\text{w}}} = \frac{1}{L} \cdot \int_0^L \dot{q}''_{\text{w}}\{x\}\, dx.$$

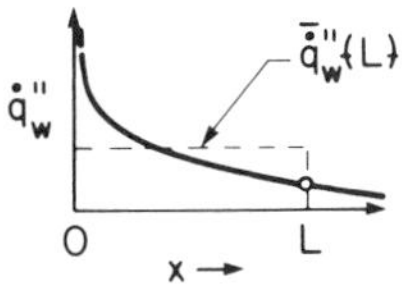

Since $\dot{q}''_{\text{w,nc}}\{x\} \sim x^{-1/4}$ (laminar BL, nat. conv.)

$$\bar{\dot{q}}''_{\text{w}} = \frac{4}{3} \cdot \dot{q}''_{\text{w}}\{L\}$$

$$(\bar{\dot{q}}''_{\text{w}})_{\text{nc}} = \frac{4}{3}\left(\frac{k\,\Delta T}{L}\right) \underbrace{\{0.517(\text{Ra}_{h,L})^{1/4}\}}_{\text{Nu}_{h,L}} \qquad \text{(Eq. (5.5-9b))}$$

$$k_{\text{air}}\{464\} \approx k\{400\}\left(\frac{464}{400}\right)^{0.8} = 78 \times 10^{-6}\left(\frac{464}{400}\right)^{0.8}$$

$$= 87.8 \times 10^{-6} \frac{\text{cal}}{\text{s} \cdot \text{cm} \cdot \text{K}}$$

$$\text{Nu}_{h,L} = 7.447 \times 10^1$$

Therefore $(\overline{\dot{q}''_{\text{w}}})_{\text{nc}} = 6.25 \times 10^{-2} \dfrac{\text{cal}}{\text{cm}^2 \cdot \text{s}}$.

$$\text{Therefore } (\overline{\dot{q}''_w})_{nc} = \left(6.25 \times 10^{-3}\frac{\text{cal}}{\text{cm}^2\text{ s}}\right)\cdot\frac{4.18\text{ J}}{1\text{ cal}}\cdot\frac{(10^2\text{ cm})^2}{(1\text{ m})^2}\cdot\frac{1\text{ kW}}{10^3\text{ W}}$$

$$= 2.62\,\frac{\text{kW}}{\text{m}^2}.$$

Conclusion

When $T_{\text{hot}} = 1589$ K, $T_{\text{cold}} = 628$ K, then:

$$\bar{\dot{q}}''_{\text{ins}} = \bar{\dot{q}}''_{\text{rad}} + \bar{\dot{q}}''_{nc}$$

$$= 7.516 + 2.616\frac{\text{kW}}{\text{m}^2} = 10.13\frac{\text{kW}}{\text{m}^2}.$$

Therefore

$$k_{\text{ins}} = \frac{\bar{\dot{q}}''_{\text{ins}}}{(\Delta T)/\text{thickness}} = \frac{10.13\dfrac{\text{kW}}{\text{m}^2}}{\left\{\dfrac{1589 - 628\text{ K}}{1.27\text{ cm}}\right\}}$$

or

$$k_{\text{ins}} = 0.134\,\frac{\text{W}}{\text{m}\cdot\text{K}} \quad \text{at } T_{\text{mean}} \approx 1110\text{ K}.$$

Therefore, for the thermal "resistance," R:

$$\text{"R"-value at 1110 K} \approx \frac{1.27 \times 10^{-2}\text{ m}}{0.134\dfrac{\text{W}}{\text{m}\cdot\text{K}}} = 0.95 \times 10^{-1}\frac{\text{m}^2\text{ K}}{\text{W}}.$$

Remark

$$k_{\text{ins}} = \left(0.134\frac{\text{W}}{\text{m}\cdot\text{K}}\right)\left(0.5778\frac{\text{BTU}}{\text{hr ft }^\circ\text{F}}\right)\left(\frac{12\text{ in}}{1\text{ ft}}\right) = 0.929\frac{\text{BTU/ft}^2\text{-s}}{(^\circ\text{F/in})}$$

(one of the common English units) at

$$T_{\text{mean}} = \tfrac{1}{2}(2400 + 670^\circ\text{F}) = 1535^\circ\text{F}.$$

Student Exercises

1. Calculate k_{ins} for the other pairs of T_h, T_w; is the resulting dependence of $k_{\text{ins}}\{T_{\text{mean}}\}$ reasonable?

2. How does k_{ins} compare to the value for "rock-wool" insulation?
3. Would this insulation behave differently under vacuum conditions?

6.9 *Catalytic Converter*

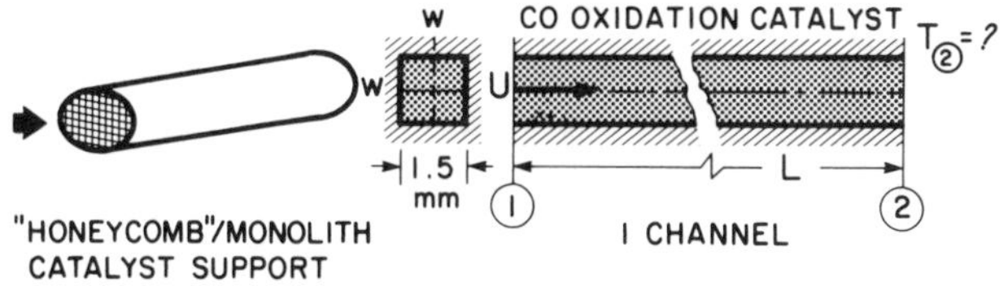

a. *Mechanism of* CO(g) *transport to the wall*
If Re < 2100 (see below), transport to the wall is by Fick diffusion of CO(g) through the prevailing mixture.

$$D_{\text{CO-mix}} \approx D_{\text{CO-N}_2} = \frac{0.216}{p}\left(\frac{T}{300}\right)^{1.73} \frac{\text{cm}^2}{\text{s}}.$$

Therefore $D_{\text{CO-mix}}(600\text{ K}, 1\text{ atm}) = \dfrac{0.216}{1}\left(\dfrac{600}{300}\right)^{1.73} = 0.684\dfrac{\text{cm}^2}{\text{s}}$

Analogous heat transfer diffusivity is $k/(\rho c_p)$ for gas mixture.

b. Discuss whether the Mass transfer Analogy Conditions (*MAC*) and Heat transfer Analogy Conditions (HAC) *are met*; implications?
Since M_{mix} and M_{CO} are close $y_{\text{CO}} \approx \omega_{\text{CO}}$; hence we will assume $\omega_{\text{CO},①} \approx 0.02$.

c. *Sc for the mixture*: $\text{Sc}_{\text{CO}} \equiv \nu_{\text{mix}}/D_{\text{CO-mix}}$

Now: $\mu_{\text{mix}} \approx \mu_{\text{air}}\left(\frac{1}{2}(700 + 500)\right)$

$\approx 3.31 \times 10^{-4}$ poise

and: $\rho \approx \dfrac{pM}{RT} = \dfrac{(1)(28.97)}{(82.06)(600)} = 5.88 \times 10^{-4}\dfrac{\text{g}}{\text{cm}^3}$

therefore $\nu \equiv (\mu/\rho)_{\text{mix}} \cong 0.563\dfrac{\text{cm}^2}{\text{s}}$

therefore $\text{Sc} \equiv \nu/D_{\text{CO-mix}} \approx \dfrac{0.563}{0.684} = 0.823.$

d. $L = ?$ We will need $\text{Re} \equiv Ud_{\text{eff}}/\nu$

Now: $$d_{\text{eff}} \equiv \frac{4A}{P} = \frac{(4)(0.15\text{ cm} \times 0.15\text{ cm})}{(4 \times 0.15\text{ cm})} = 0.15\text{ cm}$$

$$U = 10^3\text{ cm/s}$$

$$\nu \approx 0.562\text{ cm}^2/\text{s}$$

therefore $\text{Re} \approx \dfrac{(10^3)(0.15)}{(0.562)} \approx 267$ (laminar-flow regime)

For a *square channel*, Table 5.5-1 gives:

$\overline{\text{Nu}}_m\{\infty\} = 2.976$ and $C_f \cdot \text{Re} = 14.227$ (used below).

If $\omega_{\text{CO,w}} = 0$, then the mass-transfer analogy of Eq. (5.5-21a) is:

$$\frac{\omega_{\text{CO},b}\{z\}}{\omega_{\text{CO},b}\{0\}} = \exp\{-4\,\xi_m F\{\text{entrance}\} \cdot \overline{\text{Nu}}_m\{\infty\}\}$$

where (Eq. (6.5-3)):

$$\xi_m \equiv \frac{1}{\text{Re Sc}} \cdot \frac{z}{d_{\text{eff}}}.$$

We estimate $z = L$ at which $\omega_b\{L\}/\omega_b\{0\} = 0.05$.

If $$0.05 = \exp\{-X\}$$

then $$X = \ln\left(\frac{1}{0.05}\right) = 2.9957$$

therefore $$4\,\xi_m \cdot \overline{\text{Nu}}_m\{\infty\} \cdot F\{\text{entrance}\} = 2.9957.$$

Tentatively, assume $F\{\text{entrance}\} \approx 1$. Then:

$$(4) \cdot \frac{1}{(267)(0.823)} \cdot \left(\frac{L^{(0)}}{0.15\text{ cm}}\right)(2.976) = 2.9957;$$

that is,

$$\xi_m^{(0)} = \frac{2.9957}{(4)(2.976)} = 0.25166$$

(at which $F\{\text{entrance}\}$ is indeed ≈ 1). Solving for L gives: $L = 8.3$ cm (needed to give 95% CO-Conversion).

e. Discuss *underlying assumptions*, e.g.,

fully developed flow?
nearly constant thermophysical properties?
no homogeneous chemical reaction?
"diffusion-controlled" surface reaction?

f. If the catalyst were "poisoned," it would not be able to maintain $\omega_{CO,w} \ll \omega_{CO,b}$. This would cause L_{req} to exceed 8.3 cm. If catalyst were completely deactivated, then $\omega_{CO,w} \rightarrow \omega_{CO,b}$, and, of course, $L_{req} \rightarrow \infty$.

g. If heat of combustion is 67.8 kcal/mole CO, how much *heat* is *delivered to the catalyst channel per unit time? Overall CO balance* gives the CO-consumption rate/channel:

$$-\Delta\dot{m}_{CO,\,channel} = \dot{m}\,\omega_{CO,①} \cdot (\text{conversion}),$$

where

$$\dot{m} = \rho U A = (5.04 \times 10^{-4})(10^3)(0.15)^2$$
$$= 1.13 \times 10^{-2} \text{ g/s (per channel)}.$$

Moreover,

$$\omega_{CO,①} \approx 0.02; \qquad \text{conversion} = 0.95;$$

hence,

$$-\Delta\dot{m}_{CO,\,channel} = (1.13 \times 10^{-2})(0.02)(0.95) \quad \text{g/s}$$
$$= 2.156 \times 10^{-4} \text{ g/s}$$

and

$$Q = 67.8\frac{\text{kcal}}{\text{g-mole CO}} \cdot \frac{1 \text{ g-mole CO}}{28 \text{ g}} = 2.42 \times 10^3 \frac{\text{cal}}{\text{g}}$$

therefore

$$\dot{Q}_{CO\,oxid} = (2.156 \times 10^{-4})\left(2.42 \times 10^3 \frac{\text{cal}}{\text{g}}\right) = 0.522 \frac{\text{cal}}{\text{s}}.$$

The "*sensible*" *heat transfer required to keep the wall at* 500 K can be calculated from a heat balance on the 8.3 cm-long duct—i.e., once we calculate $T_b(L)$, we have:

$$\dot{Q}_{\text{sensible}} = \dot{m}\, c_p[T_b(0) - T_b(L)],$$

where

$$\dot{m} = 1.13 \times 10^{-2}\ \text{g/s},$$
$$c_p \cong 0.251\ \text{cal/(g-K)}\ \ \text{(Table 8.1-1)},$$
$$T_b(0) = 700\ \text{K}.$$

Mixing cup avg temp at duct outlet?

$$\frac{T_w - T_b(L)}{T_w - T_b(0)} = \exp\{-4\,\xi_h \cdot \overline{\text{Nu}}_h(\infty) \cdot F(\text{entrance})\}.$$

Again, we see that $F(\text{entrance}) \approx 1$ since;

$$\xi_h \equiv \frac{1}{\text{Re} \cdot \text{Pr}} \cdot \frac{L}{d_{\text{eff}}} = \frac{1}{(267)(0.706)}\left(\frac{8.3}{0.15}\right),$$
$$\xi_h = 0.293.$$

Moreover,

$$\overline{\text{Nu}}_m(\infty) = 2.976;$$

therefore

$$\frac{T_b(L) - T_w}{T_b(0) - T_w} \cong \exp\{-4(0.293)(2.976)\} = 0.0304;$$

and

$$\frac{T_b(L) - 500}{700 - 500} = 0.0304,\ \text{or}$$
$$T_b(L) = 506\ \text{K}$$

Therefore,

$$\dot{Q}_{\text{sensible}} = \dot{m}\, c_p \cdot (T_b(0) - T_b(L))$$
$$= \left(1.135 \times 10^{-2} \frac{\text{g}}{\text{s}}\right)\left(0.251 \frac{\text{cal}}{\text{g} \cdot \text{K}}\right)(700 - 506\ \text{K})$$

i.e., $\dot{Q}_{\text{sens}} = 0.552$ cal/s,

$$\dot{Q}_{\text{total}} = \dot{Q}_{\text{CO oxid.}} + \dot{Q}_{\text{sens}} = 0.522 + 0.552$$

$$= 1.07 \frac{\text{cal}}{\text{s}} \text{ per channel.}$$

h. *"Quasi-Steady" Application of These Results?*
Note that:

$$t_{\text{flow}} \equiv \frac{L}{U} \approx \frac{8.3 \text{ cm}}{10^3 \text{ cm/s}} \approx 8.3 \text{ ms}$$

and

$$t_{\text{diff}} = \frac{(\text{w}/2)^2}{D_{\text{CO-mix}}} = \frac{(0.15/2)^2}{0.684 \frac{\text{cm}^2}{\text{s}}} = 8.2 \text{ ms};$$

hence, if the characteristic period of the unsteadiness $\gg$ 8.2 ms, the previous results can be used at each flow condition.

i. *Pressure Drop $-\Delta p$?*
We have:

$$C_{\text{f}} = \frac{\bar{\tau}_{\text{w}}}{\frac{1}{2}\rho U^2} = \frac{14.227}{\text{Re}} \qquad \text{for square duct, laminar flow.}$$

Therefore

$$\bar{\tau}_{\text{w}} = \left(\frac{1}{2}\rho U^2\right) \cdot \frac{14.227}{\text{Re}} \qquad \text{(perimeter-mean shear stress),}$$

but

$$\frac{1}{2}\rho U^2 = \frac{5.043 \times 10^{-4}}{2} \frac{\text{g}}{\text{cm}^3} \left(10^3 \frac{\text{cm}}{\text{s}}\right)^2 = 2.98 \times 10^2 \frac{\text{dyne}}{\text{cm}^2}.$$

Therefore

$$\bar{\tau}_{\text{w}} = (2.98 \times 10^2) \frac{14.22}{267} = (2.98 \times 10^2)\underbrace{(5.335 \times 10^{-3})}_{C_{\text{f}}}$$

$$\bar{\tau}_{\text{w}} = 1.345 \text{ dyne/cm}^2.$$

From *overall momentum balance*:

$$4\text{w} = 4(0.15)\ \text{cm}$$

$$(-\Delta p)A_{\text{duct channel}} = \bar{\tau}_{\text{w}} \cdot (\text{Perim.}) \cdot L$$

$$-\Delta p = \tau_{\text{w}}(4\,L/\text{w})$$

$$= (1.345)(4 \times 8.3/0.15) = 2.48 \times 10^{2}\ \text{dyne/cm}^2$$

$$-\Delta p/p_{①} = 2.93 \times 10^{-5} = 0.003\%.$$

7.1 *Model Tests on Heat-Exchanger Tube Bundle*
"Prototype" Conditions:

$d_{\text{w}} = 2$ in. (diam.)
$s = 3$ in. (sep.)
$U = 10$ m/s
$p = 1$ atm
$T_{\infty} = 1600$ K
$T_{\text{w}} = 800$ K

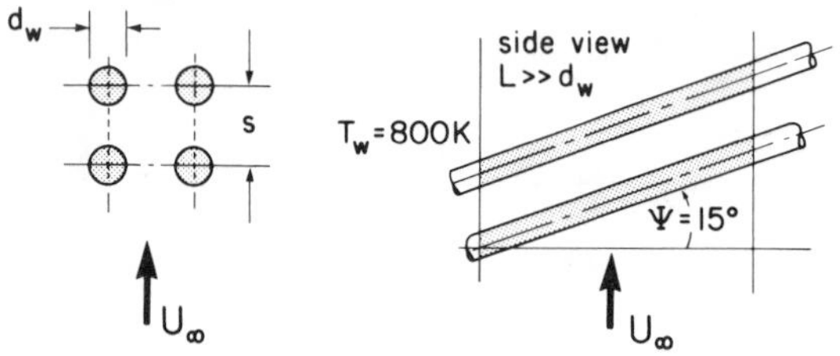

Questions:
1. Can useful "model" experiments be done?
2. What is the relation between model and full-scale heat-transfer behavior?

Note: if $\quad T_{\infty} = 1600\ \text{K}, \qquad T_{\text{w}} = 800\ \text{K}.$

Therefore $T_{\text{film}} = \frac{1}{2}(T_{\infty} + T_{\text{w}}) = \frac{1}{2}(1600 + 800) = 1200\ \text{K}.$

Therefore $\rho_{\text{film}} = \dfrac{pM}{RT_{\text{film}}} = \dfrac{(1)(28.97)}{(82.06)(1200)} = 2.94 \times 10^{-4}\ \text{g/cm}^3.$

$\mu_{\text{film}} = 4.70 \times 10^{-4}$ poise (Table 8.1-1);

therefore $\nu_{\text{film}} = 1.597\ \text{cm}^2/\text{s};$

therefore $(\text{Re})_{\text{p}} = \dfrac{Ud_{\text{w}}}{\nu} = \dfrac{(10^3\ \text{cm/s})(2 \times 2.54\ \text{cm})}{(1.6\ \text{cm}^2/\text{s})} = 3.18 \times 10^3,$

and $(\text{Pr})_{\text{p}} = 0.705$ (Table 8.1-1).

Possible "Model" Conditions (m)
Consider *small-scale model tested in water flow*;
e.g., take:

$$\left.\begin{aligned} d_{\text{m}} &= 2\ \text{mm} \\ s_{\text{m}} &= 3\ \text{mm} \end{aligned}\right\} \begin{array}{l}\text{tubes yawed @ } 15^\circ \\ \text{from } H_2O(l)\ \text{flow.}\end{array}$$

This ensures *geometric similarity.*

To ensure *dynamic similarity*, we test under conditions such that:

$$\mathrm{Re_m} = \mathrm{Re_p} = 3.18 \times 10^3;$$

i.e.,

$$\frac{U_m d_m}{\nu_m} = 3.18 \times 10^3.$$

But for water near room temperature:

$$\nu_m \approx 1.02 \times 10^2\ \mathrm{cm^2/s} \qquad (\mathrm{H_2O}(l))$$

and $\qquad d_m \approx 2$ mm. (See choice above);

$$\text{Therefore} \quad U_m = \frac{(3.18 \times 10^3)(1.02 \times 10^{-2})}{(0.2\ \mathrm{cm})} = 1.62 \times 10^2 \frac{\mathrm{cm}}{\mathrm{s}}$$

$$= 1.62\ \mathrm{m/s}.$$

For strict *thermal similarity*, we should have $\mathrm{Pr_m} = 0.706$; however: $\mathrm{Pr_m} = \mathrm{Pr}(\mathrm{H_2O}(l)) \approx 7$. We can approximately account for this[c] since we know that for an *isolated* cylinder in cross-flow the Pr-dependence of $\overline{\mathrm{Nu}}_h$ (Churchill and Bernstein, *ASME J. Heat Transfer* **99**, 300 (1977)) is:

$$\overline{\mathrm{Nu}}_h \propto \mathrm{Pr}^{1/3} \cdot \left\{1 + \left(\frac{0.4}{\mathrm{Pr}}\right)^{2/3}\right\}^{-1/4}$$

(slightly more elaborate than $\mathrm{Pr}^{1/3}$). We now assume that this Pr-dependence will also hold approximately for a tube *bank.*

Thus, if we test at the same Re, we expect:

$$\frac{\overline{\mathrm{Nu}}_{h,\mathrm{p}}}{\overline{\mathrm{Nu}}_{h,\mathrm{m}}} = \frac{\overline{\mathrm{Nu}}_h\{0.7\}}{\overline{\mathrm{Nu}}_h\{7\}} = \left(\frac{0.7}{7}\right)^{1/3} \cdot \frac{\left[1 + \left(\frac{0.4}{7}\right)^{2/3}\right]^{1/4}}{\left[1 + \left(\frac{0.4}{0.7}\right)^{2/3}\right]^{1/4}}$$

or

$$\overline{\mathrm{Nu}}_{h,\mathrm{p}} = 0.422\, \overline{\mathrm{Nu}}_{h,\mathrm{m}} \text{ at same Re,}$$

[c] Could we have alternatively tested in a fluid with the Prandtl number of air, and corrected for a difference in Reynolds' number? Which is the "safer" approach? Why?

and the scale factor is: $L_p/L_m = \dfrac{(2\,\text{in})(2.54\,\text{cm/in})}{(0.2\,\text{cm})} = 25.4.$

In the prototype:

$$(\Delta T)_p = 800\,\text{K} \qquad \text{(in air).}$$

Suppose we test with:

$$(\Delta T)_m = 10\,\text{K} \qquad \text{(in water)—(e.g., tubes } \textit{hotter} \text{ than } H_2O(l)).$$

Now,

$$\frac{\bar{\dot{q}}''_p}{\bar{\dot{q}}''_m} = \frac{\{(k/L)\cdot \overline{\text{Nu}}_h\{\text{Re, Pr}\}(\Delta T)\}_p}{\{(k/L)\cdot \overline{\text{Nu}}_h\{\text{Re, Pr}\}(\Delta T)\}_m}.$$

Therefore

$$\frac{\bar{\dot{q}}''_p}{\bar{\dot{q}}''_m} = \left(\frac{k_p}{k_m}\right)\cdot\left(\frac{\overline{\text{Nu}}_{h,p}}{\overline{\text{Nu}}_{h,m}}\right)\cdot\frac{(\Delta T)_p}{(\Delta T)_m}\cdot\frac{1}{(L_p/L_m)};$$

and, since:

$$k_p = k_{air}\{1200\,\text{K}\} = 1.88\times 10^{-4}\,\text{cal/(s}\cdot\text{cm}\cdot\text{K)},$$

and

$$k_m = k_{H_2O(l)} \approx 1.4\times 10^{-3}\,\text{cal/(s}\cdot\text{cm}\cdot\text{K) (near room temp.)},$$

therefore

$$\frac{\bar{\dot{q}}''_p}{\bar{\dot{q}}''_m} \approx \left(\frac{1.88\times 10^{-4}}{1.4\times 10^{-3}}\right)(0.422)\left(\frac{800}{10}\right)\left(\frac{1}{25.4}\right) = 0.178;$$

i.e., under these conditions: $\bar{\dot{q}}''_p \cong 0.178\,\bar{\dot{q}}''_m$.

Consider Heat-Transfer Rates per Unit Axial Length:

Since $\qquad \dot{q}'_p = \bar{\dot{q}}''_p \cdot A'_p \qquad$ and $\qquad A'_p = \pi d_p,$

(diam. in present cases)

we have: $$\frac{\dot{q}'_p}{\dot{q}'_m} = \frac{\bar{\dot{q}}''_p}{\bar{\dot{q}}''_m}\cdot\frac{L_p}{L_m} = (0.178)(25.4),$$

so that: $$\dot{q}'_p = 4.52\,\dot{q}'_m.$$

Discussion Questions

1. Are there other "model" possibilities?
2. Do the following phenomena interfere with the geometric, dynamical and/or thermal similarity:
   a. natural convection (examine $Gr_h^{1/4}/Re^{1/2}$);
   b. radiative transfer;
   c. gas compressibility; $(Ma)_p = ?$
3. Mass transfer implications? (See Fouling Problem, Section 8.2).
4. Other necessary similarity conditions (e.g., mainstream turbulence level, turbulence scale, etc.)?

Index

Numbers given in italic refer to *exercises* (at the end of each chapter, with some illustrative solutions at the end of the book).

LIBRARY
Florida Solar Energy Center
1679 Clearlake Road
Cocoa FL 32922-5703